Gas Turbine Performance

This book is dedicated to the authors' parents, wives and children.
Without the past sacrifices and efforts of Michael and Rose Walsh, and
John and Josie Fletcher, it would not have been possible.
It is for Maria, Laura and Rachel; and Mary, Robert and Samuel that we
toil today.

Gas Turbine Performance

Second Edition

Philip P. Walsh
BSc, FRAeS, CEng
Head of Performance and Engine Systems
Rolls-Royce plc

Paul Fletcher
MA (Oxon), MRAeS, CEng
Manager, Prelim Design
Energy Business
Rolls-Royce plc

Blackwell
Science

© 1998, 2004 by Blackwell Science Ltd
a Blackwell Publishing company

Editorial Offices:
Blackwell Science Ltd, 9600 Garsington Road,
  Oxford OX4 2DQ, UK
  Tel: +44 (0) 1865 776868
Blackwell Publishing Inc., 350 Main Street,
  Malden, MA 02148-5020, USA
  Tel: +1 781 388 8250
Blackwell Science Asia Pty, 550 Swanston Street,
  Carlton, Victoria 3053, Australia
  Tel: +61 (0)3 8359 1011

The right of the Author to be identified as the
Author of this Work has been asserted in
accordance with the Copyright, Designs and
Patents Act 1988.

First published 1998
Reprinted 1998, 1999, 2000 (twice)
Second edition 2004

Library of Congress Cataloging-in-Publication
Data

Walsh, Philip P.
  Gas turbine performance / Philip P. Walsh, Paul
  Fletcher. – 2nd ed.
    p. cm.
  Includes bibliographical references and index.
  ISBN 0-632-06434-X (alk. paper)
  1. Gas-turbines–Performance. I. Fletcher, Paul.
  II. Title.

TJ778.W36 2004
621.48′3–dc22
                                    2003063655

ISBN: 9780632064342

A catalogue record for this title is available from
the British Library

For further information on Blackwell Publishing,
visit our website:
www.blackwellpublishing.com

Contents

Foreword to the first edition

Sir Frank Whittle first ran his jet engine in April 1937 since when the gas turbine has had an immeasurable impact upon society. Today there are few people in the developed world whose daily life is not touched by it. The ready access to global air travel, low cost electricity, natural gas pumped across continents, and the defence of nations both in the air and at sea are but a few of the things so many of us take for granted which depend upon gas turbine power.

Those involved in the design, manufacture and operation of gas turbines over the last 60 years have made enormous strides. Today the largest turbofans produce 100 times the thrust of the Whittle and von Ohain engines, the latter having made the first flight of a gas turbine in August 1939. The challenge for the forthcoming decades will be undiminished with ever increasing demands to minimise pollution and energy consumption to preserve a better world for the next generation. There will be a continued drive for lower cost of operation, and the opening up of new applications such as mass produced automotive propulsion. The challenges are not just technical, but include adapting to a changing working environment. For example, since I joined the gas turbine industry some 35 years ago computers have revolutionised working practices. Indeed it is not possible to design today's gas turbines without the use of the very latest computational technology, whilst the authors have fully embraced 'best practice' in desk top publishing to make this textbook possible.

For those of us fortunate enough to face these challenges 'gas turbine performance' is our cornerstone. It is what our customers purchase and hence is our raison d'être. It is the functional integration, or marriage, of all elements of gas turbine technology into a product. No matter what role we may have we must understand it, and cannot afford to lose sight of it.

This book is long overdue. It is written in a clear, applied and digestible fashion and will be of benefit to engineers in all phases of their careers. It admirably describes the fundamentals of gas turbine performance and whole engine design for those who have not yet encountered them, while providing an abundance of reference material for those who are more advanced. It also describes *all* gas turbine configurations and applications. This is particularly valuable because another facet of change in our industry is that engineers are unlikely to continue their careers concentrating on only aero, or industrial and marine engines. The investment in technology is so great that it must be amortised over a wide product base, hence there are few gas turbine companies that are not in some way involved in all market sectors.

I am sure that this book will be of great benefit to you in whatever part you play in our industry. I hope your involvement with gas turbines is as enjoyable as mine has been, and is today.

Philip C. Ruffles
Director of Engineering and Technology
Rolls-Royce plc, Derby, UK

Preface

Performance is the end product that a gas turbine company sells. Furthermore, it is the thread which sews all other gas turbine technologies together. Gas turbine performance may be summarised as:

> The thrust or shaft power delivered for a given fuel flow, life, weight, emissions, engine diameter and cost. This must be achieved while ensuring stable and safe operation throughout the operational envelope, under all steady state and transient conditions.

To function satisfactorily within a gas turbine company, engineers from *all* disciplines, as well as marketing staff, *must* understand the fundamentals of performance.

The authors were motivated to write this book by experiences gained while working for three prominent gas turbine companies in the UK and the USA. These clearly showed the pressing need for a book presenting the fundamentals of performance in an applied manner, pertinent to the everyday work of those in industry as well as university based readers. The strategy adhered to in writing this book, together with some of its unique features, are described below:

- The main text contains no algebra and is laid out in a manner designed for easy reference. The latter is achieved by careful section numbering, the use of bullet points, extensive figures, charts and tables, and by deferring the more difficult concepts to the end of sections or chapters.
- Comprehensive lists of formulae, together with sample calculations are located at the end of each chapter. Formulae are presented in FORTRAN/BASIC/Spreadsheet format for ease of implementation in PC programs.
- The lead unit system employed is SI. However key unit conversions are provided on *every* figure, chart and table catering for the needs of all readers world-wide. Furthermore, a comprehensive list of unit conversions is supplied as an appendix.
- The internationally recognised aerospace recommended practice for nomenclature and engine station numbering laid down in ARP 755A is listed as an appendix, and employed throughout.
- Figures, charts, tables and formulae provide not only trends and the form of relationships, but also a *database* for design purposes. Charts are located at the end of each chapter, whereas figures are embedded in the text. Also practical guidelines for engine design are provided throughout the text.
- *All* aspects of gas turbine performance are covered, with chapters on topics not easily found in other textbooks such as transient performance, starting, windmilling, and analysis of engine test data.
- *All* gas turbine engine variants are discussed, including turbojets, turbofans, turboprops, turboshafts, auxiliary power units and ramjets.

- An introduction to gas turbine applications is provided in Chapter 1. Subsequent chapters address meeting the various requirements particular to *all* major applications such as power generation, mechanical drive, automotive, marine and aircraft installations.
- The importance of dimensionless and other parameter groups in understanding the fundamentals of gas turbine performance is emphasised throughout.
- Component performance and design is presented from an engine performance viewpoint. A comprehensive list of references is provided for those wishing to pursue detailed component aero-thermal and mechanical design issues.

This book is primarily aimed at engineers of all disciplines within the gas turbine industry, and will also be of significant value to students of mechanical and aeronautical engineering. It should also appeal to people outside the industry who have an interest in gas turbines. Experienced engineers will particularly welcome the database and list of formulae which it is hoped will make the book an invaluable reference tool.

The guidelines, charts and formulae provided should be invaluable for instructive or 'scoping' purposes, particularly where simplified forms are shown to ease implementation. Progression of projects beyond this must always be accompanied by an appropriate quality plan, however, including stringent control of the accuracy of any software produced. As such, no liability can be accepted for the consequences of any inaccuracies herein.

For the second edition all existing chapters have been reviewed, and a range of minor improvements and alterations introduced. Also, two new chapters have been added covering performance issues relating to engines 'in-service', and also the economics of gas turbines. In recent years the inexorable drive to lower operating costs has increased the need to understand both in-service performance, via issues such as health monitoring, and also the 'techno-economic' issues that determine whether or not a gas turbine project will be profitable.

The authors acknowledge the significant contribution made to this book by their wives. Mrs Mary Fletcher has made a considerable technical input, while Mrs Maria Walsh has provided secretarial support. Our thanks are extended to all of our colleagues and friends from whom we have learned so much. It is impossible to list them all here, however it would not be fitting if the late Mr Robert Chevis were not mentioned by name. He was a source of immense inspiration to both authors, as well as a fountain of knowledge. Also the support given by Mr Neil Jennings, currently the Managing Director of Rolls-Royce Industrial and Marine Gas Turbines Limited, to this project has been invaluable. We are very grateful to Mr Philip Ruffles for writing the Foreword, and to Professor John Hannis of European Gas Turbines for refereeing the manuscript so constructively. Finally our thanks are extended to Mr Christopher Tyrrell for his advice regarding preparation of the original manuscript.

Philip P. Walsh
Paul Fletcher

Gas Turbine Engine Configurations

Gas turbine engine configurations

As a precursor to the main text, this section describes gas turbine engine configurations in terms of the basic component building blocks. These components are comprehensively described in Chapter 5. Diagrams are presented of each configuration, including station numbers as per the international standard ARP 755A presented in Appendix A. Sections 3.6.5 and 6.7–6.11 discuss the thermodynamics of each configuration, and Chapter 1 covers their suitability to given applications. Section 6.2 defines the engine performance parameters mentioned.

Conventional turbojet (Fig. 1a)

Figure 1a shows a conventional single spool *turbojet* above the centre line, and one with the addition of an afterburner, convergent–divergent (con-di) intake, and con-di nozzle below.

Ambient air passes from *free stream* to the *flight intake* leading edge. As described in Chapter 5, the air accelerates from free stream if the engine is static, whereas at high flight Mach number it diffuses from the free stream, *ram* conditions. Usually, it then diffuses in the flight intake before passing through the engine intake to the compressor face resulting a small loss in total pressure.

The *compressor* then increases both the pressure and temperature of the gas. Work input is required to achieve the pressure ratio; the associated temperature rise depends on the *efficiency* level, as discussed in Chapter 3. Depending upon complexity the turbojet compressor pressure ratio ranges from 4:1 up to 25:1.

The *compressor exit diffuser* passes the air to the *combustor*. Here, fuel is injected and burnt to raise exit gas temperature to between 1100 K and 2000 K, depending upon engine technology level. The diffuser and combustor both impose a small total pressure loss.

The hot, high pressure gas is then expanded through the *turbine* where work is extracted to produce shaft power; both temperature and pressure are reduced. The shaft power is that required to drive the compressor and any engine and 'customer' auxiliaries, and to overcome engine mechanical losses such as disc windage and bearing friction. The turbine nozzle guide vanes and blades are often cooled to ensure acceptable metal temperatures at elevated gas temperatures. This utilises relatively cool air from the compression system which bypasses the combustor via *air system* flow paths which feed highly complex internal cooling passages within the vanes and blades.

On leaving the turbine the gas is still at a pressure typically at least twice that of ambient. This results from the higher inlet temperature to the turbine and the fundamental form of the temperature–entropy (T–S) diagram as described in section 3.6.5.

Downstream of the turbine the gas diffuses in the *jet pipe*. This is a short duct that transforms the flowpath from annular to a full circle at entry to the *propelling nozzle*. The jet pipe imposes a small total pressure loss. The propelling nozzle is a convergent duct that accelerates the flow to provide the high velocity jet to create the thrust. If the available expansion ratio is less than the choking value, the static pressure in the exit plane of the nozzle

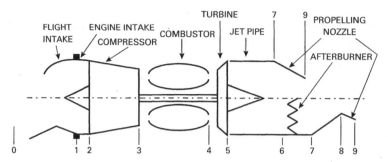

(a)   Conventional turbojet, and afterburning turbojet with con–di nozzle

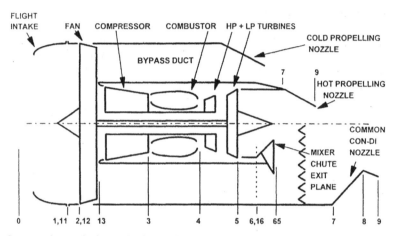

(b)   Separate jets turbofan and mixed, afterburning turbofan with con–di nozzle

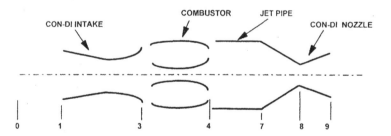

(c)   Ramjet with con-di intake and nozzle

Fig. 1   Thrust engine configurations and station numbering.

will be ambient. If it is greater than the choking value the Mach number at the nozzle will be unity (i.e. sonic conditions), the static pressure will be greater than ambient and shock waves will occur downstream. In the latter instance, the higher static pressure at nozzle exit plane relative to the intake creates thrust additional to that of the jet momentum.

In a two spool engine there are both low pressure (LP) and high pressure (HP) compressors driven by LP and HP turbines. Each spool has a different rotational speed, with the LP shaft outside of and concentric with that of the HP spool. If the spool gas paths are at different radii this arrangement necessitates short *inter-compressor* and *inter-turbine ducts*, which incur small total pressure losses.

Turbojet with afterburner and convergent–divergent nozzle (Fig. 1a)

For high flight Mach number applications an *afterburner* is often employed, which offers higher thrust from the same turbomachinery. This is also called *reheat*, and involves burning fuel in an additional combustor downstream of the jet pipe. The greatly increased exhaust temperature provides a far higher jet velocity, and the ratios of engine thrust to weight and thrust to unit frontal area are greatly increased.

To enable the jet efflux to be supersonic, and hence achieve the full benefit of the afterburner, a *convergent–divergent nozzle* may be employed. Furthermore, as described in Chapter 7, a nozzle downstream of an afterburner must be of variable area to avoid compressor surge problems due to the increased back pressure on the engine when the afterburner is lit. Usually for engines utilised in this high flight Mach number regime a *convergent–divergent intake* is also employed. This enables efficient diffusion of the ram air from supersonic flight Mach numbers to subsonic flow to suit the compressor or fan. This is achieved via a series of oblique shock waves, which impose a lower total pressure loss than a normal shock wave.

Separate jets turbofan (Fig. 1b)

A schematic diagram of a two spool *separate jets turbofan* is presented above the centre line in Fig. 1b. Here the first compressor is termed a *fan* and supplies flow to a *bypass* as well as a *core* stream. The core stream is akin to a turbojet and provides the *hot thrust*; however, the core turbines also provide power to compress the fan bypass stream.

The bypass stream bypasses the core components via the *bypass duct*, incurring a small total pressure loss. It then enters the *cold nozzle*. The total thrust is the sum of those from both the hot and cold nozzles. As described in Chapter 6, the purpose of the bypass stream is to generate additional thrust with a high mass flow rate, but low jet velocity, which improves *specific fuel consumption* (SFC) relative to a pure turbojet. However, this results in lower ratios of engine thrust to frontal area and weight.

Some turbofans have three spools, with an intermediate pressure (IP) spool as well as the HP and LP spools.

Mixed turbofan with afterburner (Fig. 1b)

This configuration is shown below the centre line in Fig. 1b. Here the two streams are combined in a *mixer* upstream of a common jet pipe with an afterburner and convergent–divergent nozzle to provide high jet velocities for supersonic flight. It is often also beneficial to mix the two streams for turbofans without afterburners, as discussed in Chapter 5.

Ramjet (Fig. 1c)

The *ramjet* is the simplest thrust engine configuration, employing no rotating turbomachinery. The ram air is diffused in a convergent–divergent intake and then passed directly to the combustor. It is accelerated to supersonic jet velocity using a convergent–divergent nozzle. As described in Chapters 1 and 6, the ramjet is only practical for high supersonic flight regimes.

Simple cycle single spool shaft power engine (Fig. 2a)

This engine configuration appears similar to a turbojet apart from the intake and the *exhaust*. The main difference is that all the available pressure at entry to the turbine is expanded to ambient to produce shaft power, apart from a small total pressure loss in the exhaust. After diffusion in the exhaust duct, the gas exit velocity is negligible. This results in turbine power

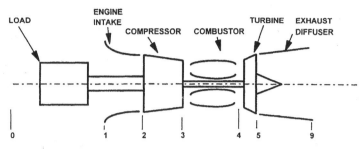

(a) Single spool shaft power engine – shown with cold end drive

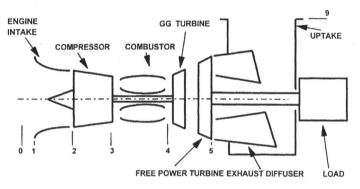

(b) Free power turbine engine – shown with hot end drive

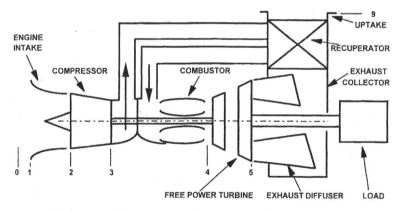

(c) Recuperated free power turbine engine – hot end drive shown

Fig. 2 Shaft power engine configurations and station numbering.

substantially greater than that required to drive the compressor, hence excess power drives the *load*, such as a propeller (*turboprop*) 'or an electrical generator (*turboshaft*)'. The gas temperature at the exhaust exit plane is typically 250 8C to 350 8C hotter than ambient, which represents considerable *waste heat* for an industrial application.

The style of the intake and exhaust varies greatly depending upon the application, though fundamentally the exhaust is normally a diverging, diffusing system as opposed to the jet pipe and nozzle employed by the turbojet for flow acceleration.

The term *simple cycle* is used to distinguish this configuration from the *complex cycles* described later, which utilise additional components such as heat exchangers or steam boilers.

Simple cycle free power turbine engine (Fig. 2b)

Here the load is driven by a *free power turbine* separate from that driving the engine compressor. This has significant impact on *off design* performance, as described in Chapter 7, allowing far greater flexibility in output speed at a power.

Gas generator

The term *gas generator* either describes the compressor and turbine combination that provides the hot, high pressure gas that enters the jet pipe and propelling nozzle for a turbojet, or the free power turbine for a turboshaft. It is common practice to use a given gas generator design for both a turbojet (or turbofan) and an *aero-derivative* free power turbine engine. Here the jet pipe and propelling nozzle are replaced by a power turbine and exhaust system; for turbofans the fan and bypass duct are removed.

Recuperated engine (Fig. 2c)

Here some of the heat that would be lost in the exhaust of a simple cycle is returned to the engine. The heat exchanger used is either a *recuperator* or *regenerator* depending on its configuration (see Chapter 5).

The compressor delivery air is ducted to the *air side* of the heat exchanger, where it receives heat from the exhaust gas passing through the *gas side*. The heated air is then ducted back to the combustor where less fuel is now required to achieve the same turbine entry temperature, which improves specific fuel consumption (SFC). Pressure losses occur in the heat exchanger air and gas sides and the transfer ducts.

Intercooled shaft power engine (Fig. 3a)

Here heat is extracted by an *intercooler* between the first and second compressors. As might be expected, rejecting heat normally worsens SFC, since more fuel must be burnt to raise cooler compressor delivery air to any given turbine entry temperature (SOT). However, intercooling improves engine power output, and potentially even SFC at high pressure ratios via reduced power absorption in the second compressor. This is due to the lower inlet temperature reducing the work required for a given pressure ratio (Chapter 6).

The intercooler rejects heat to an external medium such as sea water. The air side of the intercooler, and any ducting, impose total pressure losses.

Intercooled recuperated shaft power engine

Here both an intercooler and recuperator are employed. The increase in power from intercooling is accompanied by an SFC improvement, as the heat extraction also results in increased heat recovery in the recuperator due to the lower compressor delivery temperature.

Closed cycle (Fig. 3b)

The engine configurations described above are all *open cycle* in that air is drawn from the atmosphere, and only passes through the engine once. In a *closed cycle* configuration the *working fluid* is continuously recirculated. It may be air or another gas such as helium.

Usually the gas turbine is of intercooled recuperated configuration, as shown in Fig. 3b. However the combustor is replaced by a heat exchanger as fuel cannot be burnt directly.

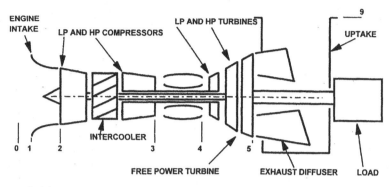

(a) Intercooled free power turbine engine – hot end drive shown

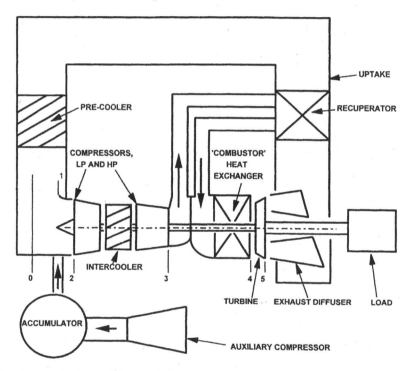

(b) Closed cycle, single spool, intercooled, recuperated shaft power engine

Fig. 3 Intercooled engine configurations and station numbering.

The heat source for the cycle may be a separate combustor burning normally unsuitable fuels such as coal, a nuclear reactor, etc.

On leaving the recuperator, the working fluid must pass through a *pre-cooler* where heat is rejected to an external medium such as sea water to return it to the fixed inlet temperature, usually between 15 8C and 30 8C. The pressure at inlet to the gas turbine is maintained against leakage from the system by an *auxiliary compressor* supplying a large storage tank called an *accumulator*. The high density of the working fluid at engine entry enables a very high power output for a given size of plant, which is the main benefit of the closed cycle. Pressure at inlet to the gas turbine would typically be around twenty times atmospheric. In addition, varying the pressure level allows power regulation without changing SFC.

Combined cycle (Fig. 4a)

Figure 4a shows the simplest *combined cycle* configuration. The gas turbine is otherwise of simple cycle configuration, but with a significant portion of the waste heat recovered in an *HRSG* (Heat Recovery Steam Generator). This is a heat exchanger with the gas turbine exhaust on the hot side, and pumped high pressure water, which forms steam, on the cold side. The first part of the HRSG is the *economiser* where the water is heated at constant pressure until it reaches its saturation temperature, and then vaporises. Once the steam is fully vaporised its temperature is increased further in the *superheater*.

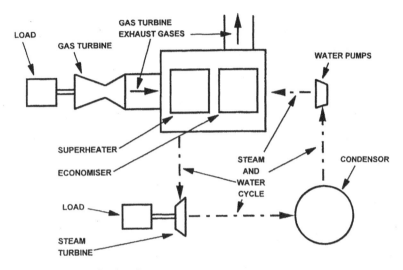

(a)   Single pressure combined cycle

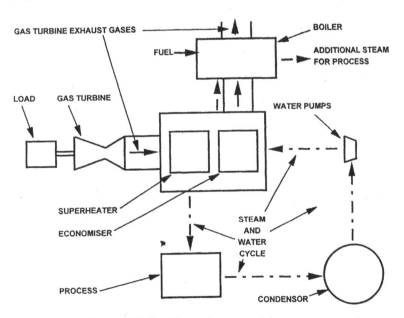

(b)   Combined heat and power (CHP) with supplementary firing

Fig. 4   Shaft power engines with bottoming cycles.

The high pressure, high temperature steam is then expanded across a *steam turbine* which provides up to an extra 45% power in addition to that from the gas turbine. On leaving the steam turbine the steam *wetness fraction* would typically be 10%. The rest of the steam is then condensed, in one of several possible ways. The most common method uses *cooling towers*, where heat is exchanged to cold water, usually pumped from a local source such as a river. When all the steam is condensed the water passes back to the pumps ready to be circulated again. Hence the steam plant is also a 'closed cycle'.

Figure 4a presents a *single pressure* steam cycle configuration. The most complex form of steam cycle used is the *triple pressure reheat*, where steam expands through three turbines in series. In between successive turbines it is returned to the HRSG and the temperature is raised again, usually to the same level as at entry to the first turbine. This cycle has the highest efficiency and specific power.

In combined cycle plant the gas turbine is often referred to as the *topping cycle*, being the hotter, and the steam plant as the *bottoming cycle*.

Combined heat and power – CHP (Fig. 4b)

There are several forms of CHP (*cogeneration*) plant, which are described below in order of increasing complexity.

In the simplest arrangments the gas turbine waste heat is used directly in an industrial process, such as for drying in a paper mill or cement works.

Adding an HRSG downstream of the gas turbine allows conversion of the waste heat to steam, giving greater flexibility in the *process* for which it may be used, such as chemical manufacture, or *space heating* in a hospital or factory.

Finally, Fig. 4b shows the most complex CHP configuration which employs *supplementary firing*. Here the simple cycle gas turbine waste heat is again used to raise steam in an HRSG, which then passes to a boiler where fuel is burnt in the *vitiated* air to raise additional steam. The boiler provides flexibility in the ratio of heat to electrical power. Once the steam has lost all of its useful heat, it passes to a condenser and pumps for re-circulation.

Aeroderivative and heavyweight gas turbines

Outside aero applications, gas turbines for producing shaft power fall into two main categories: *aero-derivative* and *heavyweight*. As implied, the former are direct adaptions of aero engines, with many common parts. The latter are designed with emphasis on low cost rather than low weight, and hence may employ such features as solid rotors and thick casings. At the time of writing both types exist up to 50 MW, with only heavyweights for powers above this.

Chapter 1
Gas Turbine Engine Applications

1.0 Introduction

This chapter provides an insight into why the gas turbine engine has been dominant in certain applications, while having only minimal success in others. Though it has undoubtedly had the greatest impact on aircraft propulsion, background information on *all* potential major applications is provided, including a description of an electrical grid system and how to evaluate marine vessel and automotive vehicle shaft power requirements. An understanding of the application is essential to appreciate fully the wider implications of gas turbine performance.

The attributes of the gas turbine are compared with other competing powerplants in relation to the requirements of each application. This discussion includes the reasons for selecting particular gas turbine configurations and cycles. High reliability and availability are prerequisites. Examples of engines and applications currently in service are provided.

Engine configurations discussed herein, such as simple or combined cycle, are described fully immediately before this chapter. Performance terms used are defined in section 6.2. Aspects of gas turbine performance which impact the choice of engine for a given application are comprehensively presented in later chapters. These include detailed cycle design, transient performance, starting performance, etc.

1.1 Comparison of gas turbine and diesel engines

The gas turbine competes with the *high and medium speed diesel* engines in all non-aero shaft power applications up to 10 MW. It is therefore logical to compare the powerplants before discussing applications individually. *Low speed diesels* are heavier and larger than high/medium speed diesels, the main difference being a higher residence time for fuel vaporisation due to a longer stroke, lower speed and indirect injection. They are used where size is not so important and can burn less refined, lower cost fuel. In most instances the attributes differ so widely from a gas turbine engine that only rarely would both be considered for a given application.

Generalised comparisons with high/medium speed diesels are presented below, with more detailed aspects relevant to particular market sectors discussed later.

- Chart 1.1 compares SFC (specific fuel consumption) and percentage power for gas turbines and diesel engines, at the power level typical of a large truck. Curves for petrol engines are also included, which are discussed in section 1.4.2. The SFC for a simple cycle gas turbine is worse than that for a diesel engine at rated power, in general the lower the rated power the larger the SFC difference, as only simpler gas turbine configurations are viable. This SFC disadvantage also increases significantly as the engines are throttled back, as the gas turbine pressure ratio and firing temperature fall.
- The recuperated gas turbine has an SFC closer to the diesel at the rated power shown on Chart 1.1 of 500 kW. For significantly higher rated powers it is comparable to the diesel. However even at higher rated powers, despite using variable power turbine nozzle guide vanes it is still worse at part load.

- Chart 1.1 also shows that the free power turbine gas turbine can deliver significantly higher torque, hence power, at low engine output speed than a diesel engine. As described in Chapter 7 a single spool gas turbine, where the load and compressor are on the same (and only) shaft, has poor part speed torque capability.
- Where the application can use the waste heat from the engine the gas turbine has a significant advantage. This is because more than 95% of the fuel energy input which is not converted to useful output power appears in the gas turbine exhaust as a single source of high grade (high temperature) heat. For the diesel engine significant proportions of the waste heat appear in a number of low grade forms, such as heat to oil.
- The gas turbine has significantly lower weight per unit of power output. For example a 5 MW *aero-derivative* gas turbine will have a specific weight of less than 1 tonne/MW, whereas a medium speed diesel would be nearer 5 tonne/MW. These values include typical packaging and 'foundations'. This advantage increases with engine output power such that above 10 MW the medium speed diesel engine is rarely a viable competitor to the gas turbine.
- The gas turbine has significantly lower volume per unit of output power. In the 5 MW example above the packaged volume of the gas turbine would approach 50% of that for the diesel engine. Again this advantage increases with power output.
- The start time to idle for a gas turbine is typically 10–60 seconds in applications where it would compete with a high speed diesel. The diesel engine has the advantage of being able to start in less than 5 seconds. The time from idle to full load can be as low as 1 second for a diesel engine or single spool gas turbine, whereas 5 seconds or longer is more typical for a gas turbine with a free power turbine.
- The gas turbine can have dual fuel capability, being able to transfer from natural gas to diesel fuel while running. This is more difficult to achieve for a diesel engine.
- The potential for low emissions of pollutants is an order of magnitude better for gas turbines. This is particularly true for NO_x as the diesel engine cannot use an exhaust catalyst to reduce (de-oxidise) the NO_x because it has excess oxygen in the exhaust. To date this advantage of the gas turbine has not noticeably increased its sales relative to the diesel as legislators have noted the macroeconomics of excluding the diesel from applications where it is the most competitive. Furthermore a diesel produces less CO_2 due to its superior SFC.
- Maintenance costs for a gas turbine are generally lower than for a diesel engine. One contributor here is the relatively low oil consumption of the gas turbine.
- The gas turbine intrinsically has a lower vibration level than a diesel engine.

It is not practical to generalise with respect to unit cost because there are so many diverse factors involved.

1.2 Power generation applications

The first gas turbine in production for electrical power generation was introduced by Brown Boveri of Switzerland in 1937. It was a standby unit with a thermal efficiency of 17%. Today the gas turbine is a major player in the huge power generation market, with orders of around 30 GW per year. This success is due partly to large reserves of natural gas which provide a cheap fuel which is rich in hydrogen, and therefore produces less carbon dioxide than liquid fuels. The other major factor is thermal efficiency, which for combined cycle powerplants approaches 60%. A final advantage is the viability of gas turbines in a very wide range of power levels, up to 300 MW per engine for simple cycle and 500 MW in combined cycle. The market is split evenly between 50 Hz areas such as much of western Europe and the former Soviet Union, and 60 Hz sectors such as North America.

1.2.1 *Major classes of power generation application*

Figure 1.1 summarises the major classes of power generation application for which the gas turbine is a candidate, including examples of actual engines used. The descriptions below refer to idents provided in Fig. 1.1.

Chart 1.2 presents thermal efficiency and *utilisation* for these applications in graphical form, again using the idents from Fig. 1.1. Utilisation is the number of hours per year that the application is typically fired. The interested reader may consult Reference 1, which is updated and reissued annually, for further information.

1.2.2 *The grid system*

Figure 1.2 illustrates a typical grid system for distribution of electrical power, showing voltages at key points. The shaft power system must operate at a constant *synchronous speed* to deliver electrical power at fixed frequency via an alternator. Frequencies are usually 50 Hz or 60 Hz depending upon the country. Reference 2 provides further background.

Until recently the trend was for grids to be supplied by a small number of large power stations. More flexible *distributed power* systems are now becoming popular however, due largely to the gas turbine's viability even at smaller sizes. Here electricity is generated locally to the consumers whether by a CHP (combined heat and power) plant, or to a lesser extent the *mid merit* power stations described later. Excess power is exported to the grid.

To illustrate the levels of power that must be transported, a city of one million people may have a peak demand of up to 2 GW. The average and peak consumption in a home for a family of four is around 1 kW and 6 kW, respectively, excluding space heating.

1.2.3 *Standby generators (idents 1, 2)*

Standby generators are employed for emergency use, where there may be a loss of main supply which cannot be tolerated. Examples include hospitals, and public buildings in areas such as Japan which may be prone to earthquakes. The power generated is used locally and the units are not connected to the grid system. Usually the fuel type is diesel. Key requirements are predominantly driven by the low utilisation, and are outlined below in order of importance:

(1) Low unit cost
(2) Often low unit weight and volume are crucial (see below)
(3) Fast start and acceleration to rated power times may be very important
(4) Thermal efficiency and emissions levels are of secondary importance.

The diesel engine is most popular, primarily due to the plethora of automotive and marine engines in the required power bracket, which reduces unit cost via high production volume. The gas turbine has made some inroads, particularly where weight and volume must be limited, for instance if the standby generator is located on the roof of an office block with limited load bearing capability.

Gas turbine engines in this sector are usually simple cycle, single spool rather than free power turbine engines. This is because the reduced number of components reduces unit cost, and because part speed torque is unimportant. Following the start sequence the engine remains at synchronous speed to ensure constant electrical frequency. Centrifugal compressor systems with pressure ratios of 5:1 to 10:1 are employed to minimise unit cost, and because the efficiency of axial flow compressors at such low flow rates is poor. The turbine blades, and usually the nozzle guide vanes, are uncooled leading to SOT (stator outlet temperature, as described in section 6.2.2) levels of typically 1100–1250 K.

Ident	Plant type	Examples of applications	Examples of engine	Power per engine (MW)
A	Microturbines	Store Small office block Restaurant	Capstone Turbec Ingersoll-Rand	0.04–0.25
1	Standby generator, simple cycle gas turbine	Office block Hospital	Yanmar AT36C, 60C, 180C Turbomeca Astazou	0.25–1.5
2	Standby generator, diesel engine	Office block Hospital	Caterpillar 352 V12 MTU 396	0.25–1.5
3	Small scale CHP, gas turbine	Hospital Small process factory	NP PGT2 Allison 501 Solar Mars Alstom Tempest	0.5–10
4	Small scale CHP, diesel or natural gas fired piston engine	Hospital Small process factory	Petter A MB 190	0.5–10
5	Large scale CHP, gas turbine	Electricity and district heating for town of up to 25 000 people. Large process factory, exporting electricity	Alstom GT10 GE LM2500 RR RB211	10–60
6	Peak lopping units, simple cycle gas turbine	Supply to grid	Alstom GT10 RR RB211 GE LM600	20–60
7	Mid merit power station, simple cycle gas turbine	Supply to grid	GE LM6000 RR Trent	30–60
8	Base load power station, gas turbine in combined cycle	Supply to grid	WEC 501F GE PG9331(FA)	50–450
9	Base load power station, coal fired steam plant	Supply to grid		200–800
10	Base load power station, nuclear powered steam plant	Supply to grid		800–2000

MTU = Motoren Turbinen Union   EGT = European Gas Turbines   NP = Nuovo Pignone
MB = Mirrlees Blackstone   WEC = Westinghouse Electric   GE = General Electric
RR = Rolls-Royce   Company (now part of Siemens)   CHP = Combined heat and power

Figure 1.1   Major classes of power generation plant. (To convert MW to hp multiply by 1341.0.)

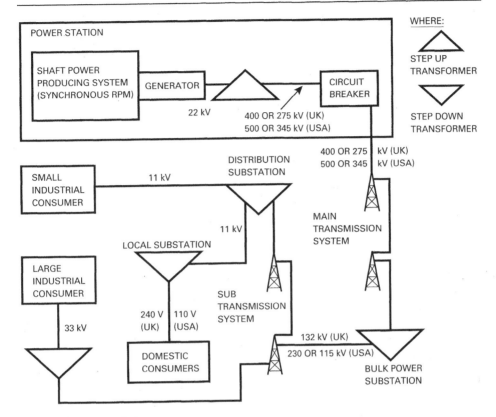

Major shaft power producing systems:
Gas turbine combined cycle plant, where gas turbine waste heat raises steam in a heat recovery steam generator (HSRG) to drive steam turbines
Coal fired boiler which raises steam to drive steam turbines
Nuclear powered boiler which raises steam to drive steam turbines

Figure 1.2   Grid for electrical power generations and distribution.

1.2.4   *Small scale combined heat and power – CHP (idents 3 and 4)*

In this application the waste heat is typically utilised in an industrial process. The heat may be used directly in drying processes or more usually it is converted by an HRSG (heat recovery steam generator) into steam for other uses. Chart 1.3 shows the presentation which manufacturers usually employ to publish the steam raising capability of an engine, specific examples of which may be found in Reference 1. Most CHP systems burn natural gas fuel. The electricity generated is often used locally, and any excess exported to the grid. The key power plant selection criteria in order of importance are:

(1)   Thermal efficiency, for both CHP and simple cycle operation. The latter becomes more significant if for parts of the year there is no use for the full exhaust heat.
(2)   Heat to power ratio is important as electricity is a more valuable commodity than heat. Hence a low ratio is an advantage as the unit may be sized for the heat requirement and any excess electricity sold to the grid.
(3)   The grade (temperature) of the heat is very important in that the process usually demands a high temperature.

(4) Owing to the high utilisation, low unit cost, start and acceleration times are all of secondary importance, as are weight, volume and part speed torque.

The attributes of the gas turbine engine best meet the above criteria, and hence it is the market leader. The diesel engine still retains a strong presence however, particularly for applications where substantial low grade heat is acceptable, or where the importance of simple cycle thermal efficiency is paramount.

Gas turbines are usually custom designed and tend to be single spool. Below 3 MW centrifugal compressors are used exclusively, with pressure ratios of between 8:1 and 15:1. This is a compromise between unit cost, and simple cycle and CHP thermal efficiencies (see Chapter 6). At the lower end of the power bracket SOT tends to be 1300–1400 K, which requires only the first stage turbine nozzle guide vanes to be cooled. At the higher end of the power bracket the first stage rotor blades may also be cooled, allowing SOT levels of up to 1450 K. This becomes viable because of the increased size of the blades, and because the increased unit cost may be supported at the higher power.

The microturbine market has emerged in recent years with a number of forecasts predicting dramatic growth. Small gas turbines of between 40 kW and 250 kW are installed in buildings, such as a store or restaurant, to generate electricity and provide space heating and hot water. A connection with the grid for import/export is usually maintained.

The very small size of microturbine turbomachinery leads to low component efficiencies and pressure ratio, hence to achieve circa 30% thermal efficiency the gas turbine must be recuperated. Otherwise the configuration is extremely simple as low unit cost is critical. Usually it comprises a single centrifugal compressor, DLE 'pipe' combustor, either a radial or two stage axial turbine and the recuperator. Another key feature is a directly driven high speed generator – the size of a gearbox to step down from the turbomachinery speed of typically 90,000 rpm to 3000/3600 rpm is impractical. This also requires power electronics to rectify the 'wild' high frequency generator output into DC, and then convert it back to 50 Hz or 60 Hz AC.

1.2.5 *Large scale CHP (ident 5)*

Here the waste heat is almost exclusively used to raise steam, which is then used in a large process application such as a paper mill, or for district heating. Again the electricity generated may be used locally or exported to the grid. The importance of performance criteria to engine selection are as for small scale CHP, except that emissions legislation is more severe at the larger engine size.

Here gas turbines are used almost exclusively. High grade heat is essential, and the weight and volume of diesel engines prohibitive at these power outputs. Furthermore the gas turbines used are often applicable to other markets, such as oil and gas, and marine, which reduces unit cost. Aero-derivative gas turbines are the most common, though some *heavyweight* engines are used. Aero-derivatives usually employ the core from a large civil turbofan as a gas generator, with a custom designed free power turbine for industrial use. Heavyweight engines are designed specifically for industrial applications and as implied are far heavier than aero-derivatives, their low cost construction employing solid rotors, thick casings, etc.

The gas turbine configuration is usually a free power turbine. While this is not necessary for CHP applications, it is essential to also allow use in oil and gas and marine. Axial flow compressors are used exclusively with overall pressure ratios between 15:1 and 25:1. The aero-derivatives are at the top end of this range as this pressure ratio level results from a civil turbofan core. This pressure ratio is a compromise between that required for optimum CHP thermal efficiency of 20:1, and the 35:1 for optimum simple cycle efficiency. These values apply to the typical SOT of between 1450 K and 1550 K. Advanced cooling systems are employed for at least both the HP turbine first stage nozzle guide vanes and blades.

1.2.6 *Applications which supply solely to a grid system (idents 6 to 10)*

Power plants supplying a grid fall into three categories:
(1) *Peak lopping* engines have a low utilisation, typically less than 10%. They are employed to satisfy the peak demand for electrical power which may occur on mid-weekday evenings as people return home and switch on a multitude of appliances.
(2) *Base load power plant* achieve as near to 100% utilisation as possible to supply the continuous need for electrical power.
(3) *Mid merit power plant* typically have a 30–50% utilisation. They serve the extra demand for electricity which is seasonal, such as the winter period in temperate climates where demand increases for domestic heating and lighting.

The considerations in selecting the type of powerplant for a base load power station are as follows.
(1) Thermal efficiency and availability are paramount.
(2) Unit cost is of high importance as the capital investment, and period of time before the power station comes on line to generate a return on the investment are large.
(3) Cost of electricity is a key factor in selecting the type of powerplant, and fuel price is a major contributor to this. Coal, nuclear and oil fired steam plants all compete with the gas turbine.

In all cases weight and volume are of secondary importance. Other specific comments are as follows.
● For base load plant, start and acceleration times are unimportant.
● For peak lopping power stations unit cost is crucial, time onto full load is very important and thermal efficiency relatively unimportant.
● Mid merit power stations are a compromise with some unit cost increase over and above peak loppers being acceptable in return for a moderate gain in thermal efficiency.

Peak loppers are mostly simple cycle gas turbines burning either diesel or natural gas, and some diesel engines are used at the lower power end. This is because the unit cost and time onto load are far lower than for other available alternatives, which involve steam plant. Both aero-derivative and heavyweight gas turbines are employed as peak loppers, either single spool or free power turbine, and with pressure ratios between 15:1 and 25:1. SOT may be as high as 1500 K, particularly where the unit is also sold for CHP and mechanical drive applications which demand high thermal efficiency.

For base load applications the gas turbine is used in combined cycle, to achieve the maximum possible thermal efficiency. It competes here with coal and nuclear fired steam plant. Historically coal fired plant had the biggest market share. In recent years the combined cycle gas turbine has taken an increasing number of new power station orders due to the availability of natural gas leading to a competitive fuel price, higher thermal efficiency and lower emissions, and the fact that the power stations may often be built with a lower capital investment. This has been supported by advances in gas turbine technology, which have increased both thermal efficiency and the feasible power output from a single engine. In particular, improvements in mechanical design have allowed SOT, and the last stage turbine stress level, to increase significantly. This particular stress is a limiting feature for large single spool engines in that as mass flow is increased at synchronous speed, so too must the turbine exit area to keep acceptable Mach number. Blade root stresses increase in proportion to AN^2 (see Chapter 5). In some countries relatively modern coal fired plant has even 'slid down the merit table' and is now only being used in mid merit applications. However in parts of the world where there is no natural gas and an abundance of coal, such as in China, coal fired steam plant will continue to be built for the foreseeable future. For nuclear power the case is complex, depending upon individual government policies and subsidies.

For base load applications above 50 MW the gas turbines are almost exclusively custom designed, single spool heavyweight configurations. The chosen pressure ratio is the optimum for combined cycle thermal efficiency at the given SOT, though as shown in Chart 6.5 this curve is relatively flat over a wide range of pressure ratios. Usually the higher pressure ratio on this flat portion is chosen to minimise steam plant entry temperature for mechanical design considerations. Engines currently in production are in the 1450–1550 K SOT range, and pressure ratios range from 13:1 to 16:1. For a number of concept engines SOT levels of 1700–1750 K are under consideration, with pressure ratios of 19:1 to 25:1. These employ advanced cycle features such as steam cooling of NGVs and blades. As indicated on Chart 1.2, these engines are targeted at a combined cycle thermal efficiency of 60%. Aero-derivative gas turbines are currently limited to around 50 MW due to the size of the largest aero engines. In this power bracket they are competitive in combined cycle, particularly at higher SOT levels.

Mid merit power stations employ simple cycle gas turbines of a higher technology level than those used for peak lopping. The higher unit cost is justified by the higher thermal efficiency, given the higher utilisation. Most engines are aero-derivative but at pressure ratios of the order of 25:1 to 35:1 for optimum simple cycle thermal efficiency. Corresponding SOT levels are 1500–1600 K.

1.2.7   *Closed cycles*

Here the working fluid, often helium, is recirculated from turbine exit to compressor entry via pre-cooling heat exchangers. Advantages of a *closed*, as opposed to *open*, cycle include the following.

- No inlet filtration requirements, or blade erosion problems.
- Reduced turbomachinery size, due to the working fluid being maintained at a high pressure and density. In addition, helium offers a high specific heat.
- The use of energy sources unsuited to combustion within an open gas turbine cycle, such as nuclear reactors or alternative fuels such as wood and coal. Helium offers a short half life for use in radioactive environments.
- A flat SFC characteristic at part power as compressor entry pressure may be modulated, preserving cycle pressure ratio and SOT.

However, few closed cycle plants have been manufactured despite numerous studies for power generation and submarine propulsion. This is because the above advantages have been offset by high unit cost, and modest thermal efficiency due to the SOT limit of around 1100 K dictated by nuclear reactor or heat exchanger mechanical integrity limits. The high unit cost results from the plant complexity and the implications of designing for very high pressures.

1.3   Industrial mechanical drive applications

Here the engine is used to drive a pump or compressor. The most prolific example is the gas and oil industry which typically orders 1 GW per year of new engines. The majority of engines are installed *onshore*, although there is an *offshore* sector where engines are located on platforms. This industry also has the need for some local *primary power generation* and *emergency power generation*. The requirements here are as per section 1.2 but the importance of low weight and volume discussed in section 1.3.2 is amplified.

1.3.1   *The gas and oil pipeline system*
Figure 1.3 shows the configuration of a natural gas pipeline system, in which gas is pumped from a well head to industrial and domestic consumers. Pipelines have diameters of typically 915 mm (36 in) to 1420 mm (56 in), and are usually underground. A notable exception is in permafrost areas where they must be raised to avoid melting the permafrost. These systems may extend over thousands of kilometres, with compression stations approximately every

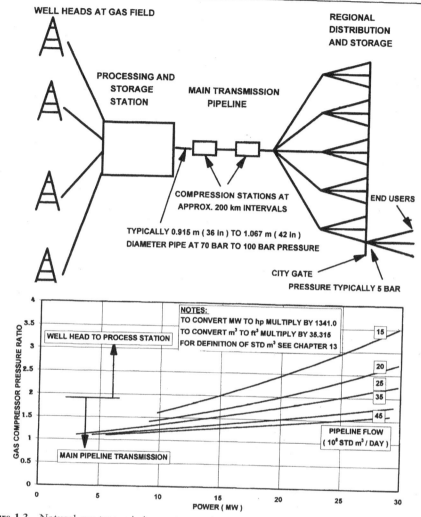

Figure 1.3 Natural gas transmission system and power requirements.

200 km. For example, pipelines run from the Alberta province of Canada to the east coast of the USA. The powerplant burns natural gas tapped off the pipeline, and drives a centrifugal compressor. Figure 1.3 also shows the typical flow rate of natural gas versus pumping power, and the pipeline compressor pressure ratio. For comparison, a family of four may consume up to 10 standard cubic metres per day in the winter period. A further use for gas turbines is to pump water into depleted natural gas fields, to increase gas extraction.

Oil pipelines are less complex. Oil is pumped from a well head to a refinery, and occasionally distillate fuels are then pumped to large industrial users. Extracting oil from the well may involve pumping gas down to raise pressure and to force oil up the extraction pipe by bubbling gas through it.

1.3.2 Engine requirements

The major power blocks required are around 6–10 MW, 15 MW and 25–30 MW. These power levels are generally beyond the practical size for a high speed diesel engine given the requirements outlined below, and hence gas turbines are used almost exclusively. In order of importance these requirements are:

(1)   Low weight, as the engines often have to be transported to remote locations, where it also may be difficult and costly to build substantial foundations
(2)   Good base load thermal efficiency, since utilisation is as near 100% as possible
(3)   Reasonable part power torque, to respond to load changes on the gas compressor. However a fast start time is not essential, and a loading rate of 2 minutes from idle to full power is typical.

For offshore well heads the engine must be located on a platform, hence the importance of low weight is amplified and low volume essential. While the gas or oil may have a high pressure as it comes out of the ground it invariably needs further pressurisation to pipe it back onshore. The engine may drive the compression unit mechanically as described above, or sometimes via an electric motor. In the latter instance a CHP arrangement supplies power for other needs, such as electricity for the drill and heat for uses such as natural gas processing or space heating.

For the middle and higher power bracket, simple cycle, free power turbine aero-derivatives best meet the above criteria, and are used almost exclusively. Pressure ratios of 20:1 to 25:1 and SOT levels of 1450–1550 K are typical, leading to thermal efficiency levels in the mid to high thirties. Engines include the Rolls-Royce RB211 and the GE LM2500, which are also utilised in power generation applications. For the lower power bracket, both custom designed industrial engines such as the Solar Mars, and aeroderivatives such as the Allison 501, are used with lower pressure ratio and SOT levels leading to thermal efficiencies in the low thirties. Reference 1 provides further details.

1.4   Automotive applications

1.4.1   *The gas turbine versus reciprocating engines*

The first gas turbine propelled automotive vehicle was the Rover JET1 produced in the UK in 1950, the design team being led by Maurice Wilks and Frank Bell. The engine had a free power turbine and produced 150 kW from a simple cycle; the vehicle fuel consumption was 5.4 km/litre (15.2 mpg). Over the ensuing decades significant effort has been spent on automotive programmes, however the diesel and petrol engines have continued to dominate, with the gas turbine only achieving a presence in specialist applications. The contributory issues are explored in this section, however there are three key reasons:

(1)   The poor part load thermal efficiency of the gas turbine, despite the use of a recuperated cycle with variable area nozzle guide vanes (see Chart 1.1). To improve thermal efficiency, ceramic turbine technology has been researched for decades, but progress towards a production standard has been frustratingly slow.
(2)   There is a relatively long acceleration time of the gas turbine gas generator spool from idle to full load.
(3)   Huge capital investment would be required in gas turbine manufacturing facilities.

These disadvantages have mostly outweighed the benefits of the gas turbine, which are:

●   Better part speed torque capability as described in Chart 1.1, which reduces the need for varying gear ratios
●   Lower weight and volume per unit power
●   Potential for significantly lower emissions.

One other use for gas turbine engines has been in thrust propelled vehicles for attempts on the world land speed record. In 1983 Richard Nobel's 'Thrust 2' achieved 1019 km/h (633 mph) using a Rolls-Royce Avon turbojet. In 1997 his 'Thrust SSC' piloted by Andrew Green exceeded the speed of sound, and set a new world land speed record of 1220 km/h (763 mph), using two Rolls-Royce Spey turbofans.

1.4.2 *The petrol engine versus the diesel engine*

Chart 1.1 includes curves of thermal efficiency and torque versus part load power for the petrol engine. Overall it has worse SFC than the diesel engine because to avoid pre-ignition its compression ratio is lower, typically 8:1 to 10:1, contrasting with 15:1 to 20:1 for diesels. Whereas the Otto cycle in a petrol engine has combustion at constant volume, producing increased pressure, in a diesel engine it is at constant pressure. Both engines can be turbo-charged to increase power, by raising inlet density and hence air mass flow. The weight and size saving can be significant, though with some expense in terms of response time, due to *turbo lag* as the turbocharger spool accelerates.

The main advantage of the petrol engine is that it has lower weight and volume, which approach those for the gas turbine at the 50 kW required for a typical family saloon. Hence petrol engines are used where fast vehicle acceleration is essential, space is at a premium, and some worsening of SFC is acceptable. Diesel engines dominate for applications such as trucks where fuel consumption is paramount due to high utilisation, engine weight and volume relative to the vehicle are low, and high vehicle acceleration is not a priority.

Ident	Vehicle class	Examples of vehicles	Engines utilised	Power at ISO (kW)
1	Family saloon car	Ford Mondeo Honda Accord Pontiac Phoenix (experimental)	4 CYL, 1.6–2.0 litre PE 4 CYL, 1.8–2.3 litre PE Allison AGT 100 GT	40–100
2	Family saloon car – hybrid electric vehicle	Volvo ECC (experimental)	Sodium sulphur Battery + gas turbine	50–60
3	Family saloon car – luxury	Jaguar XKR Mercedes Benz S Class	8 CYL, 4 litre PE 6 CYL, 3.2 litre PE	190–220
4	Supercar	Porsche 911 turbo Ferrari Testarossa	6 CYL, 3.3 litre TC PE 12 CYL Flat 5.3 litre PE	180–350
5	Formula 1 racing car	Williams FW12 Benetton B189	8 CYL, 3.5 litre PE 8 CYL, Ford HBV8 PE	500–550
6	Large truck	Scania 4 Series Ford Transcontinental H Series British Leyland Marathon T37 (experimental)	11.7 litre 6 CYL DSC12 DE Cummins NTC 355 DE Rover 2S/350R GT	300–450
7	Main battle tank	Royal Ordanance Challenger Chrysler M1 Abrams Bofors STRV 103 (experimental)	Caterpillar 12 CYL DE TL AGT-1500 GT 2 × DE, 1 × GT (boost)	900–1150

PE = Petrol engine   TC = Turbocharged   TL = Textron Lycoming   ISO = Standard ambient,
DE = Diesel engine   GT = Gas turbine   GM = General Motors   at sea level
CYL = Cylinder

To convert kW to hp multiply by 1.3410.

Figure 1.4 Major categories of automotive vehicle. Examples and engine types.

One further difference is in emissions. Though the diesel engine uses less fuel and therefore produces less CO_2, its exhaust contains more particulates and NO_x.

1.4.3 *Major classes of automotive vehicle*

Figure 1.4 provides an overview of the major automotive vehicle categories, including examples of actual vehicles and the engines utilised. Chart 1.4 uses the idents for each category defined in Fig. 1.4, and presents key facets of each vehicle type versus power required for propulsion. Reference 3 provides further information.

1.4.4 *Automotive vehicle power requirements (Formulae F1.1–F1.5)*

Figure 1.5 shows the elements which contribute to the total power requirement of an automotive vehicle, namely:

- Aerodynamic drag
- Rolling resistance, i.e. the power lost to the tyres
- Hill climb
- Acceleration

Formulae F1.1–F1.5 (at the end of the chapter) and the data provided in Chart 1.4 enable the reader to calculate approximate power requirements for a given vehicle type. Figure 1.6 provides typical coefficients for evaluating rolling resistance, and sample calculation C1.1 shows the process for a family saloon car.

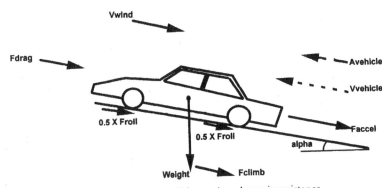

FORCES ACTING ON AUTOMOTIVE VEHICLE

F = Force,   A = Acceleration,   V = Velocity,   Fdrag = Aerodynamic resistance
Froll = Rolling resistance,   alpha = Angle of slope above horizontal
Fclimb = Component of gravitational force opposing motion = weight × sin(alpha)
Faccel = Force for vehicle acceleration
Fpropulsive and PWpropulsive = Total propulsive force and power to maintain vehicle velocity (Vvehicle)
and acceleration (Avehicle)
Fdrag = (Vwind + Vvehicle)² × drag coefficient × projected frontal area
Fpropulsive = Fdrag + Froll + Fclimb + Faccel
PWpropulsive = Fpropulsive * Vvehicle

Notes:
Forces shown as resistances
see Formulae F1.1 to F1.5

Figure 1.5 Automotive vehicle forces and power requirements.

Surface	Coefficient of rolling friction
Paving stones	0.015
Smooth concrete	0.015
Rolled gravel	0.02
Tarmacadam	0.025
Dirt road	0.05
Tracked vehicle on arable soil	0.07–0.12

Figure 1.6 Rolling resistance coefficients for radial tyres on various surfaces.

Chart 1.5 shows the relative magnitude of contributory power requirements for a truck and family saloon versus road speed. The greater contribution of rolling resistance for the truck is apparent. In both instances the power requirement at constant speed on a level road (aerodynamic drag plus rolling resistance) approximates to a cube law. This has particular significance for the family saloon where typical cruising speeds of 100 km/h and 50 km/h are at only 20% and 6% power respectively, the excess being available for acceleration and the rarely used top speed. This contrasts with the truck where the 100 km/h cruising speed is at 65% power.

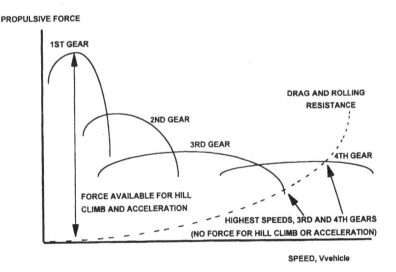

Notes:

Engine power output is propulsive force * vehicle velocity.

1st gear at high power provides high force for hill climb and acceleration, but does not allow high vehicle speed.

4th gear at high power allows high vehicle speed, but limited power is available for hill climb and acceleration.

Figure typical for piston engine, gas turbine needs less gears.

Figure 1.7 Diesel and petrol engines: the need for gearing.

1.4.5   *The need for gearing (Formulae F1.6 and F1.7)*

In almost all applications the engine speed required for the power output differs from that of the wheels, requiring some gear ratio in the transmission. In addition, for petrol and diesel engines it must be variable depending on vehicle speed as neither powerplant conforms to the approximate cube law of power versus rotational speed, as shown in Fig. 1.7. Furthermore, variable gearing enhances the piston engine's capability to provide excess power and torque at the wheel for acceleration. Chart 1.1 shows that the output torque of a piston engine falls at reduced engine rotational speed. A fixed gear ratio would mean that at low vehicle speed the engine, also being at low speed, could only produce low torque and hence power. Variable gearing enables the engine to run at high speed and power at low vehicle speeds, with high torque at the wheels.

As described in Chapter 7, the gas turbine engine with a free power turbine can readily track a cube law of output power versus rotational speed, better matching vehicle requirements. As shown in Chart 1.1, it also offers excellent torque and power at low engine output speeds for hill climb and vehicle acceleration. This is achieved by operating the gas generator at high speed and power output while the power turbine is at low speed. Hence only a small number of gears is required, which is an advantage of the gas turbine for automotive applications, notwithstanding that the gear ratios are higher.

Formulae F1.6 and F1.7 show the key interrelationships resulting from gearing. Sample calculation C1.2 shows their use and illustrates the need for gearing described above.

1.4.6   *The average and luxury family saloon (idents 1 and 3)*

For a family saloon there are several key engine requirements. In order of importance these are:

(1) Low weight and volume
(2) Fast engine acceleration, for vehicle performance
(3) Good part power SFC, down to less than 5% power where significant portions of the driving cycle occur

Despite many development programmes, such as the General Motors AGT100 highlighted in Fig. 1.4, no gas turbine has reached production. This is due to the reasons discussed in section 1.4.1. The petrol engine is most popular, with lower weight and volume relative to a diesel engine to outweigh its worse fuel consumption. Around 20% of the market is taken by diesel engines, where fuel efficiency is gained at the expense of vehicle acceleration due to the higher engine weight.

Automotive gas turbine development programmes have always used a recuperated cycle with variable power turbine nozzle guide vanes to minimise part load SFC, as described in Chapter 7. An intercooled recuperated cycle would provide further improvements, but the weight, volume and cost incurred by an intercooler is prohibitive at this engine size. The cycles always employ centrifugal compressors to suit the low air flows, and the free power turbine provides good torque at part power. Typically, at maximum power SOT is around 1200 K to avoid the need for any turbine cooling, which would be expensive and difficult for such a small engine size. As per Chart 6.5 the optimum pressure ratio for SFC of around 5:1 is used. These two parameters together provide acceptable temperatures at the power turbine and recuperator gas side inlet. In the medium term ceramic turbine blading may allow some increase in SOT and hence engine performance.

1.4.7   *The hybrid electric vehicle (ident 2)*

Severe *zero emissions* legislation is creating a significant niche market in certain parts of the world, such as in California. Pure electric vehicles have limited performance and range before battery recharging is required, even those using the most advanced battery systems such as sodium sulphur. To overcome this, a heat engine may also be fitted in one of two possible *hybrid* configurations, currently the subject of development programmes and studies.

Figure 1.8 describes the various modes in which the two hybrid electric vehicle configurations may operate:

- For a *range extender* there are two modes of operation. Mode A is that of a conventional electric vehicle, where the battery provides traction power to the wheel motors via power electronics. In mode B the heat engine drives a generator which provides power to charge the battery. In this application when operative the gas turbine runs at maximum power, where response is unimportant and the SFC difference versus piston engines is lowest.

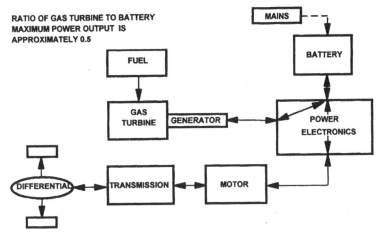

(a)   Range extender hybrid powerplant

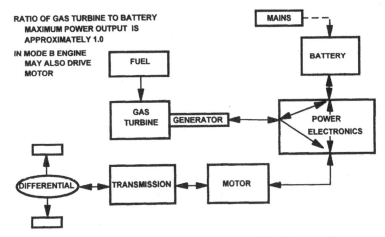

(b)   Full hybrid powerplant

Notes:
Arrows show direction of flow of power.
Differential not required if multiple wheel motors employed.
Mode A – Battery powers vehicle
Mode B – Engine charges battery

Figure 1.8   Hybrid electric vehicle powerplant configurations.

- For a *full hybrid powerplant* in mode B the engine may also provide traction power to the wheels via the generator, power electronics and motors. This would usually be at high vehicle speeds and power levels on out of town highways.

In both cases if wheel motors are utilised regenerative braking is employed where energy is recovered via the motor acting as a generator to charge the batteries. The battery is also utilised for starting the gas turbine engine during the driving cycle. References 4–7 provide a more exhaustive description of hybrid electric vehicles.

The powerplant requirements in order of importance are:

(1)   Low emissions
(2)   Low volume and weight, to accommodate the battery system, etc.
(3)   High output rotational speed, to facilitate a compact high speed directly driven generator
(4)   Good SFC at rated power

The gas turbine is the most suited to these criteria, with its Achilles' heel of poor low power SFC and acceleration time being unimportant due to the operating regime. Indeed gas turbines are being utilised in most development programmes underway.

The gas turbine configuration is usually single spool with the generator driven at fixed speed. Again a recuperated cycle is employed, but without turbine variable geometry given the low importance of part power SFC.

1.4.8   *The supercar and high speed racing car (idents 4 and 5)*

For a supercar the requirements are similar to those for a saloon but with more emphasis on vehicle performance and less on fuel consumption. Petrol engines are used exclusively due to weight and volume advantages over the diesel engine. Power turn down from full power to idle here is even greater than for a saloon and hence the gas turbine is not suitable.

The high speed racing car takes the weight and volume arguments to the extreme, and again uses petrol engines exclusively despite the lower importance of part power SFC. Occasionally gas turbine engines have been fitted to racing cars, for example Rover–BRM Le Mans entries in the 1960s for demonstration purposes. One notable achievement was that gas turbine powered cars won the Indianapolis 500 race on several occasions, before restrictions on intake orifice size made them uncompetitive.

1.4.9   *The truck (ident 6)*

The large truck has a high utilisation, as well as spending typically 80% of its driving cycle at 65–100% power. Hence its demands for a powerplant are:

(1)   Good high power SFC
(2)   Good part speed torque to accelerate the vehicle
(3)   Engine weight and volume are less important as the powerplant is a relatively small percentage of the vehicle size.

Currently, trucks are powered exclusively by diesel engines. Owing to the reduced time at low power the gas turbine is more suitable for trucks than for the saloon or supercar, and its high part load torque reduces the number of gears required from approximately 12 to only 3–5. However it has not reached production despite many development programmes world-wide, for the three prime reasons as described in section 1.4.1. In recent years further truck gas turbine programmes have been considered as a result of tougher emissions legislation. However, it is unlikely that legislation will prevent the diesel engine from being used because of the huge costs this would incur.

Gas turbine truck engine programmes to date have utilised a similar configuration and cycle to those described for the family saloon. However, if a programme were launched at the time of writing, an SOT of 1350 K would be more likely, requiring HP turbine nozzle guide vane cooling, and a pressure ratio of around 7:1 corresponding to optimum SFC. Ceramic blading, which would lead to a further increase in SOT, is being actively researched by a number of companies world-wide.

1.4.10   *The main battle tank (ident 7)*

The technical requirements for a tank powerplant in order of priority are:

(1)   Low volume due to the need for a multitude of on board systems
(2)   Excellent part speed torque for hill climb and vehicle acceleration
(3)   Good SFC, both at high power and part load
(4)   Low weight

The diesel and gas turbine have different advantages relative to the above requirements. The gas turbine's volume and weight advantages are supplemented by superior maintenance, cold starting, multifuel capabilities and quieter operation. The diesel engine offers lower SFC, but not a large cost advantage as the power level is above that of other large volume automotive applications. The diesel has the largest share of this market, but the gas turbine has a significant presence.

The most notable gas turbine application is the Abrams M1 tank shown in Fig. 1.4, of which around 11 000 have been produced. The engine configuration is again recuperated, free power turbine with variable area power turbine nozzle guide vanes. SOT is around 1470 K, requiring cooled HP turbine nozzle guide vanes and rotor blades; this is viable due to the larger engine size. An all-axial LP and an axi-centrifugal HP compressor produce a pressure ratio of over 14:1. This is above the optimum for design point SFC, but is the optimum for specific power, benefiting engine size and weight. One powerful mechanism for this is in the reduced volume of the recuperator, which is comparable in size to the rest of the engine. Again, in the future ceramic blading may allow further increases in SOT and hence engine performance.

1.5   Marine applications

Marine propulsion uses diesel engines, gas turbines, or oil or nuclear fired steam plant. Diesel engines are split into two main groups. The smaller *high and medium speed* (750 rpm to 1500 rpm) varieties burn a highly refined light diesel fuel as per marine gas turbines. Larger *low speed,* or *cathedral* diesels burn far heavier diesel oil, the low speed (120 rpm) and indirect injection not requiring rapid fuel vaporisation for combustion. While most marine propulsion uses diesel engines the gas turbine is popular in certain applications.

The first instance of naval propulsion using gas turbines was in 1947 in the UK using a Metrovick 'Gatric' engine in a modified gun boat. This was based on the F2 jet engine but with a free power turbine in the tail pipe and burning diesel. Sea trials lasted four years and convinced doubters that operation of a simple cycle lightweight engine at sea was practical. Metopolitan Vickers was later taken over by Rolls-Royce.

Another early development was the Rolls-Royce RM60 double intercooled and recuperated engine, of 4.0 MW. This had a flat SFC curve and was intended as a single engine for small ships, and as a cruise engine for larger ones. It was fitted to *HMS Grey Goose* in 1953, which became the world's first solely gas turbine propelled ship and spent four years at sea. Though mostly technically successful, the engine did not see production, being too complex for the patrol boat role and inferior to diesels as a cruise engine.

The first operational case was the use of three Bristol Engine Company (again later taken over by Rolls-Royce) Proteus engines in a fast patrol boat in 1958.

Marine propulsion system requirements differ significantly from land based units. Owing to the large vessel inertia engine acceleration time is generally not critical. Also the impact of

Ident	Vessel type	Examples of vessel	Engines utilised	Total power (MW)
1	Medium hovercraft	BHC AP1-88 Multi purpose Textron LACV-30 landing craft	4 × DEUTZ BF12L 12 CYL diesels 2 × PW ST6T GTs	1.5–3
2	Large hovercraft	BHC SR N4 passenger vessel Westamarin passenger vessel	4 × RR Proteus GTs 2 × MTU 396 diesels 2 × DEUTZ MWM diesels	3.5–10
3	Patrol boat	Souter Shipyard Wasp Bollinger Shipyard Island Class	2 × GM 16 V diesels 2 × PV diesels	2.5–4.5
4	Luxury yacht	Chritensen CXV Denison Marine Thunderbolt	2 × CAT 3412 diesels 2 × MTU 12U 396 diesels	0.5–3
5	Fast ferry	Yuet Hing Marine Catamaran Aquastrada Monohull	2 × TL TF40 GTs 1 × LM2500 GT, plus 2 × MTU 595 diesels in CODOG	6–30
6	Large merchant container	Hellenic Explorer Lloyd Nipponica	6 × diesels Boiler plus STs	20–40
7	Ultra large tanker	Sumitomo King Opama Uddevalla Nanny	Boiler plus STs Boiler plus STs	30–40
8	Attack submarine	General Dynamics Sturgeon (USN) Vickers Fleet Class (RN)	1 × PWR plus STs 1 × PWR plus STs	10–20
9	Ballistic submarine	General Dynamics Ohio Class (USN) Vickers Vanguard Class (RN)	1 × PWR plus STs 1 × PWR plus STs	40–45
10	Frigate	Yarrow Shipyard Type 23 (RN) BIW Oliver Hazard Perry Class (USN)	2 × RR SM1C GTs plus 4 × PV diesels in CODLAG 2 × GE LM2500 GTs	30–40
11	Destroyer	BIW Arleigh Burke Class (USN) RN Type 22	4 × GE LM2500 GTs 2 × RR SM1C GTs plus 2 × RR Tyne GTs in COGAG	45–75
12	Light aircraft carrier	BIW Intrepid Class (USN) Vickers Invincible Class (RN)	Boilers plus 4 × STs 4 × RR Olympus GTs	100–120
13	Large aircraft carrier	Newport News Nimitz Class (USN) Newport News J F Kennedy Class (USN)	2 × PWRs plus STs Boilers plus 4 × STs	180–220

BHC = British Hovercraft Co
RR = Rolls-Royce
TL = Textron Lycoming
PWR = Pressurised Water Reactors
BIW = Bath Iron Works

PW = Pratt & Whitney
PV = Paxman Valenta
GTs = Gas Turbines
RN = UK Royal Navy

GM = General Motors
CAT = Caterpillar
STs = Steam Turbines
USN = US Navy

Figure 1.9 Major classes of marine vessel.

current emissions legislation is negligible, particularly out at sea where pollution concentrations are low. The International Maritime Organisation (IMO) is reluctant to introduce stringent legislation which a gas turbine could meet but diesel engines could not.

1.5.1   *Major classes of marine vessel*

The major classes of marine vessel for which gas turbine engines are a candidate are summarised in Fig. 1.9. Examples of actual vessels and the engines used are provided. The gas turbine competes with the diesel engine and nuclear power plant utilising boilers and steam turbines. At the time of writing oil fired steam plant is becoming rare in new vessels, but remains in service.

Chart 1.6 presents key characteristics of these vessel classes in graphical form. The interested reader may consult References 8 and 9 for further information.

1.5.2   *Marine vessel propulsion requirements (Formulae F1.8 and F1.9)*

Figure 1.10 illustrates and quantifies the elements which comprise the total power requirement for vessel forward motion. A vessel moving through calm water creates two wave forms, one with a high water pressure at the bow, the other a reduced pressure at the stern. The energy to create this wave system is derived from the vessel via the *wave making resistance*. At high speed the wave resistance is dominant. Indeed for a given hull design a *critical hull speed* for wave making resistance is reached, where the vessel literally climbs a hill of water, with the propulsion thrust tilting upward, and it is uneconomical to go beyond this speed. The sinusoidal effect visible superimposed on the curve of wave making resistance is due to interactions of the bow and stern wave systems. The *skin friction resistance*, or *friction form resistance*, is also a major contributor to the total resistance. This is the friction between the hull and the water.

The *pressure resistance, hydrodynamic drag* or *form drag* is due to flow separations of the water around the hull creating an adverse pressure field. Any resulting eddies or vortices are in addition to the waves created by the wave making resistance.

The *air resistance* due to the drag of the vessel above the water line contributes less than 5% of the total resistance. Often for low speed vessels little effort is spent in aerodynamic profiling.

The above four resistances comprise the *naked resistance*. The *appendage resistance* must then be added to evaluate the total resistance. This is the losses incurred by rudders, bilge keels, propellers, etc. and is less than 10% of the total resistance.

Traditionally these resistances for a vessel design are evaluated by model testing in a water tank and then using non-dimensional groups to scale up the resulting formulae and coefficients to the actual vessel size. This process is complex; References 10 and 11 provide an exhaustive description. The power requirement approximates to a cube law versus vessel velocity for *displacement hulls*, which support weight by simple buoyancy. For simple calculations Formula F1.9 may be used, which shows that the resistance is also dependent on vessel *displacement* (i.e. weight). Sample calculation C1.3 illustrates how Chart 1.6 and Formula F1.9 may be used to calculate vessel power requirements. Power approaches a square law for semi planing hulls, which produce lift hydrodynamically.

1.5.3   *Engine load characteristics (Formula F1.10)*

Ship engines drive either a conventional propeller or a *waterjet* via a gearbox. The latter consists of an enclosed pump which sends a jet of water rearwards. Since both devices pump incompressible water, power versus shaft speed adheres closely to a cube law, as shown by Formula F1.10. For a propeller the vessel speed determines the shaft speed; in the absence of propeller blade *slippage* these are uniquely related. As the number of engines driving changes, engine operation moves between different possible cube laws, with slippage likely for fewer propellers driving. In the engine concept design phase cube laws are a reasonable assumption,

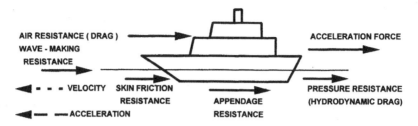

Notes:

Forces are shown as resistances

Power = total force * velocity

Total force = sum of all components

See Formulae F1.8 to F1.9

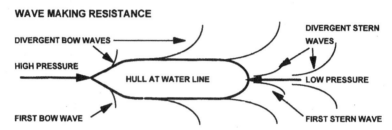

Note: Energy to create wave system is derived from vessel, hence is a resistance

(a) Forces acting on marine vessel

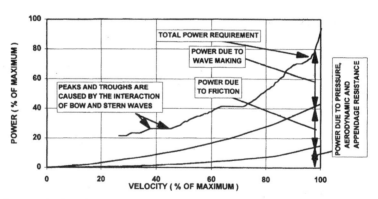

(b) Relative magnitudes of components of total power requirements (no acceleration)

Figure 1.10 Marine vessel power and force requirements.

however at the earliest opportunity the law(s) for the actual propeller or waterjet should be obtained from the manufacturer. Variable pitch propellers are often employed, which mainly affect the lowest speed characteristics.

Chart 1.7 presents power required versus ship speed for a typical displacement hull vessel adhering to Formula 1.9. Curves are presented for a twin engine, twin propeller vessel with one or two of the engines driving. With one shaft driving the power required to maintain a steady ship speed is approximately 20% higher than with two, due to the drag of the unused propeller. For waterjets this does not apply, as blocker doors are normally closed for an unused unit.

Chart 1.7 also shows engine output power versus engine output rotational speed. This also adheres to an approximate cube law and shows separate lines for one or two engines driving.

Also shown are the resulting characteristic of engine speed versus ship speed. The cube laws for power versus ship speed, and power versus engine output speed combine to make engine output speed directly proportional to ship speed. With one engine driving, however, engine output speed is almost 20% higher than with two to achieve a given ship speed, hence propeller speed, due to slippage of the loaded propeller. These multiple load characteristics must be considered when designing a gas turbine for a multi engine vessel.

1.5.4 *CODAG, CODOG, COGAG and CODLAG propulsion systems*

Figure 1.11 presents schematic diagrams of the above systems.

- In *CODAG* (COmbined Diesel And Gas turbine) systems a diesel engine provides propulsive power at low ship speed, but at high speeds the gas turbine is fired providing the relatively large additional power requirement dictated by the cube law.
- In *CODOG* (COmbined Diesel Or Gas turbine) systems the gearing is arranged such that only the diesel or gas turbine may drive the propeller or water jet at a given time.
- *COGAG* (COmbined Gas turbine And Gas turbine) and COGOG systems use a small gas turbine at low ship speed and/or a larger engine for high ship speeds.
- *CODLAG* (COmbined Diesel eLectric And Gas turbine) systems use an electric motor and diesel powered generator for low speeds. An important feature is low noise for antisubmarine work.
- *IED* (Integrated Electric Drive) or *FEP* (Full Electric Propulsion) is the subject of serious study for large naval vessels. Here the gas turbine drives a generator to provide electrical power for propulsion, ship services or, crucially, future weapons systems. A constant output speed of 3600 rpm is likely, though a gearbox or even new high power frequency converters may be employed. Such electric propulsion has long been employed on merchant vessels, but not to date on gas turbine warships.

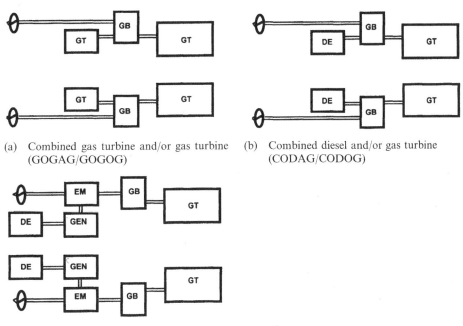

(a) Combined gas turbine and/or gas turbine (GOGAG/GOGOG)

(b) Combined diesel and/or gas turbine (CODAG/CODOG)

(c) Combined diesel–electric and gas turbine (CODLAG)

GT = Gas Turbine,   DE = Diesel Engine,
EM = Electric Motor,   GEN = Generator

Figure 1.11 Marine powerplant configurations using gas turbines.

1.5.5   *Hovercraft (idents 1 and 2)*

Hovercraft use a fan to maintain pressure under side skirts to hover above the water surface, and air propellers to provide thrust for propulsion. The ratio of power required for propulsion to that for hovering is between 5:1 and 10:1. As illustrated in Chart 1.6, these vessels are designed for high speed. They are generally used in applications such as commercial passenger ferries or landing craft where the vessel spends most of its time at maximum speed. The key powerplant criteria in order of importance are:

(1)   Low weight
(2)   Good high power SFC
(3)   Good part power SFC

The output power is below 10 MW, hence a diesel engine is practical. The diesel engine has the best high power SFC but the gas turbine the lowest weight. Consequently both powerplant types are used. The gas turbine configuration is simple cycle for minimum weight, with a free power turbine. Typically SOT levels of 1250 K are employed with pressure ratios of around 7:1, the low values partly reflecting the predominance of older engine designs in current applications.

1.5.6   *Monohull patrol boat and luxury yacht (idents 3 and 4)*

These vessels spend most of their time cruising at low speed. Because of the cube law relationship of output power to vessel speed most of the operational time is at a very low percentage output power, even allowing for twin engine vessels only cruising on one engine. The key requirement is good part load SFC; given this and the small size the diesel engine dominates.

1.5.7   *Fast ferry (ident 5)*

Fast ferries for commercial passenger transport represent a growing market. These often employ *catamaran* multiple hull configurations, and *hydrofoil* operation where a lifting surface raises much of the hull above the water at high speeds. While the vessel speed is still below that of the hovercraft it has the advantage of being able to operate in rough seas. Like the hovercraft, most of its time is at high power, and key requirements are as for the hovercraft. Low weight is particularly important given the higher speed than for other classes. Power required increases with weight even for a hydrofoil, and the fuel weight for the range must be considered when operating at high speed.

For large fast ferries the power requirement is beyond that of a diesel engine and the gas turbine dominates. There is often a CODOG arrangement with diesel engines for harbour manoeuvring. At the smaller end the gas turbine and diesel share the market.

Simple cycle gas turbines are employed, usually low weight aero-derivatives, with pressure ratios of 15:1 to 25:1. SOT is between 1450 and 1550 K with advanced cooled nozzle guide vanes and rotor blades.

1.5.8   *Large merchant container and ultra large tanker (idents 6 and 7)*

Large merchant container ships are also propelled by either oil fired boilers with steam turbines or nowadays diesel engines, giving top speeds of around 25 knots. The choice of top forward speed has varied, depending largely on fuel price. When fuel prices are low getting the cargo to market faster becomes dominant, and transatlantic carriers with speeds of up to 40 knots have been proposed using gas turbines. When fuel prices are high fuel cost considerations dictate lower speeds.

The ultra large tanker (or supertanker) is the largest vessel class at sea. It operates mostly at its relatively low maximum speed of around 15 knots. Owing to the huge size, engine weight and volume are relatively unimportant, and there is free space available beneath the crew accommodation superstructure. These vessels are almost exclusively propelled by large, slow speed cathedral diesels, though older designs used oil fired steam plant. The diesel powerplant is extremely heavy and bulky but the fuel is less refined and of far lower cost per kilowatt than that used for high speed diesels and gas turbines.

1.5.9   *Attack and ballistic submarines (idents 8 and 9)*

Owing to the elimination of refuelling, and the ability to sustain full speed under water, nuclear reactors and steam turbines are used for most modern submarines. Some lower cost, smaller attack submarines are diesel–electric.

1.5.10   *Frigate, destroyer and light aircraft carrier (idents 10, 11 and 12)*

These vessels spend most of their time on station, at low vessel speed. However substantially higher power levels are also required for sustained periods for transit to an operational zone. Hence the key powerplant criteria are:

(1)   Good part load SFC
(2)   Minimum weight, to be able to achieve high vessel speeds
(3)   Minimum volume, due to the need for many on board systems and personnel
(4)   High availability, i.e. low maintenance and high reliability

Here CODOG, COGAG and CODLAG systems predominate. The required output power at high speed would require a large number of diesels with unacceptable weight and volume, and so gas turbines are used for main engines. To achieve good SFC at cruise either diesels or smaller gas turbines are also employed. The gas turbine configuration utilised is as per fast ferries.

At the time of writing, one significant new marine gas turbine development programme is the WR21, a 25 MW class intercooled and recuperated engine funded by the US, UK and French navies. The aim is to reduce fuel usage by 30% versus existing simple cycle engines, the heat exchangers and variable power turbine nozzle guide vanes providing a very flat SFC curve to suit naval operating profiles (see Chapter 7). Rotating components are closely based on the Rolls-Royce RB211 and Trent aero turbofans.

1.5.11   *Large aircraft carrier (ident 13)*

The 'supercarrier' has the largest power requirement of any marine vessel type. The operational profile is akin to that of the other naval vessels described above.

At this power level, diesel or gas turbine installations are significantly larger than nuclear ones, especially considering the number of engines required and their fuel tanks and ducting. Hence a pressurised water nuclear reactor is employed with boilers and steam turbines. The size of the 'island' (superstructure) is reduced without engine intake and exhaust ducts, allowing deckspace for around two more aircraft for the same vessel displacement. The resulting smaller ship profile also reduces radar cross-section, and the lack of exhaust smoke and heat further reduces signatures. The elimination of engine refuelling is an advantage, though tanker support is still needed for the embarked aircraft.

1.6   Aircraft applications – propulsion requirements

The concept of using a gas turbine for jet propulsion was first patented by Guillame in France in 1921. Prior to this Rene Lorin had obtained a patent for a ramjet as early as 1908. In January 1930 Sir Frank Whittle, unaware of the earlier French patents, also obtained a patent for a turbojet in the UK. Whittle's first engine, the world's first, ran on a test bed in April 1937. The world's first flight of a turbojet propelled aircraft was the Heinkel He 178 in Germany, with Hans von Ohain's He S-3b engine, on 27 August 1939. This had been bench tested in early 1939; an earlier test in March 1937 had been hydrogen fuelled and hence not a practical engine. Whittle, dogged by lack of investment, finally got his W1A engine airborne propelling the Gloster E28/39 on 15 May 1941. The first flight of a turboprop was on 20 September 1945, the Rolls-Royce Trent powering a converted Meteor. The Trent was discontinued after five were built as Rolls-Royce concentrated on the Dart, which became the first turboprop in airline service. It should be noted that Rolls-Royce has used the name 'Trent' again in the 1990s for its latest series of large civil turbofan engines.

The gas turbine has entirely replaced the piston engine for most aircraft applications. This is in marked contrast with the automotive market discussed earlier. The difference for aircraft propulsion was that the gas turbine could deliver something the piston engine is incapable of – practical high speed aircraft, and much lower engine weight and size. For example, the thrust of the four turbofans on a modern Boeing 747 would require around one hundred World War II Merlin engines, which would then be far too heavy.

1.6.1   *Aircraft flight mechanics (Formulae F1.11–F1.16)*

Figure 1.12 shows the four forces acting on an aircraft:

(1)   The lift is due to the static pressure field around the aircraft, mainly from its wings which have a *cambered* upper surface to accelerate flow and reduce static pressure. Lift acts normal to the incident velocity, through the *centre of pressure*. As described below, for a given aircraft increasing lift usually increases drag.
(2)   The weight of the aircraft acts vertically downwards through the centre of gravity. In level flight the lift must equal the weight.
(3)   The drag acts against the direction of motion through the *centre of drag*. In level flight it acts horizontally.
(4)   The engine thrust acts along the engine centre line. In level, steady flight thrust acts very close to horizontally forwards, and must equal drag.

In steady flight the aircraft control surfaces must be set to balance any couple created by the above forces. Fuel usage moves the centre of gravity, hence large modern aircraft control fuel distribution to minimise drag. If horizontal or vertical acceleration of the aircraft is required there must be an imbalance of the above forces, as described in section 1.6.2 below.

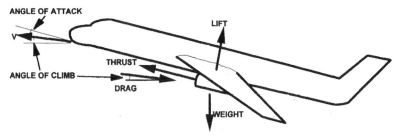

Notes:
For propeller powered aircraft, power = required thrust ∗ VTAS/propellor efficiency.
For all aircraft thrust − drag = weight ∗ sine of climb angle
Excess power to climb = Vclimb ∗ weight

In near level flight
Lift = Weight   Thrust = Drag
Acceleration requires excess thrust or lift.

Figure 1.12   Forces acting on an aircraft.

Formulae F1.11 and F1.12 show how lift and drag forces are related to incident dynamic pressure and hence equivalent and true air speeds using the *lift coefficient* and the *drag coefficient*. Lift and drag forces are proportional to equivalent air speed squared. (At altitude equivalent air speed is lower than true air speed, being that times the square root of the density ratio, as described in Chapter 2.) For a given aircraft design the lift and drag coefficients *are a function of only the angle of attack*. Their values are usually derived using computer simulation and model tests in a wind tunnel, followed by confirmatory flight testing.

For a fixed aircraft weight the lift force must be constant at all steady flight conditions. Changing the angle of attack changes both lift and drag coefficients; therefore to maintain steady flight there must be *one angle of attack for each equivalent air speed*. Hence as shown by

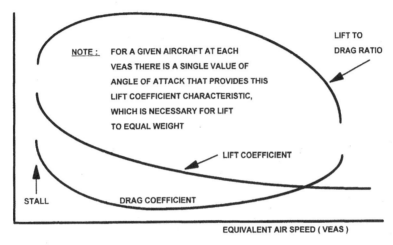

(a)   Coefficients of lift and drag; and lift to drag ratio (level, steady flight)

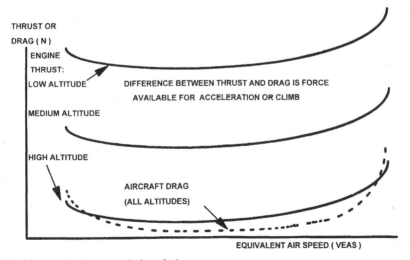

(b)   Matching engine thrust and aircraft drag

Notes:
Power increases more steeply with VEAS than does thrust.
VEAS = VTAS* SQRT ((Density at altitude)/(Density at sea level)) (This definition maintains dynamic pressure.)

Figure 1.13   Aircraft lift and drag characteristics, hence thrust requirements.

Formulae F1.13 and F1.14 for a given aircraft design *lift and drag coefficient are also a function of only equivalent air speed*, the form of the relationship also being shown in Fig. 1.13. Typical angles of attack are 158 at stall and 08 at maximum equivalent air speed.

Formula 1.14 shows that the drag coefficient comprises the following two components.

(1)   The *induced drag coefficient* is a function of the lift coefficient. It is the major contributor to total drag at low forward speeds, where a high lift coefficient and hence high angle of attack are required.

(2)   The *parasitic drag coefficient* reflects the basic drag due to the shape of the airframe and its appendages, as well as skin friction. It is the major contributor to total drag at high speeds.

The interaction of these two terms provides the characteristic shape of drag coefficient versus equivalent air speed shown in Fig. 1.13.

The *lift to drag ratio*, defined in Formula F1.15 is a measure of the efficiency of the airframe design. This is illustrated by Formula 1.16 which shows that the net thrust required for a given equivalent air speed is inversely proportional to lift to drag ratio. The form of its relationship to equivalent air speed is also shown on Fig. 1.13, being dictated by the ratio of the lift and drag coefficients. The lift to drag ratio for a subsonic transport in cruise may approach 20, whereas for a supersonic transport it will be less than 10. The value may fall to less than 5 for a fighter aircraft in combat at low altitude, and rise to 55 for a high performance glider.

Sample calculation C1.4 illustrates the use of the above formulae. References 12–14 provide a comprehensive description of aircraft flight mechanics.

1.6.2   *The flight mission and aircraft thrust requirements*

The major phases of a flight mission are *take off*, *climb*, *cruise*, *descent* and *landing*. For military aircraft *combat* must also be considered, and all aircraft must turn, albeit briefly. Figure 1.13 shows drag versus equivalent air speed, by definition (Formula F1.12) this relationship is independent of altitude. Lines of engine thrust versus equivalent air speed for low, medium and high altitude are superimposed onto Fig. 1.13. The background to the form of these lines is presented in Chapter 7. At low and medium altitudes considerable excess thrust beyond aircraft drag is available. The major phases of the flight mission are discussed below in relation to Fig. 1.13.

During takeoff, high excess thrust is available for acceleration. Typical takeoff velocity and distance for a fighter are 140 kt (0.21 Mach number) and 1.2 km, respectively. Corresponding values for a civil aircraft are up to 180 kt (0.27 Mach number) and 3 km. Takeoff is a key flight condition for engine design, with usually the highest SOT.

In order to climb, additional upwards force is required. This is achieved by maintaining a high angle of attack to increase lift coefficient. The resulting increased drag (Formula F1.14) is overcome by increasing thrust, excess being available beyond that required for steady flight. Also, a component of the thrust is directed vertically. The excess thrust available at low altitudes provides a high rate of climb, typically 500 m/min for a subsonic transport, and up to 8000 m/min for a fighter. Flight speed during climb is initially at a fixed level of equivalent air speed due to airframe structural considerations (maintaining constant dynamic pressure), and then at the limiting flight Mach number for airframe aerodynamics once achieved. At the top of climb the maximum engine thrust is just equal to the aircraft drag. This is a key sizing condition for the engine, with highest referred speeds (see Chapter 4) and hence referred air flow. It is not the highest SOT, due to the lower ambient temperature.

Aircraft usually cruise at high altitude because here the true air speed achieved for the given level of equivalent air speed is significantly higher (Formula F2.16), and because engine fuel consumption is minimised by the correspondingly lower thrust requirement. The choice of cruise altitude is complex, and depends on engine size required to achieve the altitude, the true air speed and range demanded by the market sector, etc.; Reference 12 provides an excellent description. The flight envelopes presented in Chapter 2 show the outcome of these considerations with the cruise point generally being close to the top right hand corner. The optimum altitude for cruise generally increases with the required level of flight Mach number. Over a long period at cruise required thrust may reduce by 20% of that at the top of climb, due to the reduction in aircraft fuel weight. Some aircraft therefore climb gradually as weight reduces, known as *cruise-climb*.

During descent the engines are throttled back to a *flight idle rating* and the aircraft angle of attack reduced. Both these effects reduce lift, and the flight direction is below horizontal. A component of the weight now acts in the direction of travel, supplementing the engine thrust to overcome drag. With zero engine thrust this would be *gliding*.

Turning requires centripetal force, provided by *banking* the aircraft to point the wings' lift radially inwards. To support the weight the overall lift must be increased, hence also the thrust as drag thereby increases.

The approach for landing is on a glide slope of approximately 38, with a high angle of attack and flaps set to reduce aircraft speed as far as possible to give the required lift. Typically landing speeds are between 120 kt (0.18 Mach number) and 140 kt (0.21 Mach number). Landing distances are substantially less than those required for take off, as deceleration due to *reverse thrust* or brakes and spoilers is faster than the takeoff acceleration. Most turbofan propelled aircraft employ engines with a reverse thrust capability, where the bypass air is diverted forward using either louvres in the nacelle or rearward clamshell doors.

Afterburning or *reheat* is often employed for fighter aircraft and supersonic transport. Fighters generally use it only for short durations due to the high fuel consumption. These are specific manoeuvres such as take off, transition to supersonic flight, combat or at extreme corners of the operational envelope. Generally supersonic transports such as Concorde use it for takeoff and supersonic transition.

1.6.3 *Engine configuration selection for a required flight regime (Formula F1.17)*

The key parameters of interest are:

- SFC, especially at a reasonably high thrust or power level corresponding to cruise. Other levels such as climb and descent become more important for short ranges.
- Weight and frontal area (hence engine nacelle drag), particularly for high Mach number applications.
- Cost – this can increase with engine/aircraft size, but for expendable applications such as missiles it must be as low as practical.

The gas turbine engine achieves adequate acceleration times of around 5 seconds for civil engines and 4 seconds for military, so this does not give it or a piston engine any competitive advantage.

Range factor (Formula F1.17) is the most commonly used parameter to assess the suitability of engine configurations for a required flight mission. It is the ratio of the weight of fuel and engine to the engine net thrust less pod drag (see section 5.5.4) for a range and flight speed. Clearly a low value of range factor is better. Sample calculation C1.5 shows its use for a turbofan engine. Chart 1.8 presents range factor versus flight Mach number for ranges of 1000 km and 8000 km, and a number of engine configurations including a piston engine.

It is immediately apparent why the gas turbine so readily replaced the piston engine for most aircraft propulsion: the latter is only in contention at low Mach numbers, below around 0.3. This is primarily because propulsion power requirements increase rapidly with Mach number, as shown in Fig. 1.13. The weight and frontal area of a piston engine increase far more rapidly with output power than they do for the gas turbine. The immense importance of these factors at high flight speed is quantified by the range factor diagram.

Above 0.3 Mach number the weight and frontal area considerations mean the turboprop takes over from the piston engine as the optimum powerplant. It has better fuel consumption than a turbojet or turbofan, due to a high *propulsive efficiency* (see Chapter 6), achieving thrust by a high mass flow of air from the propeller at low jet velocity. Above 0.6 Mach number the turboprop in turn becomes uncompetitive, due mainly to higher weight and frontal area. In addition, high propeller tip speeds required are a difficult mechanical design issue, and the high tip relative Mach numbers create extreme noise.

Above 0.6 Mach number the turbofan and turbojet compete, the optimum choice depending on the application. As shown by the design point diagrams in Chapter 6 the turbofan has a better SFC than the turbojet, but at the expense of worse specific thrust and hence weight and frontal area. Increasing bypass ratio provides the following engine trade offs:

- SFC improves
- The capability for reverse thrust improves
- Weight per unit thrust increases
- Frontal area per unit thrust increases (see section 5.5.4 for calculation of pod drag)
- The number of LP turbine stages to drive the fan increases rapidly
- The cost per unit of thrust increases
- Auxiliary power and bleed offtake have a more detrimental effect upon performance.

The high bypass ratio engine is most competitive at flight Mach numbers of approximately 0.8, whereas at 2.2 Mach number the ideal bypass ratio is less than 1 and a turbojet becomes increasingly competitive.

As shown in Chapter 6, above around 2.0 Mach number the specific thrust of the ramjet becomes even better than that of a turbojet, however it has poorer specific fuel consumption. The impact of this on range factor is shown on Chart 1.8. The low engine frontal area and weight resulting from the high specific thrust dominates at low range and high Mach number where the ramjet becomes the most competitive powerplant. Also applications to date requiring this flight regime have been missiles and hence the lower unit cost of the ramjet is beneficial. Furthermore Chart 2.11 shows engine ram inlet temperature ratio versus flight Mach number and altitude. For turbojet mechanical integrity the compressor delivery temperature must be kept to below approximately 950 K, hence above 2.5 flight Mach number there is very little room for compressor temperature rise.

The other possible powerplant is a rocket, which is beyond the scope of this discussion.

1.7   Shaft powered aircraft – turboprops and turboshafts

This section describes the requirements of shaft powered aircraft, while section 1.8 covers thrust propelled aircraft. The term *turboprops* usually refers to gas turbine engines which provide shaft power to drive a propeller for *fixed wing* aircraft propulsion. Those providing power for a *rotary wing* aircraft, or helicopter, are referred to as *turboshafts*.

1.7.1   *Comparison of propulsion requirements of shaft power and thrust propelled aircraft*

The *equivalent thrust* and *equivalent SFC* of a turboprop may be calculated, allowing first cut comparisons of thrust and shaft power engines for a given application. Formulae 1.18 and 1.19 provide approximate conversion factors. Furthermore these formulae may be used to convert the small amount of thrust available in a turboprop exhaust into an *equivalent shaft power*. This may be added to the delivered shaft power to get a *total equivalent shaft power*, and a corresponding SFC may be defined.

1.7.2 *Major classes of shaft powered aircraft*

Figure 1.14 presents the major classes of shaft powered aircraft together with examples of actual aircraft and the engines utilised. Chart 1.9 presents key and interesting characteristics of these aircraft classes using the idents from Fig. 1.14. Aircraft take off weight, range, maximum speed and number of seats are plotted versus required power. The interested reader may consult Reference 15 for further information.

1.7.3 *Fixed wing aircraft (idents 1, 2 and 3)*

Light aircraft are often privately owned, and used for short range transport or recreation. The business/executive turboprop is usually owned corporately to give flexibility in transporting executives. The commuter, or regional, transport turboprop is operated by commercial airlines on routes of moderate range, where the reduction in journey time offered by thrust aircraft would be of minimal benefit.

The piston engine now has only a few applications in the aircraft industry, one being for light aircraft with top speeds of less than 200 kt (0.30 Mach number). As shown by the range factors described in section 1.6.3 the piston engine is only competitive at such low flight speeds.

Ident	Aircraft type	Examples of aircraft	Engines utilised	Total shaft power (kW)
1	Light aircraft, piston engines	Piper Warrior II Beech Bonanza	1 × TL 0320-D3G flat twin 1 × TC IO 520 BB flat 6	120–220
2	Business/executive Turboprop	Piper Cheyenne 400 Cessna Caravan Dornier 228-100	2 × Garrett TPE331 1 × PW PT6A-114 2 × Garrett TPE331	500–1200
3	Commuter/regional Transport turboprop	BAE Jetstream 41 Shorts 330 BAe ATP Fokker 50	2 × Garrett TPE331 2 × PW PT6A-45R 2 × PW 126A 2 × PW 125B	1800–4000
4	Light helicopter, piston engines	Robinson R22 Schweizer 300C	1 × TL O-32-B2C flat 4 1 × TL HIO-360-DIA	120–170
5	Light helicopter, turboshaft engines	Bell-Jetranger III Bell 406	1 × Allison 250-C20J 1 × Allison 250-C30R	300–500
6	Multirole medium helicopter	Sikorsky S-70A (Black Hawk) Westland/Augusta EH101	2 × GE T700-700 3 × GE T700-401A, or 3 × RR/TM RTM322	2300–3500
7	Heavy lift helicopter	Sikorsky H53E Boeing Chinook CH-47	3 × GE T64-416 2 × TL T55-712	6500–10 000

TL = Textron Lycoming   GE = General Electric
TC = Teledyne Continental   RR = Rolls-Royce
PW = Pratt & Whitney   BAe = British Aerospace

To convert kW to hp multiply by 1.3410.

Figure 1.14 Major categories of turboprop/turboshaft aircraft.

In addition, at the low power levels required the gas turbine suffers from small scale effects, such as small blade heights and relatively thick trailing edges and fillet radii, which increasingly degrade its efficiency as size reduces.

For the flight speeds and ranges demanded by business and commuter aircraft the range factor diagrams show the turboprop to be more competitive. Also at the engine powers above 250 kW the gas turbine is clear of the worst of the small scale effects.

Engines are almost always of free power turbine configuration with a single spool or occasionally two spool gas generator. Compressors are either centrifugal or axi-centrifugal as this minimises cost, efficiency is reasonably competitive with axial compressors at such low flow levels, and because frontal area is not critical at the moderate flight speeds involved. Pressure ratio is usually in the range 7:1 to 10:1. Axial flow turbine systems are employed with SOT levels of between 1250 and 1450 K. Above 1350 K, rotor blade cooling is employed. The choice of pressure ratio reflects a compromise between a lower value reducing the cost and weight of the compression system, and a higher value improving SFC and specific power if compressor efficiency is maintained.

1.7.4   *Rotary wing aircraft (idents 4, 5, 6 and 7)*

Here the key criteria in order of importance are:

(1) Engine weight
(2) Part power SFC, as maximum power will either be sized for hot day operation, or for a multi-engine helicopter the engine failure case
(3) Rated SFC
(4) Engine frontal area is not particularly significant due to the low flight speeds and 'buried' installation.

To minimise weight some small turboshafts are single spool, which is possible because rotor pitch may be varied to change load at constant speed. For medium turboshaft helicopters the engine configuration is as per the turboprop engines described above. Levels of pressure ratio and SOT are up to around 17:1 and 1500 K respectively, the latter requiring turbine blade cooling. This pressure ratio is the optimum for specific power. At the largest engine size fully axial compressors are employed. Occasionally recuperated cycles have been considered for long range helicopters to minimise fuel weight, though none have come to fruition. This is primarily due to the increased engine cost, weight and volume, and reliability concerns. Piston engines are used only at the lowest power levels.

1.8   Thrust propelled aircraft – turbofans, turbojets and ramjets

1.8.1   *Major classes of thrust propelled aircraft*

Figure 1.15 presents the major classes of thrust propelled aircraft, together with examples of actual aircraft and the engines utilised. Chart 1.10 presents characteristics of these aircraft classes using the idents from Fig. 1.15. Again Reference 15 may be consulted for further information.

1.8.2   *Unmanned vehicle systems (ident 1)*

Unmanned vehicle systems include aircraft such as target and reconnaissance drones, decoys used by military aircraft to divert threats, and long range cruise missiles.

For expendable target drones and decoys the highest priority is minimum unit cost. A Mach number of at least 0.8 is usually required, with only a low range requirement. Single spool turbojets are usually used, often with centrifugal compressors because of their low cost and the

Ident	Aircraft type	Examples of aircraft	Engines utilised	Total thrust ISA SLS T/O (kN)
1	Unmanned Vehicle Systems (UVS)	Beech MQM 107B Target Drone IMI Delilah Decoy GD BGM-109 Tomahawk long range cruise missile	1 × MT TR160-2-097 TJET 1 × NPT 151 TJET 1 × WI F107-WR-103 TFAN	1–5
2	Business/executive jet	Swearingen SJ30 Gulfstream IV–X BAe 125 Series 800	2 × WI/RR FJ44 TFANS 2 × RR TAY 611-8 TFANS 2 × GT TFE731-5R-1H TFANS	15–120
3	Short–medium range civil transport	Fokker 100 Boeing 737-400 Airbus A320	2 × RR TAY 620 TFANS 2 × CFM56-3B-2 TFANS 2 × IAE V2500-A1 TFANS, or 2 × CFM56-5 TFANS	120–220
4	Long range civil transport	Airbus A340–500 Boeing 777	4 × RR Trent 500 TFANS 2 × PW4090 TFANS, or 2 × GE90 TFANS, or 2 × RR Trent 892 TFANS	500–1000
5	Supersonic civil transport	BAe/Aerospatiale Concorde	4 × SNECMA/RR Olympus 593 TJETS	600–700
6	Military trainer/light attack aircraft	BAe Hawk Aermachi MB-339C	1 × RR/TM Adour TFAN 1 × RR Viper TJET	20–25
7	Advanced military fighter	General Dynamics F16 Falcon Eurofighter Typhoon McDonnell Douglas F15C	1 × GE F110-GE-100 TFAN, or 1 × PW F100-PW-220 TFAN 2 × EJ 200 TFANS 2 × PW F100-PW220 TFANS	80–220
8	Ramjet propelled missiles	BAe Sea Dart (ship to air) BAe Bloodhound (gr. to air)	1 × RR ODIN 1 × RR THOR	N/A N/A

WI = Williams International   RR = Rolls-Royce   CFM = GE/SNECMA Joint Venture
MT = Microturbo   NPT = Noel Penny Turbines   IAE = International Aero Engines
IMI = Israeli Military Industries   BAe – British Aerospace   EJ – Eurojet
PW = Pratt & Whitney   GE = General Electric
GT = Garrett   TM = Turbomeca

To convert kN to lbf multiply by 224.809.

Figure 1.15 Major types of thrust propelled aircraft.

low mass flow rates. Any increased weight and frontal area is accepted. Engine pressure ratios are usually between 4:1 and 8:1 as a compromise between low values favouring weight and frontal area, and high values favouring SFC and specific thrust. Low SOT levels of around 1250 K avoid the need for turbine cooling (and also give better SFC for a turbojet). Both axial and radial turbines are used.

The long range required by cruise missiles means that they fit the turbofan regime with SFC a key issue, though engine size and cost are also important as the vehicle must be transported and is expendable. Medium bypass ratio turbofans are employed, with centrifugal compressors. Indicative cycle parameters are 1.5:1 bypass ratio, 10:1 pressure ratio and 1250 K SOT.

1.8.3   *Subsonic commercial aircraft and military trainer (idents 2, 3, 4 and 6)*

Business/executive jets and civil subsonic transports all have range and flight Mach number requirements fitting the turbofan regime. They all use multi-spool gas generators with axial flow turbomachinery (except at the smallest sizes) and sophisticated turbine blade cooling for the best SFC. The pressure ratio is selected from cycle charts to give the best cruise SFC for the given SOT.

The highest bypass ratio for an engine in production at the time of writing is 8.5:1. At ISA SLS takeoff, advanced engines utilise a fan pressure ratio of around 1.8:1, and overall pressure ratio exceeds 40:1. The corresponding SOT is around 1650 K, rising to over 1750 K on a hot day. At ISA cruise overall pressure ratio is around 10% lower, and SOT around 1400 K. The highest overall pressure ratio in the flight envelope is around 45:1 at the top of climb. For lower technology engines bypass ratio is nearer to 4:1. At ISA SLS takeoff, fan pressure ratio is approximately 1.8:1 and overall pressure ratio 25:1, with SOT around 1525 K. At cruise, pressure ratio is around 10% lower, and SOT around 1350 K.

Military trainer aircraft are again in the subsonic regime, but range requirements are shorter and unit cost very important. Here turbojets and turbofans compete.

1.8.4   *Supersonic civil transport and advanced military fighter (idents 5 and 7)*

As shown by the range factor diagrams discussed in section 1.6.3 the only engines viable here are turbojets, or turbofans with a bypass ratio of less than 1:1. Multi-spool configurations with all axial turbomachinery are used for maximum efficiency and minimum frontal area. All engines have reheat systems which are employed at key points in the operational envelope (see Chapter 5).

For the limited civil applications to date, such as Concorde, and US and Russian development programmes, afterburning turbojets have been utilised. Take off SOT exceeds 1600 K, though higher values might be chosen for more modern engine designs. Pressure ratios of around 14:1 have been employed to minimise weight, and because higher values are not practical due to the high compressor delivery temperature at high flight Mach number. In addition, this value is the optimum for a pure turbojet specific thrust around a Mach number of 1.0, and also for reheated operation at 2.2 Mach number. At this flight speed the reheat fuel is burnt at a high enough pressure that SFC is little worse than for a pure jet, though thrust is significantly higher. Studies for future applications encompass variable cycles, where higher bypass ratio minimises noise and SFC during subsonic overland flight.

Advanced military fighters use low bypass ratio afterburning turbofans, with maximum SOT exceeding 1850 K and pressure ratio around 25:1. Combustor inlet temperature approaches 900 K. Future engine designs are considering SOT levels of 2000 to 2100 K, with combustor inlet temperatures nearer 1000 K, requiring ceramic materials. Again, engine designs with variable cycles have been proposed, to achieve higher bypass ratio to improve SFC at low flight Mach number.

One other military aircraft application is for short/vertical takeoff/landing (VTOL or STOVL), operational forms of which have utilised two main approaches. The UK/US Harrier has a fixed geometry turbofan (RR Pegasus) with four rotatable propelling nozzles, two for the core stream and two for the bypass. In contrast, Russian Yakalov aircraft have used separate vertically mounted lift jets. Future variable cycle engines could be beneficial for a Harrier type approach, providing additional bypass air for jet borne flight.

1.8.5 *Ramjet propelled missiles (ident 8)*

Section 1.6.3 showed that at Mach numbers in excess of 2.5 the ramjet is the ideal powerplant. Combustion temperatures approach the *stoichiometric* value, where all oxygen is used. This ranges from 2300 to 2500 K, depending on inlet temperature and hence flight Mach number, and is feasible as there are no stressed turbine blades to consider. At these Mach numbers the only competitor engine is a rocket. Indeed as discussed in Chapter 9, starting a ramjet requires a short duration booster rocket, to accelerate the vehicle to a Mach number where operation is possible.

Air to air missiles to date have been almost entirely rocket powered, as this better suits the requirement of high thrust for a short duration. However experimental ramjet versions have been produced, particularly in France and the former USSR. Several current proposals involve ramjets, as air to air missile range requirements increase.

Surface to air missiles with ramjets have seen production, such as the UK 'Bloodhound', as range requirements are more suitable. A typical mission would be launch, climb to around 20 000 m, followed by a loiter phase and then attack. The distance covered would be around 50 km.

1.9 Auxiliary power units (APUs)

Aircraft APUs have normally fulfilled several functions in an aircraft, namely:

- Main engine starting
- Supply of cooling air for aircraft secondary systems, particularly when at ground idle in hot climates
- Supply of electrical power when main engines are shut down, including for ground checkout of aircraft systems

These functions give an aircraft self sufficiency when on the ground. In addition an APU will be required to fire up at altitude in case of main engine flame out, to power electrical systems – vital for fly by wire aircraft – and if at low flight Mach number to provide crank assistance to help restart the engines.

Until recently, new developments have been rare, but APU sophistication is now increasing to match that of recent aircraft, where APU operation is becoming less intermittent. For civil applications APU requirements may now include operation in all regions of the flight envelope, and for military aircraft advanced systems with start times as low as a second, as described in Reference 17. A current typical start time is around 6 seconds at 15 000 m. There are occasional studies on the benefits of permanent running power units which avoid compromising the design of the propulsion engines by power and bleed offtake.

Historically the main requirements for APUs have been:

(1) Low development and unit costs
(2) High reliability and maintainability
(3) Low volume and weight
(4) Good SFC

Reference 17 discusses these issues comprehensively.

1.9.1 *Gas turbines versus piston engines*

APUs for aircraft are almost exclusively simple cycle gas turbines. Power density in terms of weight and volume per unit of shaft output power are vastly superior to a piston engine, around $4.4\,kW/kg$ and $8\,MW/m^3$. This effectively makes a piston engine impractical, despite its lower unit cost. Fuel consumption becomes a secondary issue where operation is intermittent.

1.9.2 *APU power requirements of major aircraft classes*

The output power range of APUs is between $10\,kW$ and $300\,kW$, with bleed supplied requiring additional turbine power. Figure 1.16 presents specific examples of APUs employed in production aircraft.

1.9.3 *APU configurations*

For all configurations centrifugal compressors are used exclusively and often radial inflow turbines, even occasionally combined as a *monorotor* to minimise cost. SOT levels are typically 1250 to 1260 K to minimise the need for turbine cooling. Pressure ratio is generally between $4:1$ and $8:1$, though the trend is towards higher levels.

Model	Configuration	Application	Power (kW)
Turbomach T-62T-40-8	Single shaft: 1 Stage centrifugal compressor Reverse flow annular combustor 1 Stage radial turbine	Jet fuel starter General Dynamics F16 Fighter	190
Allied Signal 131-9(D)	Single shaft: 1 Stage centrifugal compressor 2 Stage axial turbine 1 Stage centrifugal load compressor	Bleed, E.G. engine start Electrical power ENV conditioning McDonnell Douglas MD90	300/100
Allied Signal 331-500B	Single shaft: 2 Stage centrifugal compressor Reverse flow annular combustor 2 Stage axial turbine 1 Stage centrifugal load compressor	Bleed, E.G. engine start Electrical power Boeing 777	850/170
APIC APS 3200	Single shaft: 1 Stage centrifugal compressor Reverse flow annular combustor 2 Stage axial turbine 1 Stage centrifugal load compressor	Electrical power Oil and fuel pumps Airbus A321	385/90

Notes:
All data is indicative.
Where two powers are shown the higher figure includes the load compressor drive power.
To convert kW to hp multiply by 1.3410.
To convert kg to lb multiply by 2.2046.
APIC – Auxilliary Power International Company.

Figure 1.16 Auxilliary power unit (APU) examples and applications.

The most common forms of APU provide high pressure air to the main engine mounted air turbine starter. These are referred to as *pneumatic APUs*. Air must usually be supplied at around five or more times ambient pressure, with the APU sized to enable hot day main engine starting. The most common pneumatic APU is a single shaft gas turbine with integral bleed. Here the engine is of single spool configuration but with the pneumatic air supply bled off from compressor delivery. This is the simplest unit and hence has the lowest cost. Also generators or pumps may be driven off the spool to provide electrical or hydraulic power.

Single shaft gas turbines driving a centrifugal load compressor, as well as application pumps or generator, are growing in popularity. This configuration has the highest power output per unit mass and volume, though is of higher cost.

A small number of APUs apply torque directly to the main engine HP shaft via its gearbox and a clutch, rather than supplying high pressure air. These are termed *jet fuel starters*. In this instance the APU is often of free power turbine configuration to provide an adequate part speed torque characteristic.

Formulae

F1.1   Automotive vehicle: Drag (kN) = fn(drag coefficient, air density (kg/m³), frontal area (m²), vehicle velocity (m/s), wind velocity (m/s))

$$Fdrag = 0.5 * RHO * Cdrag * A * (Vvehicle + Vwind)^2/1000$$

(i)   See Fig. 1.5, and also Chart 1.4 for typical drag coefficients and vehicle frontal area.

F1.2   Automotive vehicle: Rolling resistance (kN) = fn(coefficient of rolling resistance, vehicle mass (tonnes))

$$Froll = Crol * m * g$$

(i)   See Fig. 1.5, and also Fig. 1.6 for typical coefficients of rolling resistance.
(ii) $g = 9.807 \, m/s^2$.

F1.3   Automotive vehicle: Force for hill climb (kN) = fn(hill gradient (deg), vehicle mass (tonnes))

$$Fclimb = m * g * sin(alpha)$$

(i)   See Fig. 1.5.
(ii) $g = 9.807 \, m/s^2$.

F1.4   Automotive vehicle: Force for acceleration (kN) = fn(acceleration rate (m/s²), vehicle mass (tonnes))

$$Faccel = m * a$$

F1.5   Automotive vehicle: Total propulsive power requirement (kW) = fn(propulsive force (kN), vehicle velocity (m/s))

$$Fpropulsive = Fdrag + Froll + Fclimb + Faccel$$
$$PWpropulsive = Fpropulsive * Vvehicle$$

F1.6   Automotive vehicle: Engine rotational speed (rpm) = fn(vehicle velocity (m/s), gear ratio, wheel radius (m))

$$N = 60 * Vvehicle * GR/(2 * \pi * RADwheel)$$

F1.7   Automotive vehicle: Propulsive force (kN) and power (W) at wheel = fn(engine output torque (N.m), gear ratio, wheel radius (m), transmission efficiency (fraction), engine rotational speed (rpm))

Fpropulsive = TRQengine * GR * ETAtransmission/(RADwheel * 1000)
Pwpropulsive = TRQengine * ETAtransmission * N * 2 * π/60

(i)   Transmission efficiency is typically 0.88–0.93.

F1.8   Marine vessel: Propulsive power (kW) = fn(propulsive force (kN), vessel velocity (m/s))

Fpropulsive = Fwave making + Fskinfriction + Fform drag
              + Fair resistance + Fappendage + Faccel
PWpropulsive = Fpropulsive * Vvessel

(i)   See Fig. 1.10, and References 9 and 10 for formulae for the constituents of the total propulsive force.

F1.9   Marine vessel: Approximate propulsive power (kW) = fn(vessel displacement (tonnes), vessel velocity (m/s))

Ppropulsive = K1 * m^(alpha) * Vvessel^(beta)

(i)   Coefficient *K1* varies between 0.0025 and 0.0035 dependent upon hull design.
(ii)   Exponent *alpha* varies between 0.8 and 1.0 dependent upon hull design.
(iii) Exponent *beta* is approximately 3 for displacement hulls, but may be as low as 2 for semi-planing designs.
(iv)   Hence for a displacement hull of given mass and design, power versus vessel speed approximates to a cube law.

F1.10   Marine vessel: Approximate engine output power (kW) = fn(propeller/water jet rotational speed (rpm))

PW = K2 * Npropeller^3

(i)   Constant *K2* depends upon the propeller or water jet design.

F1.11   Aircraft: Lift (N) = fn(air density (kg/m^3), true air speed (m/s), lift coefficient, wing area (m^2))

Flift = 0.5 * RHO * VTAS^2 * Clift * Awing

or combining with Formula F2.16:

Flift = 0.5 * 1.2248 * VEAS^2 * Clift * Awing

F1.12   Aircraft: Drag (N) = fn(air density (kg/m^3), true air speed (m/s), drag coefficient, wing area (m^2))

Fdrag = 0.5 * RHO * VTAS^2 * Cdrag * Awing

or combining with Formula 2.16:

Fdrag = 0.5 * 1.2248 * VEAS^2 * Cdrag * Awing

F1.13 Aircraft: Lift coefficient in steady flight = fn(aircraft mass (kg), air density (kg/m^3), true air speed (m/s), wing area (m^2))

Clift = m * g/(0.5 * RHO * VTAS^2 * Awing)

or combining with Formula F2.16:

Clift = m * g/(0.5 * 1.2248 * VEAS^2 * Awing)

(i) Lift coefficient is a function of only aircraft angle of attack, or for steady flight VEAS as it will have a unique value for each angle of attack.

(ii) The lift coefficient may be up to 4 at low VEAS, falling to around 0.1 at maximum VEAS.

F1.14 Aircraft: Drag coefficient = fn(drag polar, lift coefficient)

Cdrag = Cdrag polar + Clift^2/K1

(i) The drag polar is that due to profile and friction drag.

(ii) The remaining drag is lift induced.

F1.15 Aircraft: Lift to drag ratio = fn(lift (N), drag (N))

LDratio = Lift/Drag

or:

LDratio = Clift/Cdrag

(i) Lift to drag ratio is a function of only aircraft angle of attack, or VEAS.

(ii) Typically its maximum value is between 10 and 15 at approximately 5°, falling to as low as 3 at minimum or maximum angles of attack.

F1.16 Aircraft: Required net thrust in steady flight (N) = fn(aircraft mass (kg), LDratio)

FN = m * g/LDratio

F1.17 Aircraft: Engine range factor (kg/N) = fn(engine mass (kg), thrust (N), SFC (kg/N h), range (m), true air speed (km/h), engine nacelle drag coefficient, engine frontal area (m^2), air density (kg/m^3))

Krange = ((m/FN) + ((SFC/3600) * Range/VTAS))
       /(1 − (0.5 * Cnacelle * Aengine * RHO * VTAS^2)/FN)

F1.18 Aircraft: Engine thrust (N) = fn(engine shaft power (kW)) – Approximate

FN = PW * 15

F1.19 Aircraft: Engine thrust SFC (N/kg h) = fn(engine shaft power SFC (kW/kg h)) – Approximate

SFCthrust = SFCshaft/15

Sample calculations

C1.1   **(i)**   **Calculate the power required for a typical family saloon car at ISA conditions on a tarmacadam road with no head wind at 150 km/h and**
   (ii)   **50 km/h.**
   (iii)   **Calculate the power required to accelerate from 50 km/h to 150 km/h in 15 seconds up an incline of 208.**

F1.1   Fdrag = 0.5 * RHO * Cdrag * A * (Vvehicle + Vwind)^2/1000
F1.2   Froll = Croll * m * g
F1.3   Fclimb = m * g * sin(alpha)
F1.4   Faccel = m * a
F1.5   Fpropulsive = Fdrag + Froll + Fclimb + Faccel
   PWpropulsive = Fpropulsive * Vvehicle

From Chart 1.4 for a typical family saloon Cdrag = 0.4, A = 2.2 m^2, mass = 1.25 tonnes.
From Fig. 1.6 Croll = 0.025.
From Chart 2.1 RHO = 1.225 kg/m^3.

*(i)   150 km/h = 150 * 1000/3600 = 41.67 m/s on a flat road*
Substituting values into Formulae F1.1, F1.2 and F1.5:

   Fdrag = 0.5 * 1.225 * 0.4 * 2.2 * (41.67 + 0)^2/1000
   Fdrag = 0.936 kN

   Froll = 0.025 * 1.25 * 9.807
   Froll = 0.306 kN

   Pwpropulsive = (0.936 + 0.306 + 0 + 0) * 41.67
   Pwpropulsive = 51.75 kW

*(ii)   50 km/h = 50 * 1000/3600 = 13.89 m/s on a flat road*
Repeating as for item (i) above:

   Fdrag = 0.5 * 1.225 * 0.4 * 2.2 * (13.89 + 0)^2/1000
   Fdrag = 0.104 kN

   Froll = 0.306 kN

   Pwpropulsive = (0.104 + 0.306 + 0 + 0) * 13.89
   Pwpropulsive = 5.69 kW

(iii)   Accelerating from 50 km/h to 150 km/h in 15 s up a 208 incline
Take mean of values at 50 and 150 km/h:

   Fdrag = (0.104 + 0.936)/2 = 0.52 kN

   Froll = (0.306 + 0.306)/2 = 0.306 kN

Substituting into Formulae F1.3, F1.4 and F1.5:

   Faccel = 1.25 * 100 * 1000/3600/15
   Faccel = 2.315

   Fclimb = 1.25 * 9.807 * sin(20)
   Fclimb = 4.193

   Pwpropulsive = (0.52 + 0.306 + 2.315 + 4.193) * 13.89
   Pwpropulsive = 101.87 kW

The above examples are consistent with the data shown on Chart 1.4. Note that if the engine was sized to attain the performance of item (iii) then vehicle would have a top speed of almost 200 km/h on a flat tarmacadam road. This is at the top end of the likely range for a family saloon.

C1.2 **Calculate the gear ratio for a family saloon car with a maximum vehicle speed of 210 km/h for a petrol reciprocating engine and a gas turbine engine at (i) their top speed, and (ii) at 50 km/h with the acceleration and incline as per C1.1(iii). The wheel radius is 0.29 m and 100% rotational speeds are 4500 rpm and 60 000 rpm for the petrol and gas turbine engines respectively.**

F1.6 N = 60 * Vvehicle * GR/(2 * π * RADwheel)
F1.7 Fpropulsive = TRQengine * GR * ETAtransmission
/(RADwheel * 1000)
Pwpropulsive = TRQengine * ETAtransmission * N * 2 * π/60

From the guidelines with Formula F1.7 take ETAtransmission = 0.905.

(i) Gear ratios at 210 km/h (58.33 m/s)
At maximum vehicle speed the engines will be at their 100% rotational speeds. Substituting into Formula F1.6:

4500 = 60 * 58.33 * GR/(2 * π * 0.29)
GR = 2.343   Petrol engine

60 000 = 60 * 58.33 * GR/(2 * π * 0.29)
GR = 31.23   Gas turbine

(ii) Gear ratios accelerating from 50 km/h to 100 km/h in 15 s up a 208 incline
First find engine torque at maximum speed by substituting into F1.7:

96090 = TRQengine * 0.905 * 4500 * 2 * π/60
TRQengine = 225 N m   Petrol engine

96090 = TRQengine * 0.905 * 60 000 * 2 * π/60
TRQengine = 16.9 N m   Gas turbine

From C1.1 vehicle propulsive force = 6.918 kN. Substituting into F1.7:

6.918 = TRQengine * GR * 0.905/(0.29 * 1000)
2217 = TRQengine * GR
GR = 2217/TRQengine

Substitute into the above for both petrol engine and gas turbine.

Hence for the petrol engine the gear ratio will be at a minimum when the engine is at maximum torque. From Chart 1.1 this occurs at 100% rotational speed and is 225 N m. Hence from the above the GR must be 9.85:1. This is 4.2 times that at maximum road speed.

For the gas turbine as per Chart 7.2 sheet 3 the engine may be at the full power of 96.09 kW with the gas generator at 100% speed, but the power turbine at part speed. If in this instance the power turbine is at say 23.8% speed (50/210) then, from Chart 1.1, torque is approximately 2.1 times that at 100% speed, i.e. 35.49 N m. Hence from the above the GR must be 62.47:1, this is 2 times that at maximum road speed. With the gas turbine the gear ratios are higher, but fewer gears are required.

C1.3   Calculate the power required for a displacement hull frigate at 32 knots (16.46 m/s), 15 knots (7.72 m/s) and 5 knots (2.57 m/s).

F1.9   Ppropulsive = K1 * m^(alpha) * Vvessel^(beta)

From the guidelines with Formula F1.9 take K1 = 0.003, alpha = 0.9 and beta = 3.
From Chart 1.6 take mass = 4000 tonnes. Substituting into F1.9 for 32 knots:

Ppropulsive = 0.003 * 4000^0.9 * 16.46^3
Ppropulsive = 23 349 kW

Repeating for other ship speeds gives 2409 kW and 89 kW at 15 knots and 5 knots respectively.

C1.4   Calculate the thrust required for an unmanned aircraft of 2 tonnes weight with a wing area of 10 m^2 in steady flight at the airframe maximum equivalent airspeed of 400 kt (206 m/s).

F1.12   Fdrag = 0.5 * 1.2248 * VEAS^2 * Cdrag * Awing
F1.13   Clift = m * g/(0.5 * 1.2248 * VEAS^2 * Awing)
F1.15   LDratio = Clift/Cdrag

First calculate the lift coefficient by substituting into Formula F1.13:

Clift = 2000 * 9.807/(0.5 * 1.2248 * 206^2 * 10)
Clift = 0.0755

400 kt equivalent airspeeed is the aircraft maximum flight speed and hence minimum angle of attack. From the guide lines with Formula F1.15 take lift to drag ratio to be 12.5:

12.5 = 0.0755/Cdrag
Cdrag = 0.0060

Since the aircraft is in steady flight, thrust = drag and substituting into Formula F1.12:

Fdrag = 0.5 * 1.2248 * 206^2 * 0.0060 * 10
Fdrag = 1559 N

Note: This could also have been calculated directly from F1.16.

C1.5   Calculate range factor for the turbofan of design parameters listed below for a mission of 8000 km at 0.8 Mach number at ISA 11 000 m

Engine mass = 3.5 tonnes
Engine thrust = 35 000 N
SFC = 0.065 kg/N h
Nacelle drag coefficient = 0.005
Diameters: engine = 2.5 m, intake = 2 m, propelling nozzle = 1.25 m
Engine length = 4 m

F1.17   Krange = ((m/FN) + (SFC/3600 * Range/VTAS))
                           /(1 − (0.5 * Cnacelle * Aengine * RHO * VTAS^2)/FN)
F2.5   RHOrel = RHO/1.2248
F2.15   VTAS = 1.94384 * M * SQRT(γ * R * TAMB)
F5.5.1   PodDrag = 0.5 * RHO * VTAS^2 * C * A
F5.5.2   NacelleArea = PI * L * (D.ENGINE + D.INTAKE + D.NOZZLE)/3

From Chart 2.1 RHOrel = 0.297 and TAMB = 216.7 K at ISA 11 000 m.
From the guidelines with Formula F2.15, R = 287.05 and γ = 1.4.
From the guidelines with Formula F5.5.1 Cnacelle = 0.0025.

First conduct basic calculations using Formulae F2.5 and F2.15:

$\text{VTAS} = 1.94384 * 0.8 * \text{SQRT}(1.4 * 287.05 * 216.7)$
$\text{VTAS} = 458.9\,\text{kts}$

$0.297 = \text{RHO}/1.2248$
$\text{RHO} = 0.364\,\text{kg/m}^3$

$\text{Aengine} = 4 * \text{PI} * (2.5 + 2 + 1.25)/3$
$\text{Aengine} = 24.1\,\text{m}^2$

Substituting into Formula F1.17:

$\text{Krange} = ((3500/35000) + (0.065/3600 * 8000 * 1000/458.9)$
$\qquad\qquad /(1 - (0.5 * 0.0025 * 24.1 * 0.364 * 458.9^{\wedge}2)/35000)$
$\text{Krange} = (0.1 + 0.315)/(1.0 - 0.0660)$
$\text{Krange} = 0.444\ \text{kg/N}$

This point compares favourably with Chart 1.8.

Charts

Chart 1.1   Performance of gas turbines compared with piston engines.

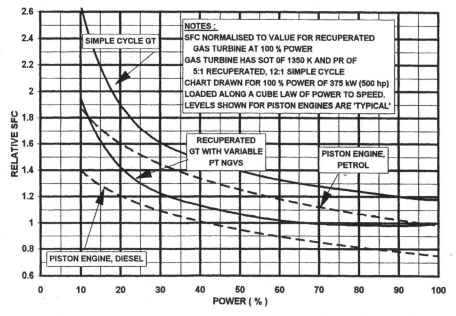

(a)   SFC versus power

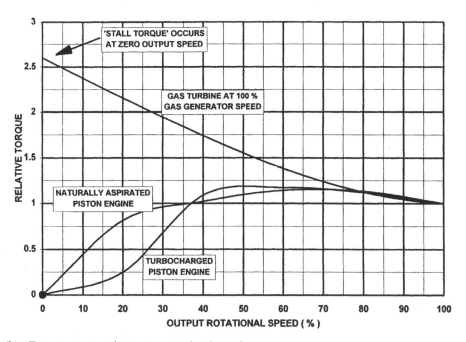

(b)   Torque versus engine output rotational speed

Chart 1.2 Characteristics of power generation plant.

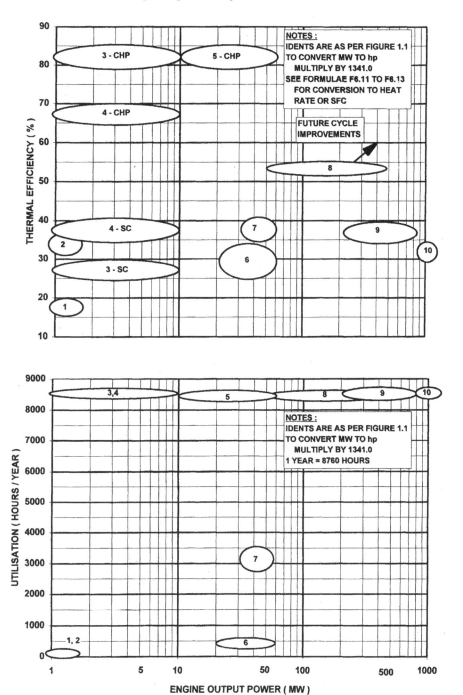

Chart 1.3   Gas turbine CHP steam production capability.

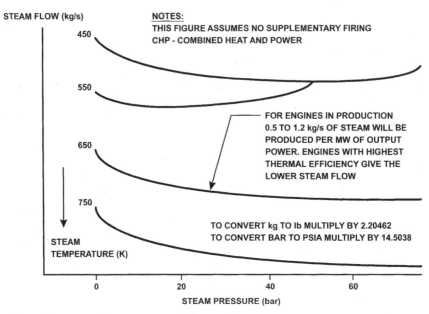

(a)   Steam flow versus steam pressure and temperature

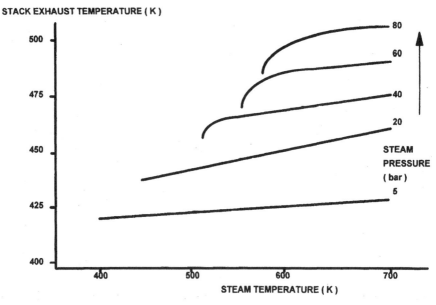

(b).   Stack exhaust temperature versus steam temperature and pressure

Chart 1.4 Automotive vehicles: leading data versus installed power.

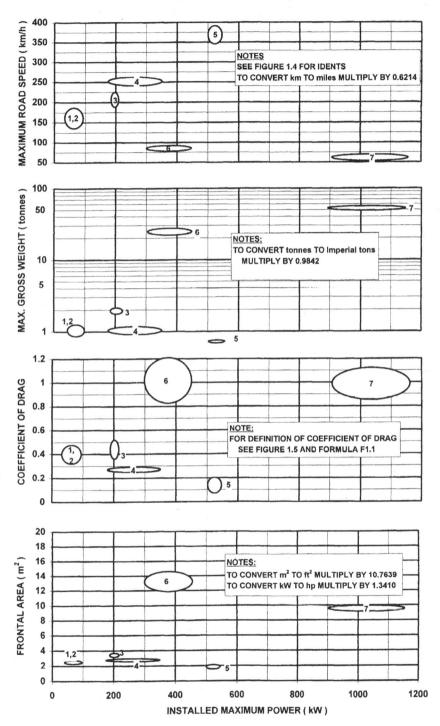

Chart 1.5   Automotive vehicles: power requirements for truck and family saloon.

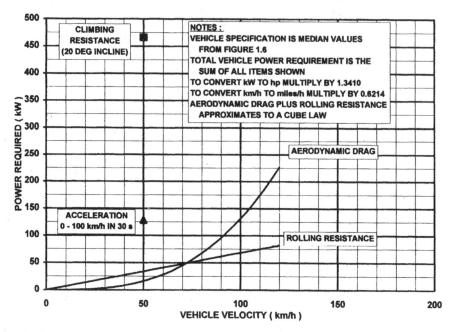

(a)   Truck

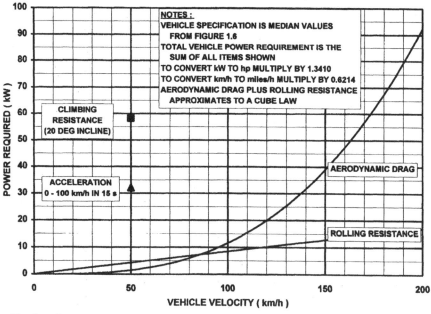

(b)   Family saloon

Chart 1.6   Marine vessels: leading data versus installed power.

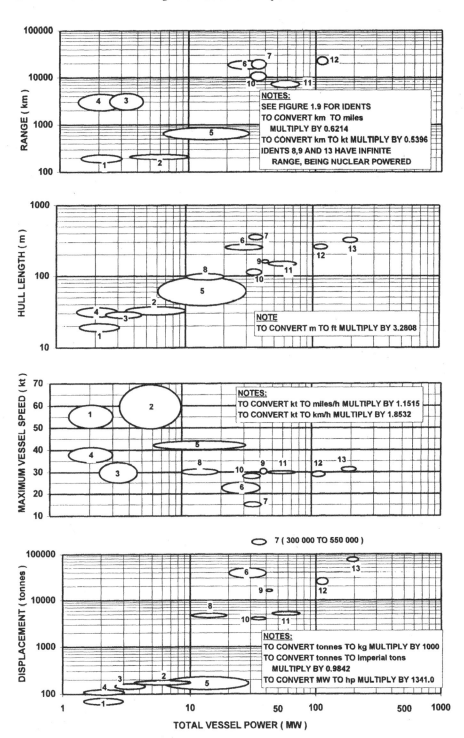

Chart 1.7   Marine engines: effect on engine power and ship speed and number of engines driving, for displacement hulls.

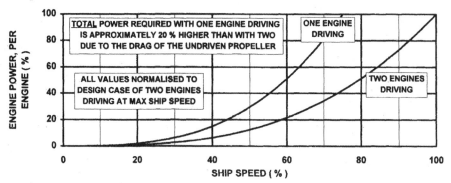

(a)   Engine power versus ship speed

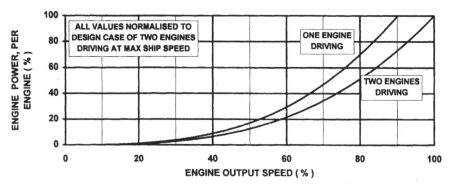

(b)   Engine power versus engine output speed

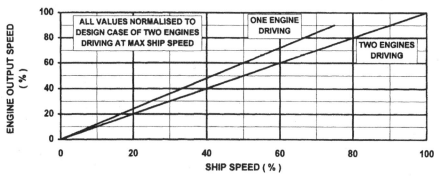

Notes:
Above figure is for two engines in a ship, each driving its own propeller.
An alternative layout is two engines per propeller in a four engine ship.
This requires low engine output speed if only one engine of a pair is on line.

(c)   Engine output speed versus ship speed

Chart 1.8   Aircraft range factor versus Mach number, for ranges of 1000 and 8000 km.

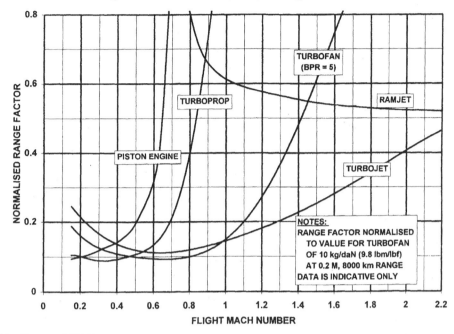

(a)   Range of 1000 km

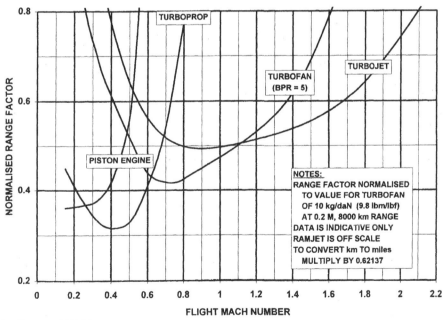

(b)   Range of 8000 km

Chart 1.9   Turboprop/turboshaft aircraft: leading characteristics.

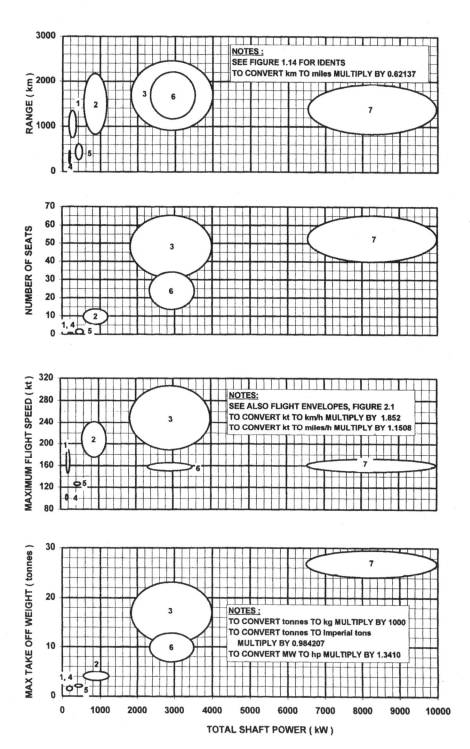

Chart 1.10   Thrust propelled aircraft: leading characteristics.

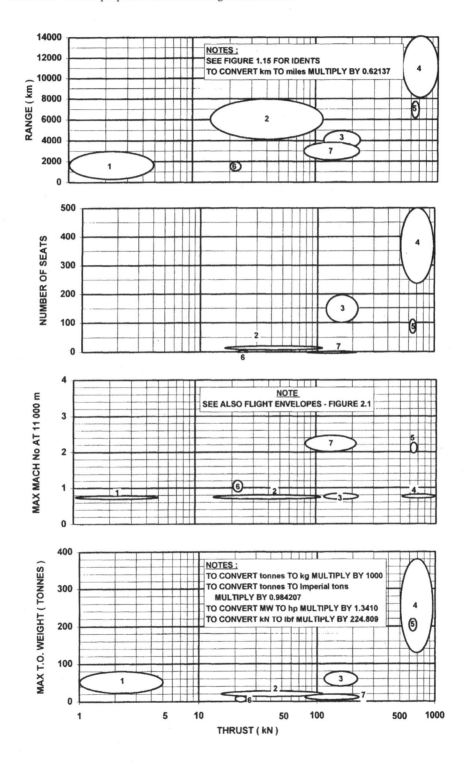

References

1. *The Diesel and Gas Turbine Worldwide Catalog*, Diesel and Gas Turbine Publications, Brookfield, Wisconsin.
2. R. Cochrane (1985) *Power to the People – The Story of the National Grid*, Newnes Books (Hamlyn, Middlesex), in association with the CEGB.
3. C. F. Foss (ed.) (1977) *Jane's All The World's Military Vehicles*, Jane's Information Group, Coulsdon, Surrey.
4. R. W. Chevis, J. Everton, M. Coulson and P. P. Walsh (1991) *Hybrid Electric Vehicle Concepts Using Gas Turbines*, Noel Penny Turbines Ltd, Birmingham.
5. A. F. Burke and C. B. Somnah (1982) Computer aided design of electric and hybrid vehicles, *International Journal of Vehicle Design* **Vol. FP2**, 61–81.
6. K. R. Pullen (1982) A case for the gas turbine series hybrid vehicle, presented at the '*Electric and Hybrid Vehicles Conference*' at the I.Mech.E., London, December 1992.
7. K. R. Pullen and S. Etemad (1995) Further developments of a gas turbine series hybrid for automotive use, presented at the *European Automobile Engineers Cooperation 5th International Congress*, Conference A, Strasbourg, 21–23 June 1995.
8. R. L. Trillo (ed.) (1995) *Jane's High Speed Marine Craft and Air Cushioned Vehicles*, Jane's Information Group, Coulsdon, Surrey.
9. R. Sharpe (ed.) (1997) *Jane's Fighting Ships*, Jane's Information Group, Coulsdon, Surrey.
10. T. C. Gillmer and B. Johnson (1982) *An Introduction to Naval Architecture*, E. & F. Spon, London.
11. K. J. Rawson and E. C. Tupper (1984) *Basic Ship Theory*, Volume 2, Longman, London.
12. P. J. McMahon (1971) *Aircraft Propulsion*, Pitman, London.
13. A. C. Kermode (1995) *Mechanics of Flight*, Longman, London.
14. J. D. Anderson, Jr. (1989) *Introduction to Flight*, 3rd edn, McGraw-Hill, New York.
15. B. Gunston (1987) *World Encylopaedia of Aero Engines*, Patrick Stephens Publishing, Wellingborough.
16. C. Rodgers (1983) Small auxiliary power unit design constraints, 19th Joint Propulsion Conference, AIAA/SAE/ASME, June 1983, Seattle.
17. C. Rodgers (1985) *Secondary Power Unit Options for Advanced Fighter Aircraft*, AIAA, New York.

Chapter 2
The Operational Envelope

2.0 Introduction

The performance – thrust or power, fuel consumption, temperatures, shaft speeds etc. – of a gas turbine engine is crucially dependent upon its inlet and exit conditions. The most important items are pressure and temperature, which are determined by the combination of ambient values and any changes due to flight speed, or pressure loss imposed by the installation.

The full range of inlet conditions that a given gas turbine engine application could encounter is encompassed in the operational envelope. This comprises:

- An *environmental envelope*, defining ambient pressure, temperature and humidity
- *Installation* pressure losses
- A *flight envelope* for aircraft engines

This chapter provides relevant data for all these, as well as useful background information. For this chapter alone tables are presented in Imperial as well as SI units, due to the wide use in industry of the former in relation to the operational envelope. All figures include conversion factors from SI to other units systems, and Appendix B provides a more comprehensive list.

2.1 The environmental envelope

The environmental envelope for an engine defines the range of ambient pressure (or pressure altitude, see section 2.1.2), ambient temperature and humidity throughout which it must operate satisfactorily. These atmospheric conditions local to the engine have a powerful effect upon its performance.

2.1.1 *International standards*

The International Standard Atmosphere (ISA) defines *standard day* ambient temperature and pressure up to an altitude of 30 500 m (100 066 ft). The term *ISA conditions* alone would imply zero relative humidity.

US Military Standard 210 (MIL 210) is the most commonly used standard for defining likely *extremes* of ambient temperature versus altitude. This is primarily an aerospace standard, and is also widely used for land based applications though with the hot and cold day temperature ranges extended. Chart 2.1 shows the ambient pressure and temperature relationships of MIL 210 and ISA, and sections 2.1.2 and 2.1.3 provide a fuller description.

For land based engines performance data is frequently quoted at the single point *ISO conditions*, as stipulated by the International Organisation for Standardisation (ISO). These are:

- 101.325 kPa (14.696 psia), sea level, ambient pressure
- 15 8C ambient temperature
- 60% relative humidity
- Zero installation pressure losses

References 1–5 include the above standards and others which are less frequently used. For the interested reader, Reference 6 provides a detailed guide to the Earth's atmosphere.

2.1.2   *Ambient pressure and pressure altitude (Formula F2.1)*

Pressure altitude, or *geo-potential altitude*, at a point in the atmosphere is defined by the level of ambient pressure, as per the International Standard Atmosphere. Pressure altitude is therefore *not* set by the elevation of the point in question above sea level. For example, due to prevailing weather conditions a ship at sea may encounter a low ambient pressure of, say, 97.8 kPa, and hence its pressure altitude would be 300 m.

Chart 2.1 includes the ISA definition of pressure altitude versus ambient pressure, and Chart 2.2 shows the relationship graphically. It will be observed that pressure falls exponentially from its sea level value of 101.325 kPa (14.696 psia) to 1.08 kPa (0.16 psia) at 30 500 m (100 066 ft). Formula F2.1 relates pressure altitude and ambient pressure, and sample calculation C2.1 shows its use.

The highest value of ambient pressure for which an engine would be designed is 108 kPa (15.7 psia). This would be due to local conditions and is commensurate with a pressure altitude of −600 m (−1968 ft).

2.1.3   *Ambient temperature (Formulae F2.2 and F2.3)*

Chart 2.1 also presents the ISA standard day ambient temperature, together with MIL 210 cold and hot day temperatures, versus pressure altitude. Chart 2.3 shows these three lines of ambient temperature plotted versus pressure altitude. Formula F2.2 shows ISA ambient temperature as a function of pressure altitude, and Formula F2.3 gives ambient pressure. Sample calculation C2.2 shows the calculation of ISA pressure and temperature.

Standard day temperature falls at the rate of approximately 6 8C per 1000 m (2 8C for 1000 ft) until a pressure altitude of 11 000 m (36 089 ft), after which it stays constant until 25 000 m (82 000 ft). This altitude of 11 000 m is referred to as the *tropopause*; the region below this is the *troposphere*, and that above it the *stratosphere*. Above 25 000 m standard day temperature rises again.

The minimum MIL 210 cold day temperature of 185.9 K (−87.3 8C) occurs between 15 545 m (51 000 ft) and 18 595 m (61 000 ft). The maximum MIL 210 hot day temperature is 312.6 K (39.5 8C) at sea level.

2.1.4   *Relative density and the speed of sound (Formulae F2.4–F2.7)*

Relative density is the atmospheric density divided by that for an ISA standard day at sea level. Chart 2.1 includes relative density, the square root of relative density, and the speed of sound for cold, hot and standard days. Charts 2.4–2.6 present this data graphically. These parameters are important in understanding the interrelationships between the different definitions of flight speed discussed in section 2.3.4.

Density falls with pressure altitude such that at 30 500 m (100 066 ft) it is only 1.3% of its ISA sea level value. The maximum speed of sound of 689.0 kt (1276 km/h, 792.8 mph) occurs on a hot day at sea level. The minimum value is 531.6 kt (984.3 km/h, 611.6 mph), occurring between 15 545 m (51 000 ft) and 18 595 m (61 000 ft).

2.1.5   *Specific and relative humidity (Formulae F2.8–F2.10)*

Atmospheric *specific humidity* is variously defined either as:

(1)   the ratio of water vapour to *dry air* by mass, or
(2)   the ratio of water vapour to *moist air* by mass

The former definition is used exclusively herein; for most practical purposes the difference is small anyway. *Relative humidity* is specific humidity divided by the saturated value for the prevailing ambient pressure and temperature.

Humidity has the least powerful effect upon engine performance of the three ambient parameters. Its effect is not negligible, however, in that it changes the inlet air's molecular weight, and hence basic properties of specific heat and gas constant. In addition, condensation may occasionally have gross effects on temperature. Wherever possible humidity effects should be considered, particularly for hot days with high levels of relative humidity. Chapter 12 discusses methods of accounting humidity effects upon engine performance.

For most gas turbine performance purposes, specific humidity is negligible below 0 8C, and also above 40 8C. The latter is because the highest temperatures only occur in desert conditions, where water is scarce. MIL 210 gives 35 8C as the highest ambient temperature at which to consider 100% relative humidity.

Chart 2.7 presents specific humidity for 100% relative humidity versus pressure altitude for cold, standard and hot days. For MIL 210 cold days specific humidity is almost zero for all altitudes. The maximum specific humidity will never exceed 4.8%, which would occur on a MIL 210 hot day at sea level. In the troposphere (i.e. below 11 000 m) specific humidity for 100% relative humidity falls with pressure altitude, due to the falling ambient temperature. Above that, in the stratosphere, water vapour content is negligible, almost all having condensed out at the colder temperatures below.

Charts 2.8 and 2.9 facilitate conversion of specific and relative humidities. Chart 2.8 presents specific humidity versus ambient temperature and relative humidity at sea level. For other altitudes Chart 2.9 presents factors to be applied to the specific humidity obtained from Chart 2.8. For a given relative humidity specific humidity is higher at altitude because whereas water vapour pressure is dependent only on temperature, air pressure is significantly lower. Sample calculation C2.3 demonstrates the use of formulae F2.8–2.10, the results from which may be compared with values from Charts 2.8 and 2.9.

2.1.6 *Industrial gas turbines*

The environmental envelope for industrial gas turbines, both for power generation and mechanical drive applications, is normally taken from Chart 2.1 up to a pressure altitude of around 4500 m (or 15 000 ft). Hot and cold day ambient temperatures beyond those of MIL 210 are often used, ±50 8C being typical at sea level. For specific fixed locations, altitude is known, and *S curves* are available defining the annual distribution of ambient temperature: these allow lifing assessments and rating selection. (The name derives from the characteristic shape of the curve, which plots the percentage of time for which a particular temperature level would be exceeded.)

The range of specific humidities for an industrial gas turbine would be commensurate with 0–100% relative humidity over most of the ambient temperature range, with some alleviation at the hot and cold extremes as discussed in section 2.1.5.

2.1.7 *Automotive gas turbines*

Most comments are as per industrial engines, except that narrowing down the range of ambient conditions for a specific application based on a fixed location is not appropriate.

2.1.8 *Marine gas turbines*

The range of pressure altitudes at sea is governed by weather conditions only, as the element of elevation that significantly affects all other gas turbine types is absent. The practice of the US

Navy, the most prolific user of marine gas turbines, is to take the likely range of ambient pressure as 87–108 kPa (12.6–15.7 psia). This corresponds to a pressure altitude variation of −600–1800 m (−1968–5905 ft).

At sea, free stream air temperature (i.e. that not affected by solar heating of the ship's decks) matches sea surface temperature, day or night. Owing to the vast thermal inertia of the sea there is a significant reduction in the range of ambient temperature that marine gas turbines encounter when on the open sea relative to land based or aircraft gas turbines. However ships must also be able to operate close to land, including polar ice fields and the Persian Gulf. Consequently for operability (if not lifing) purposes, a wide range of ambient temperature would normally be considered for a marine engine. The most commonly used ambient temperature range is that of the US Navy, which is −40–50 8C. US Navy ratings are proven at 38 8C, giving some margin on engine life. Reference 7 provides a comprehensive data base of temperatures encountered on the world's oceans.

The relative humidity range encountered by a marine gas turbine would be unlikely to include zero, due to the proximity of water. In practice values above 80% are typical. The upper limit would be commensurate with 100% relative humidity over most of the ambient temperature range, again with some alleviation at the hot and cold extremes as discussed in section 2.1.5.

2.1.9   *Aircraft engines*

The environmental envelope for aircraft engines is normally taken from Chart 2.1 up to the altitude ceiling for the aircraft. The specific humidity range is that corresponding to zero to 100% relative humidity as per Chart 2.7.

2.2   Installation pressure losses

Engine performance levels quoted at ISO conditions do not include installation ducting pressure losses. This level of performance is termed *uninstalled* and would normally be between inlet and exit planes consistent with the engine manufacturer's supply. Examples might include from the flange at entry to the first compressor casing to the engine exhaust duct exit flange, or to the propelling nozzle exit plane for thrust engines. When installation pressure losses, together with other installation effects discussed in section 6.13.5, are included the resultant level of performance is termed *installed*.

For industrial, automotive and marine engines installation pressure losses are normally imposed by *plant* intake and exhaust ducting. For aircraft engines there is usually a flight intake upstream of the engine inlet flange which is an integral part of the airframe as opposed to the engine; however for high bypass ratio turbofans there is not normally an installation exhaust duct. An additional item for aircraft engines is intake ram recovery factor. This is the fraction of the free stream dynamic pressure recovered by the installation or flight intake as total pressure at the engine intake front face.

Pressure losses due to installation ducting should *never* be approximated as a change of pressure altitude reflecting the lower inlet pressure at the engine intake flange. Whilst intake losses do indeed lower inlet pressure, exhaust losses *raise* engine exhaust plane pressure. Artificially changing ambient pressure clearly cannot simulate both effects at once.

For industrial, automotive and marine engines installation pressure losses are most commonly expressed as mm H_2O, where 100 mm H_2O is approximately 1% total pressure loss at sea level (0.981 kPa, 0.142 psi). For aircraft applications installation losses are more usually expressed as a percentage loss in total pressure ($\%\Delta P/P$).

2.2.1 Industrial engines

Overall installation inlet pressure loss due to physical ducting, filters and silencers is typically 100 mm H_2O at high power. Installation exhaust loss is typically 100–300 mm H_2O (0.981 kPa, 0.142 psi to 2.942 kPa, 0.427 psi); the higher values occur where there is a steam plant downstream of the gas turbine.

2.2.2 Automotive engines

In this instance both installation inlet and exhaust loss are typically 100 mm H_2O (0.981 kPa, 0.142 psi).

2.2.3 Marine engines

Installation intake and exhaust loss values at rated power may be up to 300 mm H_2O (2.942 kPa, 0.427 psi) and 500 mm H_2O (4.904 kPa, 0.711 psi) respectively, dependent upon ship design. Standard values used by the US Navy are 100 mm H_2O (0.981 kPa, 0.142 psi) and 150 mm H_2O (1.471 kPa, 0.213 psi).

2.2.4 Aircraft engines

For a pod mounted turbofan cruising at 0.8 Mach number, the total pressure loss from free stream to the flight intake/engine intake interface due to incomplete ram recovery and the installation intake may be as low as 0.5% $\Delta P/P$, whereas for a ramjet operating at Mach 3 the loss may be nearer 15%. For a helicopter engine buried behind filters the installation intake total pressure loss may be up to 2%, and there may also be an installation exhaust pressure loss due to exhaust signature suppression devices.

2.3 The flight envelope

2.3.1 Typical flight envelopes for major aircraft types

Aircraft engines must operate at a range of forward speeds in addition to the environmental envelope. The range of flight Mach numbers for a given altitude is defined by the flight envelope. Figure 2.1 presents typical flight envelopes for the seven major types of aircraft.

For each flight envelope the minimum and maximum free stream temperatures and pressures which the engine would experience are shown, together with basic reasons for the shape of the envelope. The latter are discussed in more detail in Chapter 1. Where auxiliary power units are employed the same free stream conditions are experienced as for the propulsion unit. The intake ram recovery is often lower, however, due both to placement at the rear of the fuselage and drag constraints on the intake design.

2.3.2 Free stream total pressure and temperature (Formulae F2.11 and F2.12)

The free stream total pressure (P0) is a function of both pressure altitude and flight Mach number. Free stream total temperature (T0) is a function also of ambient temperature and flight Mach number. Both inlet pressure and temperature are fundamental to engine performance. They are often used to *refer* engine parameters to ISA sea level static conditions, via quasi dimensionless parameter groups as described in Chapter 4. To do this the following ratios are defined:

DELTA $(\delta) = P0/101.325$ kPa   Also see Formula F2.11.

THETA $(y) = T0/288.15$ K   Also see Formula F2.12

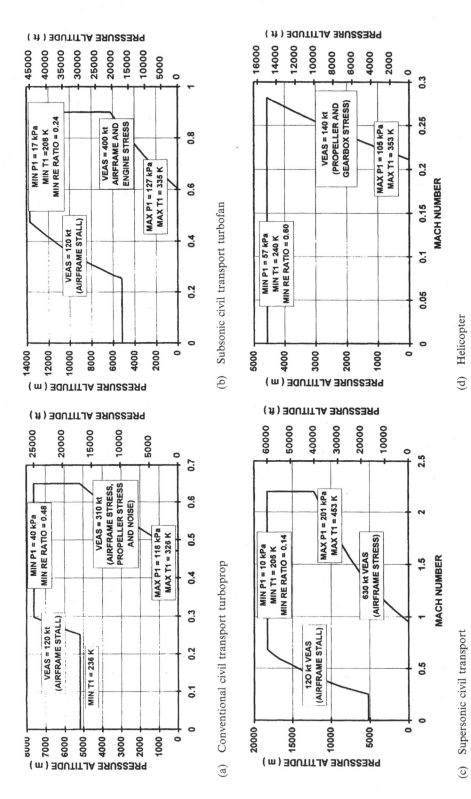

(a) Conventional civil transport turboprop

(b) Subsonic civil transport turbofan

(c) Supersonic civil transport

(d) Helicopter

Figure 2.1 Flight envelopes for the major aircraft types.

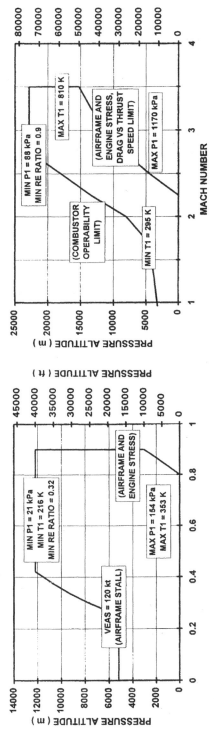

(f)   Supersonic airbreathing missile

(e)   Subsonic airbreathing missile, Drone or RPV

(g)   Advanced military fighter

Notes:

APUs have lower intake RAM recovery than propulsion engines
Pressures shown are free stream, i.e. 100% RAM recovery
To convert temperatures in K to R multiply by 1.8
To convert temperatures in K to C subtract 273.15
To convert temperatures in K to F multiply by 1.8 and subtract 459.67
To convert pressures in kPa to psia multiply by 0.145038
To convert speeds in kt to   km/h   m/s   miles/h   ft/s
multiply by   1.8520,   0.5144,   1.1508,   1.6878
Maximum temperatures shown are for MIL STD 210 Hot day
Minimum temperatures shown are for MIL STD 210 Cold day
Minimum Reynolds' Number ratios shown are for MIL STD 210 Hot day
Unducted fans would have similar flight envelope to commercial
turbofans
'RPV' = Remotely Piloted Vehicle
All numbers shown are indicative, for guidance only

Figure 2.1   *contd.*

The term *referred* is used exclusively herein, though the term *corrected* is also used, especially in the United States. For component design purposes theta and delta can also be defined using the pressure and temperature at component inlet.

Chart 2.10 presents delta based on free stream conditions versus pressure altitude and Mach number over a range that encompasses all of the flight envelopes shown in Fig. 2.1. The effect of inlet pressure losses on engine performance is additional, as discussed in section 2.2. Chart 2.11 presents theta based on free stream conditions versus pressure altitude and Mach number over a similar range for MIL 210 cold, standard and MIL 210 hot days. Sample calculation C2.4 demonstrates the use of these charts together with Formulae F2.11 and 2.12.

2.3.3   *Reynolds number ratio (Formulae F2.13 and F2.14)*

The manner in which engine performance is affected by Reynolds number is described in Chapters 4 and 7.

For any conditions in a flowing gas, the Reynolds number reflects the ratio of body forces (reflecting velocity and momentum effects) to viscous forces (causing frictional pressure losses). The Reynolds number may have a significant second order effect on engine performance at low values due to increasing viscous effects. The Reynolds number ratio shows generically how the Reynolds number varies with ram conditions. It is the value at the given operating condition divided by that at ISA sea level static (Formulae F2.13 and F2.14); Chart 2.12 shows how it varies with altitude, Mach number and ambient temperature throughout the operational envelope. Minimum Reynolds number ratios are shown on the typical flight envelopes presented in Fig. 2.1. Sample calculation C2.5 demonstrates the use of Formulae F2.13 and F2.14 in relation to Chart 2.12.

Whilst the Reynolds number ratio presented in Chart 2.12 is based upon free stream conditions, intake and compressor Reynolds number will show a corresponding variation around the operational envelope; free stream conditions are similar to engine inlet conditions. For turbines, however, the Reynolds number will additionally depend upon power or thrust level, which determines the change in turbine pressures and temperatures relative to the ram conditions. Nevertheless Chart 2.12 still provides a useful first order indication of the Reynolds number variation for these 'hot end' components.

2.3.4   *Definitions of flight speed (Formulae F2.15–F2.19)*

Traditionally aircraft speed has been measured using a pitot-static head located on a long tube projecting forward from the wing or fuselage nose. The difference between total and static pressure is used to evaluate velocity, which is shown on a visual display unit or gauge in the cockpit. The device is normally calibrated at sea level, which has given rise to a number of definitions of flight speed:

- *Indicated air speed* (VIAS) is the speed indicated in the cockpit based upon the above calibration.
- *Calibrated air speed* (VCAS) is approximately equal to VIAS with the only difference being a small adjustment to allow for aircraft disturbance of the static pressure field around the pitot-static probe.
- *Equivalent air speed* (VEAS) results from correcting VCAS for the lower ambient pressure at altitude versus that embedded in the probe calibration conducted at sea level, i.e. for a given Mach number the dynamic head is smaller at altitude. When at sea level VEAS is equal to VCAS.
- *True air speed* (VTAS) is the actual speed of the aircraft relative to the air. It is evaluated by multiplying VEAS by the square root of relative density as presented in Chart 2.5. This correction is due to the fact that the density of air at sea level is embedded in the probe calibration which provides VIAS. Both density and velocity make up dynamic pressure.

- *Mach number* (M) is the ratio of true air speed to the local speed of sound.
- *Ground speed* is VTAS adjusted for wind speed.

VEAS and VCAS are functions of pressure altitude and Mach number only. The difference between VEAS and VCAS is termed the *scale altitude effect* (SAE), and is independent of ambient temperature. Conversely, VTAS is a function of ambient temperature and Mach number which is independent of pressure altitude. The complex nature of the mathematical relationships between these different definitions of flight speed definitions is apparent from Formulae F2.15–F2.19. Reference 8 provides a comprehensive description of their derivation.

To a pilot both Mach number and ground air speed are important. The former dictates critical aircraft aerodynamic conditions such as shock or stall, whereas the latter is vital for navigation. For gas turbine engineers Mach number is of paramount importance in determining inlet total conditions from ambient static. Often, however, when analysing engine performance data from flight tests only VCAS or VEAS are available. The following charts enable one form of flight speed to be derived with knowledge of another:

- Chart 2.13 – VCAS versus pressure altitude and Mach number
- Chart 2.14 – VEAS versus pressure altitude and Mach number
- Chart 2.15 – VTAS versus pressure altitude and Mach number
- Chart 2.16 – SAE versus pressure altitude and Mach number

Sample calculation C2.6 demonstrates the interrelationships of the above flight speed definitions, using the above formulae to obtain results consistent with Charts 2.13–2.16.

Formulae

F2.1   Pressure altitude (m) = fn(ambient pressure (kPa))

PAMB > 22.633 kPa

$$ALT = 44330.48 * (1 - (PAMB/101.325)^{0.1902632})$$

If PAMB < 22.633 kPa and > 1.6 kPa

$$ALT = 6341.58 * \ln(22.63253/PAMB) + 10999.93$$

F2.2   ISA Ambient temperature (K) = fn(pressure altitude (m))

If ALT < 11 000 m

$$TAMB = 288.15 - 0.0065 * ALT$$

If ALT ⩾ 11 000 m and < 24 994 m

$$TAMB = 216.65$$

If ALT ⩾ 24 994 m and < 30 000 m

$$TAMB = 216.65 + 0.0029892 * (ALT - 24\,994)$$

F2.3   Ambient pressure (kPa) = fn(ISA ambient temp (K), pressure altitude (m))

If ALT < 11 000 m

$$PAMB = 101.325 * (288.15/TAMB)^{(-5.25588)}$$

If ALT $> 11\,000$ and $< 24\,994\,m$

  PAMB $= 22.63253/\text{EXP}(0.000157689 * (\text{ALT} - 10998.1))$

If ALT $> 24\,994\,m$ and $< 30\,000\,m$

  PAMB $= 2.5237 * (216.65/\text{TAMB})^{\wedge}11.8$

F2.4   Density of air (kg/m³) = fn(ambient pressure (kPa), ambient temperature (K))

RHO $= \text{PAMB} * 1000/(R * \text{TAMB})$

(i)   Where air is a perfect gas with a value for the gas constant R of 287.05 J/kg K.

F2.5   Relative density = fn(density (kg/m³))

RHOrel $= \text{RHO}/1.2248$

(i)   Where $1.2248\,kg/m^3$ is the density of air at ISA sea level.

F2.6   Speed of sound (m/s) = fn(ambient temperature (K))

VS $= \text{SQRT}(g * R * \text{TAMB})$

(i)   For air the gas constant R has a value of 287.05 J/kg K.
(ii) For air g may be calculated from formulae presented in Chapter 3, it is approximately 1.4 at ambient temperatures.

F2.7   Relative speed of sound = fn(speed of sound (kt))

VSrel $= \text{VS}/661.7$

(i)   Where 661.7 knots is the speed of sound at ISA sea level.

F2.8   Specific humidity (%) = fn(water content in atmosphere)

SH $= 100 *$ Mass of water vapour in a sample/Mass of dry air in the sample

(i)   The above definition is used herein.
(ii) Occasionally in other publications an alternative definition of mass of water per mass of moist air is used.

F2.9   Relative humidity (%) = fn(specific humidity (%), specific humidity if atmosphere was saturated at prevailing ambient pressure and temperature (%))

RH $= 100 * \text{SH}/\text{SHsat}$

F2.10   Specific humidity (%) = fn(relative humidity (%), ambient temperature (K), ambient pressure (kPa), saturated vapour pressure (kPa))

SH $= 0.622 * \text{PSAT} * \text{RH}/(\text{PAMB} - \text{PSAT} * (\text{RH}/100))$

where the saturated vapour pressure of water (kPa) at the ambient conditions is:

PSAT $= (1.0007 + 3.46\text{E-}05 * \text{PAMB}) * 0.61121$
    $* \text{e}^{\wedge}(17.502 * (\text{TAMB} - 273.15)/(\text{TAMB} - 32.25))$

F2.11   Delta = fn(ambient pressure (kPa), flight Mach number)

$$\delta = (PAMB/101.325) * (1 + ((\gamma - 1)/2) * M^{\wedge}2)^{\wedge}(\gamma/(\gamma - 1))$$

or

$$\delta = P1/101.325$$

(i)   For air, g may be calculated from Formula F3.7, it is approximately 1.4 at ambient temperatures.

F2.12   Theta = fn(ambient temperature (K), flight Mach number)

$$\theta = (TAMB/288.15) * (1 + ((\gamma - 1)/2) * M^{\wedge}2)$$

or

$$\theta = T1/288.15$$

(i)   For air, g may be calculated from the formulae presented in Chapter 3, it is approximately 1.4 at ambient temperatures.

F2.13   Reynolds number = fn(density (kg/m^3), dynamic viscosity (N s/m^2))

$$RE = RHO * D * V/VIS$$

(i)   D and V are a representative dimension and velocity respectively.
(ii)   VIS is given by Formula F3.30.

F2.14   Reynolds number ratio = fn(Reynolds number)

$$RERATIO = RE/67661$$

(i)   Where 67661 is the Reynolds number at the first compressor face at ISA SLS, with the representative dimensions and velocity set to unity.
(ii)   Here RE is evaluated also using unity as the representative dimension and velocity, and using engine inlet ram pressure and temperature.
(iii) Hence this term generically shows how the Reynolds number varies throughout the operational envelope.

F2.15   True air speed (kt) = fn(flight Mach number, speed of sound (m/s), ambient temperature (K))

$$VTAS = 1.94384 * M * VS$$

or

$$VTAS = 1.94384 * M * SQRT(\gamma * R * TAMB)$$

(i)   For air the gas constant R has a value of 287.05 J/kg K
(ii)   For air, g may be calculated from the formulae presented in Chapter 3, it is approximately 1.4 at ambient temperatures.

F2.16   Equivalent air speed (kt) = fn(true air speed (kt), relative density)

$$VEAS = VTAS * SQRT(RHOrel)$$

(i)   RHOrel can be evaluated from Formula F2.5.

F2.17   Equivalent air speed (kt) = fn(flight Mach number, ambient pressure (kPa))

$$VEAS = 1.94384 * M * SQRT(PAMB * 1000 * \gamma/1.2248)$$

F2.18   Calibrated air speed (kt) = fn(flight Mach number, ambient pressure (kPa), total to static pressure differential as measured by pitot (kPa))

$$\text{VCAS} = 661.478 * \text{SQRT}(2/(\gamma - 1) * (((\text{PAMB}/101.325)$$
$$* (\text{DP}/\text{PAMB}) + 1)^{\wedge}((\gamma - 1)/\gamma) - 1))$$

For M < 1: DP/PAMB may be derived from Q curve formulae:

$$\text{DP}/\text{PAMB} = (1 + (\gamma - 1)/2 * \text{M}^{\wedge}2)^{\wedge}(\gamma/(\gamma - 1)) - 1$$

For M > 1: to evaluate measured DP/PAMB a shock correction must be applied:

$$\text{DP}/\text{PAMB} = 0.7 * \text{M}^{\wedge}2 * (1.8394 - 0.7717/\text{M}^{\wedge}2 + 0.1642/\text{M}^{\wedge}4 + 0.0352/\text{M}^{\wedge}6$$
$$+ 0.0069/\text{M}^{\wedge}8)$$

F2.19   Scale altitude effect (kt) = fn(calibrated air speed (kt), equivalent air speed (kt))

$$\text{SAE} = \text{VEAS} - \text{VCAS}$$

Sample calculations

C2.1   Evaluate pressure altitude for an ambient pressure of 2.914 kPa

Since PAMB < 22.633 and >1.6 kPa then:

F2.1   ALT = 6341.58 * ln(22.63253/PAMB) + 10999.93

Substituting into F2.1:

$$\text{ALT} = 6341.58 * \ln(22.63253/2.914) + 10999.93$$
$$\text{ALT} = 24\,000\,\text{m}$$

This is as per the value presented in Chart 2.1.

C2.2   Evaluate ambient temperature and pressure for an ISA day at 5500 m

Since altitude is less than 11 000 m:

F2.2   TAMB = 288.15 − 0.0065 * ALT
F2.3   PAMB = 101.325 * (288.15/TAMB)^(−5.25588)

Substituting into Formulae F2.2 and F2.3:

$$\text{TAMB} = 288.15 - 0.0065 * 5500$$
$$\text{TAMB} = 252.4\text{K}$$

$$\text{PAMB} = 101.325 * (288.15/252.4)^{\wedge}(-5.25588)$$
$$\text{PAMB} = 50.507\,\text{kPa}$$

The above values are as per those presented in Chart 2.1.

C2.3   Calculate the mass of water vapour per kg of dry air for 50% relative humidity at a pressure altitude of 2000 m on a MIL 210 hot day

> **F2.8**   SH = 100 * Mass of water vapour in a sample/Mass of dry air in the sample

> **F2.9**   RH = 100 * SH/SHsat

> **F2.10**   SH = 0.622 * PSAT * RH/(PAMB − PSAT * (RH/100))

> PSAT = (1.0007 + 3.46E-05 * PAMB) * 0.61121 * e^(17.502 * (TAMB − 273.15)/(TAMB − 32.25))

From Chart 2.1 TAMB = 298.5 K and PAMB = 79.496 kPa.
Substituting into F2.10:

> PSAT = (1.0007 + 3.46E-05 * 79.496) * 0.61121
> * e^(17.502 * (298.5 − 273.15)/(298.5 − 32.25))
> PSAT = 0.6133 * e^(1.6664))
> PSAT = 3.1238 kPa
>
> SH = 0.622 * 3.1238 * 50/(79.496 − 3.1238 * (50/100))
> SH = 1.30%

If Chart 2.7 is looked up at 2000 m and ISA hot day, and the value multiplied by 50% then it is comparable to the above. If Chart 2.8 is looked up for 298.5 K and 50% RH, and the resulting value multiplied by the factor from Chart 2.9 for 2000 m, then it is also comparable to the above.
Substituting into F2.8:

> 1.30 = 100 * Mass water vapour/1
> Mass water vapour = 0.013 kg

C2.4   Calculate delta and theta for 11 000 m, MIL 210 cold day and 0.8 Mach number

Q curve formulae from Chapter 3:

> **F3.31**   T/TS = (1 + (γ − 1)/2 * M^2)
> **F3.32**   P/PS = (T/TS)^(γ/(γ − 1))

From Chart 2.1 TAMB = 208.0 K and PAMB = 22.628 kPa. From the guidelines with Formula F2.15, γ = 1.4.
Substituting into F3.31 and F3.32:

> T1/208.0 = (1 + (1.4 − 1)/2 * 0.8^2)
> T1 = 234.6
>
> THETA = 234.6/288.15
> THETA = 0.814

Substituting into F3.32 :

$P1/22.628 = (234.6/208.0)^{\wedge}(1.4/(1.4-1))$
$P1 = 34.480\,\text{kPa}$

$DELTA = 34.480/101.325$
$DELTA = 0.340$

If Charts 2.11 and 2.10 are looked up for the given altitude and Mach number the resultant values are comparable to the above.

C2.5   **(i) Evaluate the Reynolds number for a compressor of 50 mm blade chord with an inlet Mach number of 0.4 at ISA SLS. (ii) Evaluate approximately the Reynolds number for a MIL 210 hot day at 10 000 m, 0.8 flight Mach number**

F2.13 $RE = RHO * D * V/VIS$

F3.30 $VIS = 1.015E\text{-}06 * TS^{\wedge}1.5/(TS + 120)$

F3.1 $RHO = PS/(R * TS)$

From Chart 2.1 ISA ambient pressure and temperature are 101.325 kPa and 288.15 K.
   From Chart 3.8 at 0.4 Mach number $P/PS = 1.1166$, $T/TS = 1.032$ and $V/SQRT(T) = 7.8941$.

(i)   Calculate Reynolds number at ISA SLS
Total temperature is unchanged across the intake hence at the compressor face:

$TS = 288.15/1.032$
$TS = 279.21\,\text{K}$

$V = 7.8941 * SQRT(288.15)$
$V = 134.0\,\text{m/s}$

Approximating no loss in total pressure along the intake:

$PS = 101.325/1.1166$
$PS = 90.744\,\text{kPa}$

Substituting values into F3.1, F.3.30 and F2.13:

$RHO = 90744/287.05/279.21$
$RHO = 1.132\,\text{kg/m}^3$

$VIS = 1.015E\text{-}06 * 279.21^{\wedge}1.5/(279.21 + 120)$
$VIS = 1.18E\text{-}05$

$RE = 1.132 * 0.05 * 134.0/1.18E\text{-}05$
$RE = 4926$

(ii)   Reynolds number for a MIL 210 hot day at 10 000 m and 0.8 Mach number
From Chart 2.12, Reynolds number ratio for a MIL 210 hot day at 10 000 m and 0.8 flight Mach number is approximately 0.52, hence:

$RE = 4926 * 0.52$
$RE = 2562$

C2.6 **Calculate true air speed, Mach number, calibrated air speed and the scale altitude effect for 400 knots equivalent air speed at (i) ISA, sea level and (ii) for a MIL 210 cold day at 5000 m**

> **F2.16**   VEAS = VTAS * SQRT(RHOrel)
>
> **F2.17**   VEAS = 1.94384 * M * SQRT(PAMB * 1000 * γ/1.2248)
>
> **F2.18**   VCAS = 661.478 * SQRT(2/(γ − 1)
>   *(((PAMB/101.325) * (DP/PAMP) + 1)$^\wedge$((γ − 1)/γ) − 1))

For M < 1: DP/PAMB may be derived from Q curve formulae:

> DP/PAMB = (1 + (γ − 1)/2 * M$^\wedge$2)$^\wedge$(γ/(γ − 1)) − 1
>
> **F2.19**   SAE = VEAS − VCAS
>
> **F2.5**   RHOrel = RHO/1.2248

From Chart 2.1 PAMB = 101.325 kPa, TAMB = 288.15 K and RHOrel = 1.0 at ISA SLS.

> PAMB = 54.022 kPa, TAMB = 236.6 K and RHOrel = 0.649 at 5000 m,
>   MIL 210 cold day.

From the guidelines with Formula F2.15, γ = 1.4.

(i)   *VTAS, Mach number, VCAS and SAE at ISA, sea level*

Substituting values into F2.16, F2.17, F2.18 and F2.19:

> 400 = VTAS * SQRT(1.0)
> VTAS = 400 kt
>   = 740.8 km/h
>
> 400 = 1.94348 * M * SQRT(101.325 * 1000 * 1.4/1.2248)
> M = 0.605
>
> DP/PAMB = (1 + (1.4 − 1)/2 * 0.605$^\wedge$2)$^\wedge$(1.4/(1.4 − 1)) − 1
> DP/PAMB = 0.2805
>
> VCAS = 661.478 * SQRT(2/(1.4 − 1) * (((101.325/101.325)
>   * 0.2805 + 1)$^\wedge$((1.4 − 1)/1.4) − 1))
> VCAS = 661.478 * SQRT(5 * (1.2805$^\wedge$0.286 − 1))
> VCAS = 400 kt
>   = 740.8 km/h
>
> SAE = 400 − 400
> SAE = 0 kt

(ii)   *VTAS, Mach number, VCAS and SAE for a MIL 210 cold day at 5000 m*

Substituting values into F2.16, F2.17, F2.18 and F2.19:

> 400 = VTAS * SQRT(0.649)
> VTAS = 496.5 kt
>   = 919.55 km/h
>
> 400 = 1.94348 * M * SQRT(54.022 * 1000 * 1.4/1.2248)
> M = 0.828
>
> DP/PAMB = (1 + (1.4 − 1)/2 * 0.828$^\wedge$2)$^\wedge$(1.4/(1.4 − 1)) − 1
> DP/PAMB = 0.5679

VCAS = 661.478 * SQRT(2/(1.4 − 1) * (((54.022/101.325)
    * 0.5679 + 1)$^\wedge$((1.4 − 1)/1.4) − 1))
VCAS = 661.478 * SQRT(5 * (1.3028$^\wedge$0.286 − 1))
VCAS = 414.65 kt
    = 767.9 km/h

SAE = 400 − 414.65
SAE = −14.65 kt
    = −27.13 km/h

The above answers are consistent with Charts 2.13, 2.14, 2.15 and 2.16.

Charts

Chart 2.1 Ambient conditions versus pressure altitude.

(a) SI units: 0–15 000 m

Pressure altitude (m)	Pressure (kPa)	MIL STD 210A cold atmosphere				Standard atmosphere				MIL STD 210A hot atmosphere			
		Temp (K)	Relative density	√Relative density	Speed of sound (kt)	Temp (K)	Relative density	√Relative density	Speed of sound (kt)	Temp (K)	Relative density	√Relative density	Speed of sound (kt)
0	101.325	222.1	1.298	1.139	581.0	288.2	1.000	1.000	661.7	312.6	0.922	0.960	689.0
250	98.362	228.2	1.226	1.107	589.0	286.6	0.976	0.988	659.8	310.9	0.900	0.949	687.1
500	95.460	234.4	1.158	1.076	596.9	284.9	0.953	0.976	658.0	309.1	0.878	0.937	685.2
750	92.631	240.6	1.095	1.046	604.7	283.3	0.930	0.964	656.1	307.4	0.857	0.926	683.3
1 000	89.873	245.3	1.042	1.021	610.7	281.7	0.907	0.953	654.2	305.6	0.836	0.914	681.3
1 250	87.180	247.1	1.004	1.002	612.8	280.0	0.885	0.941	652.3	303.8	0.816	0.903	679.3
1 500	84.558	247.1	0.973	0.987	612.8	278.4	0.864	0.929	650.4	302.1	0.796	0.892	677.3
1 750	81.994	247.1	0.944	0.972	612.8	276.8	0.842	0.918	648.5	300.3	0.777	0.881	675.4
2 000	79.496	247.1	0.915	0.957	612.8	275.2	0.822	0.906	646.6	298.5	0.757	0.870	673.3
2 250	77.060	247.1	0.887	0.942	612.8	273.5	0.801	0.895	644.7	296.7	0.739	0.859	671.3
2 500	74.683	247.1	0.860	0.927	612.8	271.9	0.781	0.884	642.8	294.8	0.720	0.849	669.2
2 750	72.367	247.1	0.833	0.913	612.8	270.3	0.761	0.873	640.9	293.0	0.702	0.838	667.2
3 000	70.106	247.1	0.807	0.898	612.8	268.6	0.742	0.861	639.0	291.2	0.685	0.827	665.2
3 250	67.905	246.8	0.783	0.885	612.4	267.0	0.723	0.850	637.1	289.5	0.667	0.817	663.2
3 500	65.761	245.7	0.761	0.872	611.1	265.4	0.705	0.839	635.1	287.8	0.650	0.806	661.3
3 750	63.673	244.2	0.741	0.861	609.3	263.8	0.686	0.829	633.2	286.1	0.633	0.796	659.3
4 000	61.640	242.7	0.722	0.850	607.4	262.1	0.669	0.818	631.2	284.4	0.616	0.785	657.3
4 250	59.657	241.2	0.703	0.839	605.5	260.6	0.651	0.807	629.3	282.6	0.600	0.775	655.3
4 500	57.731	239.7	0.685	0.828	603.6	258.9	0.634	0.796	627.3	280.8	0.585	0.765	653.2
4 750	55.852	238.2	0.667	0.817	601.7	257.3	0.617	0.786	625.3	279.1	0.569	0.754	651.2
5 000	54.022	236.6	0.649	0.806	599.7	255.7	0.601	0.775	623.4	277.3	0.554	0.744	649.1
5 250	52.242	235.1	0.632	0.795	597.8	254.1	0.585	0.765	621.4	275.5	0.539	0.734	647.0
5 500	50.507	233.5	0.615	0.784	595.8	252.4	0.569	0.754	619.4	273.7	0.525	0.724	644.9
5 750	48.820	231.9	0.599	0.774	593.8	250.8	0.554	0.744	617.4	271.9	0.511	0.715	642.8

Notes:
To convert kt to m/s multiply by 0.5144.
To convert kt to km/h multiply by 1.8520.
To convert K to 8C subtract 273.15.
To convert K to 8R multiply by 1.8.

To convert K to 8F multiply by 1.8 and subtract 459.67.
Density at ISA sea level static = 1.2250 kg/m³.
Standard practice is to interpolate linearly between altitudes listed.

Chart 2.1 *contd.* (a)   SI units: 0–15 000 m

Pressure altitude (m)	Pressure (kPa)	MIL STD 210A cold atmosphere				Standard atmosphere				MIL STD 210A hot atmosphere			
		Temp (K)	Relative density	√Relative density	Speed of sound (kt)	Temp (K)	Relative density	√Relative density	Speed of sound (kt)	Temp (K)	Relative density	√Relative density	Speed of sound (kt)
6 000	47.178	230.4	0.582	0.763	591.8	249.2	0.538	0.734	615.4	270.2	0.497	0.705	640.8
6 250	45.584	228.8	0.567	0.753	589.7	247.5	0.524	0.724	613.4	268.5	0.483	0.695	638.8
6 500	44.033	227.2	0.551	0.742	587.7	245.9	0.509	0.714	611.4	266.8	0.469	0.685	636.8
6 750	42.525	225.6	0.536	0.732	585.5	244.3	0.495	0.704	609.3	265.1	0.456	0.675	634.7
7 000	41.063	224.0	0.521	0.722	583.5	242.7	0.481	0.694	607.4	263.4	0.443	0.666	632.7
7 250	39.638	222.3	0.507	0.712	581.3	241.0	0.468	0.684	605.3	261.7	0.431	0.656	630.6
7 500	38.254	220.7	0.493	0.702	579.2	239.4	0.454	0.674	603.2	259.9	0.419	0.647	628.5
7 750	36.909	219.0	0.479	0.692	577.0	237.8	0.441	0.664	601.2	258.2	0.407	0.638	626.5
8 000	35.601	217.4	0.466	0.682	574.9	236.2	0.429	0.655	599.2	256.5	0.395	0.628	624.3
8 250	34.330	215.8	0.453	0.673	572.7	234.5	0.416	0.645	597.1	254.7	0.383	0.619	622.2
8 500	33.096	214.1	0.440	0.663	570.4	232.9	0.404	0.636	595.0	252.9	0.372	0.610	620.0
8 750	31.899	212.3	0.427	0.654	568.1	231.3	0.392	0.626	592.9	251.2	0.361	0.601	617.9
9 000	30.740	210.6	0.415	0.644	565.9	229.7	0.381	0.617	590.9	249.5	0.350	0.592	615.8
9 250	29.616	209.1	0.403	0.635	563.8	228.0	0.369	0.608	588.7	247.9	0.340	0.583	613.8
9 500	28.523	208.0	0.390	0.624	562.3	226.4	0.358	0.599	586.6	246.2	0.329	0.574	611.8
9 750	27.463	208.0	0.375	0.613	562.3	224.8	0.347	0.589	584.6	244.6	0.319	0.565	609.7
10 000	26.435	208.0	0.361	0.601	562.3	223.2	0.337	0.580	582.4	242.9	0.310	0.556	607.6
10 250	25.441	208.0	0.348	0.590	562.3	221.5	0.327	0.571	580.3	241.2	0.300	0.548	605.6
10 500	24.475	208.0	0.335	0.578	562.3	219.9	0.317	0.563	578.2	239.7	0.290	0.539	603.6
10 750	23.540	208.0	0.322	0.567	562.3	218.3	0.307	0.554	576.0	238.2	0.281	0.530	601.7
11 000	22.628	208.0	0.309	0.556	562.3	216.7	0.297	0.545	573.9	236.7	0.272	0.521	599.8
11 250	21.758	208.0	0.297	0.545	562.3	216.7	0.286	0.534	573.9	235.2	0.263	0.513	597.9
11 500	20.914	208.0	0.286	0.535	562.3	216.7	0.275	0.524	573.9	233.6	0.255	0.505	595.9
11 750	20.106	208.0	0.275	0.524	562.3	216.7	0.264	0.514	573.9	232.1	0.246	0.496	593.9
12 000	19.331	208.0	0.264	0.514	562.3	216.7	0.254	0.504	573.9	231.0	0.238	0.488	592.5
12 250	18.583	208.0	0.254	0.504	562.3	216.7	0.244	0.494	573.9	230.6	0.229	0.479	592.0
12 500	17.862	208.0	0.244	0.494	562.3	216.7	0.234	0.484	573.9	230.8	0.220	0.469	592.3
12 750	17.176	208.0	0.235	0.485	562.3	216.7	0.225	0.475	573.9	231.0	0.211	0.460	592.5
13 000	16.512	207.0	0.227	0.476	560.9	216.7	0.217	0.466	573.9	231.1	0.203	0.451	592.8
13 250	15.872	205.1	0.220	0.469	558.3	216.7	0.208	0.456	573.9	231.3	0.195	0.442	593.0
13 500	15.257	202.7	0.214	0.463	555.1	216.7	0.200	0.447	573.9	231.5	0.187	0.433	593.2
13 750	14.669	200.3	0.208	0.456	551.8	216.7	0.193	0.439	573.9	231.8	0.180	0.424	593.5
14 000	14.105	197.9	0.203	0.450	548.4	216.7	0.185	0.430	573.9	232.0	0.173	0.416	593.8
14 250	13.558	195.4	0.197	0.444	545.0	216.7	0.178	0.422	573.9	232.2	0.166	0.408	594.1
14 500	13.034	193.0	0.192	0.438	541.6	216.7	0.171	0.414	573.9	232.4	0.160	0.399	594.3
14 750	12.530	190.7	0.187	0.432	538.4	216.7	0.164	0.406	573.9	232.6	0.153	0.391	594.6
15 000	12.045	188.7	0.182	0.426	535.6	216.7	0.158	0.398	573.9	232.8	0.147	0.384	594.9

Chart 2.1 *contd.*

(b) SI units: 15 250–30 500 m

Pressure altitude (m)	Pressure (kPa)	MIL STD 210A cold atmosphere				Standard atmosphere				MIL STD 210A hot atmosphere			
		Temp (K)	Relative density	√Relative density	Speed of sound (kt)	Temp (K)	Relative density	√Relative density	Speed of sound (kt)	Temp (K)	Relative density	√Relative density	Speed of sound (kt)
15 250	11.579	187.1	0.176	0.420	533.2	216.7	0.152	0.390	573.9	233.1	0.141	0.376	595.2
15 500	11.131	186.1	0.170	0.412	531.9	216.7	0.146	0.382	573.9	233.2	0.136	0.368	595.4
15 750	10.702	185.9	0.164	0.405	531.6	216.7	0.140	0.375	573.9	233.3	0.130	0.361	595.5
16 000	10.287	185.9	0.157	0.397	531.6	216.7	0.135	0.367	573.9	233.4	0.125	0.354	595.6
16 250	9.889	185.9	0.151	0.389	531.6	216.7	0.130	0.360	573.9	233.5	0.120	0.347	595.7
16 500	9.509	185.9	0.145	0.381	531.6	216.7	0.125	0.353	573.9	233.6	0.116	0.340	595.9
16 750	9.142	185.9	0.140	0.374	531.6	216.7	0.120	0.346	573.9	233.7	0.111	0.334	596.0
17 000	8.789	185.9	0.134	0.367	531.6	216.7	0.115	0.340	573.9	233.7	0.107	0.327	596.0
17 250	8.446	185.9	0.129	0.359	531.6	216.7	0.111	0.333	573.9	233.8	0.103	0.321	596.1
17 500	8.118	185.9	0.124	0.352	531.6	216.7	0.107	0.326	573.9	233.9	0.099	0.314	596.2
17 750	7.806	185.9	0.119	0.346	531.6	216.7	0.102	0.320	573.9	234.0	0.095	0.308	596.4
18 000	7.502	185.9	0.115	0.339	531.6	216.7	0.098	0.314	573.9	234.1	0.091	0.302	596.5
18 250	7.213	185.9	0.110	0.332	531.6	216.7	0.095	0.308	573.9	234.2	0.088	0.296	596.6
18 500	6.936	185.9	0.106	0.326	531.6	216.7	0.091	0.302	573.9	234.2	0.084	0.290	596.7
18 750	6.668	186.8	0.102	0.319	532.8	216.7	0.088	0.296	573.9	234.3	0.081	0.284	596.8
19 000	6.410	188.1	0.097	0.311	534.8	216.7	0.084	0.290	573.9	234.4	0.078	0.279	596.9
19 250	6.162	189.5	0.092	0.304	536.7	216.7	0.081	0.284	573.9	234.5	0.075	0.273	597.1
19 500	5.924	190.9	0.088	0.297	538.7	216.7	0.078	0.279	573.9	234.6	0.072	0.268	597.2
19 750	5.695	192.2	0.084	0.290	540.5	216.7	0.075	0.273	573.9	234.7	0.069	0.263	597.3
20 000	5.475	193.5	0.080	0.284	542.3	216.7	0.072	0.268	573.9	234.8	0.066	0.258	597.4
20 250	5.263	194.7	0.077	0.277	544.0	216.7	0.069	0.263	573.9	234.9	0.064	0.252	597.5
20 500	5.060	195.9	0.073	0.271	545.6	216.7	0.066	0.258	573.9	235.1	0.061	0.247	597.8
20 750	4.864	197.0	0.070	0.265	547.2	216.7	0.064	0.253	573.9	235.4	0.059	0.242	598.1
21 000	4.676	198.1	0.067	0.259	548.8	216.7	0.061	0.248	573.9	235.6	0.056	0.238	598.5

Notes:
To convert kt to m/s multiply by 0.5144.
To convert kt to km/h multiply by 1.8520.
To convert K to 8C subtract 273.15.
To convert K to 8R multiply by 1.8.

To convert K to 8F multiply by 1.8 and subtract 459.67.
Density at ISA sea level static = 1.2250 kg/m^3.
Standard practice is to interpolate linearly between altitudes listed.

Chart 2.1 *contd.* (b) SI units: 15 250–30 500 m

Pressure altitude (m)	Pressure (kPa)	MIL STD 210A cold atmosphere				Standard atmosphere				MIL STD 210A hot atmosphere			
		Temp (K)	Relative density	√Relative density	Speed of sound (kt)	Temp (K)	Relative density	√Relative density	Speed of sound (kt)	Temp (K)	Relative density	√Relative density	Speed of sound (kt)
21 250	4.495	199.2	0.064	0.253	550.3	216.7	0.059	0.243	573.9	236.0	0.054	0.233	598.9
21 500	4.321	200.2	0.061	0.248	551.7	216.7	0.057	0.238	573.9	236.3	0.052	0.228	599.4
21 750	4.155	201.2	0.059	0.242	553.0	216.7	0.055	0.234	573.9	236.6	0.050	0.223	599.8
22 500	3.690	203.0	0.052	0.227	555.5	216.7	0.048	0.220	573.9	237.6	0.044	0.210	601.0
22 750	3.549	202.9	0.050	0.223	555.4	216.7	0.047	0.216	573.9	237.9	0.042	0.206	601.4
23 000	3.411	202.8	0.048	0.219	555.2	216.7	0.045	0.212	573.9	238.2	0.041	0.202	601.8
23 250	3.280	202.7	0.046	0.215	555.1	216.7	0.043	0.207	573.9	238.6	0.039	0.198	602.2
23 500	3.153	202.6	0.044	0.210	554.9	216.7	0.041	0.203	573.9	238.9	0.038	0.194	602.6
23 750	3.031	202.5	0.043	0.206	554.8	216.7	0.040	0.199	573.9	239.2	0.036	0.190	603.0
24 000	2.914	202.3	0.041	0.202	554.6	216.7	0.038	0.196	573.9	239.5	0.035	0.186	603.4
24 250	2.801	202.2	0.039	0.199	554.4	216.7	0.037	0.192	573.9	239.8	0.033	0.182	603.8
24 500	2.691	202.0	0.038	0.195	554.2	216.7	0.035	0.188	573.9	240.2	0.032	0.179	604.2
24 750	2.594	201.9	0.037	0.191	554.0	216.7	0.034	0.185	573.9	240.5	0.031	0.175	604.7
25 000	2.522	201.7	0.036	0.189	553.7	216.7	0.033	0.182	573.9	240.9	0.030	0.173	605.1
25 250	2.397	201.5	0.034	0.184	553.5	217.4	0.031	0.177	574.9	241.2	0.028	0.168	605.5
25 500	2.299	201.4	0.032	0.180	553.2	218.2	0.030	0.173	575.9	241.6	0.027	0.165	606.0
25 750	2.212	201.2	0.031	0.177	553.0	218.9	0.029	0.170	576.9	241.9	0.026	0.161	606.4
26 000	2.128	201.0	0.030	0.174	552.7	219.7	0.028	0.166	577.8	242.3	0.025	0.158	606.9
26 250	2.047	200.8	0.029	0.170	552.5	220.4	0.026	0.163	578.8	242.7	0.024	0.155	607.3
26 500	1.969	200.7	0.028	0.167	552.3	221.2	0.025	0.159	579.8	243.0	0.023	0.152	607.8
26 750	1.895	200.5	0.027	0.164	552.1	221.9	0.024	0.156	580.8	243.4	0.022	0.149	608.3
27 000	1.823	200.3	0.026	0.161	551.8	222.7	0.023	0.153	581.8	243.8	0.021	0.146	608.8
27 250	1.754	200.1	0.025	0.158	551.6	223.4	0.022	0.149	582.8	244.2	0.020	0.143	609.2
27 500	1.689	200.0	0.024	0.155	551.3	224.2	0.021	0.146	583.7	244.5	0.020	0.140	609.6
27 750	1.626	199.8	0.023	0.152	551.1	224.9	0.021	0.143	584.7	244.8	0.019	0.137	610.0
28 000	1.565	199.6	0.022	0.149	550.8	225.7	0.020	0.140	585.7	245.2	0.018	0.135	610.4
28 250	1.507	199.4	0.021	0.147	550.5	226.4	0.019	0.138	586.7	245.5	0.017	0.132	610.9
28 500	1.452	199.2	0.021	0.144	550.3	227.2	0.018	0.135	587.6	245.8	0.017	0.130	611.3
28 750	1.398	199.0	0.020	0.141	550.0	227.9	0.017	0.132	588.6	246.2	0.016	0.127	611.8
29 000	1.347	198.9	0.019	0.139	549.8	228.7	0.017	0.129	589.6	246.6	0.016	0.125	612.3
29 250	1.298	198.7	0.019	0.136	549.5	229.4	0.016	0.127	590.5	247.0	0.015	0.122	612.7
29 500	1.250	198.4	0.018	0.134	549.2	230.1	0.015	0.124	591.5	247.3	0.014	0.120	613.2
29 750	1.205	198.3	0.017	0.131	549.0	230.9	0.015	0.122	592.4	247.7	0.014	0.118	613.7
30 000	1.161	198.1	0.017	0.129	548.7	231.7	0.014	0.119	593.4	248.2	0.013	0.115	614.2
30 250	1.119	197.9	0.016	0.127	548.4	232.4	0.014	0.117	594.4	248.6	0.013	0.113	614.7
30 500	1.079	197.7	0.016	0.125	548.1	233.1	0.013	0.115	595.3	248.9	0.012	0.111	615.1

Chart 2.1 *contd.*

(c) Imperial units: 0–50 000 ft

Pressure altitude (ft)	Pressure (psia)	MIL STD 210A cold atmosphere				Standard atmosphere				MIL STD 210A hot atmosphere			
		Temp (K)	Relative density	√Relative density	Speed of sound (kt)	Temp (K)	Relative density	√Relative density	Speed of sound (kt)	Temp (K)	Relative density	√Relative density	Speed of sound (kt)
0	14.696	222.1	1.298	1.139	581.0	288.2	1.000	1.000	661.7	312.6	0.922	0.960	689.0
1000	14.173	229.6	1.211	1.100	590.7	286.2	0.971	0.985	659.4	310.5	0.895	0.946	686.7
2000	13.664	237.1	1.130	1.063	600.3	284.2	0.943	0.971	657.1	308.4	0.869	0.932	684.4
3000	13.171	244.7	1.056	1.027	609.8	282.2	0.915	0.957	654.8	306.2	0.843	0.918	682.0
4000	12.692	247.1	1.007	1.004	612.8	280.2	0.888	0.942	652.5	304.1	0.818	0.905	679.6
5000	12.228	247.1	0.970	0.985	612.8	278.3	0.862	0.928	650.3	301.9	0.794	0.891	677.2
6000	11.777	247.1	0.935	0.967	612.8	276.3	0.836	0.914	647.9	299.7	0.770	0.878	674.7
7000	11.340	247.1	0.900	0.949	612.8	274.3	0.811	0.900	645.6	297.5	0.747	0.865	672.2
8000	10.916	247.1	0.866	0.931	612.8	272.3	0.786	0.887	643.3	295.3	0.725	0.851	669.7
9000	10.505	247.1	0.834	0.913	612.8	270.3	0.762	0.873	641.0	293.1	0.703	0.838	667.2
10000	10.106	247.1	0.802	0.896	612.8	268.3	0.738	0.859	638.6	290.9	0.681	0.825	664.8
11000	9.720	246.6	0.773	0.879	612.2	266.4	0.715	0.846	636.3	288.8	0.660	0.812	662.4
12000	9.346	244.8	0.749	0.865	610.0	264.4	0.693	0.833	633.9	286.7	0.639	0.799	660.0
13000	8.984	242.9	0.725	0.852	607.7	262.4	0.671	0.819	631.5	284.6	0.619	0.787	657.6
14000	8.633	241.1	0.702	0.838	605.4	260.4	0.650	0.806	629.2	282.5	0.599	0.774	655.2
15000	8.294	239.2	0.680	0.825	603.0	258.4	0.629	0.793	626.7	280.3	0.580	0.762	652.7
16000	7.965	237.4	0.658	0.811	600.7	256.4	0.609	0.780	624.3	278.2	0.561	0.749	650.1
17000	7.647	235.5	0.637	0.798	598.3	254.5	0.589	0.768	621.9	276.0	0.543	0.737	647.6
18000	7.339	233.6	0.616	0.785	595.9	252.5	0.570	0.755	619.5	273.8	0.526	0.725	645.0
19000	7.041	231.7	0.596	0.772	593.4	250.5	0.551	0.742	617.0	271.6	0.508	0.713	642.5
20000	6.753	229.8	0.576	0.759	591.0	248.6	0.533	0.730	614.7	269.6	0.491	0.701	640.0
21000	6.475	227.8	0.557	0.746	588.5	246.6	0.515	0.718	612.2	267.5	0.475	0.689	637.6
22000	6.206	225.8	0.539	0.734	585.9	244.6	0.498	0.705	609.7	265.4	0.459	0.677	635.1
23000	5.947	223.9	0.521	0.722	583.4	242.6	0.481	0.693	607.3	263.3	0.443	0.665	632.6

Notes:
To convert kt to ft/s multiply by 1.6878.
To convert kt to miles/h multiply by 1.1508.
To convert K to 8C subtract 273.15.
To convert K to 8R multiply by 1.8.

To convert K to 8F multiply by 1.8 and subtract 459.67.
Density at ISA sea level static = 0.07647 lb/ft³.
Standard practice is to interpolate linearly between altitudes listed.

Chart 2.1 *contd.* (c) Imperial units: 0–50 000 ft

Pressure altitude (ft)	Pressure (psia)	MIL STD 210A cold atmosphere				Standard atmosphere				MIL STD 210A hot atmosphere			
		Temp (K)	Relative density	√Relative density	Speed of sound (kt)	Temp (K)	Relative density	√Relative density	Speed of sound (kt)	Temp (K)	Relative density	√Relative density	Speed of sound (kt)
24 000	5.696	221.9	0.503	0.709	580.8	240.6	0.464	0.681	604.8	261.2	0.428	0.654	630.1
25 000	5.434	219.9	0.485	0.696	578.1	238.6	0.447	0.668	602.2	259.1	0.411	0.641	627.5
26 000	5.220	217.9	0.470	0.685	575.5	236.7	0.432	0.658	599.8	257.0	0.398	0.631	625.0
27 000	4.994	215.9	0.454	0.673	572.9	234.7	0.417	0.646	597.3	254.8	0.384	0.620	622.4
28 000	4.776	213.8	0.438	0.662	570.1	232.7	0.402	0.634	594.7	252.7	0.371	0.609	619.7
29 000	4.566	211.7	0.423	0.650	567.3	230.7	0.388	0.623	592.2	250.6	0.357	0.598	617.1
30 000	4.364	209.7	0.408	0.639	564.5	228.7	0.374	0.612	589.6	248.6	0.344	0.587	614.7
31 000	4.169	208.0	0.393	0.627	562.3	226.7	0.361	0.600	587.1	246.6	0.332	0.576	612.2
32 000	3.981	208.0	0.375	0.613	562.3	224.8	0.347	0.589	584.5	244.6	0.319	0.565	609.7
33 000	3.800	208.0	0.358	0.599	562.3	222.8	0.334	0.578	581.9	242.5	0.307	0.554	607.1
34 000	3.626	208.0	0.342	0.585	562.3	220.8	0.322	0.567	579.3	240.5	0.296	0.544	604.6
35 000	3.458	208.0	0.326	0.571	562.3	218.8	0.310	0.557	576.8	238.7	0.284	0.533	602.3
36 000	3.297	208.0	0.311	0.557	562.3	216.7	0.298	0.546	573.9	236.8	0.273	0.522	600.0
36 089	3.282	208.0	0.309	0.556	562.3	216.7	0.297	0.545	573.9	236.7	0.272	0.521	599.8
37 000	3.142	208.0	0.296	0.544	562.3	216.7	0.284	0.533	573.9	235.0	0.262	0.512	597.7
38 000	2.994	208.0	0.282	0.531	562.3	216.7	0.271	0.521	573.9	233.1	0.252	0.502	595.3
39 000	2.854	208.0	0.269	0.519	562.3	216.7	0.258	0.508	573.9	231.2	0.242	0.492	592.9
40 000	2.720	208.0	0.256	0.506	562.3	216.7	0.246	0.496	573.9	230.5	0.231	0.481	591.9
41 000	2.592	208.0	0.244	0.494	562.3	216.7	0.235	0.484	573.9	230.8	0.220	0.469	592.3
42 000	2.471	208.0	0.233	0.483	562.3	216.7	0.224	0.473	573.9	231.0	0.210	0.458	592.6
43 000	2.355	206.4	0.224	0.473	560.1	216.7	0.213	0.462	573.9	231.2	0.200	0.447	592.9
44 000	2.244	203.6	0.216	0.465	556.3	216.7	0.203	0.451	573.9	231.4	0.190	0.436	593.1
45 000	2.139	200.6	0.209	0.457	552.2	216.7	0.194	0.440	573.9	231.7	0.181	0.425	593.5
46 000	2.039	197.7	0.202	0.450	548.1	216.7	0.185	0.430	573.9	232.0	0.172	0.415	593.9
47 000	1.943	194.7	0.196	0.442	544.0	216.7	0.176	0.419	573.9	232.2	0.164	0.405	594.1
48 000	1.852	191.7	0.189	0.435	539.8	216.7	0.168	0.409	573.9	232.5	0.156	0.395	594.5
49 000	1.765	189.2	0.183	0.428	536.2	216.7	0.160	0.400	573.9	232.8	0.149	0.386	594.8
50 000	1.682	187.1	0.176	0.420	533.3	216.7	0.152	0.390	573.9	233.1	0.142	0.376	595.2

Chart 2.1 *contd.*

(d) Imperial units: 51 000–100 000 ft

Pressure altitude (ft)	Pressure (psia)	MIL STD 210A cold atmosphere				Standard atmosphere				MIL STD 210A hot atmosphere			
		Temp (K)	Relative density	√Relative density	Speed of sound (kt)	Temp (K)	Relative density	√Relative density	Speed of sound (kt)	Temp (K)	Relative density	√Relative density	Speed of sound (kt)
51 000	1.603	185.9	0.169	0.411	531.6	216.7	0.145	0.381	573.9	233.2	0.135	0.367	595.4
52 000	1.528	185.9	0.161	0.401	531.6	216.7	0.138	0.372	573.9	233.3	0.128	0.358	595.6
53 000	1.456	185.9	0.154	0.392	531.6	216.7	0.132	0.363	573.9	233.4	0.122	0.350	595.7
54 000	1.388	185.9	0.146	0.383	531.6	216.7	0.126	0.354	573.9	233.6	0.117	0.341	595.8
55 000	1.323	185.9	0.140	0.374	531.6	216.7	0.120	0.346	573.9	233.7	0.111	0.333	596.0
56 000	1.261	185.9	0.133	0.365	531.6	216.7	0.114	0.338	573.9	233.7	0.106	0.325	596.0
57 000	1.201	185.9	0.127	0.356	531.6	216.7	0.109	0.330	573.9	233.8	0.101	0.317	596.2
58 000	1.145	185.9	0.121	0.347	531.6	216.7	0.104	0.322	573.9	233.9	0.096	0.310	596.3
59 000	1.091	185.9	0.115	0.339	531.6	216.7	0.099	0.314	573.9	234.1	0.091	0.302	596.5
60 000	1.040	185.9	0.110	0.331	531.6	216.7	0.094	0.307	573.9	234.2	0.087	0.295	596.6
61 000	0.991	185.9	0.105	0.323	531.6	216.7	0.090	0.300	573.9	234.3	0.083	0.288	596.8
62 000	0.945	187.6	0.099	0.314	533.9	216.7	0.086	0.292	573.9	234.4	0.079	0.281	596.9
63 000	0.901	189.3	0.093	0.305	536.4	216.7	0.081	0.285	573.9	234.5	0.075	0.274	597.0
64 000	0.858	190.9	0.088	0.297	538.7	216.7	0.078	0.279	573.9	234.6	0.072	0.268	597.2
65 000	0.818	192.5	0.083	0.289	540.9	216.7	0.074	0.272	573.9	234.7	0.068	0.261	597.3
66 000	0.780	194.1	0.079	0.281	543.1	216.7	0.071	0.266	573.9	234.8	0.065	0.255	597.4
67 000	0.743	195.5	0.075	0.273	545.1	216.7	0.067	0.259	573.9	235.0	0.062	0.249	597.7
68 000	0.708	196.9	0.071	0.266	547.1	216.7	0.064	0.253	573.9	235.3	0.059	0.243	598.1
69 000	0.675	198.3	0.067	0.258	549.0	216.7	0.061	0.247	573.9	235.7	0.056	0.237	598.5
70 000	0.643	199.6	0.063	0.251	550.8	216.7	0.058	0.241	573.9	236.1	0.053	0.231	599.1
71 000	0.613	200.8	0.060	0.245	552.4	216.7	0.055	0.236	573.9	236.5	0.051	0.225	599.6
72 000	0.584	202.0	0.057	0.238	554.1	216.7	0.053	0.230	573.9	236.9	0.048	0.220	600.1
73 000	0.556	203.1	0.054	0.232	555.6	216.7	0.050	0.224	573.9	237.3	0.046	0.214	600.6
74 000	0.531	203.0	0.051	0.226	555.5	216.7	0.048	0.219	573.9	237.7	0.044	0.209	601.1

Notes:
To convert kt to ft/s multiply by 1.6878.
To convert kt to miles/h multiply by 1.1508.
To convert K to 8C subtract 273.15.
To convert K to 8R multiply by 1.8.

To convert K to 8F multiply by 1.8 and subtract 459.67.
Density at ISA sea level static = 0.07647 lb/ft³.
Standard practice is to interpolate linearly between altitudes listed.

Chart 2.1   *contd.*

(d) Imperial units: 51 000–100 000 ft

Pressure altitude (ft)	Pressure (psia)	MIL STD 210A cold atmosphere				Standard atmosphere				MIL STD 210A hot atmosphere			
		Temp (K)	Relative density	√Relative density	Speed of sound (kt)	Temp (K)	Relative density	√Relative density	Speed of sound (kt)	Temp (K)	Relative density	√Relative density	Speed of sound (kt)
75 000	0.506	202.9	0.049	0.221	555.3	216.7	0.046	0.214	573.9	238.1	0.042	0.204	601.5
76 000	0.482	202.7	0.047	0.216	555.1	216.7	0.044	0.209	573.9	238.4	0.040	0.199	602.0
77 000	0.460	202.6	0.044	0.211	555.0	216.7	0.042	0.204	573.9	238.8	0.038	0.194	602.5
78 000	0.438	202.4	0.042	0.206	554.7	216.7	0.040	0.199	573.9	239.2	0.036	0.189	603.0
79 000	0.417	202.3	0.040	0.201	554.5	216.7	0.038	0.194	573.9	239.6	0.034	0.185	603.5
80 000	0.398	202.1	0.039	0.196	554.3	216.7	0.036	0.190	573.9	240.0	0.032	0.180	604.0
81 000	0.379	201.9	0.037	0.192	554.0	216.7	0.034	0.185	573.9	240.4	0.031	0.176	604.6
82 000	0.366	201.7	0.036	0.189	553.7	216.7	0.033	0.182	573.9	240.9	0.030	0.173	605.1
83 000	0.344	201.5	0.034	0.183	553.4	217.6	0.031	0.176	575.1	241.3	0.028	0.167	605.6
84 000	0.328	201.3	0.032	0.179	553.1	218.5	0.029	0.172	576.3	241.7	0.027	0.163	606.2
85 000	0.313	201.1	0.031	0.175	552.8	219.4	0.028	0.167	577.5	242.2	0.025	0.159	606.7
86 000	0.299	200.8	0.029	0.171	552.5	220.3	0.027	0.163	578.7	242.6	0.024	0.155	607.3
87 000	0.285	200.7	0.028	0.167	552.3	221.2	0.025	0.159	579.9	243.1	0.023	0.152	607.8
88 000	0.272	200.4	0.027	0.163	552.0	222.1	0.024	0.155	581.1	243.6	0.022	0.148	608.4
89 000	0.259	200.2	0.025	0.159	551.7	223.1	0.023	0.151	582.3	244.0	0.021	0.144	609.0
90 000	0.248	200.0	0.024	0.156	551.4	223.9	0.022	0.147	583.5	244.4	0.020	0.141	609.5
91 000	0.236	199.8	0.023	0.152	551.1	224.9	0.021	0.144	584.7	244.8	0.019	0.138	610.0
92 000	0.226	199.6	0.022	0.149	550.8	225.8	0.020	0.140	585.8	245.2	0.018	0.134	610.5
93 000	0.216	199.3	0.021	0.146	550.5	226.7	0.019	0.137	587.1	245.6	0.017	0.131	611.0
94 000	0.206	199.1	0.020	0.142	550.1	227.6	0.018	0.133	588.2	246.1	0.016	0.128	611.6
95 000	0.197	198.9	0.019	0.139	549.8	228.6	0.017	0.130	589.4	246.6	0.016	0.125	612.2
96 000	0.188	198.7	0.019	0.136	549.5	229.4	0.016	0.127	590.6	247.0	0.015	0.122	612.7
97 000	0.180	198.4	0.018	0.133	549.1	230.3	0.015	0.124	591.7	247.4	0.014	0.119	613.3
98 000	0.172	198.2	0.017	0.130	548.8	231.3	0.015	0.121	592.9	247.9	0.014	0.117	613.9
99 000	0.164	197.9	0.016	0.127	548.5	232.2	0.014	0.118	594.1	248.4	0.013	0.114	614.5
100 000	0.157	197.7	0.016	0.125	548.1	233.1	0.013	0.115	595.3	248.9	0.012	0.111	615.1

Chart 2.2   Ambient presssure versus pressure altitude.

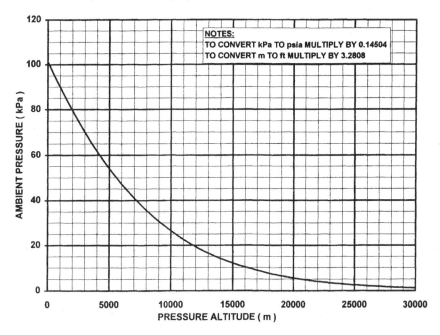

Chart 2.3   Ambient temperature versus pressure altitude.

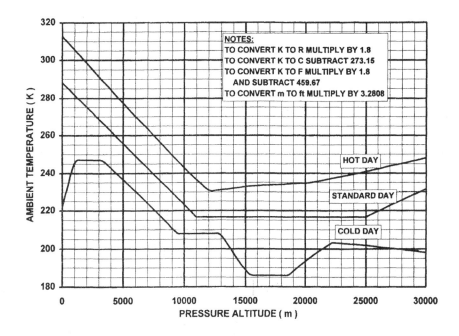

Chart 2.4   Relative density versus pressure altitude.

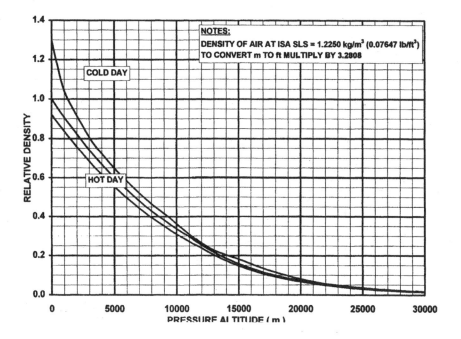

Chart 2.5   Square root of relative density versus pressure altitude.

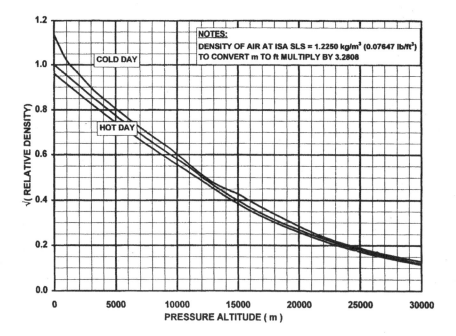

Chart 2.6　Speed of sound versus pressure altitude.

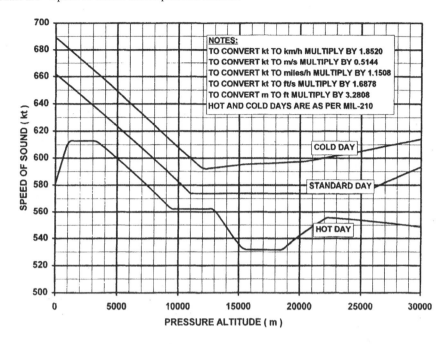

Chart 2.7　Specific humidity versus pressure altitude for 100% relative humidity.

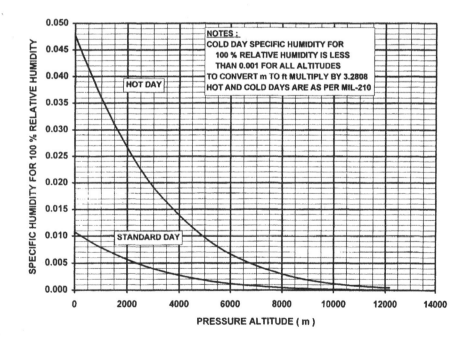

Chart 2.8   Specific humidity versus relative humidity and ambient temperature at sea level.

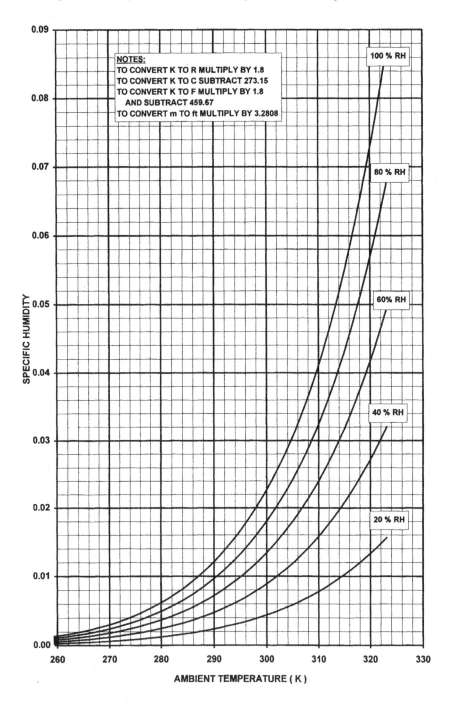

Chart 2.9 Ratio of specific humidity at altitude to that at sea level.

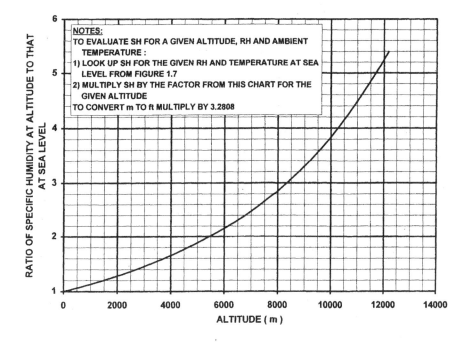

NOTES:
TO EVALUATE SH FOR A GIVEN ALTITUDE, RH AND AMBIENT
  TEMPERATURE :
1) LOOK UP SH FOR THE GIVEN RH AND TEMPERATURE AT SEA
  LEVEL FROM FIGURE 1.7
2) MULTIPLY SH BY THE FACTOR FROM THIS CHART FOR THE
  GIVEN ALTITUDE
TO CONVERT m TO ft MULTIPLY BY 3.2808

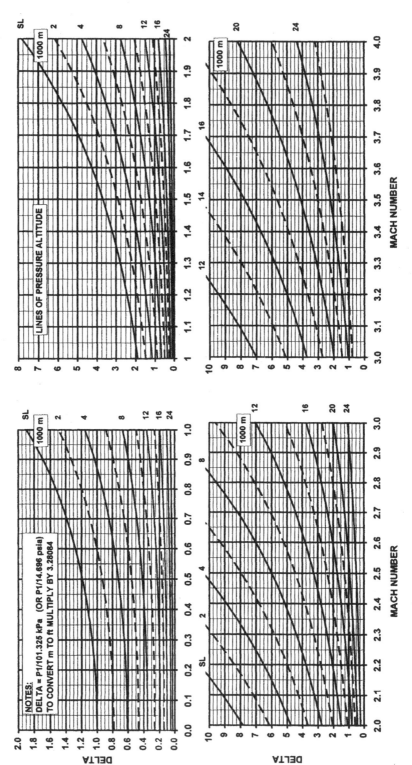

Chart 2.10 Delta versus altitude and Mach number.

Chart 2.11 Theta versus altitude and Mach number.

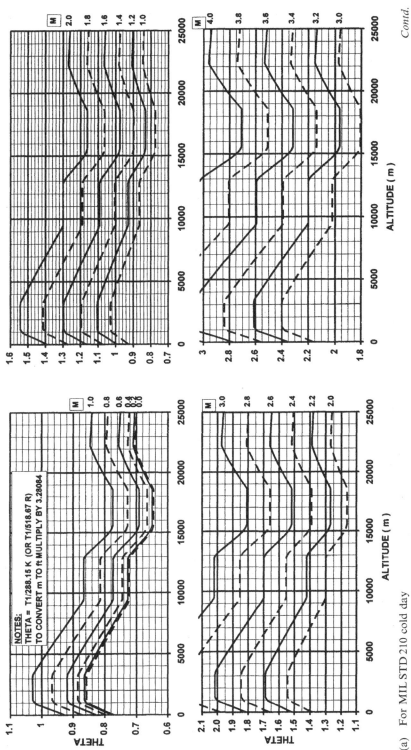

NOTES:
THETA = T1/288.15 K (OR T1/518.67 R)
TO CONVERT m TO ft MULTIPLY BY 3.28084

(a) For MIL STD 210 cold day

Contd.

Chart 2.11   *contd.*

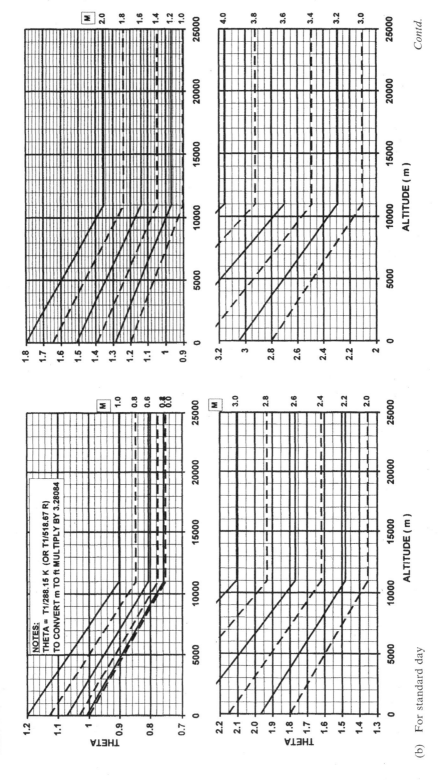

(b) For standard day

Contd.

Chart 2.11 *contd.*

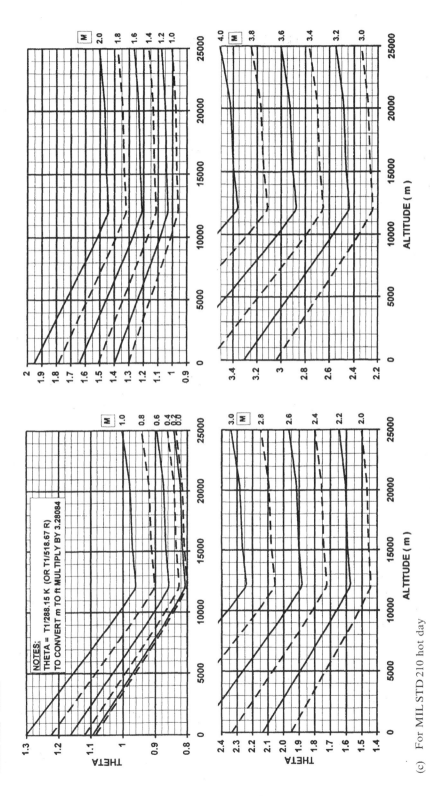

NOTES:
THETA = T1/288.15 K (OR T1/518.67 R)
TO CONVERT m TO ft MULTIPLY BY 3.28084

(c) For MIL STD 210 hot day

Chart 2.12   Reynolds number ratio versus altitude and Mach number.

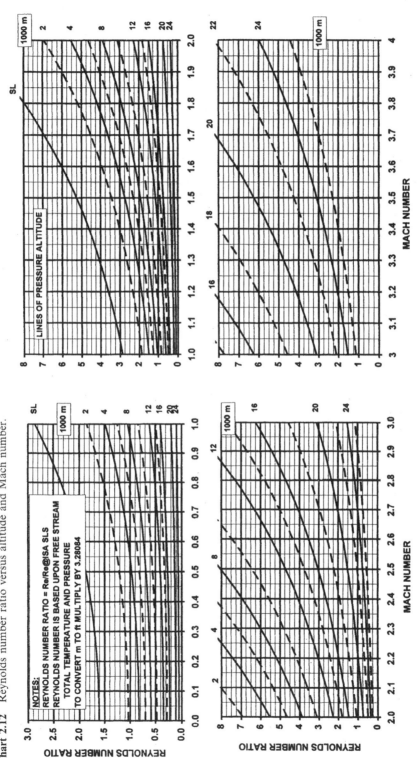

(a)   For MIL STD 210 cold day

Contd.

Chart 2.12   *contd.*

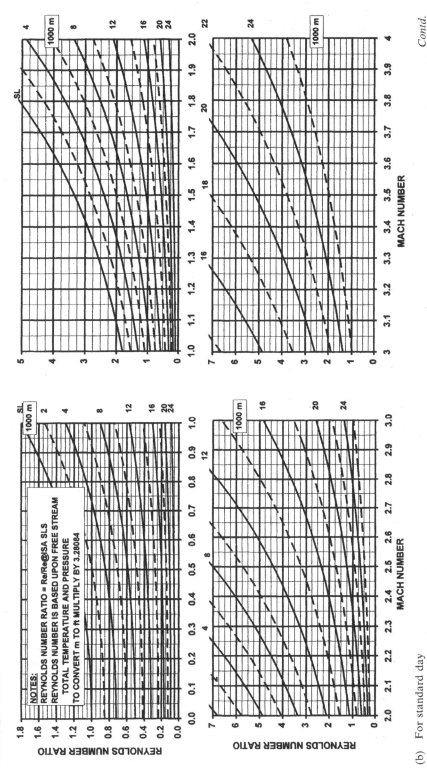

NOTES:
REYNOLDS NUMBER RATIO = Re/Re@ISA SLS
REYNOLDS NUMBER IS BASED UPON FREE STREAM
TOTAL TEMPERATURE AND PRESSURE
TO CONVERT m TO ft MULTIPLY BY 3.28084

(b)   For standard day

Contd.

Chart 2.12 *contd.*

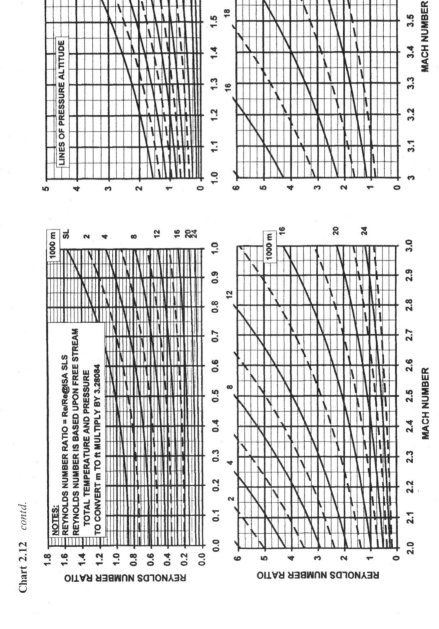

(c) For MIL STD 210 hot day

Chart 2.13   Calibrated air speed versus Mach number and altitude.

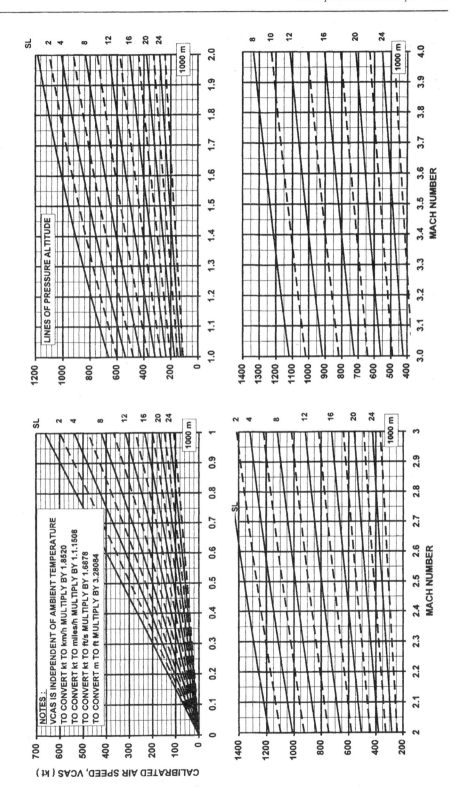

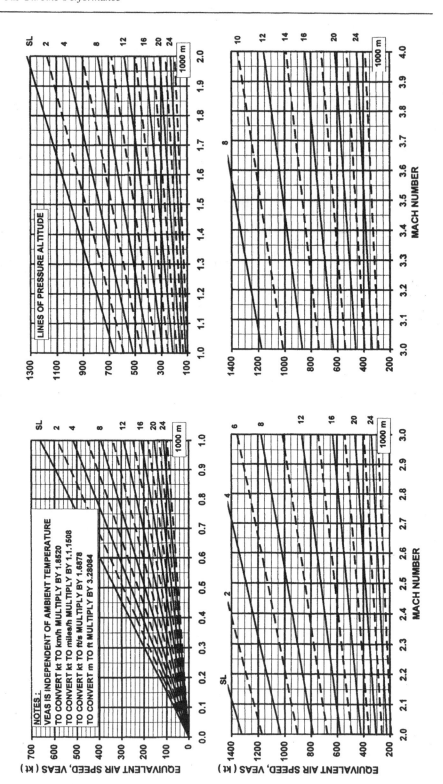

Chart 2.14 Equivalent air speed versus Mach number and altitude.

Chart 2.15 True air speed versus Mach number and ambient temperature.

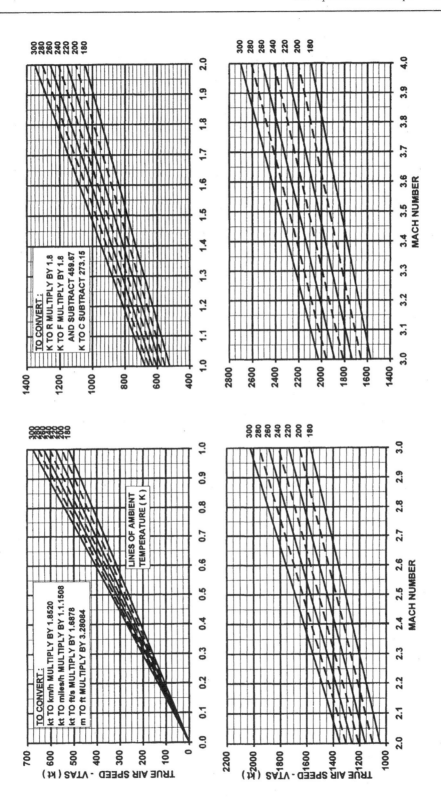

Chart 2.16   Scale altitude effect versus calibrated air speed and altitude.

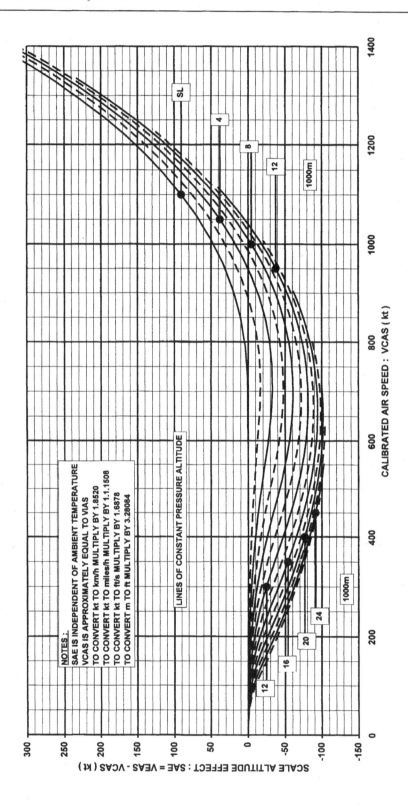

References

1. ISO (1975) *Standard Atmosphere, ISO 2533*, International Organisation for Standardisation, Geneva.
2. *Climatic Information to Determine Design and Test Requirements for Military Equipment MIL 210C*, Rev C January 1997, US Department of Defense, Massachusetts.
3. ISO (1973) *Gas Turbine Acceptance Tests, ISO 2314*, International Organisation for Standardisation, Geneva.
4. CAA (1975) *British Civil Airworthiness Requirements*, Sub-Section C1, Chapter C1–2, Civil Aviation Authority, London.
5. MOD (1968) *Defence Standard A970*, Chapter 101, UK Ministry of Defence, HMSO, London.
6. J. T. Houghton (1977) *The Physics of Atmospheres*, Cambridge University Press, Cambridge.
7. UK Meteorological Office (1990) *Global Ocean Surface Temperature Atlas (GOSTA)*, HMSO, London.
8. W. F. Hilton (1952) *High Speed Aerodynamics*, Longmans, London.

Chapter 3
Properties and Charts for Dry Air, Combustion Products and other Working Fluids

3.0 Introduction

The properties of the working fluid in a gas turbine engine have a powerful impact upon its performance. It is essential that these gas properties are accounted rigorously in calculations, or that any inaccuracy due to simplifying assumptions is quantified and understood. This chapter describes at an engineering level the fundamental gas properties of concern, and their various interrelationships. It also provides a comprehensive data base for use in calculations for:

- Dry air
- Combustion products for kerosene or diesel fuel
- Combustion products for natural gas fuel
- Helium, the working fluid often employed in closed cycles

Chapter 12 covers the impact of water content due to humidity, condensation, or injection of water or steam. Chapter 13 provides the key properties of gas turbine fuels.

3.1 Description of fundamental gas properties

Reference 1 provides an exhaustive description of fundamental gas properties. Those relevant to gas turbine performance are described below, and section 3.5 provides a data base sufficient for all performance calculations.

3.1.1 Equation of state for a perfect gas (Formula F3.1)

A *perfect gas* adheres to Formula F3.1. All gases employed as the working fluid in gas turbine engines, except for water vapour, may be considered as perfect gases without compromising calculation accuracy. When the mass fraction of water vapour is less than 10%, which is usually the case when it results from the combination of ambient humidity and products of combustion, then for performance calculations the gas mixture may still be considered perfect. When water vapour content exceeds 10% the assumption of a perfect gas is no longer valid and for rigorous calculations steam tables (Reference 2) must be employed in parallel, for that fraction of the mixture. This is described further in Chapter 12.

A physical description of a perfect gas is that its enthalpy is only a function of temperature and *not* pressure, as there are no intermolecular forces to absorb or release energy when density changes.

3.1.2 Molecular weight and the mole

The *molecular weight* for a pure gas is defined in the *Periodic Table*. For mixtures of gases, such as air, the molecular weight may be found by averaging the constituents on a *molar* (volumetric) basis. This is because a mole contains a fixed number of molecules, as described below. For example as shown by sample calculation C3.1, the molecular weight of dry air given in section 3.5.1 may be derived from the molecular weight of its constituents and their mole fractions provided in section 3.3.

A *mole* is the quantity of a substance such that the mass is equal to the molecular weight in grammes. For any perfect gas one mole occupies a volume of 22.4 litres at 0 8C, 101.325 kPa. A mole contains the Avogadro's number of molecules, 6.023×10^{23}.

3.1.3 *Specific heat at constant pressure (CP) and at constant volume (CV) (Formulae F3.2 and F3.3)*

These are the amounts of energy required to raise the temperature of one kilogramme of the gas by 1 8C, at constant pressure and volume respectively. For gas turbine engines, with a steady flow of gas (as opposed to piston engines where it is intermittent) only the specific heat at constant pressure, *CP*, is used directly. This is referred to hereafter simply as *specific heat*.

For the gases of interest specific heat is a function of only gas composition and static temperature. For performance calculations total temperature can normally be used up to Mach numbers of 0.4 with negligible loss in accuracy, since dynamic temperature (section 3.2.1) remains a low proportion of the total.

3.1.4 *Gas constant (R) (Formulae F3.4 and F3.5)*

The gas constant appears extensively in formulae relating pressure and temperature changes, and is numerically equal to the difference between *CP* and *CV*. The gas constant for an individual gas is the universal gas constant divided by the molecular weight, and has units of J/kg K. The universal gas constant has a value of 8314.3 J/mol K.

3.1.5 *Ratio of specific heats, gamma (γ) (Formulae F3.6–F3.8)*

This is the ratio of the specific heat at constant pressure to that at constant volume. Again it is a function of gas composition and static temperature, but total temperature may be used when the Mach number is less than 0.4. Gamma appears extensively in the 'perfect gas' formulae relating pressure and temperature changes and component efficiencies.

3.1.6 *Dynamic viscosity (VIS) and Reynolds number (RE) (Formulae F3.9 and F2.13)*

Dynamic viscosity is used to calculate the Reynolds number, which reflects the ratio of momentum to viscous forces present in a fluid. The Reynolds number is used in many performance calculations, such as for disc windage, and has a second-order effect on component efficiencies. Dynamic viscosity is a measure of the viscous forces and is a function of gas composition and static temperature. As viscosity has only a second-order effect on an engine cycle, total temperature may be used up to a Mach number of 0.6. The effect of fuel air ratio (gas composition) is negligible for practical purposes.

The units of viscosity of N s/m^2 are derived from N/(m/s)/m; force per unit gradient of velocity. Gas velocity varies in a direction perpendicular to the flow in the boundary layers on all gas washed surfaces.

3.2 Description of key thermodynamic parameters

The key thermodynamic parameters most widely used in gas turbine performance calculations are described below. Their interrelationships are dependent upon the values of the fundamental gas properties described above. These parameters are described further in References 1 and 3, and section 3.5 provides a data base sufficient for all performance calculations.

3.2.1   Total or stagnation temperature (T) (Formula F3.10 or F3.31)

Total temperature is the temperature resulting from bringing a gas stream to rest with no work or heat transfer. Note that here 'at rest' means relative to the engine, which may have a flight velocity relative to the Earth. The difference between the total and static temperatures at a given point is called the *dynamic temperature*. The ratio of total to static temperature is a function of only gamma and Mach number, as per Formula F3.10 or F3.31.

In general for gas turbine *performance* calculations total temperature is used through the engine, evaluated at engine entry from the ambient static temperature and any ram effect. At locations between engine components total temperature is a valid measure of energy changes. In addition, this aids comparison between predictions and test data, as it is only practical to measure total temperature. For most *component design* purposes, however, static conditions are also relevant, as for example the Mach number is often high (1.0 and greater) at entry to a compressor stator or turbine rotor blade.

Total temperature is constant for flow along ducts where there is no work or heat transfer, such as intake and exhaust systems. Total and static temperature diverge much less rapidly versus Mach number than do total and static pressure, as described below.

3.2.2   Total or stagnation pressure (P) (Formulae F3.11 or F3.32; F3.12 and F3.13)

Total pressure is that which would result from bringing a gas stream to rest without any work or heat transfer, and without any change in entropy (section 3.2.4). Total pressure is therefore an idealised property.

The difference between total and static pressure at a point is called either the *dynamic pressure*, *dynamic head* or *velocity head* (Formulae F3.12 and F3.13). The term *head* relates back to hydraulic engineering. The ratio of total to static pressure, as for temperature, is a function of only gamma and Mach number. Most performance calculations are conducted using total pressure, that at engine inlet again resulting from ambient static plus intake ram recovery.

Total pressure is not constant for flow through ducts, being reduced by wall friction and changes in flow direction, which produce turbulent losses. Both these effects act on the dynamic head; as described in Chapter 5 the pressure loss in a duct of given geometry and inlet swirl angle is almost always a fixed number of inlet dynamic heads. For this reason for performance calculations both the total and static pressure must often be evaluated at entry to ducts. Again for component design purposes both the total and static values are of interest.

Total and static pressure diverge much more rapidly versus Mach number than do total and static temperature. Calculation of pressure ratio from temperature ratio is far more sensitive to errors in the assumption of the mean gamma than the reverse calculation.

3.2.3   Specific enthalpy (H) (Formulae F3.14–F3.16)

This is the energy per kilogramme of gas relative to a stipulated zero datum. *Changes* in enthalpy, rather than absolute values, are important for gas turbine performance. Total or static enthalpy may be calculated, depending on which of the respective temperatures is used. Total enthalpy, like total temperature, is most common in performance calculations.

3.2.4   Specific entropy (S) (Formulae F3.17–F3.21)

Traditionally the property entropy has been shrouded in mystery, primarily due to being less tangible than the other properties discussed in this chapter. Section 3.6.4 shows how entropy relates to other thermodynamic properties relevant to gas turbine performance, and thereby helps overcome these difficulties.

During compression or expansion the increase in entropy is a measure of the thermal energy lost to friction, which becomes unavailable as useful work. Again, changes in entropy, rather than absolute values are of interest, as shown by Formulae F3.17 and F3.18. The former is used in conjunction with the full enthalpy polynomials discussed in section 3.3.3, and the latter is the simplified version using specific heat at mean temperature. Formulae F3.19–F3.21 provide *isentropic* versions, i.e. for zero entropy change, as described in section 3.6.4. This idealised case is used extensively in gas turbine performance calculations as is apparent from the sample calculations presented later.

3.3   Composition of dry air and combustion products

3.3.1   *Dry air*

Reference 4 states that dry air comprises the following.

	By mole or volume (%)	By mass (%)
Nitrogen (N_2)	78.08	75.52
Oxygen (O_2)	20.95	23.14
Argon (Ar)	0.93	1.28
Carbon dioxide	0.03	0.05
Neon	0.002	0.001

There are also trace amounts of helium, methane, krypton, hydrogen, nitrous oxide and xenon. These are negligible for gas turbine performance purposes.

3.3.2   *Combustion products*

When a hydrocarbon fuel is burnt in air, combustion products change the composition significantly. As shown in Chapter 13, atmospheric oxygen is consumed to *oxidise* the hydrogen and carbon, creating water and carbon dioxide respectively. The degree of change in air composition depends both on *fuel air ratio* and *fuel chemistry*. As discussed in Chapter 13 the fuel air ratio such that all the oxygen is consumed is termed *stoichiometric*.

Distilled liquid fuels such as kerosene or diesel each have relatively fixed chemistry. Properties of their combustion products can be evaluated versus fuel air ratio and temperature using unique formulae, with the fuel chemistry inbuilt. In contrast, the chemistry of natural gas varies considerably. All natural gases have a high proportion of light hydrocarbons, often with other gases such as nitrogen, carbon dioxide or hydrogen. The sample natural gas shown in section 13.1.5 is typical. Because the composition of natural gas combustion products varies, along with the fuel chemistry, unique formulae for their gas properties do not exist, hence the calculation is more complex. Sample calculation C13.1 describes how to calculate the mole and mass fractions of the constituent gases resulting from the combustion of a hydrocarbon fuel in air, and hence fundamental gas properties.

3.4   The use of CP and gamma, or specific enthalpy and entropy, in calculations

Either CP and gamma, *or* specific enthalpy and entropy, are used extensively in performance calculations. The manner of their use is described below in order of increasing accuracy and calculation complexity. This list covers all gas turbine components except for the combustor, which is discussed in section 3.6.2. Sample calculations for each method are presented later.

3.4.1 Constant, standard values for CP and gamma

This normally uses the following approximations:

- Cold end gas properties   CP = 1004.7 J/kg K, gamma = 1.4
- Hot end gas properties   CP = 1156.9 J/kg K, gamma = 1.33
- Component performance   Formulae use values of CP and gamma as above

This is the least accurate method, giving errors of up to 5% in leading performance parameters. It should only be used in illustrative calculations for teaching purposes, or for crude, 'ballpark' estimates.

3.4.2 Values for CP and gamma based on mean temperature

For formulae using CP and gamma it is most accurate to base these values on the mean temperature within each component, i.e. the arithmetic mean of the inlet and exit values. It is less accurate to evaluate CP and gamma at inlet and exit, and then take a mean value for each.

For dry air and combustion products of kerosene or diesel the formulae given for CP as a function of temperature and fuel air ratio give accuracies of within 1.5% for leading performance parameters. The largest errors occur at the highest pressure ratios.

For combustion products of natural gas Formula F3.25 gives CP for the sample natural gas composition presented in section 13.1.5. Applying this to significantly different blends of natural gas, with different combustion products, may give errors of up to 3% in leading performance parameters. To achieve the same accuracy as for kerosene and diesel, CP must be evaluated using the method described in sample calculation C13.1.

This technique is commonly used for hand calculations or personal computer programs.

3.4.3 Specific enthalpy and entropy – dry air, and diesel or kerosene

For fully rigorous calculations changes in enthalpy and entropy across components must be accurately evaluated. This improves accuracy to be within 0.25% for leading parameters at all pressure ratios. Here polynomials of specific enthalpy and entropy are utilised, obtained by integration of the standard polynomials for specific heat. In these methods, a formula for specific heat is therefore still required.

The use of specific enthalpy and entropy for performance calculations is now almost mandatory for computer 'library' routines in large companies.

3.4.4 Specific enthalpy and entropy – natural gas

For the combustion products of natural gas it is logical to use specific enthalpy and entropy only if CP is evaluated accurately. This requires the method of sample calculation C13.1, which addresses variation in fuel chemistry.

3.5 Data base for fundamental and thermodynamic gas properties

References 2, 4 and 6 provide a comprehensive coverage of fundamental gas properties, the last recognising the effects of dissociation at high combustion temperatures.

3.5.1 Molecular weight and gas constant (Formula F3.22)

Data for gases of interest are tabulated below.

	Molecular weight	Gas constant (J/kg K)
Dry air	28.964	287.05
Oxygen	31.999	259.83
Water	18.015	461.51
Carbon dioxide	44.010	188.92
Nitrogen	28.013	296.80
Argon	39.948	208.13
Hydrogen	2.016	4124.16
Neon	20.183	411.95
Helium	4.003	2077.02

Note: The universal gas constant is 8314.3 J/mol K.

Chart 3.1 shows the gas constant resulting from the combustion of leading fuel types in air plotted versus fuel air ratio. It is not possible to provide all encompassing data for natural gas due to the wide variety of blends which occur. For indicative purposes combustion of a sample natural gas, described in Chapter 13, has been used. The following are apparent:

- For *kerosene*, molecular weight and gas constant are not changed noticeably from the values for dry air up to stoichiometric fuel to air ratio.
- For *diesel* molecular weight and hence gas constant change minimally, in a linear fashion versus fuel to air ratio. For performance calculations there is negligible loss in accuracy by ignoring these small changes and using data for kerosene.
- For the sample *natural gas* molecular weight and gas constant vary linearly with fuel air ratio from the values for dry air to 27.975 and 297.15 J/kg K respectively at a fuel to air ratio of 0.05. A significant loss of accuracy will occur if this change is not accounted.

Formula F3.22 presents gas constant as a function of fuel air ratio for the above three cases.

The effect of gas fuel is more powerful than that of liquid fuels, primarily due to the constituent hydrocarbons being lighter (i.e. containing less carbon and more hydrogen); this results in a higher proportion of water vapour after combustion, which has a significantly lower molecular weight than the other constituents. A comprehensive description of how to calculate molecular weight and the gas constant following combustion of a particular blend of natural gas is provided in Chapter 13.

3.5.2 *Specific heat and gamma (Formulae F3.23–F3.25)*

Charts 3.2 and 3.3 present specific heat and gamma respectively for dry air and combustion products versus static temperature and fuel air ratio for kerosene or diesel fuels.

Chart 3.4 and Formula F3.25 show the ratio of specific heat following the combustion of the sample natural gas (Chapter 13) to that for kerosene, versus fuel to air ratio. This plot is sensibly independent of temperature. As stated earlier, specific heat is noticeably higher following the combustion of natural gas due to the higher resultant water content, which significantly impacts engine performance. Typical liquid fuel to natural gas engine performance parameter exchange rates are provided in Chapter 13.

Charts 3.5 and 3.6 show specific heat and gamma respectively versus temperature for the individual gases present in air and combustion products. The higher value for water vapour is immediately apparent. For inert gases such as helium, argon and neon specific heat and gamma do not change with temperature.

Formulae F3.23–F3.25 facilitate the evaluation of specific heat and gamma for dry air, combustion products for liquid fuel, the sample natural gas, and for each individual gas. Sample calculation C3.2 shows their application to a compressor.

3.5.3   *Specific enthalpy and specific entropy (Formulae F3.26–F3.29)*

As described in section 3.4.3, for fully rigorous calculations the changes in specific enthalpy and entropy must be evaluated using polynomials as opposed to specific heat at the mean temperature. Formulae F3.26–F3.29 provide the necessary relationships, and sample calculation C3.2 includes an illustration of their use for a compressor. A comprehensive method for calculating these properties following the combustion of any blend of natural gas is provided in Chapter 13.

Section 3.6.4 describes the temperature–entropy or 'T–S' diagram, which is frequently used for illustration.

3.5.4   *Dynamic viscosity (Formula F3.30)*

Chart 3.7 presents dynamic viscosity for dry air and combustion products versus static temperature. As stated earlier, the effect of fuel air ratio is negligible for practical purposes, and Formula F3.30 is sufficient for all performance calculations.

3.6   Charts showing interrelationships of key thermodynamic parameters

3.6.1   *Compressible flow or 'Q' curves (Formulae F3.31–F3.36)*

Compressible flow curves, commonly called *Q curves*, apply to flow in a duct of varying area with no work or heat transfer, such as intakes, exhaust systems, and ducts between compressors or turbines. They relate key parameter groups and are indispensable for rapid hand calculations, providing an instant reference for the various useful flow parameters versus Mach number. Once one parameter group relating to flow area (e.g. Mach number, or the ratio of total to static pressure or temperature) is known at a point in the duct then all the other parameter groups at that point can be evaluated.

One key phenomenon for compressible flow is *choking*, where a Mach number of 1 is reached at the minimum area along a duct. Reducing downstream pressure further provides no increase in mass flow. This is discussed in detail in Chapter 5.

It is important not to confuse compressible flow relationships with the simpler Bernoulli's equation, which only applies to *incompressible* flow such as liquid. That is however a reasonable approximation for perfect gases below 0.25 Mach number.

Owing to its immense value, tabulated Q curve data over the most commonly used Mach number range 0–1 is provided in Chart 3.8. The most useful parameter groups are also given in the charts below over a Mach number range of 0–2.5; around the highest level likely to be encountered in a convergent–divergent propelling nozzle. The values of gamma shown are 1.4 and 1.33, which are commonly used levels typical of the cold and hot ends of an engine. For calculations where higher accuracy is required, or for Mach numbers exceeding 2.5, Formulae F3.31–F3.36 should be used, with correct values for the gas properties. Calculation C3.3 illustrates the use of these formulae.

- Total to static temperature ratio versus Mach number – Chart 3.9, Formula F3.31
- Total to static pressure ratio versus Mach number – Chart 3.10, Formula F3.32
- Flow function, $W\sqrt{T}/A.P$ (Q) versus Mach number – Chart 3.11, Formula F3.33
- Flow function based on static pressure (q) versus Mach number – Chart 3.12, Formula F3.34
- Velocity function, $V/\sqrt{T}$ (i.e. based upon total temperature) versus Mach number – Chart 3.13, Formula F3.35
- Value of one dynamic head as a percentage of total pressure versus Mach number – Chart 3.14, Formula F3.36

The last chart is of particular interest in that, as described in Chapter 5, the percentage pressure loss in a particular duct is a multiple of the inlet dynamic head as a percentage of total pressure. This multiple is termed the *loss coefficient*, and has a unique value for a duct of fixed geometry and inlet swirl angle.

Some examples of the uses of Q curves are as follows:

- Calculating ram pressure and temperature at entry to an engine resulting from the flight Mach number
- Calculating the area of a propelling nozzle required when inlet total pressure, temperature and mass flow are known as well as the exit static pressure
- Calculating total pressure in a duct when the static value has been measured and mass flow, temperature and area are also known
- Calculating pressure losses in ducts

3.6.2   Combustion temperature rise charts (Formulae F3.37–F3.41)

Chart 3.15 presents temperature rise versus fuel to air ratio and inlet temperature for the combustion of kerosene. This chart is consistent with the enthalpy polynomials, and may also be used for diesel with negligible loss in accuracy. Formulae F3.37–F3.41 are a curve fit of Chart 3.15, and are sufficient for performance calculations; sample calculation C3.4 demonstrates their use. They agree closely with an enthalpy based approach whilst simplifying the process. The chart and formulae are for a fuel calorific value of 43 124 kJ/kg and a combustion efficiency of 100%. For other calorific values or efficiencies the temperature rise or fuel air ratio should be factored accordingly. Though not exact, this is a standard methodology and incurs very low error, due to fuel flow being much less than air flow and combustion efficiency being normally close to 100%.

Again a unique chart does not exist for natural gas fuel. For the sample natural gas however, a good indication will be provided by dividing the temperature rise by the specific heat ratio from Chart 3.4 and Formula F3.25. However, for rigorous calculations CP must be evaluated as per sample calculation C13.1, and then enthapy polynomials applied. It is apparent that when burning natural gas as opposed to kerosene or diesel more energy input is required for a given temperature rise. Equally, however, the higher resultant specific heat in the turbine(s) provides extra power output, and engine thermal efficiency is actually higher.

3.6.3   Isentropic to polytropic efficiency conversions for compressors and turbines (Formulae F3.42–F3.45)

Two definitions for compressor and turbine efficiency are commonly used. Isentropic and polytropic efficiency are discussed in Chapter 5. Charts 3.16 and 3.17 enable conversion between them for the standardised values of gamma of 1.4 and 1.33. For more accurate calculations Formulae F3.42–F3.45 must be used with the correct value of gamma at the average temperature through the component. Sample calculation C3.5 illustrates this approach for a compressor. As discussed, fully rigorous methods are based on enthalpy polynomials; Formulae F3.42–F3.45 show their application to polytropic efficiency.

3.6.4   Temperature entropy diagram for dry air

Most heat engine cycles are taught at university level via schematic illustration on a *temperature–entropy (T–S) diagram*. This approach becomes laborious to extend to 'real' engine effects such as internal bleeds and cooling flows, but remains a useful indication of the overall thermodynamics of a known engine cycle. Chart 3.18 presents an actual temperature–entropy diagram for dry air, complete with numbers, showing lines of constant pressure. Such a diagram is rare in the open literature.

The following are important:

- Raising temperature at constant pressure (e.g. by adding heat in a combustor) raises entropy.
- Reducing temperature at constant pressure (e.g. by removing heat in an intercooler) lowers entropy.
- Compression from a lower to a higher constant pressure line (i.e. by adding *work*) produces minimum change in temperature (i.e. requires minimum energy input) if entropy does not increase. *Isentropic* compression is an idealised process.
- In reality entropy does increase during compression, hence extra energy must be provided, beyond the ideal work required for the pressure change. This extra energy is converted to heat.
- Expansion from a higher to a lower constant pressure line produces maximum change in temperature (i.e. produces maximum work) if entropy does not increase. *Isentropic* expansion is also an idealised process.
- In reality entropy does increase during expansion, hence less work output is obtained than the ideal work produced pressure change. This 'lost' energy is retained as heat.

Entropy may be defined as *thermal energy not available for doing work*. In real compressors and turbines some energy goes into raising entropy, as some pressure is lost to real effects such as friction. The ideal work would be required or produced if entropy did not change, i.e. the process were *isentropic*. *Isentropic efficiency* is defined as the appropriate ratio of actual and ideal work, and is always less than 100%. (The term *adiabatic efficiency* is also commonly used, but is strictly incorrect. It only excludes heat transfer but not friction, and an isentropic process would have neither.)

Gas turbine cycles utilise the above processes, and rely on one other vital, fundamental thermodynamic effect:

Work input, approximately proportional to temperature rise, for a given compression ratio from low temperature is significantly lower than the work output from the same expansion ratio from higher temperature.

This is because on the T–S diagram lines of constant pressure diverge with increasing temperature and entropy.

This can be seen by considering a sample compression, heating and expansion between two lines of constant pressure, using Chart 3.18. At an entropy value of 1.5 kJ/kg K the temperature rise required to go from 100 to 5100 kPa is 500 K. If fuel is now burnt at this pressure level such that entropy increases to 2.75 kJ/kg K, and temperature to 1850 K, an expansion back to 100 kPa will achieve a temperature drop of around 1000 K. This clearly illustrates the rationale behind the Brayton cycle described in section 3.6.5.

The same fundamental effect is apparent from Formula F3.32, which gives the idealised definition of total pressure. It shows the interrelationship of pressure and temperature changes in an isentropic process, and illustrates that the temperature *difference* resulting from expansion or compression is directly proportional to the initial temperature level.

3.6.5 *Schematic T–S diagrams for major engine cycles*

Figures 3.1–3.6 show the key cycles of interest to gas turbine engineers. More detail for specific engine types is provided in the Gas Turbine Engine Configurations section and in Chapter 6.

Figure 3.1 shows the Carnot cycle. This is the most efficient cycle theoretically possible between two temperature levels, as shown in Reference 1. Gas turbine engines necessarily do not use the Carnot cycle, as unlike steam cycles they cannot add or reject heat at constant temperature.

Figure 3.2 shows the Brayton cycle. This is the basic cycle utilised by all gas turbine engines where heat is input at constant pressure. The effect of component inefficiency is shown by the non-vertical compression and expansion lines, a further difference from the ideal Carnot cycle. The form of the Brayton cycle is modified for heat exchangers and bypass flows.

Figure 3.3 presents the cycle for a turbofan. The bypass stream only undergoes partial compression, and no heating before expansion back to ambient pressure.

Figure 3.4 shows a heat exchanged cycle. Waste heat from exhaust gases is used to heat air from compressor delivery prior to combustion, thereby reducing the required fuel flow.

Figure 3.5 presents an intercooled cycle, where heat is extracted downstream of an initial compressor. This reduces the work required to drive a second compressor, and thereby increases power output.

Figure 3.6 shows a Rankine cycle with superheat. This is used in combined cycle applications, with the gas turbine exhaust gases providing heat to raise steam. Where heat is added at constant temperature during evaporation a close approximation to the Carnot cycle is achieved, the main deviation being the non-ideal component efficiencies.

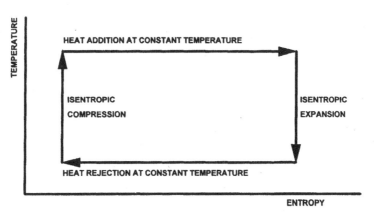

Fig. 3.1   Ideal Carnot cycle.

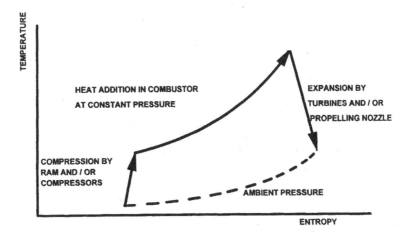

Fig. 3.2   Brayton cycle for turboshaft, turboprop, turbojet or ramjet.

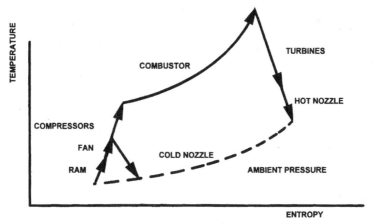

Fig. 3.3 Cycle for turbofan.

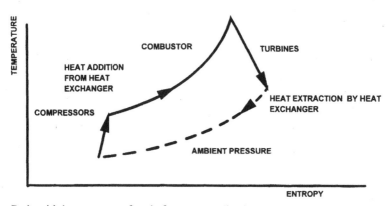

Fig. 3.4 Cycle with heat recovery for shaft power applications.

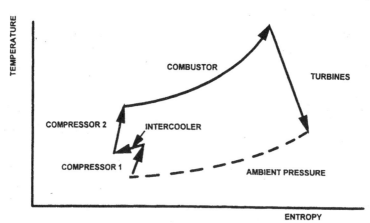

Fig. 3.5 Intercooled cycle for shaft power applications.

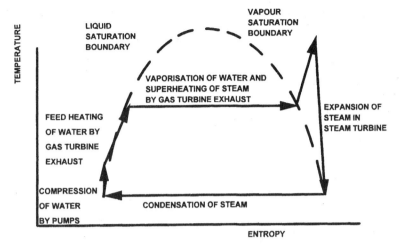

Fig. 3.6 Rankine cycle with superheat; typical steam cycle used with a gas turbine for combined cycle power generation.

Formulae

F3.1   Equation of state for perfect gas

RHO = PS/(R * TS)

F3.2   Specific heat at constant pressure (J/kg K) = fn(specific enthalpy (J/kg), static temperature (K))

CP = dH/dTS

F3.3   Specific heat at constant volume (J/kg K) = fn(specific internal energy (J/kg), static temperature (K))

CV = dU/dTS

F3.4   Gas constant (J/kg K) = fn(universal gas constant (J/kg K), molecular weight)

R = Runiversal/MW

(i)   Where the universal gas constant Runiversal = 8314.3 J/mol K.

F3.5   Gas constant (J/kg K) = fn(CP (J/kg K), CV (J/kg K))

R = CP – CV

F3.6   Gamma = fn(CP (J/kg K), CV (J/kg K))

γ = CP/CV

F3.7   Gamma = fn(gas constant (J/kg K), CP (J/kg K))

γ = CP/(CP – R)

F3.8 The gamma exponent $(\gamma-1)/\gamma$ = fn(gas constant (J/kg K), CP (J/kg K))

$(\gamma - 1)/\gamma = R/CP$

F3.9 Dynamic viscosity of dry air $(N\,s/m^2)$ = fn(shear stress (N/m^2), velocity gradient (m/s m), static temperature (K))

$VIS = Fshear/(dV/dy)$

(i)   Fshear is the shear stress in the fluid.
(ii)   V is the velocity in the direction of the shear stress.
(iii) dV/dy is the velocity gradient perpendicular to the shear stress.

F3.10 Total temperature (K) = fn(static temp (K), gas velocity (m/s), CP (J/kg K))

$T = TS + V^2/(2 * CP)$

(i)   This may be converted to the Q curve Formula F3.30 using Formulae F2.15 and F3.8.

F3.11 Total pressure (kPa) = fn(total to static temperature ratio, gamma)

$P = PS * (T/TS)^{\wedge}(\gamma/(\gamma - 1))$

(i)   *Note*: This is the definition of total pressure.

F3.12 Dynamic head (kPa) = fn(total pressure (kPa), static pressure (kPa))

$VH = P - PS$

F3.13 Dynamic head (kPa) = fn(density (kg/m^3), velocity (m/s), Mach number)

$VH = 0.5 * RHO * V^2((1 + 0.5 * (\gamma - 1) * M^{\wedge}2) - 1) * 2/(\gamma * M^{\wedge}2)$

(i)   For incompressible flow, such as that of liquids, it is sufficient to only use the first term – this is the well known Bernoulli equation.

F3.14 Specific enthalpy (kJ/kg) = fn(temperature (K), CP (kJ/kg K))

$H = H0 + \int CP\,dT$

(i)   H0 is an arbitarily defined datum. The datum is unimportant in gas turbine performance as it is changes in enthalpy that are of interest.

F3.15 and F3.16 Change in enthalpy (kJ/kg) = fn(temperature (K), CP (kJ/kg K))

For fully rigorous calculations specific enthalpy at state 1 and state 2 must be calculated from Formulae F3.26 and F3.27:

F3.15 $DH = H2 - H1$

For calculations to within 1% accuracy then CP at the mean temperature may be used as calculated from Formulae F3.23–F3.25:

F3.16 $DH = CP * (T2 - T1)$

F3.17 and F3.18   Change in entropy (J/kg K) = fn(CP (J/kg K), gas constant (J/kg K), change in total temperature and pressure)

For fully rigorous calculations $\int CP/T\,dT$ must be calculated from Formulae F3.28 and F3.29:

F3.17 $S2 - S1 = \int CP/T\,dT - R * \ln(P2/P1)$

For calculations to within 1% accuracy then CP corresponding to the mean temperature may be calculated from Formulae F3.23–F3.25:

F3.18 $S2 - S1 = CP * \ln(T2/T1) - R * \ln(P2/P1)$

F3.19–F3.21   Isentropic process formulae

For fully rigorous calculations $\int CP/T\,dT$ must be calculated from Formulae F3.28 and F3.29:

F3.19 $\int CP/T\,dT = R * \ln(P2/P1)$

For calculations to within 1% accuracy then CP corresponding to the mean temperature may be calculated from formulae F3.23–F3.25:

F3.20 $CP * \ln(T2/T1) = R * \ln(P2/P1)$

or:

$(T2/T1)^{\wedge}(CP/R) = P2/P1$

and using Formula F3.8:

F3.21 $(T2/T1)^{\wedge}((\gamma - 1)/\gamma) = P2/P1$

F3.22   Gas constant for products of combustion in dry air (J/kg K) = fn(fuel air ratio)

$R = 287.05 - 0.00990 * FAR + 1E\text{-}07 * FAR^{\wedge}2$ kerosene
$R = 287.05 - 8.0262 * FAR + 3E\text{-}07 * FAR^{\wedge}2$ diesel
$R = 287.05 + 212.85 * FAR - 197.89 * FAR^{\wedge}2$ sample natural gas

F3.23   CP For key gases (kJ/kg K) = fn(static temperature (K))

$$CP = A0 + A1 * TZ + A2 * TZ^{\wedge}2 + A3 * TZ^{\wedge}3 + A4 * TZ^{\wedge}4 \\ + A5 * TZ^{\wedge}5 + A6 * TZ^{\wedge}6 + A7 * TZ^{\wedge}7 + A8 * TZ^{\wedge}8$$

(i)   Where $TZ = TS/1000$ and the values for constants are as below.

	Dry air	O_2	N_2	CO_2	H_2O
A0	0.992313	1.006450	1.075132	0.408089	1.937043
A1	0.236688	−1.047869	−0.252297	2.027201	−0.967916
A2	−1.852148	3.729558	0.341859	−2.405549	3.338905
A3	6.083152	−4.934172	0.523944	2.039166	−3.652122
A4	−8.893933	3.284147	−0.888984	−1.163088	2.332470
A5	7.097112	−1.095203	0.442621	0.381364	−0.819451
A6	−3.234725	0.145737	−0.074788	−0.052763	0.118783
A7	0.794571	—	—	—	—
A8	−0.081873	—	—	—	—
A9	0.422178	0.369790	0.443041	0.366740	2.860773
A10	0.001053	0.000491	0.0012622	0.001736	−0.000219

(i)   Gamma may be then be calculated via Formula F3.7.

F3.24 CP for combustion products of kerosene or diesel in dry air (kJ/kg K) = fn(fuel air ratio, static temperature (K))

$$CP = A0 + A1 * TZ + A2 * TZ^\wedge 2 + A3 * TZ^\wedge 3 + A4 * TZ^\wedge 4$$
$$+ A5 * TZ^\wedge 5 + A6 * TZ^\wedge 6 + A7 * TZ^\wedge 7 + A8 * TZ^\wedge 8$$
$$+ FAR/(1 + FAR) * (B0 + B1 * TZ + B2 * TZ^\wedge 2 + B3 * TZ^\wedge 3$$
$$+ B4 * TZ^\wedge 4 + B5 * TZ^\wedge 5 + B6 * TZ^\wedge 6 + B7 * TZ^\wedge 7)$$

(i)   Where TZ = TS/1000.
A0–A8 are the values for dry air from Formula F3.23.
B0 = −0.718874,   B1 = 8.747481,   B2 = −15.863157,   B3 = 17.254096,
B4 = −10.233795,   B5 = 3.081778,   B6 = −0.361112,   B7 = −0.003919,
B8 = 0.0555930,   B9 = −0.0016079.

(ii) Gamma may be then be calculated via Formula F3.7.

F3.25 CP for combustion products of sample natural gas in dry air (kJ/kg K) = fn(CP of liquid fuel combustion products (kJ/kg K))

$$CPgas = (1.0001 + 0.9248 * FAR - 2.2078 * FAR^\wedge 2) * CPliquid$$

F3.26 Specific enthalpy for key gases (MJ/kg) = fn(temperature (K))

$$H = A0 * TZ + A1/2 * TZ^\wedge 2 + A2/3 * TZ^\wedge 3 + A3/4 * TZ^\wedge 4$$
$$+ A4/5 * TZ^\wedge 5 + A5/6 * TZ^\wedge 6 + A6/7 * TZ^\wedge 7 + A7/8 * TZ^\wedge 8$$
$$+ A8/9 * TZ^\wedge 9 + A9$$

(i)   Where TZ = TS/1000 and the values for constants are as per Formula F3.23.
(ii) If the change in enthalpy is known and the change in temperature is required, then Formulae F3.15 and F3.26 must be used iteratively.

F3.27 Specific enthalpy for combustion products of kerosene or diesel in dry air (MJ/kg) = fn(fuel air ratio, static temperature (K))

$$H = A0 * TZ + A1/2 * TZ^\wedge 2 + A2/3 * TZ^\wedge 3 + A3/4 * TZ^\wedge 4 + A4/5 * TZ^\wedge 5$$
$$+ A5/6 * TZ^\wedge 6 + A6/7 * TZ^\wedge 7 + A7/8 * TZ^\wedge 8 + A8/9 * TZ^\wedge 9$$
$$+ A9 + (FAR/(1 + FAR)) * (B0 * TZ + B1/2 * TZ^\wedge 2 + B2/3 * TZ^\wedge 3$$
$$+ B3/4 * TZ^\wedge 4 + B4/5 * TZ^\wedge 5 + B5/6 * TZ^\wedge 6 + B6/7 * TZ^\wedge 7 + B8)$$

(i)   Where TZ = TS/1000 and the values for constants are as per Formula F3.23 and F3.24.
(ii) If the change in enthalpy is known and the change in temperature is required, then Formulae F3.15 and F3.27 must be used iteratively.

F3.28 ∫ CP/T dT for key gases (kJ/kg K) = fn(temperature (K))

$$FT2 = A0 * \ln(T2Z) + A1 * T2Z + A2/2 * T2Z^\wedge 2 + A3/3 * T2Z^\wedge 3$$
$$+ A4/4 * T2Z^\wedge 4 + A5/5 * T2Z^\wedge 5 + A6/6 * T2Z^\wedge 6$$
$$+ A7/7 * T2Z^\wedge 7 + A8/8 * T2Z^\wedge 8 + A10$$

$$FT1 = A0 * \ln(T1Z) + A1 * T1Z + A2/2 * T1Z^\wedge 2 + A3/3 * T1Z^\wedge 3$$
$$+ A4/4 * T1Z^\wedge 4 + A5/5 * T1Z^\wedge 5 + A6/6 * T1Z^\wedge 6$$
$$+ A7/7 * T1Z^\wedge 7 + A8/8 * T1Z^\wedge 8 + A10$$

$$\int CP/T \, dT = FT2 - FT1$$

(i) Where T2Z = TS2/1000, T1Z = TS1/1000 and the values for constants are as per Formula F3.23.

(ii) If the change in entropy is known and the change in temperature is required then Formulae F3.17 and F3.28 must be used iteratively.

F3.29 ∫ CP/T dT for combustion products of kerosene or diesel in dry air (kJ/kg K) = fn (temperature (K))

$$
\begin{aligned}
FT2 = {}& A0 * \ln(T2Z) + A1 * T2Z + A2/2 * T2Z^\wedge 2 + A3/3 * T2Z^\wedge 3 \\
& + A4/4 * T2Z^\wedge 4 + A5/5 * T2Z^\wedge 5 + A6/6 * T2Z^\wedge 6 + A7/7 * T2Z^\wedge 7 \\
& + A8/8 * T2Z^\wedge 8 + A10 + (FAR/(1 + FAR)) * (B0 * \ln(T2) + B1 * TZ \\
& + B2/2 * TZ^\wedge 2 + B3/3 * TZ^\wedge 3 + B4/4 * TZ^\wedge 4 + B5/5 * TZ^\wedge 5 \\
& + B6/6 * TZ^\wedge 6 + B7/7 * TZ^\wedge 7 + B9)
\end{aligned}
$$

$$
\begin{aligned}
FT1 = {}& A0 * \ln(T1Z) + A1 * T1Z + A2/2 * T1Z^\wedge 2 + A3/3 * T1Z^\wedge 3 \\
& + A4/4 * T1Z^\wedge 4 + A5/5 * T1Z^\wedge 5 + A6/6 * T1Z^\wedge 6 + A7/7 * T1Z^\wedge 7 \\
& + A8/8 * T1Z^\wedge 8 + A10 + (FAR/(1 + FAR)) * (B0 * \ln(T1) + B1 * TZ \\
& + B2/2 * TZ^\wedge 2 + B3/3 * TZ^\wedge 3 + B4/4 * TZ^\wedge 4 + B5/5 * TZ^\wedge 5 \\
& + B6/6 * TZ^\wedge 6 + B7/7 * TZ^\wedge 7 + B9)
\end{aligned}
$$

∫ CP/T dT = FT2 − FT1

(i) Where T2Z = TS2/1000, T1Z = TS1/1000 and the values for constants are as per Formula F3.23 and F3.24.

(ii) If the change in entropy is known and the change in temperature is required then Formulae F3.17 and F3.29 must be used iteratively.

F3.30 Dynamic viscosity of dry air (N s/m²) = fn(static temperature (K))

VIS = 1.5105E-06 * TS^1.5/(TS + 120)

F3.31–F3.36 Q curve formulae

F3.31   T/TS = (1 + (γ − 1)/2 * M^2)

See also Formula F3.10.

F3.32   PT/PS = (T/TS)^(γ/(γ − 1))
          = (1 + (γ − 1)/2 * M^2)^(γ/(γ − 1))

F3.33   Q = W * SQRT(T)/(A * P)
          = 1000 * SQRT(2 * γ/((γ − 1) * R) * (P/PS)^(−2/γ)
          * (1 − (P/PS)^((1 − γ)/γ)))

F3.34   q = W * SQRT(T)/(A * PS)
          = (PT/PS) * Q

F3.35   V/SQRT(T) = M * SQRT(γ * R)/SQRT(T/TS)

F3.36   DP/P = 100 * (1 − 1/(P/PS))

(i)   Where T, TS = K; P, PS = kPa, A = m², W = kg/s, V = m/s, DP/P = %, R = gas constant, e.g. 287.05 J/kg K for dry air.

(ii)   DP/P is percentage pressure loss equivalent to one dynamic head.

(iii)   Formulae are for compressible flow in a duct with no work or heat transfer.

(iv)   Once one parameter group at a point in the flow is known all others may be calculated.

F3.37–F3.40   Fuel air ratio = fn(combustor inlet and exit temperatures (K))

For fully rigorous calculations:

F3.37   FAR = DH/(LHV * ETA34)

(i)   DH must be calculated from Formulae F3.15, F3.26 and F3.27.

For calculations to within 0.25% accuracy with kerosene fuel which has an LHV of 43124 kJ/kg:

F3.38A   FAR1 = 0.10118 + 2.00376E-05 * (700 − T3)
FAR2 = 3.7078E-03 − 5.2368E-06 * (700 − T3) − 5.2632E-06 * T4
FAR3 = 8.889E-08 * ABS(T4 − 950)
FAR = (FAR1 − SQRT(FAR1^2 + FAR2) − FAR3)/ETA34

For calculations to within 0.25% accuracy for diesel or kerosene fuel with an LHV other than 43124 kJ/kg:

F3.38B   FAR = F3.37 * 43124/LHV

For calculations to within 1% accuracy with the sample natural gas, CPs at the mean temperature must be evaluated from Formulae F3.24 and F3.25, and:

F3.39   FAR = F3.36 * 43124 * Cpgas/(LHV * CPliquid)

For calculations to within 5% accuracy CP may be taken as that at the mean temperature:

F3.40   FAR = CP * (T4 − T3)/(ETA34 * FHV)

F3.41   Combustor exit temperature = fn(inlet temperature (K), fuel air ratio) – iterative

T4 = 1000
START:
T4previous = T4
FARcalc = F3.37 to F3.39
IF ABS((FAR − FARcalc)/FAR) > 0.0005 THEN
T4 = (T4previous − T3) * FAR/FARcalc + T3
GOTO START:
END IF

F3.42   Compressor isentropic efficiency = fn(polytropic efficiency, pressure ratio, gamma)

ETA2 = (P3Q2$^{((\gamma - 1)/\gamma)}$ − 1)/(P3Q2$^{((\gamma - 1)/(\gamma * ETAP2))}$ − 1)

F3.43   Compressor polytropic efficiency = fn(pressure ratio, temperature ratio, gamma)

Using gamma:

ETAP2 = ln(P3Q2)$^{((\gamma - 1)/\gamma)}$/ln(T3Q2)

Using rigorous enthalpy and entropy polynomials:

ETAP2 = ln(P3Q2)/ln(P3Q2.isentropic)

(i)   P3Q2.isentropic is obtained from Formula F3.19.

F3.44 Turbine isentropic efficiency = fn(polytropic efficiency, expansion ratio, gamma)

$$\text{ETA4} = (1 - \text{P4Q5}^{\wedge}(\text{ETAP4} * (1 - \gamma)/\gamma))/(1 - \text{P4Q5}^{\wedge}((1 - \gamma)/\gamma))$$

F3.45 Turbine polytropic efficiency = fn(expansion ratio, temperature ratio, gamma)

Using gamma:

$$\text{ETAP4} = \ln(\text{T4Q5})/\ln(\text{P4Q5})^{\wedge}((\gamma - 1)/\gamma)$$

Using rigorous enthalpy and entropy polynomials:

$$\text{ETAP4} = \ln(\text{P4Q5.isentropic})/\ln(\text{P4Q5})$$

(i) P4Q5.isentropic is obtained from Formula F3.19.

Sample calculations

C3.1 Calculate the molecular weight and gas constant for dry air using the composition provided in section 3.4.1.

F3.4 R = Runiversal/MW

From the guidelines provided with F3.4, Runiversal = 8314.3 J/mol K.

Average the molecular weights of the constituents on a molar basis using the data provided in sections 3.4 and 3.6:

$$\text{MWdry air} = (78.08 * 28.013 + 20.95 * 31.999 + 0.93 * 39.948 + 0.03$$
$$* 44.01 + 0.002 * 20.183)/100$$

$$\text{MWdry air} = 28.964$$

Evaluate the gas constant using F3.4:

Rdry air = 8314.3/28.964
Rdry air = 287.05 J/kg K

C3.2 Calculate the outlet temperature and power input for a compressor of 20:1 pressure ratio, isentropic efficiency of 85%, with an inlet temperature of 288.15 K and a mass flow of 100 kg/s using:

    (i)   **constant CP of 1.005 kJ/kg K and constant $\gamma = 1.4$**
   (ii)   **CP at mean temperature across the compressor**
  (iii)   **rigorous enthalpy and entropy polynomials**
  (iv)   **calculate the error in power resulting from the first two methods.**

F5.1.2 $\text{PW2} = \text{W2} * \text{CP23} * (\text{T3} - \text{T2})$

F5.1.4 $\text{T3} - \text{T2} = \text{T2}/\text{ETA2} * (\text{P3Q2}^{\wedge}((\gamma - 1)/\gamma) - 1)$

F3.7 $\gamma = \text{CP}/(\text{CP} - \text{R})$

F5.1.3 $\text{ETA2} = (\text{H3isentropic} - \text{H2})/(\text{H3} - \text{H2})$

F3.23 $\text{CP} = 0.992313 + 0.236688 * \text{TZ} - 1.852148 * \text{TZ}^{\wedge}2 + 6.083152$
$$* \text{TZ}^{\wedge}3 - 8.893933 * \text{TZ}^{\wedge}4 + 7.097112 * \text{TZ}^{\wedge}5 - 3.234725$$
$$* \text{TZ}^{\wedge}6 + 0.794571 * \text{TZ}^{\wedge}7 - 0.081873 * \text{TZ}^{\wedge}8$$

F3.26 $H = 0.992313 * TZ + 0.236688/2 * TZ^2 - 1.852148/3 * TZ^3$
$+ 6.083152/4 * TZ^4 - 8.893933/5 * TZ^5 + 7.097112/6 * TZ^6$
$- 3.234725/7 * TZ^7 + 0.794571/8 * TZ^8$
$+ 0.081873/9 * TZ^9 + 0.422178$

F3.28 $FTZ = 0.992313 * \ln(TZ) + 0.236688 * TZ - 1.852148/2 * TZ^2$
$+ 6.083152/3 * TZ^3 - 8.893933/4 * TZ^4 + 7.097112/5 * TZ^5$
$- 3.234725/6 * TZ^6 + 0.794571/7 * TZ^7 + 0.081873/8 * TZ^8$
$+ 0.001053$

$$\int CP/T \ dT = FTZ2 - FTZ1$$

F3.19 $\int CP/T \ dT = R * \ln(P3/P2)$

where: $TZ = TS/1000$

From the guidelines with Formula F3.36, R for dry air $= 287.05$ J/kg K.

(i) Constant CP and γ
Substituting values into F5.1.4:

$T3 - T2 = 288.15/0.85 * (20^{((1.4 - 1)/1.4)} - 1)$
$T3 - T2 = 458.8$ K
$T3 = 746.95$ K

$PW2 = 100 * 1.005 * 458.8$
$PW2 = 46109$ kW

(ii) CP and γ at mean T
For pass 1 take Tmean $= T2 = 288.15$ K. From Formulae F3.23 and F3.7:

$CP = 1003.3$ J/kg K

$\gamma = 1003.3/(1003.3 - 287.05)$
$\gamma = 1.401$

$T3 = 288.15/0.85 * (20^{((1.401 - 1)/1.401)} - 1) + 288.15$
$T3 = 748.2$

Tmean $= (288.15 + 748.2)/2$
Tmean $= 518.2$

Repeat using Tmean $= 518.2$ K: $CP = 1032.9$ J/kg K, $\gamma = 1.385$, $T3 = 728.7$ K, Tmean $= 508.4$ K.
Repeat using Tmean $= 508.4$ K: $CP = 1030.9$ J/kg K, $\gamma = 1.386$, $T3 = 729.9$ K, Tmean $= 509.0$ K.
Repeat using Tmean $= 509.0$ K: $CP = 1031$ J/kg K, $\gamma = 1.3858$, $T3 = 729.7$ K, Tmean $= 508.9$ K.
Repeat using Tmean $= 508.9$ K: $CP = 1031$ J/kg K, $\gamma = 1.3858$, $T3 = 729.7$ K, Tmean $= 508.9$ K.

The iteration has converged and hence $T3 = 729.7$ K.
  Substituting into F5.1.2:

$PW2 = 100 * 1.031 * (729.7 - 288.15)$
$PW2 = 45524$ kW

(iii) Using enthalpy and entropy polynomials

First calculate $\int CP/T\,dT$ for isentropic process from F3.19, and FTZ1 at 288.15 K by substituting into F3.28:

$$\int CP/T\,dT = 0.28705 * \ln(20)$$

$$\int CP/T\,dT \text{ isentropic} = 0.859925 \,kJ/kg\,K$$

FTZ1 = 5.648475 kJ/kg K

Now solve for T3isentropic by iterating until $(\int CP/T\,dT \text{ isentropic})/(\int CP/T\,dT\ T3\text{isentropic}$ guess) is within 0.0002. Make first guess T3isentropic = 700 K, and substitute into F3.28:

FTZ2 = 6.559675 kJ/kg K

$$\int CP/T\,dT = 0.9112 \,kJ/kg\,K$$

Calculate ratio and hence second guess for T3isentropic:

$(\int CP/T\,dT \text{ isentropic})/(\int CP/T\,dT\ T3\text{isentropic guess})$
= 0.859925/0.9112 = 0.94373
T3isentropic = 700 * 0.94373 = 660.61 K

Repeat using T3isentropic guess = 660.61 K:

$(\int CP/T\,dT \text{ isentropic})/(\int CP/T\,dT\ T3\text{isentropic guess})$
= 0.859925/0.849247 = 1.01257
T3isentropic = 660.61 * 1.01257 = 668.92 K

Repeat using T3isentropic guess = 668.92 K:

$(\int CP/T\,dT \text{ isentropic})/(\int CP/T\,dT\ T3\text{isentropic guess})$
= 0.859925/0.862567 = 0.99694
T3isentropic = 768.92 * 0.99694 = 666.87 K

Repeat using T3isentropic guess = 666.87 K:

$(\int CP/T\,dT \text{ isentropic})/(\int CP/T\,dT\ T3\text{isentropic guess})$
= 0.859925/0.859295 = 1.00073
T3isentropic = 666.87 * 1.00073 = 667.36 K

Repeat using T3isentropic guess = 667.36 K:

$(\int CP/T\,dT \text{ isentropic})/(\int CP/T\,dT\ T3\text{isentropic guess})$
= 0.859925/0.860077 = 0.99982

This is within the required error band and hence T3isentropic = 667.36 K.

Now calculate DHisentropic, DH and power by substituting 288.15 K and 667.36 K into Formulae F3.26, F5.1.2 and F5.1.4:

DHisentropic = 1.10072 − 0.710724
DHisentropic = 0.389997 MJ/kg = 389.997 kJ/kg

0.85 = 389.997/DH
DH = 458.82 kJ/kg

PW2 = 100 * 458.82
PW2 = 45882 kW

To calculate T3 iterate using F3.26:

> 0.45882 = H3 − 0.710724
> H3 = 1.16953 kJ/kg

Make first guess for T3 = 700 K and substitute into F3.26 which gives H3guess = 1.135668 and DHguess = 0.424943 MJ/kg. Hence calculate error and new T3guess:

> DH/DHguess = 0.45882/0.424943 = 1.0797
> T3guess = 700 ∗ 1.0797^0.5 = 727.37 K

Repeat using T3guess = 727.37 K:

> DH/DHguess = 0.45882/0.45444 = 1.00964
> T3guess = 727.37 ∗ 1.00964^0.5 = 730.86 K

Repeat using T3guess = 730.86 K:

> DH/DHguess = 0.45882/0.458222 = 1.0013
> T3guess = 730.86 ∗ 1.0013^0.5 = 731.34 K

Repeat using T3guess = 731.34 K:

> DH/DHguess = 0.45882/0.45874 = 1.00018

This is within the target error band and hence T3 = 731.34 K.

(iv)   Calculate the errors in power and T3 of using methods (i) and (ii)
Errors in method (i):

> PW2error = (46109 − 45882)/45882 ∗ 100
> PW2error = 0.49%
>
> T3error = 746.95 − 731.34
> T3error = 15.61 K

Errors in method (ii):

> PW2error = (45524 − 45882)/45882 ∗ 100
> PW2error = −0.78%
>
> T3error = 729.7 − 731.34
> T3error = −1.64 K

Note: For this example the error in power from method (i) is actually marginally better than for CP at mean temperature. However the error in T3 of 15.6 K when using constant values for CP and γ is unacceptable for engine design purposes. Calculation of other parameters using constant values of CP and γ will also show unacceptable errors.

For similar calculations across the turbine the error in power using CP at mean temperature will tend to cancel the error in compressor power.

C3.3   **Air enters a convergent duct at plane A with a total temperature and pressure of 1000 K and 180 kPa respectively, static pressure 140 kPa and area 2 m². A short distance along the duct at plane B the duct area has reduced by 10%. Find the key flow parameters at planes A and B assuming no loss in total pressure between the two stations.**

> **F3.23**   CP = 0.992313 + 0.236688 ∗ TZ − 1.852148 ∗ TZ^2 + 6.083152 ∗ TZ^3
> − 8.893933 ∗ TZ^4 + 7.097112 ∗ TZ^5 − 3.234725 ∗ TZ^6
> + 0.794571 ∗ TZ^7 − 0.081873 ∗ TZ^8
>
> **F3.7** γ = CP/(CP − R)

F3.31–F3.35   Q curve formulae

F3.31 $T/TS = (1 + (\gamma - 1)/2 * M^2)$

See also Formula F3.10.

F3.32 $PT/PS = (T/TS)^{\wedge}(\gamma/(\gamma - 1))$
$= (1 + (\gamma - 1)/2 * M^2)^{\wedge}(\gamma/(\gamma - 1))$

F3.33 $Q = W * SQRT(T)/(A * P)$
$= 1000 * SQRT(2 * \gamma/((\gamma - 1) * R) * (P/PS)^{\wedge}(-2/\gamma)$
$* (1 - (P/PS)^{\wedge}((1 - \gamma)/\gamma)))$

F3.34 $q = W * SQRT(T)/(A * PS)$
$= (PT/PS) * Q$

F3.35 $V/SQRT(T) = M * SQRT(\gamma * R)/SQRT(T/TS)$

(i)   Where T, TS = K; P, PS = kPa, A = m^2, W = kg/s, V = m/s, DP/P = %, R = gas constant, e.g. 287.05 J/kg K for dry air.

(i)   Plane A

Derive CP and γ using F3.23 and F3.7 and the total temperature of 1000 K:

CP = 1.141 kJ/kg K

$\gamma = 1.141/(1.141 - 0.28705)$
$\gamma = 1.336$

Substituting into Q curve Formulae F3.32, F3.31, F3.33 and F3.35:

$180/140 = (1000/TS)^{\wedge}(1.336/(1.336 - 1))$
TS = 938.7 K

$1000/938.7 = (1 + (1.336 - 1)/2 * M^2)$
M = 0.623

$Q = 1000 * SQRT(2 * 1.336/((1.336 - 1) * 287.05) * (180/140)^{\wedge}(-2/1.336)$
$* (1 - (180/140)^{\wedge}((1 - 1.336)/1.336)))$
$Q = 1000 * SQRT(0.027704 * 1.285714^{\wedge}(-1.497) * (1 - 1.285714^{\wedge}(-0.25150))$
$Q = 34.1291$ kg $\sqrt{K}$/s m^2 kPa

$34.1291 = W * SQRT(1000/(2 * 180))$
W = 388.5 kg/s

$V/SQRT(1000) = 0.623 * SQRT(1.336 * 287.05)/SQRT(1000/938.7)$
V = 373.8 m/s

Note: For the above to be fully rigorous it should be repeated using CP and γ calculated using the static temperature, since the Mach number is greater than 0.4. The Q curve values in Chart 3.8, $\gamma = 1.33$ – turbines, are very close to the above, the differences being due to the small difference in γ.

(ii)   Plane B

Total temperature is unchanged as there is no work or heat transfer and area = 2 * 0.9 = 1.8 m^2. Also, since the assumption is made that there is no loss in total pressure then P = 180 kPa.

Use Formula F3.33 to determine P/PS:

$Q = 388.5 * SQRT(1000)/(1.8 * 180)$
$Q = 37.918$ kg $\sqrt{K}$/s m^2 kPa
$37.918 = 1000 * SQRT(0.027704 * (P/PS)^{\wedge}(-1.497) * (1 - (P/PS^{\wedge}(-0.25150)))$

Solving by iteration $P/PS = 1.472$. Hence $PS = 122.3$ kPa.

Substituting into Q curve Formulae F3.32, F3.31 and F3.35:

$180/122.3 = (1000/TS)^\wedge(1.336/(1.336 - 1))$
$TS = 907.4$ K

$1000/907.4 = (1 + (1.336 - 1)/2 * M^\wedge 2)$
$M = 0.779$

$V/SQRT(1000) = 0.779 * SQRT(1.336 * 287.05)/SQRT(1000/907.4)$
$V = 459.5$ m/s

Note: The same comments apply to CP and γ as for plane A.

C3.4 **(i)** **Calculate the fuel air ratio for a combustor for kerosene, and diesel with an LHV of 42 500 kJ/kg, for:**
inlet temperature T31 = 600 K
exit temperature = 1500 K
ETA34 = 99.9%
(ii) **Calculate fuel air ratio for kerosene using the approximate method and the resultant error.**

F3.38A $FAR1 = 0.10118 + 2.00376E-05 * (700 - T3)$
$FAR2 = 3.7078E-03 - 5.2368E-06 * (700 - T3) - 5.2632E-06 * T4$
$FAR3 = 8.889E-08 * ABS(T4 - 950)$
$FAR = (FAR1 - SQRT(FAR1^\wedge 2 + FAR2) - FAR3)/ETA34$

F3.38B $FAR = F3.37 * 43124/LHV$

F3.40 $FAR = CP * (T4 - T3)/ETA34/LHV$

(i) *FAR for kerosene and diesel using rigorous method*
Substituting values for kerosene into Formula F3.38A:

$FAR1 = 0.10118 + 2.00376E-05 * (700 - 600)$
$FAR1 = 0.103184$

$FAR2 = 3.7078E-03 - 5.2368E-06 * (700 - 600) - 5.2632E-06 * 1500$
$FAR2 = -0.004711$

$FAR3 = 8.889E-08 * ABS(1500 - 950)$
$FAR3 = 0.000049$

$FAR = (0.103184 - SQRT(0.103184^\wedge 2 - 0.004711) - 0.000049)/0.999$
$FAR = 0.02612$

Substituting values for diesel into F3.38B:

$FAR = 0.02612 * 43124/42500$
$FAR = 0.0265$

The FAR value for kerosene is in agreement with Chart 3.15.

(ii) *Using approximate method for kerosene*
Look up CP at the mean temperature of 1050 K from Chart 3.5 using a guessed FAR of 0.02. This gives a value of 1.189 kJ/kg K.
Substituting values into F3.40:

$FAR = 1.189 * (1500 - 600)/0.999/43124$
$FAR = 0.0248$

Error in approximate method:

$$FARerror = (0.0248 - 0.0265)/0.0265$$
$$FARerror = -6.4\%$$

Note: Even if CP were calculated accurately using F3.24 then the error would still be large. This is mainly because the large temperature rise means that there is a significant error incurred by not using the fact that $DH = \int CP\,dT$. F3.38A is a curve fit of $DH = \int CP\,dT$ for the products of kerosene combustion.

C3.5 For the compressor operating point of C3.2 derive the corresponding polytropic efficiency using CP and γ at mean temperature.

F3.42 $ETA2 = (P3Q2^{\wedge}((\gamma - 1)/\gamma) - 1)/(P3Q2^{\wedge}((\gamma - 1)/(\gamma * ETAP2)) - 1)$

From C3.2 Tmean = 508.9 K, CP = 1031 J/kg K, $\gamma = 1.3858$, P3Q2 = 20 : 1, ETA2 = 0.85. Substituting into F3.42:

$$0.85 = (20^{\wedge}((1.3858 - 1)/1.3858) - 1)/(20^{\wedge}((1.3858 - 1)/(1.3858 * ETAP2)) - 1)$$
$$0.85 = 1.3025/(20^{\wedge}(0.27840/ETAP2) - 1)$$
$$20^{\wedge}(0.27840/ETAP2) - 1 = 1.5325$$
$$\ln(20) * 0.27840/ETAP2 = \ln(2.53235)$$
$$ETAP2 = 0.8976$$

This is very similar to the data presented in Chart 3.16. The minor difference is due to the difference in γ.

Charts

Chart 3.1   Gas constant, R, for combustion products of kerosene, diesel and natural gas versus fuel air ratio.

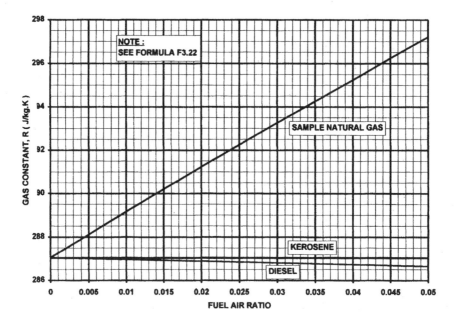

Chart 3.2 Specific heat, CP, for kerosene combustion products versus temperature and fuel air ratio.

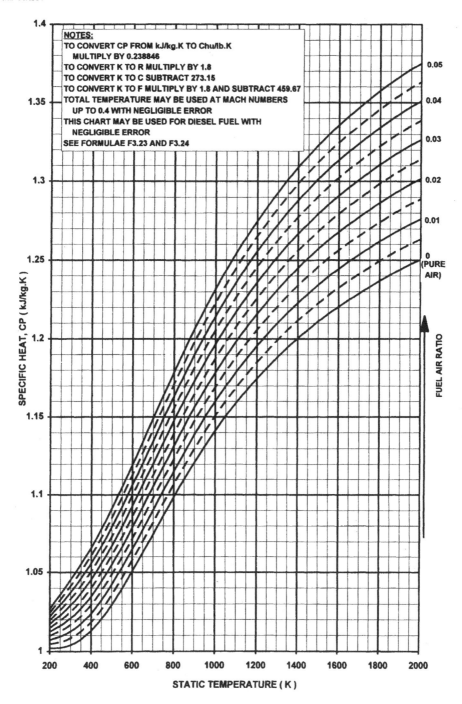

NOTES:
TO CONVERT CP FROM kJ/kg.K TO Chu/lb.K
    MULTIPLY BY 0.238846
TO CONVERT K TO R MULTIPLY BY 1.8
TO CONVERT K TO C SUBTRACT 273.15
TO CONVERT K TO F MULTIPLY BY 1.8 AND SUBTRACT 459.67
TOTAL TEMPERATURE MAY BE USED AT MACH NUMBERS
    UP TO 0.4 WITH NEGLIGIBLE ERROR
THIS CHART MAY BE USED FOR DIESEL FUEL WITH
    NEGLIGIBLE ERROR
SEE FORMULAE F3.23 AND F3.24

Chart 3.3 Gamma for kerosene combustion products versus temperature and fuel air ratio.

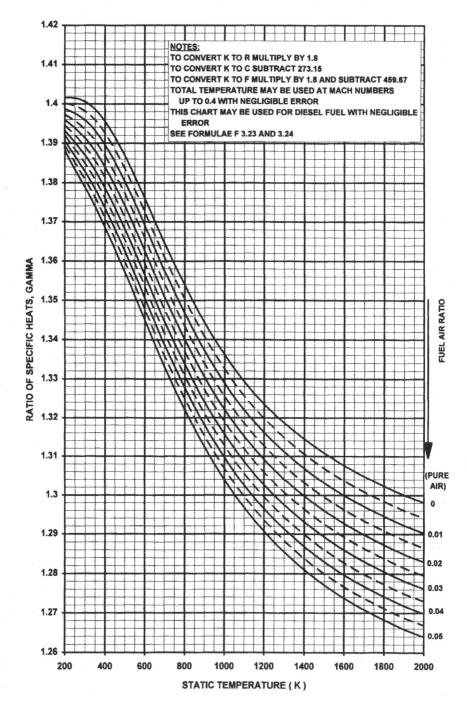

NOTES:
TO CONVERT K TO R MULTIPLY BY 1.8
TO CONVERT K TO C SUBTRACT 273.15
TO CONVERT K TO F MULTIPLY BY 1.8 AND SUBTRACT 459.67
TOTAL TEMPERATURE MAY BE USED AT MACH NUMBERS
    UP TO 0.4 WITH NEGLIGIBLE ERROR
THIS CHART MAY BE USED FOR DIESEL FUEL WITH NEGLIGIBLE
    ERROR
SEE FORMULAE F 3.23 AND 3.24

Chart 3.4   Specific heat, CP, for typical natural gas combustion products relative to kerosene versus fuel air ratio.

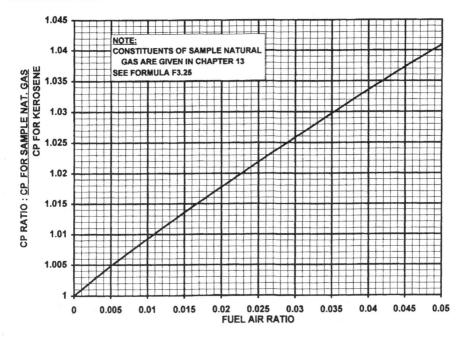

Chart 3.5   Specific heat, CP, for the constituents of air and combustion products versus temperature.

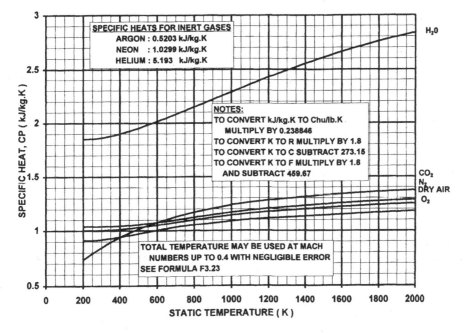

Chart 3.6   Gamma for the constituents of air and combustion products versus temperature.

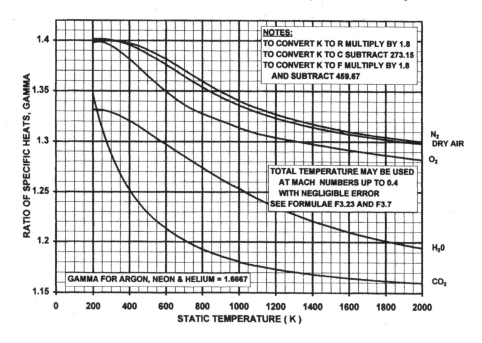

Chart 3.7   Dynamic viscosity versus temperature for pure air and kerosene combustion products.

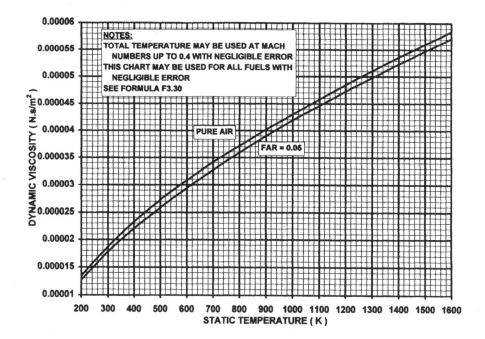

Chart 3.8 Q curve data for dry air or kerosene combustion products for Mach numbers 0.0–1.0.

(a) Gamma = 1.4 – compressors

Mach No.	P/PS (−)	(P − PS)/P %	T/TS (−)	$V/\sqrt{T}$ $m/\sqrt{K}\,s$	q $kg\sqrt{K}/m^2\,kPa\,s$	Q $kg\sqrt{K}/m^2\,kPa\,s$
0.01	1.0001	0.0070	1.0000	0.2005	0.6983	0.6983
0.02	1.0003	0.0280	1.0001	0.4010	1.3967	1.3963
0.03	1.0006	0.0630	1.0002	0.6014	2.0951	2.0938
0.04	1.0011	0.1119	1.0003	0.8018	2.7937	2.7906
0.05	1.0018	0.1748	1.0005	1.0022	3.4924	3.4863
0.06	1.0025	0.2516	1.0007	1.2025	4.1914	4.1808
0.07	1.0034	0.3422	1.0010	1.4027	4.8906	4.8738
0.08	1.0045	0.4467	1.0013	1.6028	5.5900	5.5651
0.09	1.0057	0.5649	1.0016	1.8029	6.2899	6.2543
0.10	1.0070	0.6969	1.0020	2.0028	6.9901	6.9414
0.11	1.0085	0.8424	1.0024	2.2027	7.6907	7.6259
0.12	1.0101	1.0015	1.0029	2.4024	8.3918	8.3077
0.13	1.0119	1.1741	1.0034	2.6019	9.0933	8.9866
0.14	1.0138	1.3600	1.0039	2.8013	9.7955	9.6622
0.15	1.0158	1.5592	1.0045	3.0005	10.4982	10.3345
0.16	1.0180	1.7715	1.0051	3.1996	11.2015	11.0031
0.17	1.0204	1.9970	1.0058	3.3984	11.9055	11.6678
0.18	1.0229	2.2353	1.0065	3.5971	12.6102	12.3283
0.19	1.0255	2.4865	1.0072	3.7955	13.3157	12.9846
0.20	1.0283	2.7503	1.0080	3.9937	14.0219	13.6363
0.21	1.0312	3.0267	1.0088	4.1917	14.7290	14.2832
0.22	1.0343	3.3155	1.0097	4.3895	15.4370	14.9252
0.23	1.0375	3.6165	1.0106	4.5869	16.1458	15.5619
0.24	1.0409	3.9297	1.0115	4.7841	16.8557	16.1933
0.25	1.0444	4.2547	1.0125	4.9811	17.5665	16.8191
0.26	1.0481	4.5915	1.0135	5.1777	18.2783	17.4391
0.27	1.0520	4.9400	1.0146	5.3740	18.9913	18.0531
0.28	1.0560	5.2998	1.0157	5.5701	19.7053	18.6610
0.29	1.0601	5.6709	1.0168	5.7658	20.4206	19.2625
0.30	1.0644	6.0530	1.0180	5.9611	21.1370	19.8575
0.31	1.0689	6.4460	1.0192	6.1561	21.8546	20.4459
0.32	1.0735	6.8497	1.0205	6.3508	22.5735	21.0273
0.33	1.0783	7.2638	1.0218	6.5451	23.2938	21.6018
0.34	1.0833	7.6882	1.0231	6.7390	24.0154	22.1690
0.35	1.0884	8.1227	1.0245	6.9325	24.7384	22.7290
0.36	1.0937	8.5670	1.0259	7.1257	25.4628	23.2814
0.37	1.0992	9.0210	1.0274	7.3184	26.1887	23.8263
0.38	1.1048	9.4844	1.0289	7.5107	26.9162	24.3633
0.39	1.1106	9.9570	1.0304	7.7026	27.6452	24.8925
0.40	1.11661	10.4386	1.0320	7.8941	28.3757	25.4137
0.41	1.12271	10.9289	1.0336	8.0851	29.1080	25.9268
0.42	1.1290	11.4278	1.0353	8.2756	29.8418	26.4316
0.43	1.13551	11.9349	1.0370	8.4657	30.5774	26.9280
0.44	1.14221	12.4502	1.0387	8.6553	31.3148	27.4160
0.45	1.14911	12.9733	1.0405	8.8445	32.0539	27.8955
0.46	1.15611	13.5040	1.0423	9.0331	32.7949	28.3663
0.47	1.16341	14.0420	1.0442	9.2213	33.5377	28.8283
0.48	1.17081	14.5872	1.0461	9.4089	34.2824	29.2815
0.49	1.17841	15.1393	1.0480	9.5960	35.0290	29.7259
0.50	1.18621	15.6981	1.0500	9.7826	35.7777	30.1613

Q is flow function $W\sqrt{T}/A.P.$
q is static flow function $W\sqrt{T}/A.PS.$
V is velocity, m/s.
PS is static pressure, kPa.
TS is static temperature, K.

W is flow, kg/s.
P is total pressure, kPa.
T is total temperature, K.
To convert m/s to ft/s multiply by 3.28084.
To convert $kg\sqrt{K}/s\,m^2\,kPa$ to $lb\sqrt{K}/s\,in^2\,psia$
multiply by 0.009806.

Chart 3.8   *contd.*

(a)   Gamma = 1.4 – compressors

Mach No.	P/PS (−)	(P − PS)/P %	T/TS (−)	V/√T m/√K s	q kg √K/m² kPa s	Q kg √K/m² kPa s
0.51	1.1942	16.2633	1.0520	9.9687	36.5283	30.5876
0.52	1.2024	16.8346	1.0541	10.1542	37.2810	31.0049
0.53	1.2108	17.4119	1.0562	10.3392	38.0358	31.4130
0.54	1.2194	17.9950	1.0583	10.5236	38.7927	31.8119
0.55	1.2283	18.5835	1.0605	10.7075	39.5517	32.2016
0.56	1.2373	19.1772	1.0627	10.8908	40.3130	32.5821
0.57	1.2465	19.7759	1.0650	11.0735	41.0765	32.9532
0.58	1.2560	20.3794	1.0673	11.2556	41.8422	33.3150
0.59	1.2656	20.9873	1.0696	11.4372	42.6103	33.6675
0.60	1.2755	21.5996	1.0720	11.6181	43.3807	34.0106
0.61	1.2856	22.2159	1.0744	11.7984	44.1534	34.3443
0.62	1.2959	22.8361	1.0769	11.9781	44.9286	34.6687
0.63	1.3065	23.4598	1.0794	12.1572	45.7062	34.9836
0.64	1.3173	24.0869	1.0819	12.3357	46.4863	35.2892
0.65	1.3283	24.7171	1.0845	12.5135	47.2689	35.5854
0.66	1.3396	25.3502	1.0871	12.6907	48.0541	35.8723
0.67	1.3511	25.9860	1.0898	12.8673	48.8418	36.1498
0.68	1.3628	26.6242	1.0925	13.0432	49.6321	36.4180
0.69	1.3748	27.2647	1.0952	13.2184	50.4251	36.6769
0.70	1.3871	27.9072	1.0980	13.3930	51.2208	36.9265
0.71	1.3996	28.5515	1.1008	13.5669	52.0192	37.1669
0.72	1.4124	29.1975	1.1037	13.7402	52.8204	37.3982
0.73	1.4254	29.8448	1.1066	13.9127	53.6243	37.6203
0.74	1.4387	30.4932	1.1095	14.0846	54.4310	37.8333
0.75	1.4523	31.1427	1.1125	14.2558	55.2406	38.0372
0.76	1.4661	31.7930	1.1155	14.4263	56.0531	38.2322
0.77	1.4802	32.4438	1.1186	14.5961	56.8685	38.4182
0.78	1.4947	33.0950	1.1217	14.7652	57.6868	38.5954
0.79	1.5094	33.7464	1.1248	14.9337	58.5081	38.7637
0.80	1.5243	34.3978	1.1280	15.1014	59.3324	38.9233
0.81	1.5396	35.0491	1.1312	15.2683	60.1597	39.0743
0.82	1.5552	35.7000	1.1345	15.4346	60.9901	39.2167
0.83	1.5711	36.3504	1.1378	15.6002	61.8237	39.3505
0.84	1.5873	37.0000	1.1411	15.7650	62.6603	39.4760
0.85	1.6038	37.6488	1.1445	15.9291	63.5001	39.5930
0.86	1.6207	38.2966	1.1479	16.0925	64.3431	39.7019
0.87	1.6378	38.9431	1.1514	16.2551	65.1893	39.8025
0.88	1.6553	39.5883	1.1549	16.4170	66.0387	39.8951
0.89	1.6731	40.2320	1.1584	16.5782	66.8914	39.9797
0.90	1.6913	40.8740	1.1620	16.7386	67.7475	40.0564
0.91	1.7098	41.5142	1.1656	16.8983	68.6068	40.1253
0.92	1.7287	42.1524	1.1693	17.0573	69.4696	40.1865
0.93	1.7479	42.7886	1.1730	17.2154	70.3357	40.2401
0.94	1.7675	43.4225	1.1767	17.3729	71.2052	40.2862
0.95	1.7874	44.0540	1.1805	17.5296	72.0782	40.3249
0.96	1.8078	44.6830	1.1843	17.6855	72.9547	40.3563
0.97	1.8285	45.3095	1.1882	17.8407	73.8347	40.3806
0.98	1.8496	45.9331	1.1921	17.9951	74.7182	40.3978
0.99	1.8710	46.5540	1.1960	18.1487	75.6052	40.4080
1.00	1.8929	47.1718	1.2000	18.3016	76.4959	40.4114

Q is flow function W√T/A.P.
q is static flow function W√T/A.PS.
V is velocity, m/s.
PS is static pressure, kPa.
TS is static temperature, K.

W is flow, kg/s.
P is total pressure, kPa.
T is total temperature, K.
To convert m/s to ft/s multiply by 3.28084.
To convert kg √K/s m² kPa to lb √K/s in² psia multiply by 0.009806.

Chart 3.8   *contd.*

(b)   Gamma = 1.33 – turbines

Mach No.	P/PS (−)	(P − PS)/P %	T/TS (−)	V/√T m/√K s	q kg √K/m²kPa s	Q kg √K/m²kPa s
0.01	1.0001	0.0066	1.0000	0.1954	0.6806	0.6806
0.02	1.0003	0.0266	1.0001	0.3908	1.3613	1.3609
0.03	1.0006	0.0598	1.0001	0.5862	2.0420	2.0408
0.04	1.0011	0.1063	1.0003	0.7815	2.7229	2.7200
0.05	1.0017	0.1661	1.0004	0.9768	3.4038	3.3982
0.06	1.0024	0.2390	1.0006	1.1721	4.0850	4.0752
0.07	1.0033	0.3252	1.0008	1.3673	4.7663	4.7508
0.08	1.0043	0.4245	1.0011	1.5624	5.4479	5.4248
0.09	1.0054	0.5368	1.0013	1.7575	6.1297	6.0968
0.10	1.0067	0.6622	1.0017	1.9525	6.8119	6.7668
0.11	1.0081	0.8006	1.0020	2.1473	7.4944	7.4344
0.12	1.0096	0.9519	1.0024	2.3421	8.1772	8.0994
0.13	1.0113	1.1160	1.0028	2.5368	8.8605	8.7616
0.14	1.0131	1.2929	1.0032	2.7313	9.5442	9.4208
0.15	1.0150	1.4824	1.0037	2.9257	10.2283	10.0767
0.16	1.0171	1.6845	1.0042	3.1199	10.9130	10.7292
0.17	1.0194	1.8990	1.0048	3.3140	11.5982	11.3780
0.18	1.0217	2.1259	1.0053	3.5080	12.2840	12.0228
0.19	1.0242	2.3651	1.0060	3.7017	12.9704	12.6636
0.20	1.0269	2.6164	1.0066	3.8953	13.6574	13.3001
0.21	1.0297	2.8798	1.0073	4.0887	14.3451	13.9320
0.22	1.0326	3.1550	1.0080	4.2819	15.0335	14.5592
0.23	1.0356	3.4420	1.0087	4.4749	15.7226	15.1814
0.24	1.0389	3.7406	1.0095	4.6677	16.4125	15.7986
0.25	1.0422	4.0506	1.0103	4.8602	17.1032	16.4104
0.26	1.0457	4.3721	1.0112	5.0525	17.7947	17.0167
0.27	1.0494	4.7047	1.0120	5.2446	18.4871	17.6174
0.28	1.0532	5.0483	1.0129	5.4364	19.1804	18.2121
0.29	1.0571	5.4028	1.0139	5.6279	19.8747	18.8009
0.30	1.0612	5.7680	1.0149	5.8192	20.5699	19.3834
0.31	1.0655	6.1437	1.0159	6.0102	21.2661	19.9595
0.32	1.0699	6.5298	1.0169	6.2009	21.9633	20.5291
0.33	1.0744	6.9260	1.0180	6.3913	22.6616	21.0920
0.34	1.0791	7.3323	1.0191	6.5814	23.3610	21.6481
0.35	1.0840	7.7484	1.0202	6.7712	24.0615	22.1971
0.36	1.0890	8.1741	1.0214	6.9607	24.7632	22.7390
0.37	1.0942	8.6092	1.0226	7.1498	25.4660	23.2736
0.38	1.0995	9.0536	1.0238	7.3386	26.1701	23.8008
0.39	1.1051	9.5070	1.0251	7.5270	26.8755	24.3204
0.40	1.1107	9.9693	1.0264	7.7151	27.5821	24.8324
0.41	1.1166	10.4403	1.0277	7.9029	28.2901	25.3365
0.42	1.1226	10.9196	1.0291	8.0902	28.9994	25.8327
0.43	1.1288	11.4072	1.0305	8.2772	29.7101	26.3210
0.44	1.1351	11.9029	1.0319	8.4638	30.4222	26.8010
0.45	1.1416	12.4064	1.0334	8.6500	31.1357	27.2729
0.46	1.1483	12.9174	1.0349	8.8358	31.8507	27.7364
0.47	1.1552	13.4359	1.0364	9.0212	32.5672	28.1915
0.48	1.1623	13.9615	1.0380	9.2062	33.2853	28.6382
0.49	1.1695	14.4941	1.0396	9.3908	34.0049	29.0762
0.50	1.1769	15.0335	1.0413	9.5749	34.7262	29.5056

Use for diesel fuel incurs negligible error.
Q is flow function W√T/A.P.
q is static flow function W√T/A.PS.
V is velocity, m/s.
PS is static pressure, kPa.
TS is static temperature, K.

W is flow, kg/s
P is total pressure, kPa.
T is total temperature, K.
To convert m/s to ft/s multiply by 3.28084.
To convert kg √K/s m² kPa to lb √K/s in² psia
multiply by 0.009806.

Chart 3.8   *contd.*

(b)   Gamma = 1.33 – turbines

Mach No.	P/PS (−)	(P − PS)/P %	T/TS (−)	V/√T m/√K s	q kg √K/m² kPa s	Q kg √K/m² kPa s
0.51	1.1845	15.5793	1.0429	9.7586	35.4490	29.9263
0.52	1.1923	16.1315	1.0446	9.9419	36.1735	30.3382
0.53	1.2003	16.6898	1.0463	10.1247	36.8997	30.7413
0.54	1.2085	17.2539	1.0481	10.3070	37.6277	31.1354
0.55	1.2169	17.8237	1.0499	10.4889	38.3573	31.5206
0.56	1.2255	18.3989	1.0517	10.6703	39.0888	31.8969
0.57	1.2343	18.9794	1.0536	10.8512	39.8221	32.2641
0.58	1.2432	19.5648	1.0555	11.0316	40.5572	32.6222
0.59	1.2524	20.1550	1.0574	11.2116	41.2941	32.9713
0.60	1.2618	20.7497	1.0594	11.3910	42.0330	33.3113
0.61	1.2714	21.3488	1.0614	11.5700	42.7738	33.6421
0.62	1.2813	21.9520	1.0634	11.7484	43.5166	33.9638
0.63	1.2913	22.5591	1.0655	11.9264	44.2613	34.2763
0.64	1.3016	23.1700	1.0676	12.1038	45.0081	34.5797
0.65	1.3121	23.7842	1.0697	12.2807	45.7568	34.8739
0.66	1.3228	24.4018	1.0719	12.4570	46.5077	35.1590
0.67	1.3337	25.0224	1.0741	12.6328	47.2607	35.4349
0.68	1.3449	25.6458	1.0763	12.8081	48.0158	35.7017
0.69	1.3563	26.2719	1.0786	12.9829	48.7730	35.9594
0.70	1.3680	26.9004	1.0809	13.1570	49.5325	36.2080
0.71	1.3799	27.5311	1.0832	13.3306	50.2941	36.4476
0.72	1.3921	28.1639	1.0855	13.5037	51.0580	36.6781
0.73	1.4045	28.7984	1.0879	13.6762	51.8242	36.8996
0.74	1.4171	29.4346	1.0904	13.8481	52.5926	37.1122
0.75	1.4300	30.0723	1.0928	14.0194	53.3634	37.3158
0.76	1.4432	30.7112	1.0953	14.1902	54.1365	37.5106
0.77	1.4567	31.3511	1.0978	14.3604	54.9120	37.6965
0.78	1.4704	31.9919	1.1004	14.5300	55.6899	37.8737
0.79	1.4844	32.6333	1.1030	14.6990	56.4702	38.0421
0.80	1.4987	33.2753	1.1056	14.8673	57.2530	38.2019
0.81	1.5133	33.9176	1.1083	15.0351	58.0383	38.3531
0.82	1.5281	34.5600	1.1109	15.2023	58.8260	38.4958
0.83	1.5433	35.2023	1.1137	15.3689	59.6163	38.6300
0.84	1.5587	35.8445	1.1164	15.5349	60.4092	38.7558
0.85	1.5745	36.4863	1.1192	15.7002	61.2047	38.8734
0.86	1.5905	37.1275	1.1220	15.8649	62.0027	38.9826
0.87	1.6069	37.7681	1.1249	16.0290	62.8034	39.0838
0.88	1.6236	38.4077	1.1278	16.1925	63.6068	39.1768
0.89	1.6406	39.0464	1.1307	16.3553	64.4128	39.2619
0.90	1.6579	39.6839	1.1337	16.5175	65.2216	39.3391
0.91	1.6756	40.3201	1.1366	16.6791	66.0331	39.4085
0.92	1.6936	40.9548	1.1397	16.8400	66.8473	39.4701
0.93	1.7120	41.5879	1.1427	17.0003	67.6643	39.5241
0.94	1.7307	42.2193	1.1458	17.1600	68.4842	39.5706
0.95	1.7497	42.8488	1.1489	17.3190	69.3069	39.6097
0.96	1.7692	43.4763	1.1521	17.4773	70.1324	39.6414
0.97	1.7890	44.1016	1.1552	17.6350	70.9608	39.6659
0.98	1.8091	44.7247	1.1585	17.7921	71.7921	39.6833
0.99	1.8297	45.3454	1.1617	17.9485	72.6264	39.6937
1.00	1.8506	45.9636	1.1650	18.1042	73.4636	39.6971

Use for diesel fuel incurs negligible error.
Q is flow function W√T/A.P.
q is static flow function W√TS/A.PS.
V is velocity, m/s
PS is static pressure, kPa.
TS is static temperature, K.

W is flow, kg/s
P is total pressure, kPa.
T is total temperature, K.
To convert m/s to ft/s multiply by 3.28084.
To convert kg √K/s m² kPa to lb √K/s in² psia
multiply by 0.009806.

Chart 3.9   Q curves: total to static temperature ratio versus Mach number.

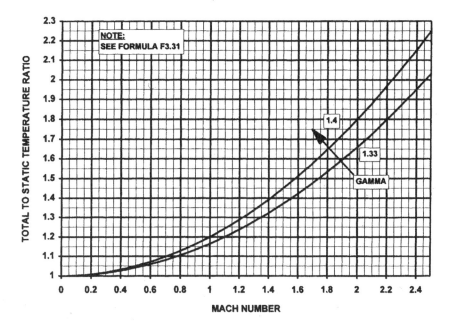

Chart 3.10   Q curves: total to static pressure ratio versus Mach number.

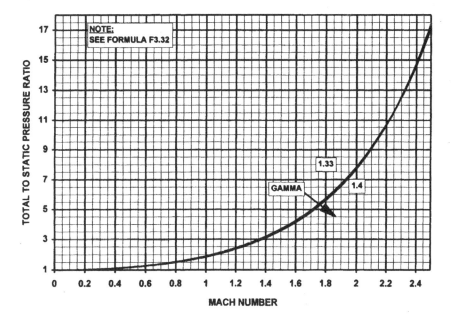

Chart 3.11   Q curves: total flow function Q versus Mach number, $Q = W\sqrt{T}/AP$.

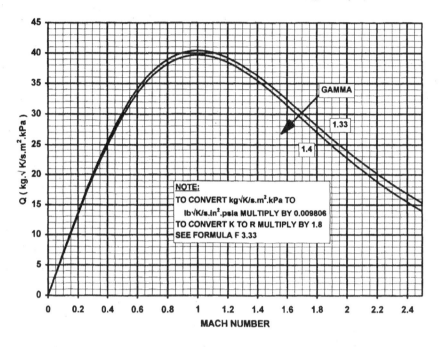

Chart 3.12   Q curves: static flow function q versus Mach number, $q = W\sqrt{T}/APS$.

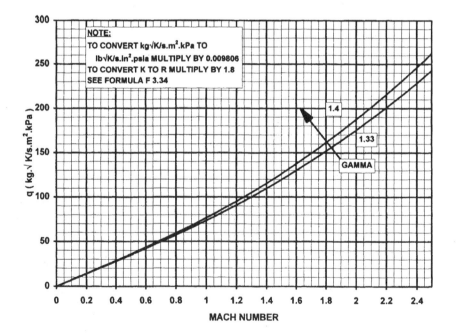

Chart 3.13   Q curves: velocity function $V/\sqrt{T}$ versus Mach number.

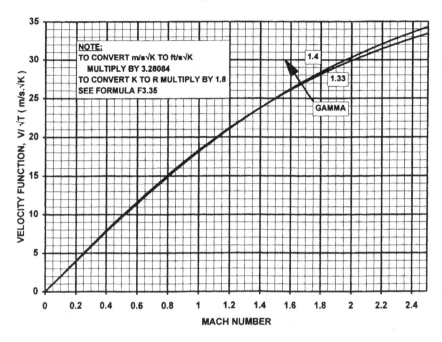

Chart 3.14   Q curves: dynamic head as a percentage of total pressure versus Mach number.

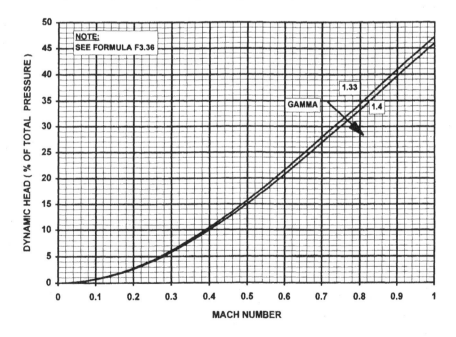

Chart 3.15 Combustion temperature rise versus fuel air ratio and inlet temperature for kerosene fuel.

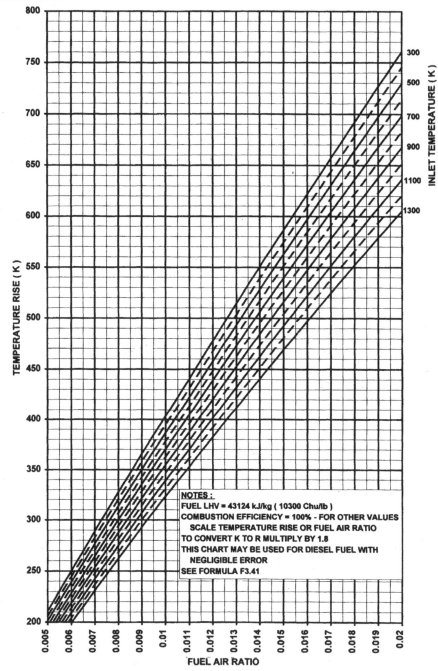

(a)   Fuel air ratio = 0.005–0.02

Chart 3.15 *contd.*

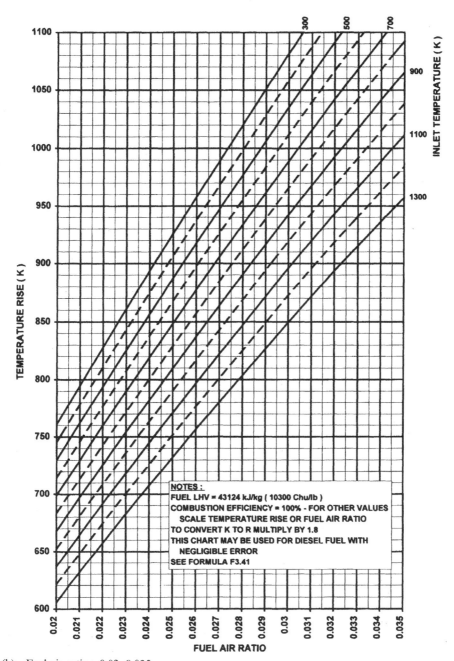

(b)   Fuel air ratio = 0.02–0.035

Chart 3.16   Isentropic efficiency versus polytropic efficiency and pressure ratio for compressors (gamma = 1.4).

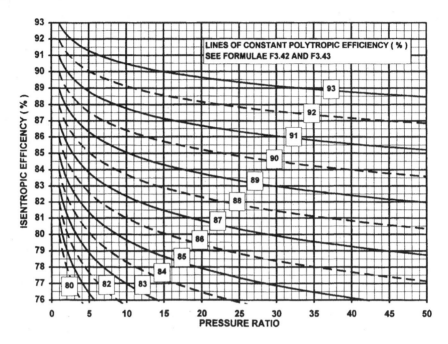

Chart 3.17   Isentropic efficiency versus polytropic efficiency and expansion ratio for turbines (gamma = 1.33).

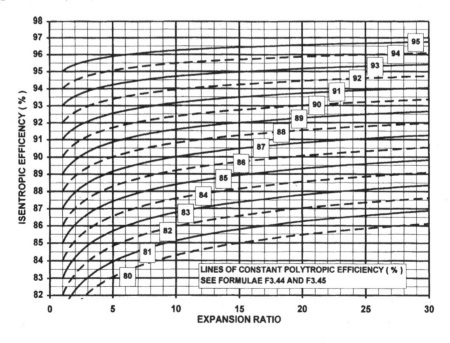

Chart 3.18 Temperature – entropy ($T - S$) diagram for dry air.

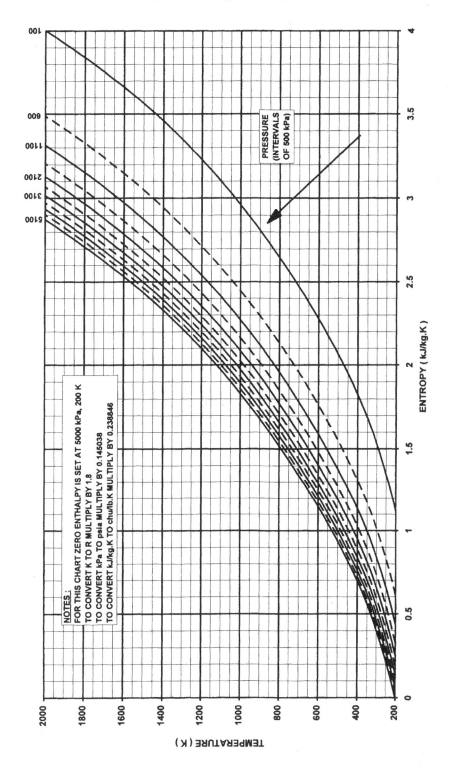

References

1. G. F. C. Rogers and Y. R. Mayhew (1967) *Engineering Thermodynamics Work and Heat Transfer*, Longmans, Harlow.
2. Y. R. Mayhew and G. F. C. Rogers (1967) *Thermodynamic and Transport Properties of Fluids*, Basil Blackwell, Oxford.
3. B. S. Massey (1968) *Mechanics of Fluids*, Van Nostrand Reinhold, Wokingham.
4. A. M. Howatson, P. G. Lund and J. D. Todd (1972) *Engineering Tables and Data*, Chapman and Hall, London.
5. US Department of Commerce (1965) *JANAF Thermochemical Tables*, PB 168370, US Clearinghouse for Federal Scientific and Technical Information, Washington DC.
6. S. Gordon and B. J. McBride (1994) *Computer Program for Calculation of Complex Equilibrium Compositions*, NASA Reference Publication 1311, NASA, Washington DC.

Chapter 4
Dimensionless, Quasidimensionless, Referred and Scaling Parameter Groups

4.0 Introduction

The importance of *dimensionless*, *quasidimensionless*, *referred* and *scaling* parameter groups to all aspects of gas turbine performance cannot be over emphasised. Understanding and remembering the form of the parameter group relationships allows 'on the spot' judgements concerning the performance effects of changing ambient conditions, scaling an engine, a change of working fluid, etc. All gas turbine performance calculations rely to some degree on these parameter groups, which are used in two main ways:

(1) Rigorous representation of *component* characteristics
(2) First order approximation of *overall engine* steady state and transient performance

This chapter provides a tabular quick reference for the major parameter groups, and explains their background. The application of these groups is described extensively in later chapters covering components, off design performance, transient performance, starting, and test data analysis. This chapter includes a brief introduction to these descriptions. It also discusses second-order 'real engine effects' which parameter groups do not account.

4.1 The importance of parameter groups

Many variables are required to describe numerically engine performance throughout the operational envelope. This is accentuated where linear scales of the engine are considered, or working fluids other than dry air. For instance, the steady state mass flow of a turbojet of a given design is a function of eight parameters, as shown later.

The *Buckingham PI theorem* described in Reference 1 reduces the large number of parameters to a smaller number of dimensionless parameter groups. In these groups the parameters are multiplied together and each raised to some exponent, possibly negative or non-integer. The results greatly simplify understanding, and graphical representation, of engine performance.

For example the Buckingham PI theorem may be applied to the mass flow for a given design of turbojet. The parameter *group* for mass flow is then a function of only three other parameter groups, rather than of eight parameters.

Inlet mass flow is a function of:	Dimensionless group for inlet mass flow is a function of:
1 Ambient temperature	1 Dimensionless group for engine speed
2 Ambient pressure	2 Flight Mach number
3 Flight Mach number	3 Dimensionless group for viscosity (has only a second-order effect, and is often ignored for initial calculations)
4 Engine rotational speed	
5 Engine diameter (scale factor)	
6 Gas constant of working fluid	
7 Gamma for working fluid	
8 Viscosity of working fluid	

The Buckingham PI theorem may be applied even more easily to individual components such as compressors and turbines. A simple illustration is that the mixed out temperature from two flow streams depends on the *ratios* of inlet flows and temperatures, i.e. two parameter groups rather than four parameters.

4.2   Tables of parameter groups and description

4.2.1   *Table of parameter groups*

Chart 4.1 presents the parameter groups for overall engine performance, while Chart 4.2 presents corresponding groups for components. These may be derived from first principles by applying the Buckingham PI theorem discussed above. Reference 2 gives an example for a compressor.

4.2.2   *Dimensionless groups*

Otherwise called *non-dimensional groups* or *full dimensionless groups*, these contain all variables affecting engine or component performance, including engine linear scale and fluid properties. This form is of interest if different working fluids, such as helium in a closed cycle, are to be considered. Column 1 of Charts 4.1 and 4.2 presents dimensionless groups for main engine and component parameters.

4.2.3   *Quasidimensionless groups*

Otherwise called *semi-dimensional groups*, these have the specific gas constant, gamma, and the engine diameter omitted. This suits the most common situation of an engine or component design of fixed linear scale, using dry air as the working fluid; i.e. only operational condition and throttle setting are to be considered. Quasidimensionless parameter groups are often confusingly referred to as non-dimensional. Whilst this does not normally affect the validity of the engineering answers it should be noted that these groups *do* have dimensions; e.g. for mass flow $W\sqrt{T}/P$ has units of $kg\sqrt{K}/kPa\,s$. Column 2 of Charts 4.1 and 4.2 presents quasi-dimensionless groups for main engine and component parameters.

4.2.4   *Referred or corrected groups*

Referred or corrected parameter groups are directly proportional to quasidimensionless groups and hence they are interchangeable in usage. The difference is the substitution of theta (θ) and delta (δ) for engine or component inlet pressure and temperature as defined in Chapter 2, where:

delta (δ) = inlet pressure/101.325 kPa
theta (θ) = inlet temperature/288.15 K

As outlined in Chapter 2, overall engine performance is frequently referred to standard inlet conditions of 101.325 kPa, 288.15 K. The referred parameters take the values that the basic parameters would have at ISA sea level static conditions. The units are those of the basic parameter, e.g. kg/s for mass flow. The resulting groups are presented in column 3 of Charts 4.1 and 4.2.

Chapter 2 shows the variation of delta and theta with altitude, ambient temperature and flight Mach number.

4.2.5   *Scaling parameter groups*

These are the dimensionless groups with only the working fluid properties omitted. Their use is of particular value in the concept design of new engines. They enable the performance effects of linearly scaling an existing engine, or matching differentially scaled existing compressors and turbines, to be quickly assessed. Column 4 of Charts 4.1 and 4.2 presents scaling parameter groups for the main engine and component parameters.

4.2.6   *Combining parameter groups*

Further parameter groups may be derived by combining existing groups. For example the group for fuel air ratio presented in Chart 4.1 may be obtained by dividing the groups for fuel flow and engine mass flow.

4.3   Examples of applications

This section provides some examples of the multitude of applications of parameter groups. This serves as a prelude to the comprehensive descriptions in later chapters.

4.3.1   *Component characteristics*

The compressor and turbine characteristics described in Chapter 5 rigorously define the component's performance. The component characterisation is simplified dramatically by using the parameter groups as opposed to the larger number of base parameters. Section 4.7.1 illustrates why this representation is valid.

For a component of fixed geometry the characteristic is unique. To a first order, changing physical inlet conditions does not change the component characteristic. This is of crucial importance in the application to overall engine performance. For a compressor, for example, defining referred speed and referred mass flow fixes pressure ratio and efficiency; an *operating point* may then be plotted onto the characteristic. Once two suitable parameter groups are defined then all others are fixed.

4.3.2   *Engine steady state off design performance*

Figure 4.1 illustrates a referred parameter performance representation of a turbojet. To first-order accuracy, for an engine of fixed geometry, one such figure will fully define engine performance at all ambient temperatures and pressures, flight Mach numbers and throttle settings. It may be seen that:

- If the propelling nozzle is choked then once one referred parameter is fixed all others have a unique value.
- If the propelling nozzle is not choked then a second referred parameter group, usually flight Mach number, must be specified to define all other groups.

This figure is an invaluable tool for making 'on the spot' judgements such as during engine tests, or discussing the impact of an extreme operating point on engine design. For example, if the control system is governing to a constant referred speed, and ambient pressure is reduced at a given flight Mach number, then it is immediately apparent that fuel flow, gross thrust and engine mass flow will reduce in proportion whereas SOT is unaffected, as their referred parameter groups must remain unchanged. Conversely, if Mach number were increased with the nozzle choked, again governing to constant referred speed then fuel flow, gross thrust, mass flow and SOT would increase. This is because again the parameter groups remain unchanged but P1 and T1 increase. For mass flow the effect of the increase in P1 outweighs that of T1, as T1 is square rooted and increases far less with increasing Mach number. Sample calculation C4.1 illustrates this for a turbojet run at different ambient condtions.

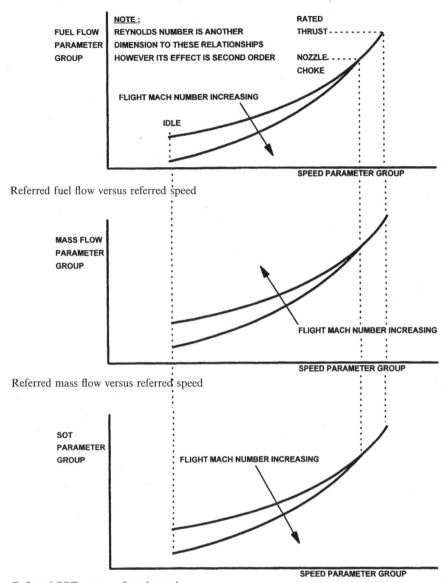

Referred fuel flow versus referred speed

Referred mass flow versus referred speed

Referred SOT versus referred speed

Figure 4.1   Turbojet parameter group relationships.

Similar relationships apply to turbofans. For engines delivering shaft power, output speed depends on the driven load, which need not behave non dimensionally. Relative to Fig. 4.1 this requires another axis in the graphical representation. Chapter 7 provides a more comprehensive description for all engine configurations.

Gross thrust parameter

Variation in flight altitude changes ambient pressure. With an unchoked propelling nozzle at the same flight Mach number, the gross thrust, momentum drag and hence net thrust are proportional to engine inlet pressure, as shown in Chart 4.1. However with a choked nozzle there is also pressure thrust (as per Formula F6.3), since expansion only to a Mach number of 1 at the nozzle exit leaves a higher static pressure there than the ambient value. In this instance changes in ambient pressure must be accounted via the *gross thrust parameter* shown in Chart 4.1.

4.3.3 *Comparison of sets of engine test data*

Over a series of engine tests, for example before and after introducing a component modification, ambient conditions may vary widely. By correcting all sets of data to standard day conditions using the referred parameter groups, the data may be compared on an 'apples with apples' basis.

4.3.4 *Scaling engine and component designs*

Engine scaling is particularly common in the heavyweight power generation industry where engines with a single spool driving the load directly are employed. Designs for 60 Hz running at 3600 rpm are often linearly scaled up by a factor of 1.2 to drop rotational speed to the 3000 rpm demanded by 50 Hz applications. Scaling of existing components for use in new engine projects is common practice throughout the gas turbine industry.

Use of the scaling parameter groups in column 4 of Charts 4.1 and 4.2 is straightforward. For example, if all geometry of a given engine design is linearly scaled by a factor of two, then to first-order accuracy:

- At a particular non-dimensional operating condition the value of the scaling parameter groups will be unchanged.
- Mass flow will increase by a factor of four.
- Thrust or output power will increase by a factor of four.
- Rotational speed will decrease by a factor of two (note this maintains key stress parameters constant, such as blade tip and disc rim speeds).
- SFC remains unchanged.

The effect on other parameters may be seen by reference to Chart 4.1. The impact of linearly scaling components is apparent from the scaling parameter groups in Chart 4.2. Sample calculation C4.2 shows a linear scaling exercise for a turbojet.

In practice, mechanical design issues and wider cost/benefit considerations mean that changes in engine size often deviate from pure linear scaling, as discussed below.

The impact on engine weight of 'scaling' is not obvious. Theoretically weight would change with the cube of the linear scale factor. Indeed for large industrial heavyweight engines, where weight is not an important issue, an exponent of around 3.0 is common. However, in general items such as casing wall thicknesses, trailing edge radii, etc. often do not change as an engine is scaled up. Furthermore, for aero engines where the design must be inherently lightweight, the 'scaling up' process would include all possible steps to minimise the weight such as reviewing disc thickness (which does change stresses), retaining axial gaps, etc. This leads to weight being proportional to the linear scale factor to a power of less than 2.5. Indeed for very small engines the exponent may approach 2.0 because of the strong influence of accessories. Clearly the thrust, or power, to weight ratio decreases as an engine is scaled up if the exponent exceeds 2.0.

The impact on shaft acceleration of linearly scaling is included in Chart 4.1 via the parameter group for NU, and is worthy of further discussion. Here, the effect of scaling on shaft inertia and the speed range to be accelerated through must be considered in addition to the DI term. Theoretically acceleration times would increase in proportion to the linear scale factor. In fact they do not change significantly, since though theoretically shaft inertia increases with the fifth power of the linear scale factor, in reality it approaches the fourth power, for similar reasons to those discussed for scaling weight. The other results of the scaling process are as follows.

- The torque available for acceleration increases with the cube of the linear scale factor as shown by its scaling parameter given in Chart 4.1.
- The speed range to be accelerated through reduces in direct proportion to the linear scale factor.

In practice the viability of scaling a particular engine design by a scale factor of greater than 1.5 is tenuous. A large engine of complex configuration scaled down will often be too expensive. Equally a small engine of simple configuration scaled up will be too inefficient. This arises because the quantity of fuel used during the life of a large engine justifies a higher initial unit cost to achieve low SFC, and much of the manufacturing cost is fixed by the number of operations, independent of engine size. Furthermore, it is impractical to scale axial turbomachinery much below the point beyond which *actual* tip clearances must remain fixed due to manufacturing limitations. This would lead to increased *relative* tip clearances, which would have a powerful impact on component performance.

4.3.5   *Other working fluids*

In closed cycles working fluids other than air, such as helium, are used. To a first-order the dimensionless groups presented in Charts 4.1 and 4.2 enable the effect on leading component, and engine, performance parameters to be evaluated. Helium has a far larger specific heat and gas constant than air, as shown in Chapter 3. This results in very high specific powers, which can be seen from the non-dimensional group for shaft power in Chart 4.1.

Another situation where the full dimensionless groups are of benefit is in dealing with high water content, due to humidity, or steam or water injection, as described in Chapter 12.

4.3.6   *Engine transient performance*

Most of the foregoing has covered steady state performance, however parameter groups may also be applied to transient performance problems where parameters are changing with time. The turbojet example of Fig. 4.1 may again be used for illustration, where under transient operation the following are true to first-order accuracy:

- When the propelling nozzle is choked then two groups, as opposed to one for steady state operation, must be fixed to give a unique value of all others.
- When the propelling nozzle is not choked then two (as opposed to one) parameter groups must be fixed as well as flight Mach number (or any third group) to fix all other groups.

The application of the above to transient performance is discussed further in Chapter 8. However one illustration is worthwhile at this point. Transiently, if the propelling nozzle is choked and referred fuel flow is scheduled against referred speed, then all other groups will follow a unique 'trajectory' during transients. Hence the compressor working line will be the same for all transients, and so to a first order will be compressor surge margin. Indeed engine control strategies during transient operation are invariably based upon parameter group relationships.

The parameter groups presented in Chart 4.1 for parameters such as engine gain, time constant and unbalanced shaft torque enable a fundamental understanding of gas turbine transient performance.

4.4   Second-order effects – steady state performance

This section sets out the various phenomena which have a second-order effect upon engine matching, and which therefore affect the parameter group relationships. Although these effects may be ignored if first-order accuracy only is required, it is always advisable to make some assessment of the likely errors.

When a rigorous analysis must be pursued then *all* effects must be fully accounted by the methods described in Chapter 7. This invariably requires complex computer codes.

4.4.1   *P1 effects – Reynolds number*

The non-dimensional group for the effect of viscosity on engine performance is Reynolds number, as shown in Charts 4.1 and 4.2. Below *critical Reynolds number* for a given engine design viscosity has a second-order detrimental effect on engine performance, reducing component efficiencies and flow capacities. In this instance Reynolds number is actually another axis in the parameter group relationships, as indicated in Fig. 4.1. These Reynolds number effects are often called *P1 effects*, as P1 has the most impact on its value. For an engine of fixed geometry Reynolds number reduces with falling inlet pressure, and its effect is therefore most pronounced for high altitude operation, as shown in Chapter 2. If alternatively an engine is scaled down in size then Reynolds number will also reduce.

The methods presented in Chapter 5 provide a reasonable adjustment of referred and corrected parameters for these effects. The critical Reynolds number level reflects a change in component flow behaviour corresponding to a transition from turbulent to laminar flow, which causes increasing flow separation and hence pressure losses.

4.4.2   *T1 effects*

Gas properties (Cp and γ) vary with temperature and fuel air ratio, as shown in Chapter 3. Furthermore, at a non-dimensional operating point fuel air ratio varies with inlet temperature, as shown in Chart 4.1. When plotting dimensionless groups it is normal to use gas properties for one ambient temperature for the cold end of the engine, and gas properties for a fixed firing temperature for the hot end. This introduces a second-order error when using these relationships at other operational temperatures.

An additional result of changing engine inlet temperature may be a second-order effect on engine geometry. Changing inlet temperature changes mechanical speed if $N/\sqrt{T}$ is held constant. This changes blade and disc stresses, and hence physical growths. This then changes tip and seal clearances, and also blade untwist, all of which affect component performance. One powerful instance of a large change in inlet temperature is for rig tested components where the ambient rig inlet temperatures may have been much lower than those encountered in an engine.

One approach to overcoming these inaccuracies due to P1 and T1 effects is to raise theta or T1 and delta or P1 by exponents other than exactly 0.50 or 1.00. These new exponents may be derived from testing or more rigorous engine modelling.

4.4.3   *Variable geometry features*

The parameter group relationships for an engine or component are only unique for a given geometry. Variable compressor or turbine vanes, and variable propelling nozzles, change geometry and hence change the parameter group relationships. If compressor vanes are scheduled versus $N/\sqrt{T}$ then the compressor may be regarded as a 'black box' with a single characteristic, and non-dimensional behaviour is preserved. Turbine vanes and nozzle areas are not usually scheduled this way, however, and are additional items that must be defined along with throttle setting to describe a non-dimensional operating point.

Engines frequently incorporate handling bleeds at low power. If their switch points are not scheduled versus $N/\sqrt{T}$ they will significantly detract from non-dimensional behaviour. Similarly installation bleed offtakes should be considered.

4.4.4   *Heat exchangers*

Land based engines may incorporate intercooling between the compressors and/or heat recovery upstream of the combustor. In practice neither of these processes is completely non-dimensional.

To a first order, engines with recuperators or regenerators do adhere to parameter group relationships. However these units frequently also have a variable power turbine nozzle, to improve part load SFC. In this instance plotting overall engine performance requires additional lines corresponding to nozzle schedules, such as lines of constant temperature ratio. With this extension dimensionless groups remain a very useful tool.

For intercoolers, if sink temperature does not neatly follow ambient temperature then again the representation must be extended with additional lines of constant sink temperature.

4.4.5   *Inlet and exit conditions*

For a given engine design non-standard inlet and exit conditions may cause the engine to deviate from its normal non-dimensional behaviour. Examples might include:

- Different installation inlet and exhaust losses due to change of application, filter blockage, etc.
- Flow distortion at first compressor inlet due to cross wind or aircraft pitch and yaw

4.4.6   *External power offtake from thrust engines*

As stated earlier, the parameter groups relationships of an engine delivering shaft power require two, rather than one, parameter groups to be fixed at a flight Mach number to define an operating point. This effect extends to small power offtakes from thrust engines, such as that for providing electrical or hydraulic power; a minor adjustment must be made to the parameter relationships for each value of the power offtake parameter group.

4.4.7   *Humidity and water or steam injection*

The effect of water vapour on engine performance can be significant, because gas properties change. Chapter 2 shows how specific humidity varies with ambient conditions and relative humidity. To a first order the effect is small, and may be evaluated using the full dimensionless groups presented in Charts 4.1 and 4.2.

Chapter 12 shows how engine performance is affected by larger water concentrations as in water and steam injection, and describes methods for modelling these effects.

4.5   Second-order effects – engine scaling

Certain 'real' effects are encountered if an engine or component is linearly scaled down. As mentioned, dimensions such as tip clearances, fillet radii, trailing edge thickness, surface finish, etc. cannot be maintained in scale beyond a certain point. If an engine or component is scaled to a size below this threshold then a second-order loss in performance will occur relative to the level suggested by the scaling parameter groups.

More fundamental difficulties are encountered when scaling combustors. These are discussed in Chapter 5.

4.6   Second-order effects – transient performance

For transient performance there are additional secondary phenomena which cause deviation from non-dimensional behaviour, as summarised below. Chapter 8 provides a more comprehensive description.

4.6.1   *Heat soakage*

For steady state operation of a gas turbine there is negligible net heat transfer between the gas path and the engine carcass. During engine transients, however, heat is transferred as the carcass soaks to a new temperature. The significance of heat soakage on referred parameter relationships depends upon the thermal mass and hydraulic diameter of each component, as well as the speed of the transient manoeuvre and the size of the temperature change.

4.6.2   *Volume packing*

For steady state operation the mass flow entering a component at a given instant is equal to that leaving it. For transient operation, however, this is not the case since the density of the flow is changing with time. This phenomenon has a second-order effect on dimensionless group relationships during transient manoeuvres, and is particularly significant for large volumes such as heat exchangers.

4.6.3   *Geometry changes*

During a transient manoeuvre minor engine geometry changes may occur, such as tip clearances increasing during an acceleration due to casing thermal growths being faster than the discs. Since this is a change of engine geometry it will effect component behaviour and hence the non-dimensional relationships.

4.7   Why components and engines adhere to the parameter group relationships

For the advanced reader this section provides a physical description of why component and engines behaviour may be represented by parameter group relationships.

4.7.1   *Basic component behaviour*

Parameter groups reflect the fundamental fluid dynamic processes within an engine component. This may be simply illustrated by considering the operation of an unchoked compressor at an operating point where $W\sqrt{T}/P$ and $N/\sqrt{T}$ are set, and with fluid properties and geometry fixed:

- At a given level of flow parameter group $W\sqrt{T}/P$, the Mach number of the local air relative to stationary vanes is fixed. This is because Q, the flow parameter group $W\sqrt{T}/A.P$, is a unique function of Mach number as shown by the Q curves discussed in Chapter 3, and A is fixed by the compressor geometry.
- At a given level of speed parameter group $(N/\sqrt{T})$, the Mach number of a blade relative to the local air is fixed, as N reflects blade speed and $\sqrt{T}$ the speed of sound.
- With Mach numbers fixed for both the air and the rotating blades, incidence angles onto blades and vanes are fixed (by similar velocity triangles), and hence so are pressure loss coefficients and work input.
- At a given level of Mach number, the ratio of dynamic to total pressure is fixed (as shown by the Q curves discussed in Chapter 3).
- With pressure loss coefficients fixed, as well as the ratio of dynamic to total pressure, blade and vane pressure losses become fixed fractions of their inlet total pressure. With work already fixed this fixes overall pressure ratio.
- With work and pressure ratio fixed, efficiency is also fixed and component performance has been defined.

For the unchoked compressor fixing the parameter groups for speed and flow fixes all others. If the compressor is choked then, as described in Chapter 3, changes in pressure ratio and $W\sqrt{T}/P$ become independent, and pressure ratio must be specified instead to ensure unique conditions in all stages. As mentioned there are various suitable pairs of parameter groups which fix all others.

4.7.2 *Extension to engine matching*

This section describes why, for a single spool turbojet with choked propelling nozzle, fixing one parameter group fixes all the others and hence the component operating points. Put briefly, the characteristics of turbines and nozzles give flow sizes that depend on their expansion ratios, and engine operation gives expansion ratios that depend on flow sizes. The flow parameter $W\sqrt{T}/P$ reflects flow size because the group $W\sqrt{T}/A.P$ reflects Mach number, and for a fixed Mach number the shorter form is simply the latter form multiplied by flow area A.

- If the propelling nozzle expansion ratio is greater than 1.86 the propelling nozzle will be choked. Otherwise the flow parameter group, $W\sqrt{T}/P$ will vary uniquely versus nozzle expansion ratio alone. (As described in Chapter 5, nozzles have a discharge coefficient, but this too is a unique function of expansion ratio.)
- The turbine must have a flow parameter group $W\sqrt{T}/P$ given by its characteristic, and therefore dependent on its non-dimensional speed and expansion ratio.
- The fact that $W\sqrt{T}/P$ for both the turbine and propelling nozzle are unique defines the expansion ratio the turbine must have in the engine, at any one operating point. Consequently the operating point is unique and turbine power and speed can only change by varying inlet temperature (fuel flow). This gives a unique trajectory of turbine and nozzle flow parameter groups versus speed (in practice at higher powers both turbines and nozzles become choked anyway).
- Hence, the compressor pressure ratio at each speed will also be fixed. The turbine $W\sqrt{T}/P$ downstream of it is fixed, and fuel flow varies uniquely with speed, hence the $\sqrt{T}$ also: this fixes compressor exit W/P. Compressor operation will be confined to a unique *running line*, with higher speeds at higher fuel flows.
- When inlet pressure and temperature vary these effects hold, with parameters such as speed expressed by their appropriate parameter groups such as $N/\sqrt{T}$.

For engine configurations without a choked propelling nozzle a unique running line may also be obtained if additional parameters are fixed. These may be flight Mach number for an unchoked propelling nozzle, or output speed law for a shaft power engine. As noted earlier, various effects cause engines to deviate from ideal non-dimensional behaviour, and parameter groups provide a first-order treatment only. Chapter 7 describes off design engine matching in more detail.

Sample calculations

C4.1 A turbojet at maximum rating, ISA, sea level has the following performance:

W1 = 5 kg/s	SOT = 1200 K	P3 = 500 kPa
WF = 27.5 kg/h	A9 = 0.02 m^2	T3 = 500 K
FN = FG = 2.75 kN	SFC = 0.01 kg/N h	N = 28 000 rpm

At maximum rating the propelling nozzle is always choked, and the control system governs the engine to constant referred speed. To first-order accuracy derive the above parameters at maximum rating for a MIL 210 cold day at 11 000 m, 0.8 Mach number.

F6.3   FN = FG − FRAM
F2.15   VTAS = 1.94384 * M * SQRT(γ * R * TAMB)

(i) Evaluate referred parameter groups at ISA SLS
At ISA SLS THETA = 1.0, DELTA = 1.0, hence the values of the referred parameter groups in Chart 4.1 are as per the absolute values above. Since the propelling nozzle is choked the gross thrust parameter, rather than the group for referred gross thrust, must be used:

FGparameter = (2750/(0.02 * 101325) + 1)/(101325/101325)
FGparameter = 2.357

(ii) Evaluate performance data for MIL 210 cold day at 11 000 m, 0.8 Mach number

From sample calculation, C2.4 for a MIL 210 cold day at 11 000 m and 0.8 Mach number THETA = 0.814, DELTA = 0.340 and PAMB = 22.628 kPa. Also from F2.15 VTAS = 231.3 m/s.

As per section 4.3.2, since the propelling nozzle is choked and one referred parameter group is fixed (referred speed), then all other parameter groups will have the same values as at ISA SLS. Hence substituting values into the referred parameter groups:

5 = W1 * SQRT(0.814)/0.34
W1 = 1.88 kg/s

1200 = SOT/0.814
SOT = 977 K

500 = P3/0.340
P3 = 170 kPa

500 = T3/0.814
T3 = 407 K

28000 = N/SQRT(0.814)
N = 25 262 rpm

27.5 = WF/(SQRT(0.814) * 0.34)
WF = 8.44 kg/h

Note: It is assumed that ETA34 is as per ISA SLS.
Evaluate net thrust using gross thrust parameter and momentum drag:

2.357 = (FG/(0.02 * 22628) + 1)/(0.34 * 101325/22628)
FG = 1171 N

FRAM = 1.88 * 231.3
FRAM = 435 N

FN = 1171 − 435
FN = 736 N

SFC = 8.44/736
SFC = 0.015 kg/N h

C4.2   Derive performance parameters at the ISA SLS maximum rating for the engine in C4.1 if all dimensions are linearly scaled by a factor of 1.5.

Using the scaling parameter groups from Chart 4.1 with THETA = 1.0, DELTA = 1.0 and DI the diameter of the original engine:

5/DI^2 = W1/(DI * 1.5)^2
W1 = 11.25 kg/s

28 000 * DI = N * (1.5 * DI)
N = 18 667 rpm

27.5/DI^2 = WF/(1.5 * DI)^2
WF = 61.88 kg/h

2750/DI^2 = FN/(1.5 * DI)^2
FN = 6188 N

The gas path temperatures and pressures and SFC are unchanged by scaling.

Charts

Chart 4.1 Engine parameter groups.

Performance parameter	Dimensionless group	Quasidimensionless group	Referred parameter	Scaling parameter
Temperature at station n (Tn)	$\dfrac{CP*(Tn/T1-1)}{\gamma*R}$	$\dfrac{Tn}{T1}$ or $\dfrac{TSn}{T1}$	$\dfrac{Tn}{\theta}$ or $\dfrac{TSn}{\theta}$	$\dfrac{Tn}{\theta}$ or $\dfrac{TSn}{\theta}$
Pressure at station n (Pn)	$\dfrac{CP*((Pn/P1)^{(\gamma-1)/\gamma}-1)}{\gamma*R}$	$\dfrac{Pn}{P1}$ or $\dfrac{PSn}{P1}$	$\dfrac{Pn}{\delta}$ or $\dfrac{PSn}{\delta}$	$\dfrac{Pn}{\delta}$ or $\dfrac{PSn}{\delta}$
Mass flow (W)	$\dfrac{W*\sqrt{(T1*R)}}{DI^2*P1*\sqrt{(\gamma)}}$	$\dfrac{W*\sqrt{(T1)}}{P1}$	$\dfrac{W*\sqrt{(\theta)}}{\delta}$	$\dfrac{W*(\theta)}{DI^2*\delta}$
Rotational speed (N)	$\dfrac{N*DI}{\sqrt{(\gamma*R*T1)}}$	$\dfrac{N}{\sqrt{(T1)}}$	$\dfrac{N}{\sqrt{(\theta)}}$	$\dfrac{DI*N}{\sqrt{(\theta)}}$
Fuel flow (WF)	$\dfrac{WF*FHV*\sqrt{(R)}*ETA31}{CP*DI^2*P1*\sqrt{(T1*\gamma)}}$	$\dfrac{WF*FHV*ETA31}{P1*\sqrt{(T1)}}$	$\dfrac{WF*FHV*ETA31}{\delta*\sqrt{(\theta)}}$	$\dfrac{WF*FV*ETA31}{DI^2*\delta*\sqrt{(\theta)}}$
Fuel air ratio (FAR)	$\dfrac{FAR*FHV*ETA31}{CP*T1}$	$\dfrac{FAR*FHV*ETA31}{T1}$	$\dfrac{FAR*FHV*ETA31}{\sqrt{(\theta)}}$	$\dfrac{FAR*FHV*ETA31}{\sqrt{(\theta)}}$
Shaft power (PW)	$\dfrac{PW}{\gamma*DI^2*P1*\sqrt{(\gamma*R*T1)}}$	$\dfrac{PW}{P1*\sqrt{(T1)}}$	$\dfrac{PW}{\delta*\sqrt{(\theta)}}$	$\dfrac{PW}{DI^2*\delta*\sqrt{(\theta)}}$
Shaft power SFC (SFC)	$\dfrac{SFC*FHV*\gamma*R*ETA31}{CP}$	$SFC*FHV*ETA31$	$SFC*FHV*ETA31$	$SFC*FHV*ETA31$
Specific power (SPW)	$\dfrac{SPW}{\gamma*R*T1}$	$\dfrac{SPW}{T1}$	$\dfrac{SPW}{\theta}$	$\dfrac{SPW}{\theta}$

Chart 4.1 *contd.*

Performance parameter	Dimensionless group	Quasidimensionless group	Referred parameter	Scaling parameter
Gross thrust (FG)	$\dfrac{FG}{\gamma * DI^2 * P1}$	$\dfrac{FG}{P1}$	$\dfrac{FG}{\delta}$	$\dfrac{FG}{DI^2 * \delta}$
Momentum drag (FRAM)	$\dfrac{FRAM}{\gamma * DI^2 * P1}$	$\dfrac{FRAM}{P1}$	$\dfrac{FRAM}{\delta}$	$\dfrac{FRAM}{DI^2 * \delta}$
Gross thrust parameter (FG)	$\dfrac{FG/(A9 * PAMB) + 1}{DI^2 * \gamma * P1/(PAMB)}$	$\dfrac{FG/(A9 * PAMB) + 1}{P1/(PAMB)}$	$\dfrac{FG/(A9 * PAMB) + 1}{P1/(PAMB)}$	$\dfrac{FG/(A9 * PAMB) + 1}{DI^2 * P1/(PAMB)}$
Thrust SFC (SFC)	$\dfrac{SFC * FHV * \sqrt{(\gamma * R)} * ETA31}{CP * \sqrt{(T1)}}$	$\dfrac{SFC * FHV * ETA31}{\sqrt{(T1)}}$	$\dfrac{SFC * FHV * ETA31}{\sqrt{(\theta)}}$	$\dfrac{SFC * FHV * ETA31}{\sqrt{(\theta)}}$
Specific thrust (SFG)	$\dfrac{SFG}{\sqrt{(\gamma * R * T1)}}$	$\dfrac{SFG}{\sqrt{(T1)}}$	$\dfrac{SFG}{\sqrt{(\theta)}}$	$\dfrac{SFG}{\sqrt{(\theta)}}$
Gas velocity at station n (Vn)	$\dfrac{Vn}{\sqrt{(\gamma * R * T1)}}$	$\dfrac{Vn}{\sqrt{(T1)}}$	$\dfrac{Vn}{\sqrt{(\theta)}}$	$\dfrac{Vn}{\sqrt{(\theta)}}$
Density at station n (RHOn)	$\dfrac{RHOn * R * T1}{P1}$	$\dfrac{RHOn * T1}{P1}$	$\dfrac{RHOn * \theta}{\delta}$	$\dfrac{RHOn * \theta}{\delta}$
Shaft torque (TRQ)	$\dfrac{TRQ}{(\gamma * DI^3 * P1)}$	$\dfrac{TRQ}{P1}$	$\dfrac{TRQ}{\delta}$	$\dfrac{TRQ}{DI^3 * \delta}$
Shaft rate of acceleration (NU)	$\dfrac{NU * J}{\gamma * DI^3 * P1}$	$\dfrac{NU}{P1}$	$\dfrac{NU}{\delta}$	$\dfrac{NU * J}{DI^3 * \delta}$

Chart 4.1 *contd.*

Performance parameter	Dimensionless group	Quasidimensionless group	Referred parameter	Scaling parameter
Shaft time constant (TC)	$\dfrac{TC * J * \sqrt{(R * T1)}}{DI^4 * P1 * \sqrt{(\gamma)}}$	$\dfrac{TC * \sqrt{(T1)}}{P1}$	$\dfrac{TC * \sqrt{(\theta)}}{\delta}$	$\dfrac{TC * J * \sqrt{(\theta)}}{DI^4 * \delta}$
Shaft gain (K)	$\dfrac{K * J * CP * \sqrt{(T1)}}{DI * FHV * ETA34 * \sqrt{(\gamma * R)}}$	$K * \sqrt{(T1)}$	$K * \sqrt{(\theta)}$	$\dfrac{K * J * \sqrt{(\theta)}}{DI}$
Compressor efficiency (ETA2)	ETA2	ETA2	ETA2	ETA2
Turbine efficiency (ETA4)	ETA4	ETA4	ETA4	ETA4
Work parameter (ΔH/T)	ΔH/T	ΔH/T	ΔH/T	ΔH/T
Reynolds number (RE)	$\dfrac{P1 * Vn * DI}{R * T1 * VIS}$	$\dfrac{P1 * Vn}{T1 * VIS}$	$\dfrac{\delta * Vn}{\theta * VIS}$	$\dfrac{\delta * Vn * DI}{\theta * VIS}$

Note: Reynolds number is another axis in the non-dimensional performance representation. See section 4.4.1.

Chart 4.2 Component parameter groups.

Performance parameter	Dimensionless group	Quasidimensionless group	Referred parameter	Scaling parameter
Mass flow (W)	$\dfrac{W*\sqrt{(T_{IN}*R)}}{DI^2*P_{IN}*\sqrt{(\gamma)}}$	$\dfrac{W*\sqrt{(T_{IN})}}{P_{IN}}$	$\dfrac{W*\sqrt{(\theta)}}{\delta}$	$\dfrac{W*\sqrt{(\theta)}}{DI^2*\delta}$
Rotational speed (N)	$\dfrac{N*DI}{\sqrt{(\gamma*R*T_{IN})}}$	$\dfrac{N}{\sqrt{(T_{IN})}}$	$\dfrac{N}{\sqrt{(\theta)}}$	$\dfrac{DI*N}{\sqrt{(\theta)}}$
Shaft power (PW)	$\dfrac{PW}{\gamma*DI^2*P_{IN}*\sqrt{(\gamma*R*T_{IN})}}$	$\dfrac{PW}{P_{IN}*\sqrt{(T_{IN})}}$	$\dfrac{PW}{\delta*\sqrt{(\theta)}}$	$\dfrac{PW}{DI^2*\delta*\sqrt{(\theta)}}$
Shaft torque (TRQ)	$\dfrac{TRQ}{(\gamma*DI^3*P_{IN})}$	$\dfrac{TRQ}{P_{IN}}$	$\dfrac{TRQ}{\delta}$	$\dfrac{TRQ}{DI^3*\delta}$
Compressor efficiency (ETA2)	ETA2	ETA2	ETA2	ETA2
Turbine efficiency (ETA4)	ETA4	ETA4	ETA4	ETA4
Work parameter ($\Delta H/T$)	$\Delta H/T$	$\Delta H/T$	$\Delta H/T$	$\Delta H/T$
Reynolds number (RE)	$\dfrac{P_{IN}*Vn*DI}{R*T_{IN}*VIS}$	$\dfrac{P_{IN}*Vn}{T_{IN}*VIS}$	$\dfrac{\delta*Vn}{\theta*VIS}$	$\dfrac{\delta*Vn*DI}{\theta*VIS}$
Stage loading	$\Delta H/U^2$	$\Delta H/U^2$	$\Delta H/U^2$	$\Delta H/U^2$
Velocity ratio	V_A/U	V_A/U	V_A/U	V_A/U
Mach number	M	M	M	M

References

1. B. S. Massey (1971) *Units, Dimensional Analysis and Physical Similarity*, Van Nostrand Reinhold, London.
2. H. Cohen, G. F. C. Rogers and H. I. H. Saravanamuttoo (1996) *Gas Turbine Theory*, 4th edn, Chapter 4, Section 5, Longmans, Harlow.

Chapter 5
Gas Turbine Engine Components

5.0   Introduction

There are many excellent textbooks available which comprehensively describe the design of gas turbine components. This chapter does not attempt to repeat these works, but instead takes a radically different approach. Information is provided which is not readily available in traditional textbooks, and which is particularly pertinent to whole engine performance; the extensive coverage of off design component performance is a good example of this approach. The objectives of this chapter are as follows.

(1)   To enable the reader to derive realistic levels for component performance parameters, such as efficiency, for use in engine design point performance calculations. Most previous textbooks simply give values for the reader to use 'blindly' in sample calculations.
(2)   To enable the reader to conduct basic sizing of each component in parallel with design point calculations. This includes guidelines on component selection for a given duty such as whether an axial or centrifugal compressor should be employed, the number of turbine stages, etc. Hence the reader may to a first order sketch out the whole engine design resulting from a design point calculation, rather than just input numbers with no idea as to whether the resultant performance is practical.
(3)   To provide key information to enable off design modelling of each component, as well as guidelines for practical considerations which limit off design operation. This encompasses steady state, transient, windmilling and start performance.

For each component a design point section is provided covering the first two of the above, and a separate section deals with off design operation to cover the third. It should be recognised that for the design point issues a text such as this can only ensure that a component or engine design is in 'the right ball park'. The extensive references provided enable further honing of these first pass component performance levels and sizing. Unfortunately component design is highly complex and for detailed design gas turbine company proprietary design rules and computer codes built up over decades are required.

Unless stated, the component performance and basic sizing guidelines provided are for ISA sea level static at the maximum rating. Engine design point and off design calculations utilising the component performance levels, methodology and formulae presented herein are provided in Chapters 6 and 7 respectively. The sample calculations in this chapter concentrate on how the data base provided may be used to evaluate first pass component performance levels and sizes.

5.1   Axial compressors – design point performance and basic sizing

The purpose of a *compressor* is to increase the total pressure of the gas stream to that required by the cycle while absorbing the minimum shaft power possible. For aero applications diameter and weight are also key design issues. A *fan* is the first, usually single stage compressor on a bypass engine or turbofan and has distinct design features, as discussed in sections 5.5 and 5.6. For multiple stage fans the difference is less and the term *LP compressor* is more frequently used.

Axial flow compressors have a greater number of design parameters than any other gas turbine component. Hence providing basic sizing guidelines is challenging and those presented here can only allow very first-order *scantlings* to be derived for a given design point. References 1–5 provide a comprehensive description of axial flow compressor design. The reasoning behind choosing an axial flow or centrifugal flow compressor for a given application is discussed in sections 5.3.6 and 5.3.7.

5.1.1   *Configuration and velocity triangles*

Figure 5.1 illustrates the typical blading configuration for an axial flow compressor. One *stage* comprises a row of *rotor blades* followed by a row of *stator vanes*. A number of stages, with the rotors on a common shaft, form a compressor. Often an additional row of *outlet guide vanes* (OGVs) are required downstream of the last stator row to carry structural load, or remove any residual swirl prior to the flow entering the downstream duct. Also as discussed in section 5.2 *variable inlet guide vanes* (VIGVs) may be employed. These are a row of stator vanes whose angle may be changed by control system action to improve off design operation. Some of the stator rows may also be of variable angle and these are referred to as *variable stator vanes* (VSVs). As a first-order rule 1 stage of VIGVs or VSVs is required per each additonal compressor stage beyond 5 to provide a satisfactory part speed surge line. This ratio will be reduced if handling bleed valves are available.

Figure 5.2 shows typical rotor blade and stator vane sections at the *pitch line*, i.e. the mean of hub and tip. It also shows the changes in leading parameters through the stage, and defines *incidence*. Figure 5.3 shows velocity triangles for the pitch line blading at the design speed and pressure ratio, as well as near to surge and in choke. The rotor blades convert the shaft power input into enthalpy in the form of increased static temperature, absolute velocity and hence total temperature. However as shown by Fig. 5.3 within all blade and vane rows relative velocity does decrease, and hence static pressure increases. Power input is the product of mass

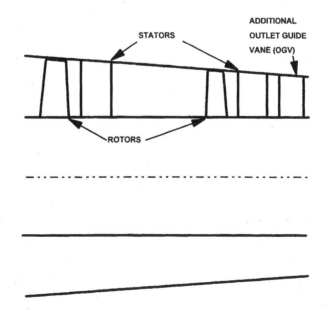

Fig. 5.1   The axial compressor annulus.

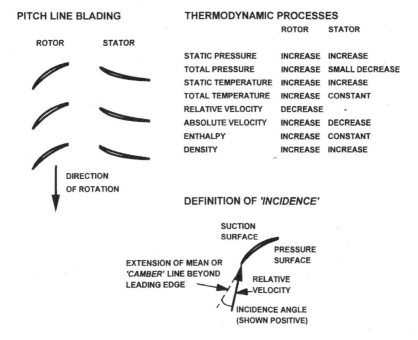

Fig. 5.2   Axial compressor blading and thermodynamics.

flow, blade speed and the change in gas whirl velocity (Formulae F5.1.1 and F5.1.2; the former is the *Euler work*, and can be seen to be simply force times velocity). For the stator vanes there is no work or heat transfer, just friction and turbulent mixing losses. Here the flow is merely diffused, with velocity exchanged for a further increase in static pressure. Owing to the adverse static pressure gradient through the rotor and stator the pressure ratio achievable in a single stage is limited to avoid flow separation and reversal.

5.1.2   *Scaling an existing compressor design*

If an existing compressor design is linearly scaled then to a first order the following are apparent from the scaling parameters presented in Chapter 4:

* Rotational speed change is inversely proportional to the linear scale factor.
* Flow change is proportional to the linear scale factor squared.
* Pressure ratio and efficiency are unchanged.
* Blade speeds and velocity triangles are unchanged.

If scaling 'down' results in a small compressor then Reynolds number effects must be considered. Methods for accounting Reynolds number when scaling are provided in section 5.2. Also in this instance it may not be possible to scale all dimensions exactly, such as tip clearance or trailing edge thickness leading to a further second-order loss in flow, pressure ratio and efficiency at a speed.

5.1.3   *Efficiency (Formulae F5.1.3–F5.1.4)*

As defined by formula F5.1.3, *isentropic efficiency* is the ideal specific work input, or total temperature rise, for a given pressure ratio divided by the actual. Isentropic efficiency is

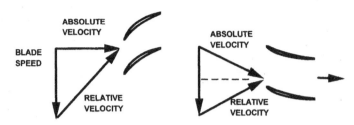

(a)   Design operating point

Notes:

For stators vane angles match air absolute inlet and outlet angles.
For rotors blade angles match air relative inlet and outlet angles.

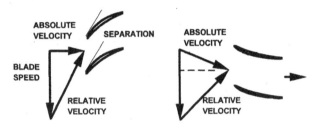

(b)   Operation close to surge

Notes:

Axial velocity is reduced due to lower flow.
Rotor blade is stalled with flow separation from suction surface due to high positive incidence onto
    rotor blades.

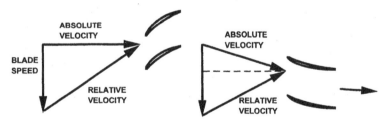

(c)   Operation close to choke

Notes:

Axial velocity increased due to higher flow.
Throat at inlet to rotor blade is choked.
Negative incidence onto rotor blades.

Fig. 5.3   Axial compressor velocity triangles.

sometimes wrongly referred to as *adiabatic efficiency*; the definition of isentropic is adiabatic
plus reversible, i.e. both heat transfer and friction are excluded. Formula F5.1.4, and the sample
calculations in Chapter 6, show its application in design point analysis. One fundamental point,
as explained in Chapter 3, is that *the total temperature rise, and hence power input, to sustain a
given pressure ratio are proportional to inlet total temperature*.

    Polytropic efficiency is defined as the isentropic efficiency of an infinitesimally small step in
the compression process, such that its magnitude would be constant throughout. As described
in Reference 1 it accounts for the fact that the inlet temperature to the back stages of a

compressor is higher, and hence more work input is required for the same pressure rise. Chart 3.16 and Formula F3.42 show the interrelationship between polytropic and isentropic efficiencies.

Polytropic efficiency is not used directly in design point calculations. However it is important in that it enables compressors of different pressure ratio to be compared on an 'apples with apples' basis. Those of the same technology level, average stage loading, and geometric design freedom such as for frontal area, will have *the same polytropic efficiency regardless of pressure ratio*. However, as apparent from Chart 3.16, *isentropic efficiency falls as pressure ratio is increased for the same polytropic efficiency*.

Polytropic efficiency for axial flow compressors improves as size and technology level available for detailed design increase, loading and hence pressure ratio per stage is reduced and geometric constraints such as achieving very low frontal area are relaxed. Chart 5.1 presents typical levels of polytropic efficiency versus average stage loading (section 5.1.4). The highest line is applicable to large industrial heavyweight engines or civil turbofans designed by companies with decades of experience and state of the art design tools, while the lowest line is typical of RPV engines. Owing to geometric constraints of minimising frontal area, supersonic engines will be between 1 and 3 points lower than the highest line.

5.1.4   *Guide to basic sizing parameters*

Guidelines for key parameters for setting the outline annulus geometry, or scantlings, of an axial flow compressor are presented below.

Mean inlet Mach number
This is the mean Mach number at the compressor face calculated using Q curves together with the known inlet flow, pressure, temperature and front face area. While it is desirable, particularly for aero-engines, to have a high inlet Mach number to minimise frontal area this leads to high relative velocities at the first stage blade tip, and hence inefficiency. Values between 0.4 and 0.6 are common, the highest level being for aero-engines in supersonic applications.

Tip relative Mach number
The highest tip relative Mach number will occur on the first stage. Unless IGVs are employed then inlet absolute gas velocity will usually be axial and may be considered to be constant across the annulus. Hence tip relative Mach number can be evaluated by drawing the velocity triangle and knowing mean inlet Mach number and tip speed.

Conservative and ambitious design levels are 0.9 and 1.3 respectively. The latter requires high diffusion relative to the blade to achieve subsonic conditions, which increases pressure losses. VIGVs may be employed to reduce these levels.

Stage loading (Formulae F5.1.5 and F5.1.6)
Loading is a measure of how much work is demanded of the compressor or stage. As shown in Chapter 4, it is a dimensionless group and is the enthalpy increase per unit mass flow of air, divided by the blade speed squared. Formula F5.1.5 shows its definition for a single stage and Formula F5.1.6 shows average loading for a multi-stage compressor as used in Chart 5.1. Efficiency improves as loading is reduced, but more stages are required for a given pressure ratio.

Apart from supersonic aero-engines, loading along the pitch line should be between 0.25 and 0.5 for all stages. The lowest values are generally only viable for LP compressors on multiple spool engines. For supersonic flight engines pitch line loading may be as high as 0.7, some loss in efficiency being accepted in return for reduced number of stages. For a first pass it is reasonable to use a constant value for all stages. For further iterations it may be varied through the stages as part of achieving acceptable rim speeds, first stage tip relative Mach number, hub loading, etc. Common design practice is to reduce it through the compressor, or occasionally allow it to rise up to the mid stages before reducing again.

Stage loading can also be calculated at radial positions other than the pitch line. A key design issue is its value at the hub of the first stage where it is at its highest due to the lower blade speed. Here to maintain acceptable diffusion rates a value of 0.6 would be conservative and 0.9 ambitious.

Rotational speed

This must be set to keep other parameters discussed within target levels, while also being acceptable for turbine design. The turbine is often the dominant factor due to its high temperature and stress levels. For single spool engines directly driving a generator the speed must be either 3000 rpm or 3600 rpm.

Pressure ratio, number of stages and spools

Chart 5.2 shows the range of LP compressor pressure ratios which have been accomplished by a given number of stages. Invariably the stage pressure ratio falls from front to rear, due to increasing temperature. The achievable pressure ratio for a given number of stages is governed by many factors, however the most important are achieving satisfactory part speed surge margin and good efficiency. As described in section 5.2 the front stages of a multi-stage axial flow compressor are pushed towards stall at low speed. The higher the number of stages, and pressure ratio per stage, the worse is this effect. To deal with this variable geometry such as VIGVs and VSVs, or handling bleed valves, must be introduced as described in section 5.2. Furthermore as per Chart 5.1 the higher the overall pressure ratio in a given number of stages, and hence loading, the lower the efficiency.

For HP compressors lower pressure ratios than those shown on Chart 5.2 are achievable. This is because loading for a given pressure ratio and blade speed is proportional to inlet temperature. Formula F5.1.5 may be used to estimate this effect.

Splitting compression between two spools has a number of advantages as described in Chapter 7. First, the part speed matching and surge line issue is eased. This means that the same pressure ratio may be achieved in fewer stages, and with less variable geometry. Secondly employing a higher rotational speed for the rear stages enables them to have a lower loading, and to be on a lower pitch line for the same loading hence alleviating hub tip ratio concerns described below. However these advantages must be balanced against the added layout complexity incurred.

Hub tip ratio

This is the ratio of hub and tip radii. At high values of hub tip ratio, tip clearance becomes a more significant percentage of the blade height. As described in section 5.2 this leads to reduced efficiency and surge margin. At low hub tip ratios disc and blade stresses become prohibitive and secondary flows become powerful.

To balance these two effects hub tip ratio should be greater than 0.65 for the first stage. For back stages on high pressure ratio compressors values may be as high as 0.92.

Hade angle

Hade angle is that of the inner or outer annulus line to the axial. For industrial engines a falling tip line and zero inner hade angle is a good starting point, as it allows some commonality of discs and root fixings reducing cost. Conversely, for aero-engines a rising hub line and zero outer hade angle will minimise loading (see below) hence minimising the number of stages and weight. This also simplifies the mechanical design for achieving good tip clearance control. As design iterations progress it may be necessary to move away from these starting points to other arrangements, such as constant radius pitch line, to achieve acceptable levels for other key design parameters.

A hade angle of up to 10° may be used for the outer annulus design, but preferably less than 5°. The inner annulus line hade angle should be kept to less than 10°.

Axial velocity and axial velocity ratio (Formula F5.1.7)

The axial component of velocity at any point through the compressor may be evaluated from Q curves. *Axial velocity ratio or Va/U* is the axial velocity divided by the blade speed on the pitch line.

To a first order the axial component of velocity is normally kept constant throughout the compressor. Hence the annulus area decreases from front to rear due to increased density, and axial Mach number reduces due to the increase in temperature. Axial velocity ratio would normally be between 0.5 and 0.75 for all stages. It is often at the lower end of this range for the last stage to achieve acceptable exit Mach number (see below).

Aspect ratio (Formula F5.1.8)

Aspect ratio is defined as height divided by vane or blade chord. Both axial and true chord are used. Where weight is important high aspect ratio blading is desirable but at the expense of reduced surge margin and more blades, leading to higher cost.

Typical design levels are 1.5–3.5, based upon axial chord, the lower values being more prevalent for HP compressors and for small engines where mechanical issues dominate.

Blade gapping

The axial gap between a blade row and its downstream stator row must be large enough to minimise the vibratory excitation due to the upstream bow wave and also to avoid *clipping* in the event of surge moving the tip of the rotor blade forward. Conversely it should be minimised for engine length and weight considerations. Typically the gap is set to 20% of the upstream chord.

Rim speed and tip speed

Rim speed is primarily constrained by disc stress limitations and is usually of most concern for the rear stage where it will be at its highest. Tip speed impacts both blade and disc stresses. Often compressor limits are not a major driver on rotational speed selection, as turbine requirements dominate. Limits depend upon geometry, material and temperature. For titanium LP compressors the rim speed may be as high as 350 m/s, and tip speeds up to 500 m/s. For the HP rear stages nickel alloy discs are required, allowing a 350 m/s rim speed, and tip speeds of 400 m/s with titanium blades.

Exit Mach number and swirl angle

These values must be minimised to prevent excessive downstream pressure loss. If this requires more turning than is practical in the last stator then an additional row of OGVs must be considered. Mach number should not be higher than 0.35 and ideally 0.25. Exit swirl should ideally be zero but certainly less than 10°.

Surge margin (Formula F8.5)

Design point target surge margins for the major engine applications are presented in Chapter 8.

Pitch/chord ratio – DeHaller number and diffusion factor (Formulae F5.1.9 and F5.1.10)

Remaining within limiting values of these prevents excessive pressure losses caused by flow diffusion and potential separation. The DeHaller number is simply the ratio of row exit to inlet velocity, and should be kept above 0.72. The diffusion factor is more elaborate, and is an empirical reflection of the effect of blade spacing (pitch/chord) on the peak blade surface velocity. The limiting maximum value is 0.6 for the pitch line, or 0.4 for rotor tip sections.

5.1.5 *Applying basic efficiency and sizing guidelines*

The first pass design of an axial compressor design is highly iterative. Sample calculation C5.1 shows how first pass efficiency level and scantlings may be derived from the above.

5.1.6 *Blading design*

References 6 and 7 describe the design process for two compressors now in production.

5.2 Axial flow compressors – off design performance

5.2.1 *The compressor map*

Once the compressor geometry has been fixed at the design point then the *compressor map* may be generated to define its performance under all off design conditions. The form of a map, sometimes called the *characteristic* or *chic*, is presented in Fig. 5.4. Pressure ratio and isentropic efficiency are plotted versus referred flow for a series of lines of constant referred speed. The *surge line* is discussed later in section 5.2.6. For each referred speed line there is a maximum flow which cannot be exceeded, no matter how much pressure ratio is reduced. This operating regime is termed *choke*. Velocity triangles at three operating points at the design referred speed are described in section 5.1.

Ignoring second-order phenomena such as Reynolds number effects, for a fixed inlet flow angle and no rotating/tertiary stall or inlet distortion the following apply:

- For a fixed compressor geometry the map is unique.
- The operating point on the compressor map is primarily dictated by the components surrounding it as opposed to the compressor itself.
- Each operating point on the map has a unique velocity triangle (with velocity expressed as Mach number).
- Pressure ratio, CP.dT/T and efficiency are related by Formulae F5.1.3 and F5.1.4, and any two out of the three parameters may be used as the ordinates for the map. In fact any combination referred or full dimensionless groups will be suitable if they define flow, pressure ratio and temperature rise.

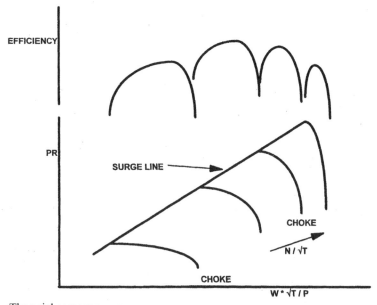

Fig. 5.4 The axial compressor map.

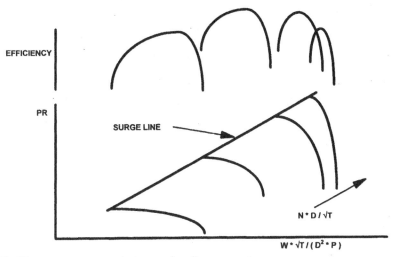

Fig. 5.5 The compressor map in terms of scaling parameters.

The aerodynamic design methods to produce a map for given compressor geometry are complex and involve the use of large computer codes. References 8 and 9 describe the methodology.

5.2.2 *Impact on the map of linearly scaling an existing compressor design*

Section 5.1.2 discusses the impact on design point performance of linearly scaling a compressor. The whole map, plotted in terms of referred parameters as per Fig. 5.4, may be scaled in a similar fashion. Figure 5.5 shows the compressor map plotted in terms of the scaling parameters described in Chapter 4. To a first order, this map is unique for any linear scale of a compressor design.

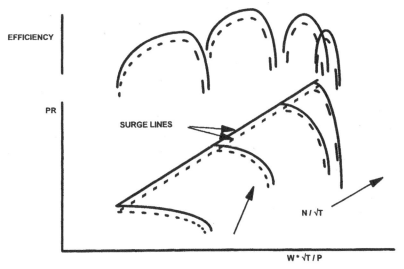

Note:
Dashed lines show effect of Reynolds numbers less than the critical value.

Fig. 5.6 The compressor map: effect of Reynolds number.

5.2.3   *Reynolds number (Formula F2.13) and T1 effects*

When Reynolds number falls below the *critical value* viscous flow effects have a second-order effect leading to lower flow, pressure ratio and efficiency at a speed. Low values may occur due to ambient conditions or due to linearly scaling a compressor to a smaller size. Reynolds number is in fact a fourth dimension to the map as shown in Fig. 5.6. Formulae F5.2.1–F5.2.2 show the form of corrections to data read from a map to account for Reynolds number.

As inlet temperature changes then the compressor geometry, and hence its map, may be modified due to thermal expansion and changing air properties. Differential radial growths between the discs/blades can cause tip clearance to change. Normally T1 effects are small and usually ignored. One important exception is HP compressor rig to engine differences, where the faster engine speed due to higher inlet temperature (than rig ambient) will change stress related growths.

5.2.4   *Change in the working fluid*

If the working fluid is not simply dry air, such as when humidity is present, then the full dimensionless parameters presented in Chapter 4 must be invoked. When the map is plotted in terms of dimensionless parameters as per Fig. 5.7, with the same stipulations as for the referred parameter map stated in section 5.2.1, then it is unique for all linear scales and working fluids. In practice the map based upon dry air is usually utilised with the change in gas properties being accounted as shown in Fig. 5.7. This is also described in Chapter 12.

5.2.5   *Loading compressor maps into engine off design performance models – beta lines*

To facilitate loading a compressor map into an engine off design performance computer model *beta lines* are employed. These are arbitrary lines, drawn approximately equi-spaced and

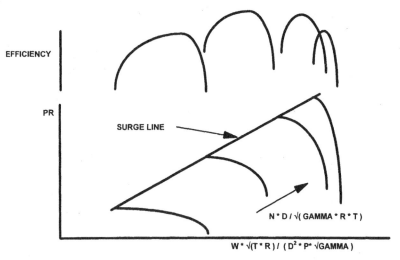

Notes:
To use full non-dimensional groups for effects such as humidity in off design performance models.
  Load map in terms of referred parameter groups for dry air as per Fig. 5.4.
  Multiply referred speed by square root of ratios of dry gamma and R to prevailing values.
  Look up referred map with the adjusted referred speed and beta.
  Multiply each group output from the map by the ratios of the prevailing gamma and R to the dry datum with appropriate exponents as per Chart 4.1.

Fig. 5.7   The compressor map in terms of full dimensionless groups.

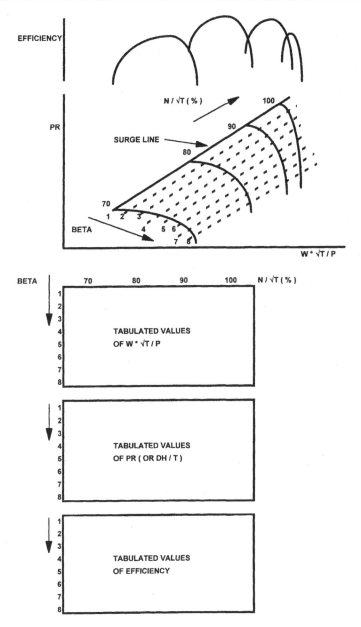

Fig. 5.8 The compressor map and beta lines.

parallel to the surge line, on the map. The map may then be tabulated as shown in Fig. 5.8. Beta serves simply as an array address, and for a plot of pressure ratio versus flow it avoids the problem of horizontal and vertical portions of constant speed lines.

The engine off design performance program can then use these tables to obtain consistent values of referred flow, pressure ratio and efficiency at given levels of referred speed and beta. The use of beta lines in the whole engine off design performance calculation process is described in Chapter 7. Maps for engine starting models utilise alternative variables, as described in section 5.2.11, to assist in model convergence.

5.2.6 *Surge, rotating stall, and locked stall*

At a given speed aerofoil rows may *stall*, that is to say the flow separates from the suction surface, as pressure ratio and hence incidence increase as shown in Figs 5.2 and 5.3. For an aerofoil the point of stall is defined as the incidence at which the aerofoil loss coefficient reaches double its minimum value. In a multi-stage compressor stalled operation can be acceptable. For instance at low speeds following start up the front stages may well be stalled during normal operation, but steady state operation is possible as the rear stages are unstalled and stabilise the flow against the pressure gradient. However if the stall becomes severe, or is entered suddenly, a number of unacceptable flow regimes can result.

Surge can occur throughout the speed range if the surrounding components force the compressor operating point up a speed line such that the pressure ratio is increased to the surge

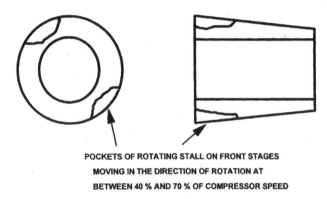

POCKETS OF ROTATING STALL ON FRONT STAGES
MOVING IN THE DIRECTION OF ROTATION AT
BETWEEN 40 % AND 70 % OF COMPRESSOR SPEED

(a) Compressor aerodynamics

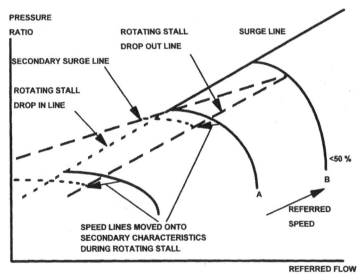

(b) Compressor map

Notes:
Rotating stall cannot drop in at higher speeds than A, nor exist above speed B.
Flow, pressure ratio and efficiency are around 20% lower when in rotating stall.

Fig. 5.9 Axial compressor rotating stall.

line value as per Fig. 5.4. It is the point where blade stall becomes so severe that the blading can no longer support the adverse pressure gradient, and with a lower pressure rise now being produced the flow instantaneously breaks down. The result is a loud bang with part of the flow reversing through the compressor from high to low pressure. In an engine a flame will often be visible at the engine intake and exhaust as combustion moves both forwards and rearwards from the combustor. If action is not taken immediately to lower the working line and hence *recover* from surge, such as by opening bleed valves or reducing fuel flow, then the compressor flow will re-establish itself and then surge again. The surge cycle would continue at a frequency of between five and ten times a second eventually leading to engine damage. Changes in parameters during surge, as well as methods of surge detection, are discussed in Chapter 8, the most dramatic sign being a step decrease in compressor delivery pressure.

Rotating stall or *secondary stall* consists of single, or a number of, stall pockets on the front stages rotating at between 40 and 70% speed in the direction of rotation, as shown in Fig. 5.9. The mechanism of movement is that the blade passage circumferentially ahead of a stalled

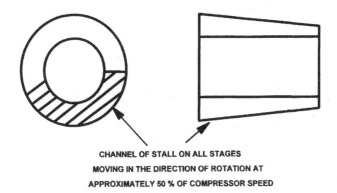

CHANNEL OF STALL ON ALL STAGES
MOVING IN THE DIRECTION OF ROTATION AT
APPROXIMATELY 50 % OF COMPRESSOR SPEED

(a) Compressor aerodynamics

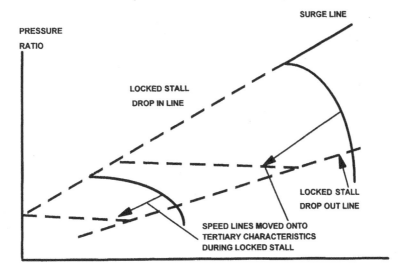

(b) Compressor map

Note:
Flow, pressure ratio and efficiency are around 50% lower when in locked stall.

Fig. 5.10 Axial compressor locked or tertiary stall.

one receives additional flow and moves away from stall. The passage behind the stalled one receives less flow and *it* stalls, deflecting more flow into the first passage which then recovers. For a well designed compressor rotating stall will not occur above 50% speed. Steady state operation in rotating stall is undesirable due to deteriorated compressor and hence engine performance, and the possibility of inducing destructive high cycle blade vibration. Figure 5.9 also illustrates the rotating stall region on a compressor map. If the working line crosses the *drop in* line then it will be in rotating stall. To recover it must be depressed to below the *drop out* line which is considerably lower. While operating in the rotating stall regime the compressor exhibits *secondary characteristics* due to the modified aerodynamics where flow, pressure ratio and efficiency may be reduced by up to 20%. There is a *secondary surge line* which crosses the high speed surge line at the highest speed at which rotating stall drop in may occur. This is significantly lower than the high speed surge line and may be encountered when driving up in speed while in rotating stall. It is often difficult to detect rotating stall from changes in engine performance parameters, unless its onset causes a downstream compressor to surge; this is discussed further in Chapter 8.

Locked stall or *tertiary stall* may occur at low engine speeds following a surge. In this instance instead of the flow recovering and then surging again a channel of stall rotating at approximately 50% engine speed in the direction of rotation remains. This is different from rotating stall in that the stalled section is present over the full axial length of the compressor as opposed to just the front stages as shown in Fig. 5.10. A *tertiary characteristic* is created, again due to modified aerodynamics. Speed lines are almost horizontal on the map. Figure 5.10 shows that the locked stall 'drop out' line is substantially below the surge line. While operating in locked stall referred flow, pressure ratio and efficiency at a referred speed reduce by around 50%. It is characterised by the engine running down while the turbine entry temperature is rapidly rising and the engine must be shut down immediately to avoid damage. Locked stall is also discussed further in Chapter 8.

5.2.7   *Operation of multi-stage compressors*

Each individual stage in a multi-stage compressor has its own unique map. Normally all these are *stacked* together to form an overall map which is more convenient for engine performance. References 8 and 9 show the stacking technique. Figure 5.11 shows how at low and high speeds the operating points for front, mid and rear stages vary on their individual maps. The front stages are pushed towards surge at low speed, due to the flow being restricted by the rear stages which move towards choke. At high speed the situation is reversed with the front stages in choke and the rear stages moving towards surge. These effects occur because the rear stages' referred flow increases strongly as speed increases.

Figure 5.11 also shows how extracting inter stage bleed at low speed alleviates surge concerns due to the extra flow passing through the front stages. However at high speeds the rear stages are starved of flow and move more towards surge.

5.2.8   *Effect of inlet flow angle – VIGVs*

As stated in section 5.2.1 the compressor map is only unique for a fixed value of inlet flow angle. In most instances the inlet flow is axial, however VIGVs are sometimes employed to change the inlet flow angle to modify the map in certain key operating ranges.

Figure 5.12 shows the impact of VIGVs on a compressor map. At low speed they move referred speed lines approximately horizontally on the compressor map; they are scheduled *closed* (high rotative swirl angle) to reduce flow at a speed, and more importantly move the surge line to the left. As described in Chapter 7 the compressor working line often migrates towards surge as an engine is throttled back. VIGVs provide a mechanism to mitigate this by raising the part speed surge line. It is important to note that to a first order the working line in terms of pressure ratio versus flow is unaffected by their setting.

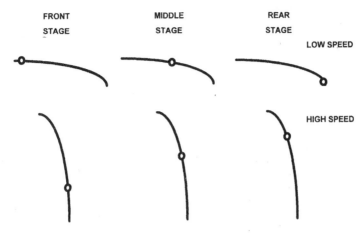

(a) With no interstage bleed

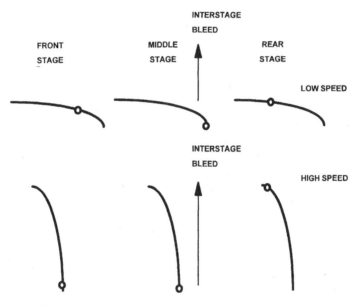

(b) With interstage bleed

Note:
Lines are of constant referred speed, plotted as pressure ratio versus referred flow.

Fig. 5.11 Axial compressor stage matching.

At high speed the effect of VIGVs on referred speed lines is more 'diagonal'. Here the VIGVs and VSVs are fully *open*, being axial or providing small negative incidence, as it is important that the compressor passes as much flow as possible to maximise output power or thrust. Only a small improvement in surge line may occur since the rear rather than the front stages control surge.

VIGVs and VSVs are mainly required to allow a compressor to have an acceptable low speed surge line with all the stages on one shaft. Different feasible schedules only have a second-order impact on compressor efficiency and the relationships between leading engine referred parameters. For example, SFC versus thrust or power is virtually unchanged by their

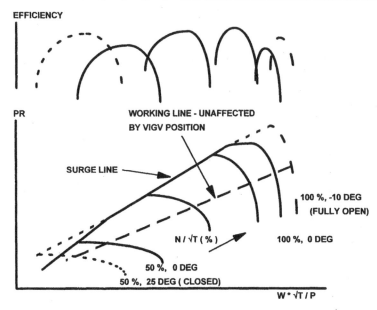

Fig. 5.12   Axial compressor map – effect of VIGV angle.

movement, though a higher power or thrust is attainable. The only other significant effect is that the compressor speed at which they both occur does change significantly.

5.2.9   *Handling bleed valves*

When *bleed valves* downstream of a compressor are opened *the compressor map is not affected* but the working line shows a step change downwards as shown in Fig. 5.13. Bleed valves may be used to maintain acceptable part speed surge margin instead of, or as well as, VIGVs. The choice between VIGVs or handling bleed valves is complex. Bleed valves have lower cost, are lighter and generally more reliable than variable vanes. However they incur a far more

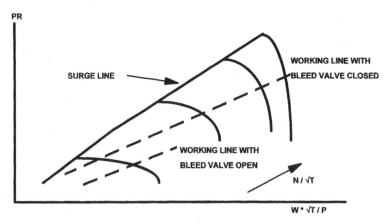

Note:
The compressor map is unchanged.

Fig. 5.13   Axial compressors: effect of downstream handling bleed.

severe SFC penalty since the bleed valve flow, which can be up to 25% of the mainstream and has had considerable work input, is either dumped overboard or into a bypass duct.

The effect of interstage bleed valves located part way along a multi-stage compressor is to change the internal compressor geometry, not just the boundary conditions imposed upon it, hence *the map itself changes* when they are opened. Ideally during rig tests the compressor map should be evaluated with varying interstage bleed levels, and interstage bleed then used as an extra variable when loading it into an engine off design performance model. Opening interstage bleed valves improves the overall surge line at part speed, however it may deteriorate it at high speed. In both instances the working line is lower.

5.2.10   *Inlet pressure and temperature distortion*

Inlet distortion, which is spatial variation of inlet pressure or temperature, can significantly affect the overall compressor map. The most important effect is a reduction in the surge line. The *method of parallel compressors* is employed to evaluate this. Here the exit pressure and temperature are considered to be constant circumferentially. The map is then applied to two parallel streams as described below.

For aircraft engines in cross winds or at high angles of attack the inlet flow may be distorted circumferentially, leading to sectors where inlet pressure is significantly lower than the average.

The *DC60 coefficient* is usually employed to quantify the degree of inlet pressure distortion. This is the difference between the average total pressures in the most distorted 60° sector and the full 360° intake, divided by the average inlet dynamic head (Formula F5.2.3). Worst values in the operational envelope are:

- −0.2 for a civil subsonic transport
- −0.9 for a military fighter aircraft
- Less than −0.1 is usual for industrial, marine and automotive engines

These values, together with knowledge of the average inlet dynamic head enable the depressed value of inlet pressure in the worst 60° sector to be evaluated. The compressor outlet pressure circumferential profile is considered constant. Hence the 60° sector where inlet pressure is depressed must operate at a higher working line than the average and the additional surge margin required to allow for inlet distortion may be determined. Figure 5.14 illustrates this.

Inlet temperature distortion may occur due to a number of reasons such as poor test bed design, or ingestion of thrust reverser exhaust or another engine's exhaust. Again the method of parallel compressors may be used to determine additional surge margin required. In this instance it is the inlet capacity in the 120° sector with the lowest temperature which is used for one stream, and the mean temperature in the remaining sector used for the second sector. This gives rise to a *TC120* coefficient.

5.2.11   *Peculiarities of the low speed region of the map*

Idle will usually occur in speed range 40–70%. However as described in Chapters 9 and 10 operation below this is important for both starting and windmilling. Figure 5.15 illustrates some of the key features peculiar to this region of the map.

At zero rotational speed the compressor behaves as a cascade of vanes. There is no work input and any flow is accompanied by a pressure drop. Pressure loss varies as for flow in a duct as described in section 5.12, and total temperature is unchanged.

At low rotational speeds there is a region where the machine operates as a *paddle* in that there is work input and a temperature rise, but a pressure drop. There is also a region of the speed line where the machine is behaving as a compressor in that there is a pressure and temperature rise resulting from the work input. These two modes of operation are encountered

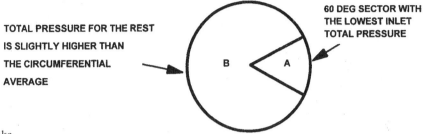

Intake

Note:
Static pressure is considered constant across the whole intake.

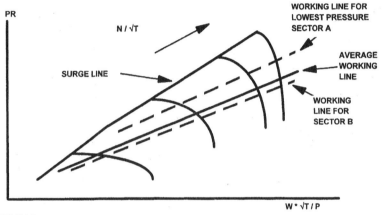

Compressor map

Notes:
Exit pressure considered to be circumferentially constant.
The 608 sector with the lowest inlet pressure of all possible sectors has traditionally been used to indicate
  compressor behaviour.
A 908 sector is more often considered for modern compressors.

Fig. 5.14   Effect of compressor inlet distortion.

during starting and windmilling. It is also theoretically possible to operate as a turbine with
work output and a temperature and pressure drop.

It is not possible to use the standard definition of efficiency (Formula F5.1.3) in the low
speed region as, when the compressor acts as a paddle, efficiency becomes negative and
produces a discontinuity. To load maps into starting and windmilling models $N/\sqrt{T}$ and beta
lines should still be employed but the flow, pressure ratio and efficiency maps are replaced with
$W.T/N.P$, along with $CP.DT/N^2$ and $E.CP.DT/N^2$. To produce the revised map the existing version is easily translated to this form, as the groups are simple combinations of the
existing ones. It is then plotted and extrapolated to low speed and low work, knowing that
zero speed must coincide with zero work.

5.2.12   *Effect of changing tip clearance*

Tip clearance is the radial gap between the rotor blades and casing and is usually in the range
1–2% rms steady state, and greater values transiently (Formula F5.2.4). If modified it is a
change in compressor geometry and hence the map is changed. Tip clearance has a particularly
powerful effect on small compressors where it is a more significant percentage of blade height.

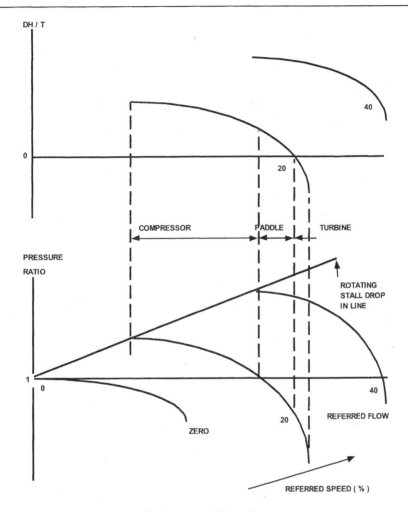

Fig. 5.15 The axial compressor map in the sub idle region.

Typically a 1% increase in rms tip clearance reduces efficiency by approximately 1–2%. Perhaps more importantly the surge line is also deteriorated. The amount depends upon the particular compressor design and must be determined by utilising design codes or more accurately via rig test. The exchange rate will be in the range of a 1% increase in rms tip clearance reducing the surge line by between 2% and 15% of surge margin (Formula F8.5).

5.2.13 *Applying factors and deltas to a map*

Often during the concept design phase a compressor map may be required for predicting engine off design performance, but it may not yet have been generated by the compressor aerodynamic prediction codes. Common practice is to use a map from a similar compressor design and apply 'factors' and 'deltas' (Formula F5.2.5) to align its design point to that required. This should not be confused with linearly scaling a compressor in that it is a technique to provide only an approximate map shape for early engine off design performance. A similar technique is also used to align an engine off design performance model and test data, as described in Chapter 11.

5.2.14   *The compressor rig test*

When a new compressor has been designed it may be tested on a rig prior to being built into an engine. This allows the compressor geometry to be optimised in a controlled environment, often before the rest of the engine hardware is available. There are so many design parameters involved with an axial flow compressor that unless the design is well within previous experience a rig test is essential.

The typical rig configuration is shown in Fig. 5.16. The compressor is driven by an electric motor which is controlled to a specified speed. Measurements are taken allowing flow, pressure ratio and efficiency to be calculated. The exit valve is then closed with the compressor speed maintained, forcing the pressure ratio to be increased and the flow decreased. This process is repeated until the surge line is encountered. A similar procedure is then followed for a number of speed lines. For each throttle setting varying the speed will produce a unique working line, akin to compressor operation within an engine.

5.2.15   *Flutter*

Flutter is the excitation of a blade or/and disc natural frequency due to compressor aerodynamics. *Choke flutter* occurs where the compressor is operating heavily in choke, with the excitation being due to the flow regime associated with very high local Mach numbers and high negative incidence. It is this phenomenon which usually imposes an upper limit on the referred speed to which a compressor may be operated.

Stall flutter can occur at any engine speed when the compressor is close to the surge line with the excitation being due to the flow unsteadiness associated with heavily stalled flow at high positive incidence.

5.3   Centrifugal compressors – design point performance and basic sizing

The changes in key parameters through the rotor and stator of a centrifugal compressor are similar to those for an axial flow compressor described earlier. However flow is changed from an axial to radial direction in the centrifugal *impeller*, and this is followed by a *radial diffuser*. The increasing diameter provides a far greater area ratio and hence diffusion in both than may be achieved in an axial flow stage. A significantly higher pressure ratio is attainable in a single stage than for an axial flow compressor, over 9:1.

References 1, 4, 10 and 11 describe centrifugal compressor design in more detail. References 12 and 13 provide details for actual designs.

5.3.1   *Configuration and velocity triangles*

Figure 5.17 shows the configuration of a centrifugal compressor. The impeller inlet is called the *inducer*, or *eye*, and the outlet the *exducer*. The impeller has a tip clearance relative to a *stationary shroud*, and has seals relative to a *back plate*. The impeller vanes at the exducer may be radial, or for higher efficiency at the expense of frontal area *backswept*. In the *vaneless space* the flow is in free vortex (whirl velocity varies inversely with radius) until the leading edge of the *diffuser vanes*. Often in turbochargers for reciprocating engines no diffuser vanes are employed, however this is rare for gas turbine engines due to the efficiency penalty.

On leaving the diffuser the flow will have a high degree of swirl, typically around 50°, and so usually it flows around a bend into a set of *axial straightener vanes* before entering the combustion system. However if a single pipe combustor is employed then immediately after the diffuser the flow passes into a *scroll*, which is a single pipe rather than an annular passage.

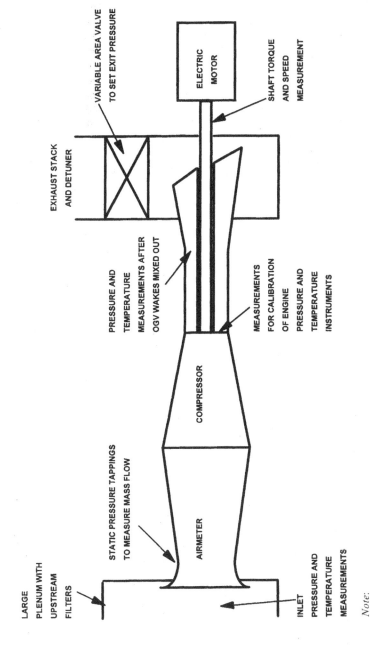

Note:
Measurement of both shaft power input and temperature rise produces 'shaft' and 'gas path' efficiency levels respectively. Shaft efficiency will include disc windage. Gas path may not, if heated air exhausts separately.

Fig. 5.16 Compressor rig layout.

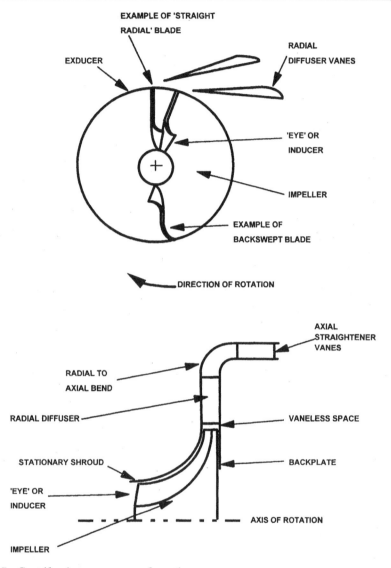

Fig. 5.17 Centrifugal compressor configuration.

Figure 5.18 shows velocity triangles for both radial and backswept vanes. The required work input is defined by Formulae F5.3.1 as well as F5.1.2 and F5.1.4 as per an axial flow compressor. Ideally for radial vanes relative velocity at the exducer would be radial, and the whirl component of absolute velocity equal to rim speed. However in reality as shown in Figure 5.18 some *slip* occurs. As shown by Formula F5.3.2, slip is defined as the ratio of the whirl component of absolute velocity to blade speed. Formula F5.3.3 gives an empirical expression for predicting slip factor based on the number of vanes.

Backsweep dramatically reduces the absolute Mach number out of the impeller, hence reducing pressure loss in the vaneless space and diffuser and improving efficiency. However due to the lower whirl velocity less work, and hence pressure ratio, is achieved in a given diameter. For high pressure ratios above around 5:1 backsweep is essential to avoid excessive pressure losses due to high Mach numbers at the diffuser leading edge.

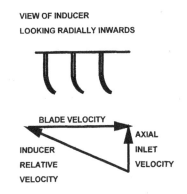

(a)   Inlet

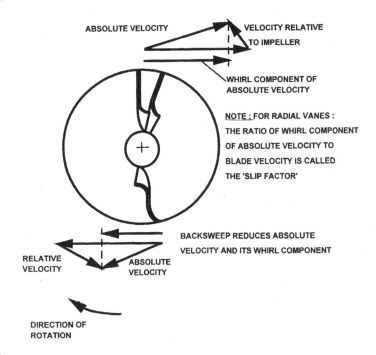

(b)   Exit

Fig. 5.18   Centrifugal compressor – impeller velocity triangles.

5.3.2   *Scaling an existing centrifugal compressor design*

The comments in section 5.1.2 for an axial flow compressor are all equally applicable to a centrifugal flow compressor.

5.3.3   *Efficiency*

The definitions of isentropic and polytropic efficiencies in section 5.1.3 are equally applicable to a centrifugal flow compressor. However polytropic efficiency is best correlated versus the parameter *specific speed* as opposed to loading. Specific speed is peculiar to radial turbomachinery, the most common definition being presented in Formula F5.3.4. It relates back to hydraulic engineering.

Chart 5.3 shows polytropic efficiency versus specific speed. The lower line is for low technology level, zero backsweep, a low diffuser radius ratio and a small size, while the upper is for the converse. This chart may be used to estimate centrifugal compressor efficiency for initial design point calculations. The optimum specific speed for efficiency is around 0.75. Hence once the mass flow rate and pressure ratio required for a given design are set, the inlet volumetric flow rate and enthalpy change may be calculated, and the rotational speed required to achieve this optimum specific speed may be derived. As a design progresses it may be found that the rotational speed has to be changed because of the inability of keeping all of the other aerodynamic and mechanical design limits acceptable, or to suit the turbine designer's needs. Specific speed would then be moved away from the optimum with a consequent loss of efficiency.

5.3.4   Guide to basic sizing parameters

Guidelines for key parameters for setting the scantlings of a centrifugal compressor are presented below. Many of the parameters are common to axial flow compressors and hence their definition is as presented in section 5.1.

Mean inlet Mach number
The mean inlet Mach number into the inducer should be in the range 0.4–0.6.

Inducer tip relative Mach number
Inducer tip relative Mach number values of 0.9 and 1.3 are conservative and ambitious respectively. For a centrifugal rear stage of an axi-centrifugal compressor even lower values may be inevitable.

Rotational speed
This must be set to maximise efficiency by optimising specific speed while keeping other parameters discussed herein within target levels, while being acceptable for turbine design. It is unusual for single spool engines with centrifugal compressors to drive a generator directly with no intermediary gearbox. This is because at the size for which they are practical the optimum speed for performance is significantly higher than 3600 rpm. One exception is for hybrid electric vehicle engines, where high speed alternators are being considered.

Pressure ratio and number of stages
The pressure ratio achieved for a given rim speed, backsweep angle and efficiency may be calculated from Formula F5.3.5. Formula F5.3.6 allows rim speed to be calculated for a given work input. These two formulae, together with rotational speed, allow basic impeller geometry to be defined.

The highest pressure ratio possible from a single stage is approximately 9:1, and from two stages up to 15:1. Owing to ducting difficulties it is unusual to use more than two centrifugal stages in series. If the two stages are on the same spool then necessarily the second stage will end up at a lower specific speed than the optimum for efficiency. Centrifugal compressors used as driven equipment for industrial processes differ in having a lower pressure ratio per stage to promote wide flow range, and hence may use many stages in series.

Backsweep
For maximum efficiency a backsweep angle of up to 40° is practical. However this will result in an increased diameter for a given mass flow and pressure ratio.

Inducer hub tip ratio and blade angle
Hub tip ratio must be large enough to ensure that the hub is of sufficient size for manufacture and to allow suitable bearing and nose bullet designs. Hence the lower limit is set by either impeller vane manufacturing capability or the shaft mechanical design. Its upper limit is governed by inducer tip relative Mach number, if the shaft has no upstream axial stages.

Values should ideally be in the range 0.35–0.5, with 0.7 as an absolute upper limit. The inducer tip blade angle should not exceed 60°.

Rim speed and exducer exit temperature
Exducer rim speed should not exceed around 500 m/s for aluminium and 625 m/s for titanium. Owing to temperature considerations aluminium is acceptable for LP compressors for pressure ratios of up to 4.5 : 1.

Exducer height
This is initially set to achieve a target relative velocity ratio from inducer tip to exit of around 0.5–0.6. Ideally this should be optimised by rig testing.

Impeller length
Typical impeller length may be derived using the *length parameter* defined by Formula F5.3.7. For good efficiency this should be in the range 1.1–1.3.

Vaneless space radius ratio
The vaneless space allows free vortex diffusion, which though relatively slow gives some reduction in Mach number prior to the diffuser vane leading edge. However if it is too long then the required overall diameter will increase. The diffuser vane leading edge to impeller tip radius ratio should be at least 1.05; lower values risk mechanical damage due to excitation.

Radial diffuser exit to impeller tip radius ratio
The radius ratio required to achieve a given level of diffuser area ratio depends on the number of vanes used. A lower limit for the number of vanes is normally set by the requirement to pass bolts or services through the vanes. A high radius ratio provides improved efficiency at the expense of frontal area and weight, guidelines are as follows.

Turbojets and turbofans	1.3–1.5
Turboprops	1.4–1.7
Industrial, marine and automotive	1.7–2.2

Diffuser radial to axial bend radius ratio
As described earlier, for some engine configurations the flow is turned from radial to axial and then straightened with vanes prior to the downstream component. Bend pressure loss reduces as the bend radius ratio is increased. This improves efficiency, but leads to a larger diameter compressor.

The *bend parameter* defined by Formula F5.3.8 should be between 0.4 and 1.5. The lower values are for aero thrust engines and the higher ones for industrial, marine and automotive applications.

Exit Mach number and swirl angle
Where a bend and axial straighteners are employed, then exit Mach number and swirl angle should be less than 0.2 and 10° respectively. If a bend and axial straighteners are not employed then the swirl angle is that coming out of the diffuser vanes which will be of the order of 50°. This is only acceptable if using a scroll outlet duct as described in section 5.12.

5.3.5   *Applying basic efficiency and sizing guidelines*

Sample calculation C5.2 illustrates the application of the basic efficiency and sizing guidelines presented herein.

5.3.6   *Centrifugal compressors versus axial flow compressors*

Axial flow and centrifugal flow compressors are compared here in a qualitative fashion. The following section shows the mass flow and pressure ratio ranges for the major applications to which each is best suited.

An axial flow compressor has the following advantages.

- Frontal area is lower for a given mass flow and pressure ratio. For example at a pressure ratio of 5:1, and the same mass flow, an axial compressor would have a diameter of about half that for a centrifugal compressor.
- Weight is usually less because of the lower resulting engine diameter.
- For mass flow rates greater than around 5 kg/s the axial flow compressor will have a better isentropic efficiency, the magnitude of this advantage increases with mass flow rate.
- Owing to manufacturing difficulties there is a practical upper limit of around 0.8 m on the diameter of the centrifugal impeller, and hence mass flow and pressure ratio capability.

The centrifugal compressor has the following advantages.

- Over 9:1 pressure ratio is achievable in a single stage. For an axial flow compressor this may take between six and twelve stages depending upon design requirements and constraints. Even higher values are possible for centrifugal compressors, but are not normally competitive due to falling efficiency.
- Centrifugal compressors are significantly lower in unit cost for the same mass flow rate and pressure ratio.
- At mass flow rates significantly less than 5 kg/s the isentropic efficiency is better. This is because in this flow range axial flow compressor efficiency drops rapidly as size is reduced due to increasing relative levels of tip clearance, blade leading and trailing edge thicknesses and surface roughness with fixed manufacturing tolerances.
- The centrifugal compressor is significantly shorter for a given flow and pressure ratio. This advantage increases with pressure ratio up to the point where a second centrifugal stage is required.
- Exit Mach number will usually be lower from a centrifugal compressor, hence reducing pressure loss in the downstream duct.
- Centrifugal compressors are less prone to foreign object damage (FOD) than axial flow compressors. This advantage is amplified at very small sizes where the axial flow compressor rotor may consist of a 'blisk' (bladed disk), as opposed to the more conventional drum and separate blading. This construction is required to overcome manufacturing difficulties and to reduce unit cost. In this instance if FOD occurs then the complete blisk must be replaced rather than just individual blades.
- Centrifugal compressors have significantly higher low speed surge and rotating stall drop in lines than multi-stage axial compressors. This is due to a combination of the fundamental low speed aerodynamics, and because the axial flow multi-stage effects described in section 5.2 are not present. Hence for single stage centrifugal compressors with pressure ratios of up to 9:1, usually no handling bleed valves or variable inlet guide vanes or/and variable stator vanes are required to avoid low speed surge problems. This further enhances the cost effectiveness of the centrifugal compressor as well as aiding its relative weight.
- Centrifugal compressor surge lines are less vulnerable to high tip clearance than for axial compressors since, as explained, the pressure rise is not all manifested as a pressure difference across each blade.

In summary, axial flow compressors dominate where low frontal area, low weight and high efficiency are essential and are the only choice at large sizes. Conversely, centrifugal compressors dominate where unit cost is paramount, and at small size.

5.3.7   *Mass flow ranges suited to axial and centrifugal compressors*

The table below shows referred mass flow ranges at ISA SLS maximum rating suited to axial or centrifugal compressors for the major engine applications. References 14 and 15 enable the reader to examine the compressor configurations for engines in production.

	Aircraft engines (kg/s)	Industrial, marine and automotive engines (kg/s)
Predominantly centrifugal	<1.5	<5
Centrifugal or axial depending upon requirements	1.5–10	5–15
Predominantly axial	>10	>15

For thrust engines, axial compressors dominate down to very low mass flows of around 1.5 kg/s. The premium of higher cost is warranted due to the importance of low frontal area and weight to minimise drag at high flight speeds. For aircraft shaft power engines centrifugal compressors are competitive up to 10 kg/s since flight speeds are lower. For industrial, marine and automotive engines low frontal area and weight are of less importance and to minimise unit cost centrifugal compressors are competitive to higher mass flows. For auxiliary power units centrifugal compressors are almost exclusively used as, in this instance, unit cost is the main driver, and the engine weight and frontal area are small in relation to the aircraft.

For the mass flow range where both axial flow or centrifugal flow compressors may be suitable then a concept design phase should address designs for both. Also here axi-centrifugal compressors may be considered where a number of axial flow stages are followed by a centrifugal stage.

5.4   Centrifugal compressors – off design performance

All of the items in section 5.2 discussing the off design operation of axial flow compressors also apply to centrifugal compressors. Only items worthy of further comment are discussed here.

5.4.1   *Effects of changing tip clearance*

Formula F5.4.1 shows the effect of tip clearance between the impeller and the stationary shroud. Flow recirculates from high to low pressure regions, absorbing additional work. Efficiency falls but pressure ratio shows little change. Excessive tip clearance should be considered in the design phase if considered likely. The effect on the surge line is much less than for axial compressors.

5.4.2   *Aerofoil stall, surge, rotating stall and tertiary stall*

As shown by Fig. 5.19, the low speed surge line, as well as flow range of centrifugal compressors are greater than for axial flow compressors. This is due to both aerodynamics and

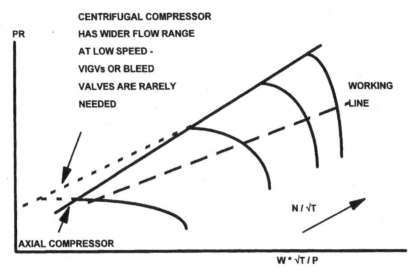

Fig. 5.19   Comparison of centrifugal and axial compressor maps.

because they are not exposed to the stage matching difficulties of multi-stage axial flow compressors. Also centrifugal compressors are not as prone to rotating stall, and are very unlikely to suffer from tertiary stall.

5.4.3   *Flutter*

Owing to its construction the centrifugal compressor is far less prone to flutter than an axial flow compressor.

5.5   Fans – design point performance and basic sizing

Fan is the term given to the first compressor in a turbofan engine. The term reflects the fact that it has a high flow and low pressure ratio compared with core compressors. Immediately downstream of the fan the flow is split into the *cold* or *bypass*, and the *hot* or *core* streams.

This section discusses single stage fans. Multi-stage fans are effectively axial flow compressors and hence the design guidelines presented in section 5.1 are applicable. Multi-stage fans will be at the top of the band shown on Chart 5.2 for pressure ratio derived from a given number of stages. This is because multi-stage fans will generally be applicable to high flight Mach number military aircraft and hence weight must be minimised. Also, as described in Chapter 7, the bypass ratio increases as the engine is throttled back, greatly improving the part speed matching problems described in section 5.2.

5.5.1   *Configuration*

Fans are always axial flow as the diameter and downstream ducting required for a centrifugal stage are prohibitive. The typical arrangement of a single stage fan is shown in Fig. 5.20. The rotor blade is followed by *fan tip* and *fan root* stators. These are usually downstream of the *splitter*, and their aerodynamic design is often compromised by the requirements for structural duty, and to allow services such as oil pipes to cross the gas path. The terms *fan tip* and *fan root* are commonly used to describe the bypass and core streams respectively.

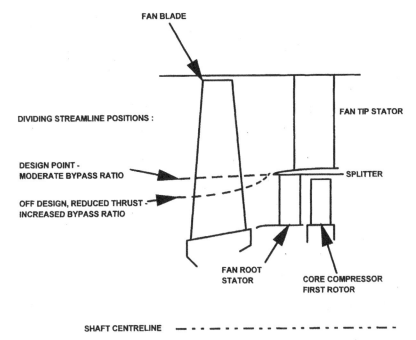

Fig. 5.20 Typical single stage fan configuration.

5.5.2 Scaling an existing fan

The comments in section 5.1.2 for an axial flow compressor are all equally applicable to a fan.

5.5.3 Efficiency

Chart 5.4 shows typical levels of polytropic efficiency versus pitch line loading for single stage fans. A band is presented indicating the variation due to size, technology level and degree of compromise in meeting the good design guidelines presented below. If *clappers* or *snubbers* are used to provide acceptable blade vibration characteristics the efficiency level from Chart 5.4 must be reduced by around 1.5% points. For first pass predictions, Chart 5.4 should be applied separately to the pitchline for the fan tip and fan root, with loading calculated for each from Formula F5.1.5. The root pressure ratio should initially be assumed equal to the tip for low bypass ratio RPV engines, and around 80% of it for the highest bypass ratio civil engines.

As soon as design iterations move to the point where computers are being used for fan design then a radial distribution of work, and hence loading, is applied. Design practices vary widely from company to company depending upon culture and experience base. At this stage significantly different efficiencies for both streams may emerge.

5.5.4 Bypass ratio

Chapter 6 shows how generally increasing bypass ratio improves turbofan SFC but deteriorates specific thrust. There are a number of practical considerations that dictate the upper limit to bypass ratio for a given engine design:

- Engine frontal area increases and hence so do weight and pod drag. Formulae F5.5.1 and F5.5.2 show how to evaluate pod drag. Cost may also increase.

- The number of fan turbine stages increases rapidly. This is because as bypass ratio increases the fan tip speed must be held approximately constant and hence its rotational speed must reduce. For a given core size, fan turbine diameter is fixed and hence its blade speed reduces. This coupled with the fact that the fan turbine specific work must increase, as the ratio of fan flow to fan turbine flow has increased, means that its loading (see section 5.5.5) would be unacceptably high. This would lead to low efficiency unless the number of stages are increased. To date it has proved impractical to put a gearbox between the fan turbine and fan, as for a large turbofan it would have to transmit around 50 MW.
- Cabin air and aircraft auxiliary power offtakes have a greater effect on SFC and specific thrust.
- The perimeter for sealing the thrust reverser when not operational increases, leading to higher leakage.

The above leads to long range civil turbofans having bypass ratios of between 4 and 6. However recently the GE90 has moved to between 8 and 9. Shorter range turbofans typically have a bypass ratio of between 1 and 3, though modern designs are tending to higher values to reduce noise and allow commonality with long range aircraft engines. For supersonic military engines bypass ratio is usually between 0.5 and 1 to minimise frontal area.

5.5.5   *Guide to basic sizing parameters*

The parameters utilised are similar, and calculated in a similar fashion, to those for axial flow compressors defined in section 5.1. As described above the guidelines presented here are for single stage fans.

Inlet Mach number
Inlet Mach number is usually between 0.55 and 0.65, the highest values are typical for military applications. These values are higher than for axial flow compressors due to the need to minimise fan frontal area, and because higher tip relative Mach numbers are acceptable as described below.

Tip relative Mach number
Fans will invariably be transonic at the tip. This is because for a turbofan to be viable it must have a high mass flow in the minimum frontal area, and high Mach numbers are viable as there are no downstream stages. Values between 1.4 and 1.8 are common with tip blade angle being less than 65°.

Stage loading (Formula F5.1.6)
Pitch line loading is higher than for multi-stage compressors. Chart 5.4 shows the typical range and relationship to efficiency.

Rotational speed
This must be set to keep other parameters discussed herein within target levels, while also being acceptable for turbine design. For large turbofans, speeds between 3000 and 3600 rpm are often compatible with these constraints, which facilitates an industrial derivative with the fan turbine driving the load.

Pressure ratio
The maximum pressure ratio achievable from a single stage fan is around 1.9. This is significantly higher than that attainable from the first stage of a multi-stage core compressor for the reasons already discussed. This will apply to the top of climb where the fan operates at its highest referred speed in the operational envelope. Hence at cruise the maximum pressure ratio from a single stage will be between 1.7 and 1.8.

The optimum fan pressure ratio for turbofan cycles is presented in Chapter 6. It will be apparent that a pressure ratio of 1.8 is suitable for medium to high bypass ratios at 0.8 flight Mach number. As shown by Charts 5.20 and 5.21 if the hot and cold streams are mixed prior to a common propelling nozzle then the optimum fan pressure ratio is reduced and a single stage is applicable to even lower bypass ratios.

As discussed in section 5.5.3, pressure ratio at the fan root relative to that at the fan tip varies depending upon design practice. For initial studies assuming the guidelines provided is a good starting point.

Hub tip ratio
Hub tip ratio is minimised to achieve the smallest frontal area for a given mass flow rate. The lower limit is dictated by ensuring that there is sufficient disc circumference for blade fixing, and to achieve an acceptable level of secondary losses.

These result in hub tip ratio for medium to high bypass ratio single stage fans being between 0.3 and 0.4.

Hade angle
The hade angle guidelines presented for axial flow compressors are generally applicable to fans, though for the highest pressure ratios higher hade angles are likely at the hub.

Axial velocity and axial velocity ratio (Formula F5.1.7)
Most comments and definitions apply as per axial compressors. The value of axial velocity ratio would normally be between 0.5 and 0.8 for all stages.

Aspect ratio (Formula F5.1.8)
Blade aspect ratio at the pitch line based upon axial chord should be between 2.0 and 2.5 for fans without clappers. If a clapper must be employed to ensure satisfactory blade vibration characteristics then it should be in the range 3.5–2.5. Fan stator aspect ratio will be in the same range as for LP compressors unless they have a structural duty or are carrying services. In this case aspect ratio may be as low as 2.0.

Rim speed and tip speed
For mechanical integrity, rim and tip speeds should be less than 180 m/s and 500 m/s respectively, for fans in the hub tip ratio range 0.3–0.4. If higher hub tip ratios are used then these values may be increased.

Exit Mach number and swirl angle
As described in section 5.13, bypass duct Mach number must be between 0.3 and 0.35 as a compromise between acceptable engine frontal area and duct pressure loss. Usually the fan tip will be of the same diameter as the bypass duct outer wall, hence there is no diffusion between the two and fan exit Mach number must be equal to that of the bypass duct. Occasionally the fan tip diameter may be smaller leading to an exit Mach number of up to 0.4. Fan stator exit swirl should ideally be zero.

Surge margin (Formula F8.5)
Design point target surge margins are presented in Chapter 8.

Pitch/chord ratio – DeHaller number and diffusion factor
Comments and definitions are as per axial compressors. The DeHaller number should be kept above 0.72. Depending on technology level, the limiting maximum diffusion factor values may slightly exceed those for axial compressors of 0.6 for the pitch line or 0.4 for tip sections.

5.5.6   *Application of basic sizing guidelines*

The sizing process for a fan is similar to that for an axial flow compressor presented in sample calculation C5.1.

5.6   Fans – off design performance

All of the items in section 5.2 discussing the off design operation of axial flow compressors also apply to fans. Only items worthy of further comment are discussed here.

5.6.1   *Change in bypass ratio at part speed and multiple fan maps*

As a turbofan is throttled back the swallowing capacity $(W\sqrt{T}/P)$ of the first core compressor reduces at a faster rate than that of the cold stream propelling nozzle. This results in bypass ratio increasing as the engine is throttled back. As shown in Fig. 5.20 the stream line curvature through the fan is changed significantly. This leads to *multiple maps* or characteristics; i.e. *there is a different map for each bypass ratio.*

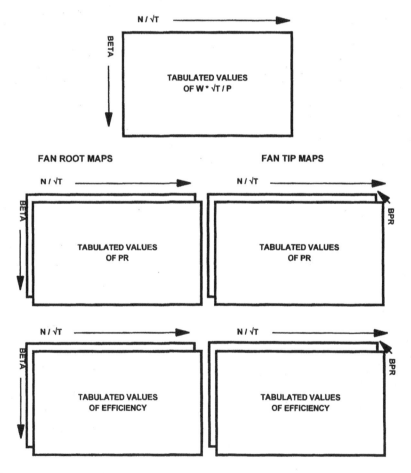

Note:
Fan root and fan tip maps are repeated at intervals of 0.5 bypass ratio.

Fig. 5.21   Fan maps required for rigorous off design modelling.

Furthermore, the fan tip map will usually be different from the fan root map. Hence for rigorous off design modelling a series of maps is required for both as a function of bypass ratio.

5.6.2 *Loading fan maps into off design performance models*

Beta lines and the manner in which a compressor map is loaded into an engine off design performance model are described in section 5.2. Figure 5.21 shows how, for rigorous modelling, fan maps are loaded. Total fan inlet referred flow is tabulated against referred speed and beta as for a compressor. However a series of maps for efficiency and pressure ratio, at discrete intervals of bypass ratio, are loaded for both the fan tip and fan root. The engine off design performance model must first interpolate for bypass ratio, and then for referred speed and beta.

For initial off design modelling a single map as for a compressor may be used for all bypass ratios assuming equal tip and root pressure ratio and efficiency. As a first improvement a map, or series of maps versus bypass ratio, may be used for the tip only. Fan root efficiency and pressure ratio are then evaluated by applying factors and deltas, scheduled versus referred speed, based on the fan design computer code.

5.7   Combustors – design point performance and basic sizing

Combustion systems are the least amenable of all gas turbine components to analysis. While significant steps have been made in recent years in improving design methodology, particularly via 'computational fluid dynamics', or 'CFD', much of the design process still relies upon empirically derived design rules. Hence a significant combustion system rig test programme is essential both before and in parallel with an engine development programme. This rig testing must address not only design point and above idle off design operation, but also the extremely challenging phenomena encountered during starting such as ignition, light around and relight.

The efficiency and basic sizing guidelines presented in this section are representative of all combustion systems except for afterburners and ramjets. These special cases are described in sections 5.21 and 5.22. References 14–19 comprehensively describe the fundamentals of combustor design. The chemistry of combustion, and the range of fuels encountered, are described in Chapter 13.

5.7.1   *Configurations*

Figures 5.22 and 5.24 shows the major features of an *annular* combustion system comprising:

- A compressor exit diffuser to reduce the Mach number of the air before it reaches the combustor
- Primary, secondary and tertiary injector holes through the combustor wall, these are often *plunged* (rounded) to improve CD and jet positional stability. Mach number through the holes is of the order of 0.3 to provide sufficient *penetration* of the jets into the combustor
- A slow moving recirculating 'primary zone' to enable the fuel injected to be mixed sufficiently with the air to facilitate combustion and flame stabilisation
- A secondary zone where further air is injected and combustion is completed
- A tertiary zone where the remaining air is injected to quench the mean exit temperature to that required for entry to the turbine, and to control the radial and circumferential temperature traverse
- Wall cooling systems
- Fuel *injectors* or *burners*
- Ignition system

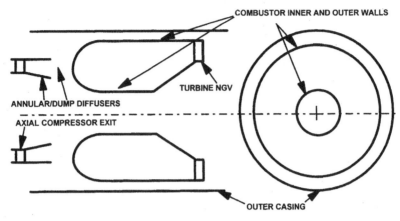

(a)   Forward flow annular

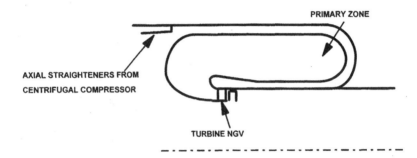

(b)   Reverse flow annular

Fig. 5.22   Annular combustor configurations.

The annular combustor is used almost exclusively for aircaft engines due to its low frontal area and weight for a given volume. It is usually forward flow, but when employing a centrifugal compressor reverse flow is often favoured. This is because the higher diameter of the centrifugal compressor enables the turbine to be arranged 'underneath' (radially inboard of) the combustor, hence reducing engine length.

Early aircraft engines employed a number of cans within an annulus. Figure 5.23 shows the arrangement of such a *cannular* combustion system. However due to its higher diameter and weight this configuration has now been superseded by annular systems. Also the *inter-connectors* between pots required for 'light around' after ignition in one or two cans added a further weight penalty, and were also a mechanical integrity concern. For industrial engines frontal area and weight are not such significant issues, and some still employ this arrangement. It allows one can to be independently rig tested, reducing the size and cost of the rig test facility, and also to be independently changed out during maintenance. For small industrial engines for minimum cost a *single pipe combustor* may be employed, also shown in Fig. 5.23. This is particularly suited to a scrolled exit from a centrifugal compressor.

The requirement for *dry low emissions* or *DLE*, has created further complication to the combustor configuration. This is discussed further in section 5.7.8.

The fuel supply system, fuel injector or burner and the ignition system are each large subjects in their own right. They are not described further here as the objective of this chapter is to enable the outline geometry of a components to be derived to first-order accuracy during early engine concept design. These systems are described comprehensively in References 14–18.

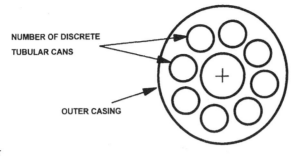

(a) Cannular

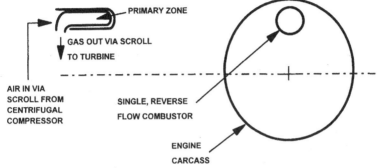

(b) Single pipe

Fig. 5.23 Cannular and single pipe combustor configurations.

5.7.2 *Scaling an existing combustor design and non-dimensional performance*

The combustor is the least amenable of all gas turbine components to scaling. Should this be attempted then for the same inlet temperature, inlet pressure, and temperature rise the following applies:

- Flow change is proportional to the linear scale factor squared.
- Fuel air ratio is unchanged.
- Air and gas velocities are unchanged.
- Percentage pressure loss is unchanged.
- Combustor loading and combustor intensity are inversely proportional to the scale factor, whereas residence time is directly proportional to it. (Loading, intensity and residence time are defined in sections 5.7.3 and 5.7.6 respectively.)

The change in the last three parameters modify the efficiency, ignition, stability, etc. characteristics of the combustor. These may be held constant by scaling only the diameters, but not the length. However in this instance the amount of cooling air per unit surface area is reduced for scale factors less than one. Also for both cases the velocities entering the primary zone are the same, but the radial distance from wall to wall is changed. Hence the relative penetration is modified, changing the aerodynamics and hence fuel mixing and flame stabilisation. From this it is apparent that *purely linearly scaling a combustor is not practical*. However, experience learned at one size will be of immense benefit as the basis for design at another size.

Furthermore, combustor loading and combustor intensity, as well as fuel injector functionality, are dependent upon the absolute level of inlet pressure and temperature. *Hence unlike other components combustor performance is significantly modified when at the same non-dimensional operating point but with different absolute values of inlet pressure and temperature.*

5.7.3   *Combustion efficiency (Formulae F5.7.1 and F5.7.2)*

Combustion efficiency is the ratio of fuel burnt in the combustor to the total fuel input (Formula F5.7.1). In the early years of gas turbine engineering much empirical rig testing showed that it could be correlated versus *combustor loading* and fuel air ratio. Chart 5.5 shows generic data sufficient for concept design. Owing to the one for one exchange rate of combustion efficiency with SFC usually other compromises are made to ensure that design guidelines are met such that the curve for an unconstrained design from Chart 5.5 is achieved.

Combustor loading (Formula F5.7.2) may be considered as a measure of the difficulty of the combustor design duty. For efficiency correlations loading is calculated using the total air flow and can volume (not including the outer annuli), as this reflects the entire combustion process. It is apparent from Chart 5.5 that a low value of loading improves combustion efficiency. The chart is characterised by the knee point occurring at a loading value of $50 \, \text{kg/s} \, \text{atm}^{1.8} \, \text{m}^3$ for an unconstrained design, above this efficiency falls rapidly. As design point mass flow and temperature increase then the flame tube volume must be increased to maintain a given value of loading, and hence efficiency. However the dominant term is combustor pressure due to it being raised to the exponent 1.8. As inlet pressure increases the required volume for a given loading level decreases rapidly. In some companies loading is defined as the reciprocal of Formula F5.7.2.

Combustor volume should initially be set to achieve a loading value of less than $10 \, \text{kg/s} \, \text{atm}^{1.8} \, \text{m}^3$ at the sea level static maximum rating condition. This provides an efficiency of greater than 99.9% for an unconstrained design, and should ensure respectable combustion stability characteristics as discussed in section 5.8. During later concept design iterations this may have to be modified if the required volume is impractical or, conversely, if off design efficiency is poor. Efficiencies of less than 90% anywhere in the operational envelope are unlikely to be tolerable.

5.7.4   *Pressure loss*

Compressor exit Mach number will be of the order of 0.2–0.35. This must be reduced in the combustor entry diffuser to between 0.05 and 0.1 around the can, otherwise can wall pressure loss will be unacceptably high. Design point performance of the combustor entry diffuser is described in section 5.13.

The combustor *cold loss* is due to the dump of air being injected through the wall. Good designs would have a value of between 2 and 4% of total pressure at the design point depending upon geometric constraints. For high flight Mach number, aero-engines Mach number outside the can may be higher than desired to minimise frontal area. In this instance cold pressure loss may be as high as 7%.

In addition there is a *fundamental* or *hot loss* in the combustion section of the flame tube. Flow in a duct with heat transfer is called *Raleigh flow* and the fundamental thermodynamics dictate that there is a pressure loss associated with the heat release; reduced density increases velocity, requiring a pressure drop for the momentum change. Reference 1 describes this phenomenon and shows the loss in dynamic head versus combustor Mach number and temperature ratio. With the typical combustor Mach number of 0.025 design point hot loss is around 0.05% and 0.15% for temperature ratios of 2 and 4 respectively.

5.7.5   *Combustor temperature rise*

Charts and formulae for combustion temperature rise as a function of inlet temperature, fuel air ratio and fuel type are provided in Chapter 3.

5.7.6   *Guide to basic sizing parameters*

Guidelines for generating first pass scantlings for a combustor are presented below.

Loading

Combustor volume must be derived by considering loading (F5.7.2) at a number of operational conditions. The guidelines provided here are again based upon the total can volume (not including the outer annuli) and mass flow. At the sea level static maximum rating loading should be less than $10\,kg/s\,atm^{1.8}\,m^3$, and preferably less than $5\,kg/s\,atm^{1.8}\,m^3$. For industrial engines greater volume is practical and values as low as $1\,kg/s\,atm^{1.8}\,m^3$ may be attainable.

The highest loading value in the operational envelope will usually occur at idle at the highest altitude, lowest flight Mach number and the coldest day. Ideally loading at this condition should be less than $50\,kg/s\,atm^{1.8}\,m^3$, to ensure acceptable efficiency and weak extinction margin. At worst it should be less than $75\,kg/s\,atm^{1.8}\,m^3$ or $100\,kg/s\,atm^{1.8}\,m^3$ for constrained or unconstrained designs respectively.

Furthermore, for aero-engines to achieve combustor relight loading must be less than $300\,kg/s\,atm^{1.8}\,m^3$ when windmilling at the highest required altitude and lowest Mach number. Combustor inlet conditions while windmilling may be derived from the charts presented in Chapter 10. Typical restart flight envelopes are provided in Chapter 9.

Combustor volume must be the largest of the three values derived from the above guidelines.

Combustion intensity

As defined by Formula F5.7.3, *combustion intensity* is a measure of the rate of heat release per unit volume. As for loading it is another measure of the difficulty of combustion and a low value is desirable. At the sea level static maximum rating it should be less than $60\,MW/m^3\,atm$. This is readily achievable for industrial engines but can be a challenge for aero-engines. Combustor volume must be sized to ensure that the guidelines for both loading and intensity are satisfied.

Residence time

Residence time is that taken for one air molecule to pass through the combustor, and may be calculated from Formula F5.7.4. It should be a minimum of 3 ms for conventional combustors.

Local Mach numbers and combustion system areas

Design guidelines for local Mach numbers and equivalence ratios are presented in Fig. 5.24. The Mach number in the inner and outer annuli prior to the primary zone injector ports should be of the order of 0.1, leading to lower levels further along the annuli. Hence the area of each annulus may be derived for given inlet conditions using Q curves. Low annulus Mach number is essential to maintain a level of Mach number for the injector ports of circa 0.3, since a ratio of injector port to annulus Mach number of greater than 2.5 is required for good coefficient of discharge. The injector port Mach number of 0.3 is a compromise between minimising pressure loss while achieving good penetration. Unless the ports are angled it is reasonable to assume that half of the air entering through the primary ports joins the upstream primary zone, and half the downstream secondary zone.

The flow regime in the primary zone is complex with the most usual being the double toroid shown in Fig. 5.24. This is essential to mix the fuel and air properly, and to provide a region of slow velocity in which the flame may be stabilised. The mean axial Mach number leaving the primary zone must be of the order 0.02–0.05. Despite heat release it is acceptable to use Q curves to evaluate flame tube area at this plane using the known mass flow (derived using fractions as per the next section), pressure and the stoichiometric temperature described below.

After the secondary zone air flow has been introduced the Mach number within the flame tube may rise to around 0.075–0.1. Finally, the tertiary air is introduced and the flow is accelerated along the turbine entry duct to about 0.2 at the nozzle guide vane leading edge.

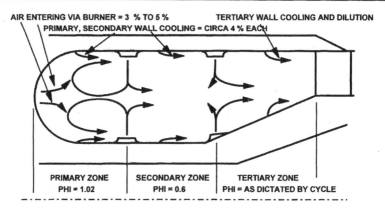

(a)   Stoichiometry

Notes:

50% of primary port flow enters primary zone and 50% secondary zone.

50% of secondary port flow enters secondary zone and 50% tertiary zone.

Primary wall cooling air takes part in secondary combustion, etc.

Primary zone flow for combustion will be 25–45% to give PHI = 1.02. The percentage increases with combustor exit temperature. Secondary port percentage flows may then be calculated to give PHI = 0.6.

The tertiary dilution will be the balance of the air available for cooling and exit temperature traverse control.

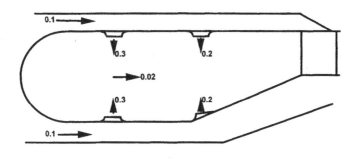

(b)   Combustor Mach numbers

Notes:

Mass flows may be derived using values of PHI given in (a) and design point fuel flow, pressures and temperatures are as per performance design point. Combustor areas are then derived from Q curves and the Mach number guidelines given in (b).

Primary exit Mach number is based upon primary mass flow only.

Temperature is compressor delivery for annuli/ports, but stoichiometric after primary zone.

Fig. 5.24   Combustor design guidelines.

Fuel air ratios and equivalence ratios

Equivalence ratio is the local fuel to air ratio divided by the corresponding stoichiometric value (Formula F5.7.5). *Stoichiometric* fuel to air ratio is that where the fuel is sufficient to burn with all the air and may be calculated from Formula F5.7.6. Equivalence ratio guidelines for sea level static maximum rating for the primary and secondary zones are 1.02 and 0.6 respectively. These guidelines enable the amounts of air required in the primary and secondary zones to be evaluated. They will give a temperature of around 2300 K in the primary zone, and 1700 K in the secondary.

The primary zone usually needs to be marginally richer than stoichiometric at the design point to avoid weak extinction at low power. In addition a small percentage of air may be introduced for wall cooling which will not take part in the combustion process until the secondary zone.

The remaining air is introduced in the tertiary zone where the dilution reduces the temperature down to the level required for turbine entry. With careful placement, tertiary dilution holes can be used to control the traverse (discussed below) to address nozzle guide vane and turbine blade oxidation and creep concerns.

Outlet temperature distributions

Figure 5.25 shows the circumferential and radial temperature distributions at the outlet plane of an annular combustor. For a given combustor design these distributions are quantified by two terms.

The *OTDF* (Overall Temperature Distribution Factor), defined by Formula F5.7.7, is the ratio of the difference between the peak and mean temperature in the outlet plane, to

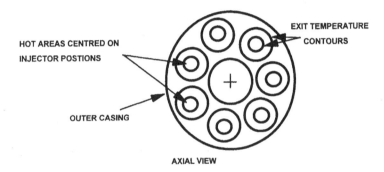

(a)   Circumferential temperature distribution – OTDF (overall temperature distribution factor)

Note:
OTDF is outlet peak temperature minus outlet mean temperature divided by mean combustor temperature rise.

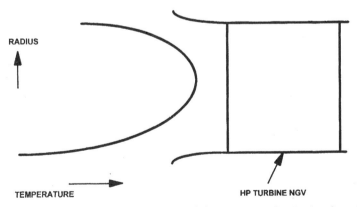

(b)   Radial temperature distribution – RTDF (radial temperature distribution factor)

Note:
RTDF is circumferentially meaned outlet peak temperature minus outlet mean temperature divided by mean combustor temperature rise.

Fig. 5.25   Combustor exit temperature profile – OTDF and RTDF.

the combustor mean temperature rise. It cannot be predicted and hence must be measured on a rig or engine. A rig may utilise traverse gear, in an engine thermal paint is applied to the turbine nozzle guide vanes. Early quantification is essential in a development programme as the peak temperatures strongly affect turbine nozzle guide vane life. OTDF should be controlled to less than 50% and ideally less than 20%.

The *RTDF* (Radial Temperature Distribution Factor), also defined by Formula F5.7.7, is analogous to OTDF but uses circumferentially meaned values. This parameter determines turbine rotor blade life since due to their rotation they experience the circumferential average of the temperatures in any given radial plane. RTDF should be controlled to less than 20%.

5.7.7   *Applying basic efficiency and sizing guidelines*

Sample calculation C5.3 illustrates the application of the basic sizing guidelines presented herein.

5.7.8   *Dry low emissions combustion systems for industrial engines*

Low emissions of NO_x, CO and unburned hydrocarbons have become essential for combustion systems. This is particularly true for industrial engines where *dry low emissions*, or *DLE*, has become mandatory for many applications. The term dry relates to the fact that no water or steam is injected into the combustor to lower flame temperature and hence NO_x. For these land based engines legislation is requiring emissions of NO_x and CO to be simultaneously less than between 42 and 10 vppm depending on geographical location (volume parts per million), over a wide operating range. Conventional combustors produce around 250 vppm.

Virtually all design solutions for industrial engines pre-mix the fuel and air outside the combustor, and then burn the homogeneous mixture inside it. This is essential since local high or low temperature regions inside the combustor will produce large amounts of NO_x or CO respectively. This is far more challenging for diesel fuel than natural gas due to the lower *auto-ignition delay time* at a *temperature and pressure*. Chart 5.6 shows the resulting levels of NO_x and CO versus temperature resulting from combustion of the homogeneous mixture. To achieve low emissions of both NO_x and CO simultaneously at base load the primary zone must burn weak at around 1850 K, much less than the conventional combustor temperature of around 2300 K. This then gives a fundamental problem as the engine is throttled back because weak extinction occurs at around 1650 K, giving negligible operating range. This is discussed further in section 5.8 where practical design solutions such as variable geometry, series and parallel fuel staged systems are described. The latter two systems involve additional fuel injection points, switched depending on power level. Their only impact on overall engine performance at both design point and off design is a small increase in the combustor entry diffuser or/and wall pressure loss, though they add significant control complexity.

It is beyond the scope of this book to fully describe DLE design solutions, however References 16–19 give a good introduction. Crudely parallel staged systems will have the same length as a conventional combustor, but an increase in area is desirable. For series staged systems the area required is approximately the same as for a conventional combustor, but twice the length is necessary. In some instances, such as that described in Reference 18, this has led to a conventional annular combustor being replaced by a number of radial pots to achieve the required length while retaining the original distance between the compressor and turbine.

5.8   Combustors – off design performance

5.8.1   *Efficiency and temperature rise*

Chart 5.5 may be used to determine efficiency for engine off design performance models, with a chosen curve digitised so that the model can interpolate along it using loading evaluated from the known inlet conditions and combustor volume. Formula F5.7.8 presents a polynomial fit for the unconstrained design which is able to meet all the key design guidelines (section 5.7). In fact, fuel air ratio is a third dimension to Chart 5.5 but its effect is small and depends on the combustor design; no generic chart can be prepared and it may be ignored for early models.

Again the combustor temperature relationships described in Chapter 3 are applicable to all off design conditions.

5.8.2 Pressure loss

Cold and hot pressure loss may be derived from Formulae F5.7.9 and F5.7.10. The constants may be derived at the design point where percentage pressure loss as well as inlet and outlet parameters are known.

5.8.3 Combustor stability

If fuel is injected correctly into a well designed combustor then stability is primarily a function of velocity, absolute pressure and temperature. A low velocity aids flame stability, while high inlet pressure and temperature promote combustion by creating a closer density of air and fuel molecules or higher molecular activity. These three variables are all included in the loading parameter (velocity indirectly). For stability correlations loading is calculated using only the primary zone air flow and volume, as this is where combustion begins. In fact this will not be significantly different from that calculated using the total can volume and mass flow. Equivalence ratio is derived using the total fuel flow and primary zone air flow.

Chart 5.7 shows a generic *combustor stability loop* of primary zone equivalence ratio versus loading. There is a loading value of around $1000\,kg/s\,atm^{1.8}\,m^3$ beyond which combustion is not practical, this is primarily driven by velocity. As loading is reduced the flammable equivalence ratio band increases. Rich and weak extinction fuel air ratios may also be plotted versus primary zone exit velocity, as opposed to loading, so there are then families of curves for absolute pressure and temperature.

The fraction of combustor entry air entering the primary zone is constant for off design operation, hence primary zone fuel air ratio may be derived from knowing total combustor inlet mass flow and fuel flow. Rich extinction is rarely encountered in an engine as over-temperature of other components would normally precede it. However weak extinction is a threat, and since the exact curve is highly dependent upon the individual combustor design it must be determined by rig test. The levels in Chart 5.7 are a reasonable first indication, however.

A further instability called *rumble* can occur at weak mixtures. It is characterised by a 300–700 Hz noise generated by the combustion process.

5.8.4 Weak extinction versus ambient conditions and flight Mach number

As shown on Chart 5.7, for industrial, marine and automotive engines loading only increases marginally as the engine is throttled back to idle. Primary zone equivalence ratio typically falls from 1 to around 0.4 at idle, and additionally around 30 to 50% underfuelling relative to steady state occurs during a decel. Chart 5.7 shows that weak extinction is around 0.25 equivalence ratio, hence even at idle the permissible underfuelling would be around 40%. The decel schedule is set to prevent weak extinction, which is then not usually a threat given a well designed system and anyway such a broad permissible band.

For aircraft engines high altitudes provide a more severe off design condition for weak extinction. The typical variation in loading and fuel to air ratio for a turbofan at key operating conditions are also illustrated on Chart 5.7. The worse case is usually a decel to just above idle at the highest altitude and lowest Mach number, however depending upon the idle scheduling this worst case can occur at an intermediate altitude. In contrast to an industrial engine, loading does increase significantly and hence great care must be taken to ensure that the stability loop is satisfactory throughout the operational envelope.

5.8.5   *Starting and restarting – ignition, light around and relight*

Chapter 9 describes the phases of starting and restarting. After dry cranking, fuel must be metered to the combustor and then ignited. Usually igniters are located in two positions, once ignition has been achieved the rest of the burners, or cans, must *light around*. A typical ignition loop is shown on Chart 5.7, again for an individual combustor it must be determined by rig test. Light off occurs with primary zone equivalence ratios in the range 0.35–0.75, depending partly on the loading, and immediate combustion efficiency is around 60–80%.

As described in Chapter 9 for aircraft engines the capability to *relight* within the *restart* envelope is essential. This is a particular challenge at high altitude and low flight Mach number where loading is high due to the low inlet pressure and temperature. As stated in section 5.7, designing the combustor to have a loading of less than $300 \, \text{kg/s atm}^{1.8} \, \text{m}^3$ when windmilling at this flight condition is essential. Also it is vital to measure altitude relight performance in the rig test programme at the earliest opportunity.

5.8.6   *The combustor rig test*

Figure 5.26 shows a typical combustor rig. Air enters through a venturi measuring section as described in Chapter 11. It is compressed, and if necessary heated to provide the inlet pressure and temperature per the engine condition being tested. It then passes into the combustor test section where the fuel is burned, and leaves via a diffuser and throttle valve. For cannular systems a single can may be tested reducing the size of the rig facility. For a new design of combustor a rig test is mandatory prior to any development engine testing, and as a minimum must establish and develop:

- Combustion efficiency versus loading and fuel air ratio
- Combustor cold loss pressure coefficient by flowing the rig without fuel being metered
- Combustor rich and weak extinction boundaries
- Combustor ignition boundaries
- Combustor wall temperatures using thermal paint and/or thermocouples
- OTDF and RTDF using traversing thermocouple rakes or thermal paint
- Emissions levels using a cruciform probe with a good coverage of sampling points at the exhaust

If the rig cannot achieve full engine pressure then it must be set up to the same inlet $W\sqrt{T}/P$ as the engine condition under consideration. However since the absolute pressure, and hence loading are different then care must be taken in interpreting results. Quartz viewing windows are of tremendous value. Also cold tests using water and air in perspex models of the combustor are an invaluable tool in deriving satisfactory aerodynamics.

5.8.7   *Industrial dry low emissions systems*

Section 5.7.7 introduced industrial engine DLE systems and described the increased likelihood of weak extinction at part power due to the primary zone being operated premixed and lean at base

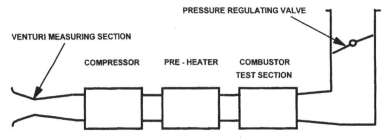

Fig. 5.26   Combustor rig test facility.

load. To overcome this either *variable geometry* must be employed, or the fuel must be *staged*. In addition a conventional fuel injector is required for starting and low power operation.

In variable geometry systems the amount of air entering the primary zone is reduced as the engine is throttled back, retaining a temperature of around 1850 K. The remaining air is spilled to the secondary zone.

In *parallel fuel staged systems* there are a large number of burners in the primary zone. As the engine is throttled back some are switched off retaining a burn temperature of around 1850 K local to those that are still operative. In *series fuel staged systems* the primary zone is fuelled to around 1850 K, and the secondary zone fuelled a little lower at base load. As the engine is throttled back fuel is metered to the primary zone to maintain 1850–1900 K allowing a safe margin versus weak extinction, and the remaining fuel is spilled to the secondary zone. The secondary zone can be operated to significantly lower exit temperatures at part power without weak extinction due to the heat of the primary zone upstream of into it. Figure 5.27 illustrates these part power temperature profiles. Another method is to employ a *diffusion flame* (i.e. conventional rich burning) pilot burner to provide stability and a premixed main burner, with a variable fuel split between them.

It is clear from this commentary that DLE combustion systems introduce another dimension to off design and transient engine performance, as well as control system design.

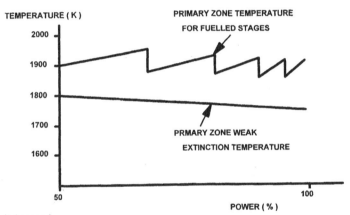

(a)   Parallel staged

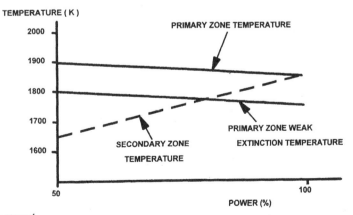

(b)   Series staged

Note:
Primary zone weak extinction temperature rises marginally at part power due to reduced T31 and P31.

Fig. 5.27   Dry low emissions combustion – part power temperature trends.

5.9   Axial flow turbines – design point performance and basic sizing guidelines

A turbine extracts power from the gas stream to drive either engine compressors or, in the case of a power turbine, a load such as a propeller or electrical generator. References 1, 4 and 20 comprehensively describe axial turbine design. Sections 5.11.6 and 5.11.7 describe why an axial or radial turbine is best suited to individual applications. Section 5.15 describes turbine blade and disc cooling.

5.9.1   *Configuration and velocity triangles*

Figure 5.28 presents the configuration of a single stage axial turbine. The stage comprises a row of *nozzle guide vanes* (*NGVs*) followed by a row of *rotor blades* mounted on a *disc*. *Shrouded* blades have reduced clearance losses and are often interlocked, providing mechanical damping. However, the shroud creates increased stress levels. For a multi-stage turbine the blading is arranged sequentially in an annulus with the discs connected via conical drive features forming the *drum*.

Figure 5.29 shows the pitch line NGV and blade aerofoils together with inlet and outlet velocity triangles, the variation of key thermodynamic parameters through the stage is also annotated. High temperature and pressure gas usually enters the first stage NGVs axially at less than 0.2 Mach number and is then accelerated by turning it, which reduces flow area. The mean NGV exit Mach number may be between 0.75 to supersonic. There is no work or heat transfer, and only a small loss in total pressure due to friction and turbulent losses. Total temperature remains unchanged, except by addition of any cooling air, while static pressure and temperature reduce due to the acceleration.

Power is extracted across the rotor via the change in whirl velocity; as for a compressor the Euler work is this times the blade velocity. Total temperature and total pressure are reduced.

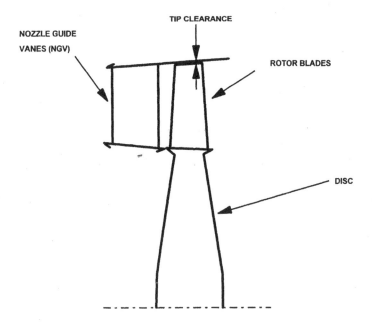

Note:
Figure 5.29 shows blading details.

Fig. 5.28   Axial turbine configuration.

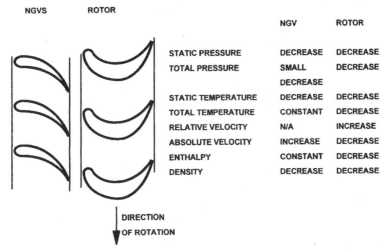

(a) Pitchline blading

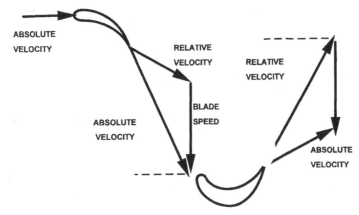

(b) Velocity triangles for design operating point

Notes:
Rotor relative inlet and outlet gas angles are close to blade angles.
Stator absolute inlet and outlet gas angles are close to vane angles.

Fig. 5.29 Axial turbine blading and velocity triangles.

Relative velocity increases, and relative total temperature remains constant. Power may be calculated via Formulae F5.9.1 and 5.9.2 which are similar to those for a compressor.

5.9.2 *Scaling an existing turbine*

All of the comments in section 5.1.2 regarding linearly scaling a compressor are equally applicable to a turbine. In addition exit swirl angle is unchanged.

5.9.3 *Efficiency (Formulae F5.9.3 and F5.9.4)*

As defined by Formulae F5.9.3 and F5.9.4, *isentropic efficiency* is the actual specific work output, or total temperature drop, for a given expansion ratio divided by the ideal.

As for a compressor *polytropic efficiency* is defined as the isentropic efficiency of an infinitesimally small step in the expansion process, such that it is constant throughout. As described in Reference 1, it accounts for the fact that the inlet temperature to the back stages of a multistage turbine is lower, and hence less work output is achieved for the same pressure drop. Though polytropic efficiency is not used directly in design point calculations it is important in that it enables comparison of turbines of different expansion ratio on an 'apples with apples' basis. Those of the same technology level, with similar geometric design freedom with respect to frontal area and expansion ratio required per stage, will have the same polytropic efficiency regardless of overall expansion ratio. Formulae and charts for conversions between these two efficiency types are provided in Chapter 3, Formula F3.44 and Chart 3.17.

Chart 5.8 based upon that in Reference 21 is commonly referred to as a Swindell or Smith chart. It shows contours of constant isentropic efficiency versus loading (Formula F5.9.5) and axial velocity ratio (Formula F5.9.6). As well as being an excellent comparator for different design options the chart may be used to give first-order judgement on the efficiency attainable for a given design. The following should be noted.

- The chart provided is for the highest technology level in terms of 3D orthogonal aerodynamic design, large blading such as for a big engine LP or power turbine (capacity greater than 10 kg K/s kPa), no cooling air affecting gas path aerodynamics, no windage, 50% reaction zero tip clearance and no other geometric compromises.
- In a practical design which has all the above merits the highest efficiency level attainable would be 95%.
- At the other extreme for low technology blading around three points should be debited from the values from Chart 5.8.
- For low capacity (around 0.1 kg K/s kPa) then levels from Chart 5.8 should be further debited by approximately three percentage points, with the loss increasing more rapidly at the bottom end of the size range.
- Values between the above two datum levels will be attained for intermediate technology levels, or where some of the other key design parameters described later cannot be set at their optimum level due to geometric or mechanical constraints.
- Cooling air also lowers the attainable efficiency levels. To a first order, for each percent of rotor blade cooling air the values from Chart 5.8 should be debited as below. These values are based on the performance model, assuming that the cooling air does no work in the blade row (section 5.15).

1.5% per 1% of suction surface film cooling
0.5% per 1% of rotor shroud cooling by upstream injection
0.5% per 1% of trailing edge cooling
0.25% per 1% of leading edge or pressure surface cooling

- Where applicable the exchange rates are approximately half of the above for NGVs.
- Non zero tip clearance is usually inevitable, and lowers efficiency levels as discussed in section 5.10.8.

5.9.4 *Guide to basic sizing parameters*

Inlet Mach number
To minimise pressure losses in upstream ducting and to ensure that the gas will accelerate at all points along the NGV surface the mean inlet Mach number to the first stage should ideally be less than 0.2. It may be higher for subsequent stages.

Blade inlet hub relative Mach number
This should be less than 0.7 to ensure that there is acceleration relative to the blade all the way through the blade passage. Should diffusion occur then it may lead to separation and increased pressure loss. NGV exit angle will be between 65° and 73°.

Rotational speed

This must be set to maintain rim speed, tip speed and AN^2 within the limits acceptable for mechanical integrity, while optimising efficiency via the stage loading and axial velocity ratio. It must also be a suitable compromise with the driven equipment speed requirements.

Stage loading (Formula F5.9.5), expansion ratio and number of stages

As for the axial flow compressor, stage loading is a non-dimensional parameter which is a measure of the difficulty of the duty of the stage. For most engines a pitch line value of 1.3–2 is typical with the higher values being on the front stages. These result in expansion ratios per stage of between 2:1 and 3:1. The highest expansion ratio practical from a single stage with any acceptable level of efficiency is 4.5:1, this pushes the hade angle guidelines to the limit. The number of stages is a compromise between achieving low loadings and good efficiency, or high loadings and low cost and weight. Small and expendable RPV engines will have the highest loadings.

Axial velocity ratio (Formula F5.9.6)

This is the ratio of the axial velocity to the blade speed, also known as *flow coefficient* or *Va/U*. Axial velocity at any point in the annulus may be evaluated using Q curves knowing the area, mass flow, total temperature and pressure. It may be assumed to be constant across the annulus. For a given stage loading the corresponding pitch line axial velocity ratio for optimum efficiency is apparent from the correlation presented in Chart 5.8. However if frontal area is paramount then a larger value may be chosen.

Hade angle

This is the angle of the inner or outer annulus wall to the axial. These angles are normally kept to less than $15°$ to avoid flow separation.

Hub tip ratio

This should be greater than 0.5 to minimise secondary losses, but less than 0.85 due to the increased impact of tip clearance as the blade height is reduced. These values are also commensurate with realistic stress levels.

Aspect ratio (Formula F5.1.8)

Aspect ratio, as defined for an axial flow compressor based upon axial chord should ideally be between 2.5 and 3.5, however it may be as high as 6 for LP turbines.

Axial gap

To avoid blade vibration difficulties this should be approximately 0.25 times the upstream axial chord.

Reaction (Formula F5.9.7)

This is the ratio of the static pressure or static temperature drop across the rotor to that across the total stage. For best efficiency pitch line reaction should be around 0.5, however for cases where blade temperature is borderline with respect to creep or oxidation then it may go as low as 0.3. This will increase the NGV exit and blade inlet relative velocities, reducing the static temperature and hence also the blade metal temperature. It will also reduce the rearwards axial thrust load which the bearing must react. Hub reaction should ideally always be greater than 0.2.

AN^2

This is the product of the annulus area mid-way along the rotor blade, and the blade rotational speed squared. As shown in Reference 20, blade stress is proportional to AN^2. It is a key mechanical parameter with respect to blade creep life for HP stages and disc stress for LP stages. The allowable AN^2 with respect to creep life must be derived from material creep curves where stress is plotted against life for lines of constant metal temperature. It may be necessary for a value as low as $20E06\,rpm^2\,m^2$ for a low technology, uncooled small industrial HP turbine, but conversely due to lower temperatures may be allowed to rise to up to $50E06\,rpm^2\,m^2$ for the last stage of a high technology heavyweight powergen engine. The allowable AN^2 for disc stress depends also on rim speed, discussed below.

Rim speed
For disc stress, rim speed must be limited to around 400 m/s for HP turbines. For the last stage of an LP turbine, designed using the upper limit to AN^2 of $50E06\,rpm^2\,m^2$ the rim speed must be limited to around 350 m/s.

Final stage exit Mach number
The final stage exit Mach number should be around 0.3. The highest allowable is 0.55, above which dramatic breakdown in flow may occur in the downstream diffusing duct such as an exhaust, jet pipe or inter-turbine duct. A new design should always be in the lower portion of this band as the engine will almost certainly require some further uprate which will bring with it higher flow and hence exit Mach number.

Final stage turbine exit swirl angle
This should be less than $20°$ and ideally $5°$ on the pitch line to minimise downstream duct pressure loss as described in section 5.13.

5.9.5   *Applying basic efficiency and sizing guidelines*

Sample calculation C5.4 illustrates the application of the basic efficiency and sizing guidelines presented herein.

5.10   Axial flow turbines – off design performance

5.10.1   *The turbine map*

Once the turbine geometry has been fixed at the design point then the *turbine map* may be generated to define its performance under all off design conditions. The most common form of map, sometimes called the *characteristic* or *chic*, is presented in Fig. 5.30. Capacity (referred flow), efficiency and exit swirl angle are plotted for lines of constant referred speed versus the *work parameter* (dH/T or CP.dT/T). For each referred speed line there is a maximum flow capacity which cannot be exceeded no matter how much CP.dT/T is increased. This operating regime is termed *choke*. For the map shown in Fig. 5.30 the choking capacity is the same for all referred speed lines. This is usually the case when choking occurs in the NGV, should it occur in the rotor blades then these lines separate out with choking capacity reducing marginally as referred speed is increased due to decreased density in the rotor throat. *Limiting output* or *limit load* is the point on the characteristic beyond which no additional power results from an increased expansion ratio. Here the shock wave moves from the rotor throat to its trailing edge, hence its aerodynamics are not affected by downstream pressure.

Ignoring second-order phenomena such as Reynolds number effects, and for a fixed inlet flow angle the following applies.

- For a fixed turbine geometry the map is unique.
- The operating point on the turbine map is dictated by the components surrounding it as opposed to the turbine itself.
- Each operating point on the map has a unique velocity triangle, expressed as Mach number.
- Expansion ratio, CPdT/T and efficiency are related by Formulae F5.9.3 and F5.9.4, hence in fact any two of the three parameters may be used as the ordinates for the map.

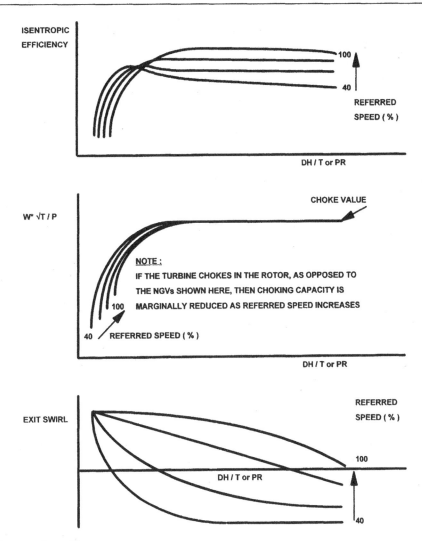

Fig. 5.30 The turbine map.

The aerodynamic design methods to produce a map for given turbine geometry are complex and involve the use of large computer codes. References 21 and 23 describe the methodology.

5.10.2 *Impact on the map of linearly scaling a turbine design*

Sections 5.1.2 and 5.2.2 discuss the impact on a compressor map of linearly scaling the compressor hardware. The same rules apply to a turbine map, i.e. if Fig. 5.30 is plotted in terms of the scaling parameters presented in Chapter 4, then to a first order it is unchanged when the design is linearly scaled.

If scaling 'down' results in a small turbine then it may not be possible to scale all dimensions exactly, such as tip clearance or trailing edge thickness leading to a further loss in capacity pressure ratio and efficiency at a speed. In addition Reynolds number effects must be considered, as described below.

5.10.3   *Reynolds number and inlet temperature effects*

As for the compressor map shown in Fig. 5.6 Reynolds number is strictly also a fourth dimension for a turbine map. Capacity and efficiency are both marginally reduced at a referred speed and CPdT/T. However due to the high pressures and temperatures in a turbine, Reynolds number rarely falls below the critical value to then have an effect. Formulae F5.10.1–F5.10.3 show corrections to the map for Reynolds number effects.

As for a compressor, changes in turbine geometry due to changes in absolute temperature have only a tertiary effect and are usually ignored.

5.10.4   *Change in the working fluid*

When the map is plotted in terms of dimensionless parameters, as shown for a compressor in Fig. 5.7, then to a first order, and for a fixed inlet flow angle, it is unique for all linear scales and working fluids. The turbine map will normally be generated in terms of referred parameters as per Fig. 5.30 using gas properties for dry air. In reality these properties will be modified by the presence of combustion products, and possibly by humidity or water or steam injection. Chapter 3 describes how new gas properties may be derived.

Most engine off design performance models use a map for dry air and deal with any change in gas properties as described for a compressor in Fig. 5.7 and Chapter 12.

5.10.5   *Loading turbine maps into engine off design performance models*

Figure 5.31 shows how the turbine map is digitised, and then arranged in three tables which are loaded into the engine off design performance model. The use of this map in such models is described in Chapter 7. Maps for engine starting models utilise alternative variables, as described in section 5.10.7, to assist in model convergence.

5.10.6   *Effect of inlet flow angle – variable area NGVs*

As stated in section 5.10.1, the turbine map is only unique for a fixed value of inlet flow angle. Changes will cause a second-order reduction of capacity and efficiency at a referred speed. This is in marked contrast with the compressor where the presence of inlet swirl is very powerful. This is because for a turbine the first blade row is the NGV which has a rounded leading edge tolerant to incidence variation and the throat is at the trailing edge as opposed to the leading edge. Also the flow is accelerating within the NGV passage and will quickly *reattach* if there is any separation such that the NGV exit flow angle is unchanged.

Variable area NGVs (VANs) are occasionally employed on LP or power turbines for recuperated cycles to maintain high turbine gas path temperatures, and hence heat recovery, at part power. The operating mechanism to pivot the NGVs is expensive and complex being in a far higher temperature environment than compressor VIGVs or VSVs. They are not practical for HP turbines due to the extreme temperatures and extensive cooling requirements. Each NGV angle represents a unique geometry and hence has its own turbine map. Hence a suite of turbine maps as per Fig. 5.30 must be loaded into an off design performance program, one for each VAN angle. The use of such maps and the control system scheduling of VAN angle is discussed in Chapter 7.

5.10.7   *Peculiarities of the low speed region of the map*

At low speed during starting or windmilling the turbine will not normally show abnormal modes of operation, such as the 'paddle' phenomena described for compressors. Generally it always acts as turbine, apart from at zero speed where it behaves as a cascade with a pressure drop but no change in total temperature.

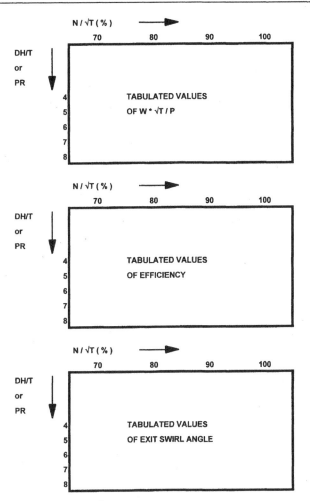

Fig. 5.31 Turbine map representation.

Near zero work the capacity slope becomes very steep on a conventional map, and the definition of efficiency becomes tenuous. To overcome these difficulties alternative parameter groups are used for loading maps into starting and windmilling models. The groups $N/\sqrt{T}$ and $CP.DT/N^2$ are used to read the map, with $W.T/N.P$ and $E.CP.DT/N^2$ returned from it. To produce the revised map the existing version is easily translated to this form, as the groups are simple combinations of the existing ones. It is then plotted and extrapolated to low speed and low work, knowing that zero speed must coincide with zero work.

5.10.8 *Effect of changing tip clearance*

Tip clearance is the radial gap between the rotor blades and casing. Its ratio to blade height must be set in the range of 1–2% depending upon layout design and size. This is larger than for axial flow compressors since the transient thermal growths are greater for a turbine. A 1% reduction in rms tip clearance (Formula F5.2.4) will reduce efficiency by around 1%. This amount reflects shrouded blades which have *tip fences* to extract work from any overtip leakage gas. Shroudless blades have a simple gap and hence the effect will be larger. It may be reduced by using *squealers*, where a thin portion of the blade stands proud and is abraded during engine running in to produce the lowest achievable clearance.

5.10.9   *Applying factors and deltas to a map*

Often during the engine concept design phase a turbine map may be required for predicting off design performance, but it will not yet have been generated by the turbine design codes. As for compressors, common practice is to apply factors and deltas to a map from a similar turbine design, as described by Formula F5.10.4, to align its design point to that required. This should not be confused with linearly scaling a turbine, only the map shape is being used to enable early engine off design performance modelling. A new turbine aerodynamic design is still required.

5.10.10   *The turbine rig test*

Turbine rig tests, prior to engine testing, are only carried out for the highest technology engines. This is because of the cost and complexity of the rig to deliver representative inlet conditions, which requires a large heater and compressor with independent control. The turbine output power is absorbed by a water brake or dynamometer, hence referred speed may be held constant and an outlet throttle valve varied to map the speed line.

5.11   Radial turbines – design

In the *radial turbine* flow is changed from a radially inwards to axial direction. This allows a far greater area ratio and hence expansion ratio than may be achieved by only changing gas angles and the annulus lines for an axial flow stage.

References 1, 4 and 24 provide further details of radial turbine design. References 25–27 provide details for actual designs.

5.11.1   *Configuration and velocity triangles*

Figure 5.32 presents a typical blading configuration for a radial turbine. The stage comprises a ring of nozzle guide vanes (NGVs), followed by a bladed disc called the *wheel*. In contrast to an axial flow turbine the flow enters the NGVs in a mostly radial direction. The turbine entry duct geometry employed to achieve this primarily depends upon combustor type. For instance if an annular combustor is employed then the annular turbine entry duct must turn from axial to radial shortly upstream of the NGVs. Often radial turbines are employed in small industrial engines where a single can combustor is utilised, requiring a scroll and hence some tangential velocity is present at NGV entry. Radial turbines for automotive turbochargers often omit the NGVs and generate tangential velocity at rotor inlet via the effect of the scroll alone.

Figure 5.32 also shows the inlet and outlet velocity triangles, the manner in which key thermodynamic parameters change through the NGV and rotor blades is as per Fig. 5.29 for an axial turbine. Formulae F5.11.1 gives the Euler work, and F5.9.2 applies equally to radial turbines.

The gas is accelerated through the NGVs by both the reduction in area due to the lower exit radius, and by turning the flow from radial to between 65° and 80° to it. The mean exit Mach number may be between 0.6 and supersonic, the latter applying to very high expansion ratio designs. There is no work or heat transfer, and only a small loss in pressure due to friction and turbulent losses. Total temperature remains unchanged, while static pressure and temperature reduce due to the acceleration.

Work is extracted across the rotor via a change in swirl velocity, which produces torque. Achieving these velocities requires a drop in total pressure and produces a drop in total temperature. As well as expanding the gas the rotor turns the flow from radial to axial at exit.

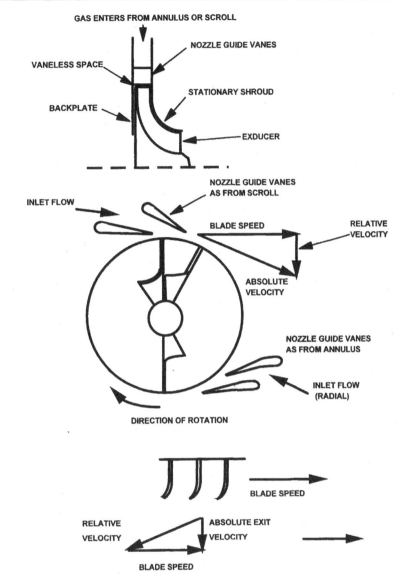

GAS ENTERS FROM ANNULUS OR SCROLL

NOZZLE GUIDE VANES

VANELESS SPACE

STATIONARY SHROUD

BACKPLATE

EXDUCER

NOZZLE GUIDE VANES
AS FROM SCROLL

INLET FLOW

BLADE SPEED

RELATIVE
VELOCITY

ABSOLUTE
VELOCITY

NOZZLE GUIDE VANES
AS FROM ANNULUS

INLET FLOW
(RADIAL)

DIRECTION OF ROTATION

(a)

BLADE SPEED

RELATIVE
VELOCITY

ABSOLUTE EXIT
VELOCITY

BLADE SPEED

(b)   View radically inwards onto exducer

Fig. 5.32   Radial turbine configuration and velocity triangles.

5.11.2   *Scaling an existing design*

The comments in section 5.9.2 for an axial flow turbine are all equally applicable to a radial flow turbine.

5.11.3   *Efficiency*

The definitions of isentropic and polytropic efficiencies defined in section 5.9.3, and via Formulae F5.9.3–F5.9.4 are equally applicable to a radial turbine. As for a centrifugal compressor, and in contrast to an axial flow turbine, efficiency is correlated versus the parameter *specific speed* described in section 5.3. The most common definition is presented in Formula F5.11.2. Both *total to total* and *total to static* efficiencies are considered, the latter using the exit static pressure on the basis that the exit dynamic head is lost. Though total to total efficiency is appropriate for cycle calculations, high exit Mach number will inevitably increase downstream pressure losses hence total to static efficiency is a fair comparitor of turbine designs.

Chart 5.9 presents total to total isentropic efficiency for given NGV exit angles versus specific speed. This chart is for a high technology level, large size (0.5 kg $\sqrt{\text{K}}$/s kPa) and other scantlings designed to the guidelines provided in section 5.11.4. For the smallest size (0.05 kg $\sqrt{\text{K}}$/s kPa), designed without 3D aerodynamic codes, up to 3% points must be deducted from the levels shown. Chart 5.9 may be used to estimate radial turbine efficiency for design point calculations.

The optimum specific speed for efficiency is around 0.6. Hence once the mass flow rate and expansion ratio required for a given design are set, the exit volumetric flow rate and enthalpy change may be calculated, and also the rotational speed required to achieve this optimum specific speed for efficiency may be derived. As a design progresses other constraints may cause the rotational speed to be changed, moving specific speed away from the optimum with a consequent loss of efficiency.

5.11.4   *Guide to basic sizing parameters*

Guidelines for key parameters for designing the scantlings of a radial turbine are presented below. Many of the parameters are common to other turbomachinery and hence their definitions are as presented earlier.

Inlet Mach number
To minimise pressure losses in upstream ducting, and to ensure that the gas will accelerate at all points along the NGV surface, this should ideally be less than 0.2.

Rotational speed
This must be set to maintain wheel rim speed within the limits acceptable for mechanical integrity, while optimising efficiency via specific speed. It must also be a suitable compromise with the driven equipment speed requirements.

Specific speed
As for centrifugal compressors specific speed is a non-dimensional parameter against which efficiency can be correlated. Chart 5.9 shows the optimum specific speed for turbine efficiency. Figure 5.33 shows typical geometries resulting from the guidelines presented at low and high specific speeds.

Expansion ratio, number of stages
The highest expansion ratio practical from a single stage with any acceptable level of efficiency is around 8 : 1. Two radial turbines in series are rarely considered seriously due to the complexity of the inter-turbine duct and because in small engines where they are most common there is rarely sufficient expansion ratio. One common layout is a single stage radial turbine driving a high pressure ratio gas generator compressor, followed by an axial free power turbine driving the load.

Wheel inlet tip speed and diameter
Wheel inlet tip speed is calculated from Formula F5.11.3. Hence tip diameter may be derived once rotational speeed has been set.

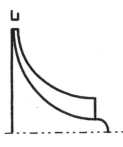   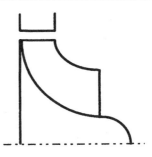

(a)   Specific speed = 0.25   (b)   Specific speed = 1.2

Note:
Values of specific speed shown are dimensionless.

Fig. 5.33   Effect of specific speed on radial turbine geometry.

NGV height
Chart 5.10 shows the optimum value of the ratio of NGV height to rotor inlet diameter for efficiency, versus specific speed. It increases with specific speed reflecting the higher volumetric flow rate, and should always be greater than 0.04 to avoid excessive frictional losses.

Rotor exit tip diameter
Chart 5.10 also shows the optimum ratio of rotor exit to inlet tip diameters for efficiency versus specific speed. It increases with specific speed reflecting the increasing ratio of specific work to volumetric flow rate. It must be less than 0.7 to avoid unfavourable velocity ratios.

Rotor exit hub tip ratio and length
The ratio must be less than 0.4 to minimise the impact of tip clearance. Rotor length may be evaluated using the impeller length parameter (F5.3.7). It should be in the range 1.0–1.3 for radial turbine rotors.

Vaneless space radius ratio
This should be of the order of 1.10 to avoid blade excitation.

NGV radius ratio and exit angle
The ratio of the NGV outer to inner radii will be between 1.35 and 1.45. The optimum NGV exit angle for efficiency may be taken from Chart 5.9.

Wheel rim speed
Formula F5.11.3 enables the blade tip speed for given duty to be calculated. For mechanical integrity the wheel rim speed should be less than 600 m/s. However the velocity may rise to 800 m/s at the blade tip if the wheel back plate is 'scalloped', i.e. it is cut away between blades.

Final stage exit Mach number
For a good design this should be around 0.3. The highest allowable value is 0.55, above which dramatic breakdown in flow may occur in the downstream diffusing duct. As for axial flow turbines new designs should be at the lower end of this range to provide future uprate capability.

Final stage turbine exit swirl angle

This should be less than 20° (and ideally 5°) on the pitch line to minimise downstream duct pressure loss as described in section 5.13.

5.11.5   *Applying basic efficiency and sizing guidelines*

Sample calculation C5.5 illustrates the application of the basic efficiency and sizing guidelines presented herein.

5.11.6   *Radial flow turbines versus axial flow turbines*

Radial and axial flow turbines are compared here in a qualitative fashion. Section 5.11.7 shows the capacity and expansion ratio ranges for the major applications to which each is suited.
  An axial flow turbine has the following advantages.

- It can be designed for a very large range of loadings, around 1–2.2, with a large variation in size, speed and efficiency depending on the requirements.
- For a highly loaded design it has a lower frontal area for a given mass flow and pressure ratio.
- For a highly loaded design weight is lower.
- For capacities greater than around 0.05 kg $\sqrt{\text{K}}$/s kPa the axial flow turbine will have a better isentropic efficiency, this advantage increases with capacity.
- If the required expansion ratio is such that more than one radial turbine is required then the inter-turbine duct is complex. This leads to multi-stage radial turbines rarely being considered. There are no such problems with axial flow turbines.
- Manufacturing (forging) difficulties may limit the viable diameter of the radial turbine wheel to around 0.6 m, and hence impact capacity and expansion ratio capability.

  The radial turbine has the following advantages.

- Radial turbines are capable of up to 8:1 expansion ratio in a single stage. For an axial flow turbine this will require at least two stages.
- Radial turbines are significantly lower in unit cost for the same capacity and expansion ratio.
- At small size, i.e. capacities of less than 0.05 kg $\sqrt{\text{K}}$/s kPa, the isentropic efficiency is better. As with compressors this is because in this capacity range axial flow turbine efficiency drops rapidly as size is reduced due to increasing relative levels of tip clearance, blade leading and trailing edge thicknesses and surface roughness with fixed manufacturing tolerances. However this capacity range corresponds to extremely small gas turbine engines which are comparatively rare.
- It has a shorter length than two axial stages, but similar to one.

In summary, axial flow turbines dominate where low frontal area, low weight and high efficiency are essential and are the only choice at large sizes. Conversely, radial turbines are competitive where unit cost is paramount, and at small size.

5.11.7   *Capacity ranges suited to axial and radial flow turbines*

The table presented below shows capacity ranges suited to axial or radial flow gas generator turbines for the major engine applications. References 28 and 29 enable the reader to examine the turbine configurations for engines in production.

	Aircraft engines (kg $\sqrt{K}$/s kPa)	Industrial, marine and automotive engines (kg $\sqrt{K}$/s kPa)
Predominantly radial	<0.05	<0.15
Radial or axial depending upon requirements	0.05–0.1	0.15–0.5
Predominantly axial	>0.1	>0.5

For thrust engines axial turbines dominate down to very low capacities. The higher cost is warranted due to the importance of low frontal area and weight to minimise drag at high flight speeds. For aircraft shaft power engines radial turbines are competitive to the top of the capacity bands shown because flight speeds are lower. For industrial, marine and automotive engines low frontal area and weight are of lower importance and, to minimise unit cost, radial turbines are competitive to higher capacities.

5.12   Radial turbines – off design performance

All the items in section 5.10 discussing the off design operation of axial flow turbines also apply to radial turbines.

5.13   Ducts – design

The components discussed to date have all involved work or heat transfer. A variety of *ducts* are required which merely pass air between these components, and into or out of the engine. The latter ducts have a more arduous duty for aero thrust engines, as *intakes* must diffuse free stream air from high flight Mach number with minimum total pressure loss, and *propelling nozzles* must accelerate hot exhaust gas to produce thrust. The modelling of intake and nozzles is usually combined with that of their corresponding exit and entry ducts, hence the descriptions of all duct varieties are combined in this section.

Within ducts, *struts* are often required to provide structural support or to allow vital services such as oil flow or cooling air to cross a duct. Duct pressure losses cannot be treated lightly, and for certain engine types such as supersonic aero and recuperated automotive engines, they are critical to the success of the engine project.

Some fundamentals of duct flow are discussed prior to describing duct performance and basic sizing. The importance of Q curves cannot be over-emphasised. References 1 and 30–36 provide further information.

5.13.1   *Subsonic flow in a duct with area change but no work or heat transfer*

The majority of gas turbine ducts have subsonic flow. Figure 5.34 shows schematically the effect of area change on leading parameters for subsonic flow in a duct with no work or heat transfer. Reducing the area accelerates the flow, and reduces static pressure and temperature. Total temperature is unchanged along the duct, however there is a small loss in total pressure due to friction. Conversely, when area is increased then velocity decreases, and static temperature increases. Static pressure will also increase if the area changes gradually to form an effective *diffuser*; otherwise in a sudden expansion the velocity will be *dumped* and dissipated as turbulence. Again total temperature is unchanged and there is a loss in total pressure.

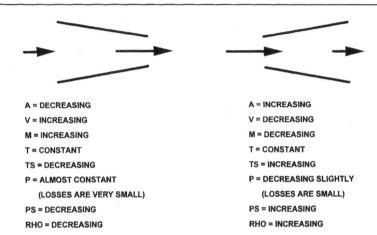

A = DECREASING	A = INCREASING
V = INCREASING	V = DECREASING
M = INCREASING	M = DECREASING
T = CONSTANT	T = CONSTANT
TS = DECREASING	TS = INCREASING
P = ALMOST CONSTANT	P = DECREASING SLIGHTLY
(LOSSES ARE VERY SMALL)	(LOSSES ARE SMALL)
PS = DECREASING	PS = INCREASING
RHO = DECREASING	RHO = INCREASING

Fig. 5.34 Subsonic flow in a duct with no work or heat transfer.

The fact that there is a loss in total pressure means that the use of Q curves is an approximation. If care is taken to use the most appropriate local value of pressure then there is negligible error in utilising them and it is universal practice in the gas turbine industry to do so. Hence *once any Q curve parameter group is known at a point in the duct, then all others may be derived* via the charts, tables or formulae provided in Chapter 3.

5.13.2 *Supersonic flow in a duct with area change but no work or heat transfer*

The only gas turbine engine ducts where flow is supersonic are aero-engine propelling nozzles, and supersonic aircraft engine intakes. Figure 5.35 shows the impact of varying area when the flow is supersonic, which is opposite to that for subsonic flow described above. Reducing area now causes the velocity to reduce as opposed to increase. Again total temperature is unchanged along the duct, there is a small percentage reduction in total pressure and *Q curves can be applied*.

Convergent nozzles
In a convergent nozzle flow accelerates to the *throat* which is the exit plane. If total pressure divided by ambient is less than the choking value derived from Q curves then flow is subsonic

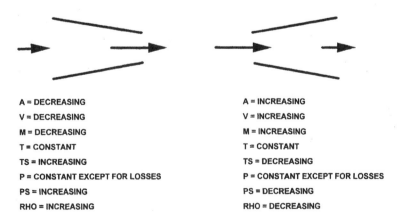

A = DECREASING	A = INCREASING
V = DECREASING	V = INCREASING
M = DECREASING	M = INCREASING
T = CONSTANT	T = CONSTANT
TS = INCREASING	TS = DECREASING
P = CONSTANT EXCEPT FOR LOSSES	P = CONSTANT EXCEPT FOR LOSSES
PS = INCREASING	PS = DECREASING
RHO = INCREASING	RHO = DECREASING

Fig. 5.35 Supersonic flow in a duct with no work or heat transfer.

at the throat. Also in this instance *static pressure in the throat plane is ambient*. However, if the ratio of total pressure to ambient is greater than the choking value derived from Q curves then flow is sonic (Mach number of 1) at the throat. Here the nozzle is choked and static pressure in the throat plane is derived from the total pressure and the choking pressure ratio. It is higher than ambient and there are shock waves downstream of the nozzle.

Guidance on where *convergent–divergent*, or *con–di*, as opposed to just *convergent* propelling nozzles are employed is provided later in this section.

Con–di propelling nozzles

A con–di nozzle initially converges to the throat and then diverges. Figure 5.36 shows total to static pressure ratio and Mach number distributions. At fixed inlet total pressure, four levels of ambient static pressure at exit are applied, reducing from line A to line D. For each line the inlet total temperature and duct geometry is unchanged. Practical con–di nozzles for thrust engines are designed such that *they always run over full (see below), hence only lines C and D are real considerations*. However lines A and B are described to aid understanding.

For line A the flow accelerates as area is reduced to the throat where it is still subsonic. After the throat the flow then decelerates until the exit. Hence the duct has acted as a *venturi*.

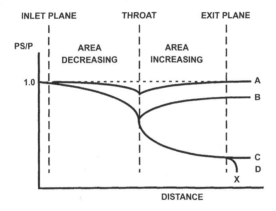

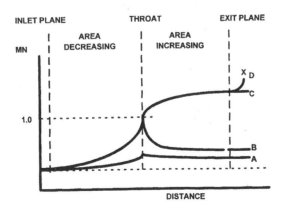

Notes:

X = shock down to subsonic flow and ambient static pressure.

For explanation of lines A, B, C, D see text.

Fig. 5.36 Flow in a convergent–divergent duct at various exit pressure levels.

Total temperature along the duct has remained constant and apart from a small pressure loss so has total pressure. At the exit plane the static pressure is ambient and hence if the small total pressure loss is known, total to static pressure ratio can be calculated. Knowing this together with the exit area and duct total temperature then mass flow *may be calculated from Q curves*.

For line C the exit static pressure is significantly lower. Here the flow accelerates to the throat, *and* through the diverging section. The Mach number is 1 at the throat which is choked, and supersonic at exit. In this case where the static pressure in the exit plane is ambient the con–di nozzle is said to be running *full*. Again *Q curves apply* and mass flow may be calculated from the exit plane total to static pressure ratio, the exit area and duct total temperature. It is apparent from Q curves that, due to the higher total to static pressure ratio at the exit plane, mass flow is significantly higher than for line A.

For line D the exit pressure is lower than for line C. Again *Q curves apply* and in this instance the total to static pressure ratio, and hence other Q curve parameter groups, along the duct are the same as for line C. This includes the exit plane and hence static pressure of the flow is higher than ambient, and the flow shocks down to ambient pressure outside the duct. Mass flow is unchanged from line C, i.e. the con–di nozzle is choked and reducing exit pressure further will not change this situation. In this instance the nozzle is said to be running *over full*.

Line B has an exit pressure between that of A and C such that the flow accelerates to, and after the throat with a Mach number of 1 at it. However, part way along the diverging section the flow shocks from being supersonic to subsonic. The flow accelerates to the shock wave and decelerates after it to the exit. *Q curves apply before and after the shock wave, but not across it.* Across the shock wave the following parameter changes occur.

- Mach number changes from supersonic to subsonic.
- Total temperature is unchanged.
- Total pressure reduces.
- Static temperature increases.
- Static pressure increases.
- Mass flow is unchanged.

Reference 1 describes how to calculate these changes for normal shock waves, and oblique shock wave systems, the latter having a lower total pressure loss.

Con–di intakes

For supersonic aircraft the flow must diffuse from supersonic to low subsonic speeds in the intake. To give any acceptable level of efficiency the intake must be of con–di configuration. Here flow is equivalent to line B of Fig. 5.36, but in reverse. Supersonic flow enters the convergent section and diffuses. The geometry is set such that a series of oblique shock waves occur near the throat, this is more efficient than a normal shock. The flow then continues to diffuse in the divergent section to the compressor face. As for a con–di nozzle *Q curves apply before and after the shock wave, but not across it.* A con–di intake is the only practical gas turbine duct where shock waves occur within the engine.

5.13.3   *Configurations*

For each gas turbine duct type there a large number of potential geometries depending upon the application, and individual design companies' culture and experiences. There are far too many to describe them all here. However to provide a flavour for the geometries encountered, and the aerodynamic and mechanical design challenge involved, Fig. 5.37 presents the most common configuration for each duct type.

That shown for an industrial engine intake is the most common for hot end drive. There is usually a large *plenum* upstream of a *flare*. A *snow hood* is arranged such that air is taken from

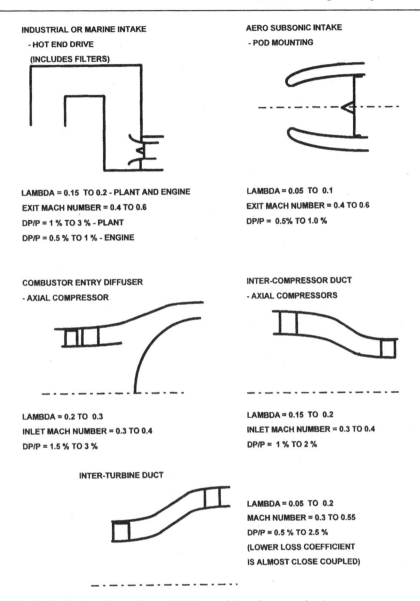

INDUSTRIAL OR MARINE INTAKE
- HOT END DRIVE
(INCLUDES FILTERS)

LAMBDA = 0.15 TO 0.2 - PLANT AND ENGINE
EXIT MACH NUMBER = 0.4 TO 0.6
DP/P = 1 % TO 3 % - PLANT
DP/P = 0.5 % TO 1 % - ENGINE

AERO SUBSONIC INTAKE
- POD MOUNTING

LAMBDA = 0.05 TO 0.1
EXIT MACH NUMBER = 0.4 TO 0.6
DP/P = 0.5% TO 1.0 %

COMBUSTOR ENTRY DIFFUSER
- AXIAL COMPRESSOR

LAMBDA = 0.2 TO 0.3
INLET MACH NUMBER = 0.3 TO 0.4
DP/P = 1.5 % TO 3 %

INTER-COMPRESSOR DUCT
- AXIAL COMPRESSORS

LAMBDA = 0.15 TO 0.2
INLET MACH NUMBER = 0.3 TO 0.4
DP/P = 1 % TO 2 %

INTER-TURBINE DUCT

LAMBDA = 0.05 TO 0.2
MACH NUMBER = 0.3 TO 0.55
DP/P = 0.5 % TO 2.5 %
(LOWER LOSS COEFFICIENT
IS ALMOST CLOSE COUPLED)

Fig. 5.37   Major duct configurations and design point performance levels.

ambient vertically upwards, filters and silencers are located in the vertical *downtake*. If the engine is arranged for cold end drive then a *radial intake* is employed where a *plug* surrounds the output shaft.

Subsonic aero-engines usually employ a pod mounting with the flight intake diffusing from leading edge to the engine intake leading edge. From here there is acceleration along the *nose bullet* as the flow area transitions from circular to annular before the compressor face. At high flight speeds there is also some diffusion upstream of the flight intake with the flow moving from a narrow stream tube to fill the intake front face. Conversely, when stationary flow accelerates to the flight intake leading edge from both behind and in front of the leading edge, hence to avoid flow separation the leading edge must be rounded.

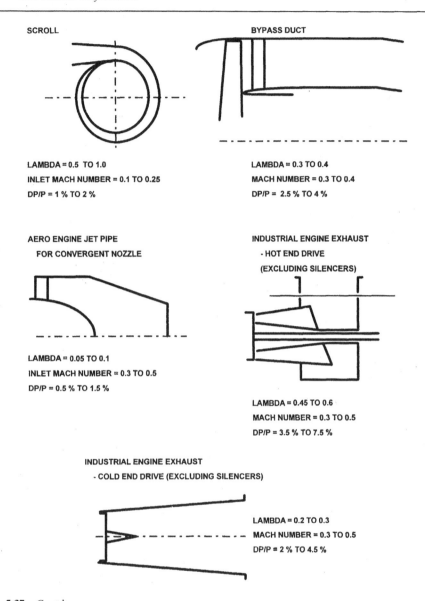

SCROLL

LAMBDA = 0.5 TO 1.0
INLET MACH NUMBER = 0.1 TO 0.25
DP/P = 1 % TO 2 %

BYPASS DUCT

LAMBDA = 0.3 TO 0.4
MACH NUMBER = 0.3 TO 0.4
DP/P = 2.5 % TO 4 %

AERO ENGINE JET PIPE
FOR CONVERGENT NOZZLE

LAMBDA = 0.05 TO 0.1
INLET MACH NUMBER = 0.3 TO 0.5
DP/P = 0.5 % TO 1.5 %

INDUSTRIAL ENGINE EXHAUST
- HOT END DRIVE
(EXCLUDING SILENCERS)

LAMBDA = 0.45 TO 0.6
MACH NUMBER = 0.3 TO 0.5
DP/P = 3.5 % TO 7.5 %

INDUSTRIAL ENGINE EXHAUST
- COLD END DRIVE (EXCLUDING SILENCERS)

LAMBDA = 0.2 TO 0.3
MACH NUMBER = 0.3 TO 0.5
DP/P = 2 % TO 4.5 %

Fig. 5.37 *Contd.*

As described above, supersonic flight intakes must be con–di, and Chapter 7 explains how variable area is advantageous for off design operation. A two-dimensional (rectangular) intake is typical, with variable throat area, and with the capability to dump, or draw in, some flow to or from overboard downstream of the throat.

Inter-compressor ducts normally have a reduction in mean line radius and accelerating flow. For inter-turbine ducts the converse is true. The combustor entry duct shown in Fig. 5.37 is annular, followed by a dump feeding an annular combustor.

Figure 5.37 shows a *scrolled* turbine entry duct typical of a single can feeding either a radial or axial flow turbine. A scroll from a centrifugal compressor exit to a single can is of similar geometry but with flow in the reverse direction.

A turbofan bypass duct is usually of constant cross-sectional area, and hence flow Mach number. Aero-engine propelling nozzles may be convergent, or con–di depending upon the application. If an afterburner (section 5.21) is employed then the propelling nozzle *must be of variable throat area design.*

Industrial engine exhausts for cold end drive engines are usually long conical diffusers. This is not practical for hot end drive due to the output shaft, hence that shown in Fig. 5.37 is most commonly employed. There is a short conical diffuser after which the flow dumps into a collector box. It then is ducted to atmosphere via a vertical *uptake* incorporating any silencers.

5.13.4   *Scaling existing duct designs*

If an existing duct design is linearly scaled then its loss coefficient, as defined below, is unchanged. Hence the design point percentage total pressure loss will only be different if the duct inlet dynamic head, or Mach number, is different for the scaled application.

5.13.5   *Duct pressure loss*

In design point calculations total pressure loss is applied as a percentage of inlet total pressure via Formula F5.13.1. Duct percentage pressure loss is a function of only:

- Duct geometry – this is accounted by a *loss coefficient,* usually called *lambda*
- Inlet swirl angle
- Inlet Mach number or dynamic head. Formula F5.13.2 defines dynamic head

The loss coefficient lambda, as defined by Formula F5.13.3, is the ratio of the difference between inlet and outlet total pressure, to the inlet dynamic head. Hence lambda is the fraction of dynamic head lost in the duct whatever the level of Mach number: *its magnitude is a function of only duct geometry and inlet swirl angle.* Formula F5.13.4 gives total pressure loss as a function of lambda. The variation of lambda with inlet swirl angle is discussed in section 5.14. Apart from turbine exit ducts most ducts have constant inlet swirl of zero degrees, and hence lambda is then a function of only duct geometry. Once duct geometry has been set, and lambda has been determined, then percentage total pressure loss only varies with inlet dynamic head and hence Mach number.

Traditionally inlet Mach number is used to judge the severity of the inlet conditions. Chart 3.14 shows dynamic head plotted versus Mach number, as defined by Q curves. Formula F5.13.5 is also of great benefit, expressing inlet dynamic head divided by inlet total pressure, as a function of inlet total to static pressure ratio. Hence percentage pressure loss may be calculated via Formula F5.13.6. As described in Chapter 3, inlet total to static pressure ratio may be determined once any Q curve parameter is known.

The value of lambda for a given geometry must initially be determined from experience, and by using commercially available correlations such as Reference 36. At later stages of an engine project perspex models may be tested in a cold flow rig test facility and the predicted lambda confirmed empirically.

Guidelines for design point values of lambda, inlet Mach number, and hence percentage total pressure loss for the major gas turbine duct types are provided in Fig. 5.37. These are suitable for initial engine design point performance calculations. Generally ducts which are diffusing, as opposed to accelerating, have higher loss coefficients. This is because the flow is more prone to separate, due to the adverse static pressure gradient, incurring significantly higher turbulent losses which overshadow the wall friction losses from which both suffer. If struts are present then they will typically increase the loss coefficients shown by between 5 and 10%, or more if significant incidence or turning occur. Ducts for thrust aero-engines will tend to be towards the higher end of the Mach number range to minimise engine frontal area. It will also be noted from

Fig. 5.34 that engine intakes are treated differently in that the dynamic head at the duct exit as opposed to entry is used (the lambdas shown for intakes are relative to exit dynamic head). This is because at ISA SLS the duct entry Mach number may be very low.

For duct geometries outside those presented in Fig. 5.37 an estimate of lambda may be made by combining the building blocks listed below. If more than one of these features is used in series then the lambda applies to the dynamic head entering each individual section.

- Sudden expansion: Lambda is a function of area ratio as per Formula F5.13.7.
- Dump: This is a sudden expansion to infinity and from Formula F5.13.7 lambda = 1.
- Large step contraction: Lambda = 0.5 based on exit dynamic head. If a radius is employed at the point of contraction then this may be reduced significantly.
- Flow in a pipe of constant cross-sectional area such as a bypass duct: The lambda due to friction may be found from Formula F5.13.8, the value of friction factor may be found from a 'Moody chart' as provided in Reference 30.
- Conical diffusers: Lambda can be found from Chart 5.11 for a range of area ratios. An included angle of 6° is optimum.
- Conical nozzles: For included cone angles of between 15° and 40° lambda is between 0.15 and 0.2 depending upon area ratio.
- Other accelerating or decelerating passages: Lambda can be found from Reference 36.

5.13.6 *Aero-engine intakes – ram recovery factor and efficiency*

The term *ram recovery factor* is commonly used for aero-engine intakes as an alternative to using percentage pressure loss (Formula F5.13.9, note that other terms are also used). This is applied to any ducting supplied as part of the airframe, upstream of the engine/aircraft interface at the engine front flange. For subsonic intakes typical design point percentage pressure loss levels are derived from recovery factor via Formula F5.13.10, and the data provided on Fig. 5.37. For supersonic intakes the design point ram recovery levels shown on Fig. 5.37 include pressure loss across the shock system, as well as subsonic diffusion in the downstream section of the intake. The methodology for deriving design point levels is described in section 5.14.3. In all instances the ram recovery factor includes pressure loss in the free stream upstream of the flight intake leading edge, as well as in the intake itself.

Another term used for aero-engine intakes is the *intake efficiency* as defined in Formula 5.13.11. It calculates an ideal total temperature at exit from the intake based on an isentropic compression from ambient static to intake exit total pressure, and divides this by the actual temperature difference between ambient static and free stream total. The ideal total temperature is a purely theoretical parameter because total temperature is constant along all ducts where there is no work or heat transfer. However, as the vehicle is doing work upstream to compress the inlet air and develop the free stream total temperature the concept of efficiency has some valididity. This book uses ram recovery as opposed to intake efficiency because it is easier to use and, more importantly, easier to measure on a perspex model or engine test.

5.13.7 *Additional design point considerations for aero-engine propelling nozzles*

Section 5.13.1 describes the basic functionality of an aero-engine propelling nozzle. For convergent nozzles thrust is determined by Formula F5.13.12 or F5.13.13 depending upon whether the nozzle is choked or unchoked respectively. In both instances exit velocity is calculated from Q curve Formula 5.13.14. When choked there is additional pressure force due to static pressure in the exit plane being greater than the ambient pressure acting upon the equal area at the front of the engine. Con–di nozzles are designed to run 'over full', hence in this instance Formula 5.13.12 for a choked nozzle does indeed apply.

As defined by Formula F5.13.15, propelling nozzle *coefficient of discharge* (*CD*) is *effective area* (that available for the mainstream flow to pass through) divided by the geometric area. Any *blockage* is due to aerodynamic separation at the wall. Chart 5.13 shows typical levels of

CD versus nozzle expansion ratio for a range of cone half angles and diameter ratios for convergent nozzles. For a good design with low cone angle and diameter ratio and the likely design point expansion ratio of 2:1 to 4:1 CD varies between 0.95 and 0.97. For con–di nozzles it is not possible to generalise and each design must be individually assessed.

As shown by Formula F5.13.14, propelling nozzle exit velocity, and static pressure if choked, are calculated using Q curves. Actual velocity is slightly lower than that calculated as there is some friction and flow non-uniformity. The *coefficient of thrust* (*CX*) or *coefficient of velocity* (*CV*), defined by Formulae F5.13.16 and F5.13.17, are used to account for this. CX is used herein as it is the most commonly used in industry. Chart 5.14 shows how CX varies with nozzle expansion ratio for convergent propelling nozzles. This plot is sensibly independent of nozzle cone angle and diameter ratio. For the likely design point expansion ratio range of 2:1 to 4:1 CX is greater than 0.98. For con–di nozzles there is additional gross thrust loss because of additional flow non-uniformity due to wall cooling, and flow angularity. For the latter reason the included angle of the divergent section of the nozzle must be less than 30° to minimise the component of velocity perpendicular to the axis. This leads to a long heavy nozzle, and CX will be around 0.95–0.97 at high flight Mach numbers, depending on whether or not the cone angle of the walls can be adjusted via variable geometry.

For a typical engine, Chart 5.12 shows the ratio of gross thrust with a con–di nozzle, to that with a convergent nozzle (Formula F5.13.18) versus expansion ratio. The gross thrust shown for the con–di nozzle assumes it is just running full. This is optimistic in that for off design reasons discussed in section 5.14, and to keep its diameter equal to the intake and main engine, con–di nozzles are designed to run over full, and hence less flow acceleration is achieved. It is apparent from Chart 5.12 that at a nozzle expansion ratio of 4:1 the convergent nozzle is 5% worse off; this will be a significantly greater difference for net thrust as momentum drag is unchanged. This is around the value that offsets the additional weight and cost of the con–di nozzle, these items are significant in that most supersonic aero-engines employ an afterburner and hence the propelling nozzle must be of variable throat area. Chart 5.12 also shows typical propelling nozzle expansion ratio versus flight Mach number for turbofans, turbojets and ramjets. Hence a con–di nozzle will generally be selected for engine applications in aircraft which operate much above Mach 1. For a ramjet the lowest flight Mach number is of the order of 2 and hence a con–di nozzle is universally employed.

5.13.8 *Basic sizing parameters*

Owing to the vast array of gas turbine duct geometries it is not possible to give basic sizing guidelines for all of them here. References 31–35 describe actual designs for a range of duct types. Initial sketches may be made using the following generic guidelines, together with the data presented earlier:

- Size upstream component exit area for a suitable duct inlet Mach number with respect to pressure loss.
- Size duct exit area to give a suitable inlet Mach number for the downstream component using the guidelines provided in this chapter.
- The 'swan neck duct parameter' (Formula F5.13.19) for inter-compressor and inter-turbine ducts should be limited to around 4 for area ratios around 1.1, rising to around 8 for area ratios of 2.
- Centrifugal compressor exit, or turbine entry duct scrolls are normally designed for constant angular momentum.
- Owing to the conflict of minimising engine frontal area and weight, while maintaining acceptable pressure loss, design point bypass duct Mach number is rarely designed outside the range defined in Fig. 5.37. Hence bypass duct area is easily derived.
- Convergent propelling nozzle cone half angle and diameter ratio should be in the range shown on Chart 5.13.

- For industrial engine exhausts, Mach number at the exit flange should be less than 0.05 to minimise the dump pressure loss. For turboprops this may be as high as 0.25 provided that the exhaust is orientated to give some gross thrust (see Chapter 6). Hence in either case the exit area can be evaluated.
- As per Chart 5.11 industrial engine conical diffuser exhaust systems should have a cone included angle as close to 6° as possible within the installation space constraints. Owing to length, and hence weight constraints, conical diffusers in aero-engines employ a cone included angle of 15°–25°.
- For diffusers there is little additional static pressure recovery in going beyond an area ratio of 2:1, and none in going beyond 3:1.

5.13.9   *Applying basic pressure loss and sizing guidelines*

Sample calculation C5.6 illustrates the application of the pressure loss and sizing guidelines presented herein.

5.14   Ducts – off design performance

5.14.1   *Loss coefficient lambda*

Once the duct geometry has been fixed by the design process then the characteristic of lambda versus inlet swirl angle is fixed. The only exception to this rule is if dramatic flow separation occurs such that the *effective geometry* is significantly modified.

Inlet swirl is usually constant throughout the operational envelope for ducts downstream of compressors or fans. This is because, in general, the last component is a stator which will have a constant exit flow angle, unless it is operated so severely off design that it stalls. Hence it is usually only after turbines where there is any significant variation in swirl angle at off design conditions. In general exit swirl angle only changes dramatically at off design conditions for the last turbine in a turboshaft engine, where exhausting to ambient produces larger changes in expansion ratio. Exit swirl angle changes may be even larger in power generation as the power turbine must operate synchronously, hence changes of up to 30° between base load and synchronous idle are typical. It is essential to account for this in performance modelling, as well as in the aerodynamic and mechanical design of the duct. The latter is of particular concern for high cycle fatigue if vanes are present which may be aerodynamically excited.

Chart 5.15 shows the typical variation in lambda with inlet swirl angle for duct types which commonly occur downstream of turbines. The optimum swirl angle is of the order of 15°. Also lambda rises rapidly for higher swirl angles for the hot end drive configuration of industrial engine exhaust shown in Fig. 5.14.

An improvement is to model the strut loss separately, as a 'bucket' of lambda versus inlet swirl angle. This will be non-symmetrical if the strut leading edge angle is not zero, as incidence and turning losses will not be minimised simultaneously.

5.14.2   *Pressure loss – all ducts except aero-engine intakes*

As for the design point, pressure loss at off design may be found from Formula F5.13.4 with the loss coefficient being determined as per section 5.14.1. This requires the duct area also to be input into the engine off design performance model such that with the known flow conditions $W\sqrt{T}/AP$ may be calculated. Total to static pressure ratio may then be found via Q curve Formulae F3.32 and F3.33 so that percentage pressure loss is calculated via Formula F5.13.6. Solving for total to static pressure ratio involves iteration and hence is cumbersome.

For a given geometry it can be shown that $(W\sqrt{T}/P)^2$ is approximately proportional to inlet dynamic head divided by inlet pressure (Formula 5.14.1). To reduce computation in off design engine performance models it is common practice to use formula F5.14.2 as opposed to F5.13.6

to compute duct pressure loss. The *pseudo loss coefficient*, or *alpha*, is directly proportional to lambda and all the rules described earlier apply equally to it. For a given duct geometry alpha is calculated from lambda at the design point via Formula F5.14.3. Hence in engine off design performance models, total pressure loss may be easily calculated from Formula 5.14.2 once inlet conditions are known, without recourse to the iteration described above. However, often Mach number values are required for information, and such simplification is not possible. Mach number must then be calculated iteratively from the duct inlet conditions and area.

Generally duct inlet Mach numbers, and hence percentage pressure loss, reduce as an engine is throttled back. Exceptions occur when the downstream capacity does not fall, such as for bypass ducts as described in Chapter 7, and combustor entry ducts.

5.14.3 *Ram recovery factor – aero-engine intakes*

For subsonic intakes, ram recovery at off design conditions is calculated in the same fashion as for other ducts using either lambda or alpha. However for supersonic intakes there is additional loss of total pressure across the shock system. Formula F5.14.4 is a first pass working rule for the pressure ratio across the shock. The pressure loss in the downstream section must be derived as per section 5.13.5 and the two values multiplied together to give an overall exit pressure. If needed, the overall ram recovery factor can then be calculated from Formula F5.13.9. At a flight Mach number of 2, typically 8–10% of free stream total pressure will be lost in the intake system.

5.14.4 *Specific features of propelling nozzles*

Propelling nozzle CD and CX at off design conditions may be derived from Charts 5.13 and 5.14. In engine off design performance models these may be loaded in tabular form and linear interpolation employed for a known value of propelling nozzle expansion ratio. Alternatively a polynomial fit may be utilised.

For variable area nozzles the control schedule must also be included in the engine off design performance model such that area can be derived for a given operating point.

5.15 Air systems, turbine NGV and blade cooling – design point performance

5.15.1 *Configuration*

An engine *air system* comprises a number of air flow paths parallel to the main gas path. For each of these air is extracted part way through the compressors, either via slots in the outer casing, or at the inner through axial gaps or holes in the drum. The air is then transferred either internally through a series of *orifices* and *labyrinth finned seals*, or externally via *pipes* outside the engine casing. The earlier the extraction point, the lower the performance loss as less work has been done on the air. However the extraction point must be of sufficient pressure for the air to be at higher pressure than the main gas path prior to joining at its destination, after allowance for losses through the air system. The *source* and *sink* pressures are the static pressure in the gas path at the points of extraction and return respectively. For early approximations there needs to be a pressure ratio of at least 1.3. Reference 37 describe the fundamentals of parallel gas flow paths or *networks*.

An engine air system will consist of some, if not all, of the following components.

- Turbine *disc cooling* and *rim sealing* requires a radially outward flow up each disc face.
- *Bearing chamber sealing* is required such that oil does not escape into the engine. Air must flow through finned seals into the bearing chamber, and then through an air–oil separator to overboard.

- *Leakage* occurs from high to low pressure air system flow paths. While every effort is made to minimise this using mechanical seals it is not possible to eradicate it.
- *Thrust balance* pistons may be required to reduce part of a spool axial load to reduce the thrust bearing duty. They comprise two air system flows of different static pressures on each side of a rotating disc. Occasionally an additional, or increased, air system flow is required to accomplish this.
- *Engine auxiliary cooling* may be required for aircraft engines, flowing over the accessory location on the engine casing, and usually to overboard. For industrial, marine and automotive engines the auxiliaries are usually cooled by a fan drawing air through the enclosure and hence an engine air system flow path is not required.
- *Handling bleeds*: as described in section 5.2 these may be required to manage compressor surge margin at part power.
- *Customer bleed extraction* may be required for functions such as cooling plant systems or aircraft cabin pressurisation. As described in Chapter 6, this is accounted as installation loss and hence is not included in the engine uninstalled performance.

In addition to the above general air system flows further flow paths are required for high technology engines for turbine *NGV and blade cooling*.

5.15.2 *Magnitudes of general air system flows*

The impact of the air system on overall engine performance is very powerful, and must be accurately accounted. The total percentage of engine inlet mass flow extracted before the combustor may be as low as 2% for a simple RPV engine, but up to 25% for a high technology aero or industrial engine.

An estimate of the station for extraction may be made from a first pass engine performance design point using the rule for source and sink pressures given in section 5.15.1. Typical magnitudes of air system flows are summarised below, each expressed as a percentage of the engine inlet flow.

- Turbine disc cooling and rim sealing: for HP turbines around 0.5% per disc face is required. For LP or power turbines the disc sealing requirement reduces to 0.25%, however if a low technology rim seal is employed then 0.5% must again be used to prevent the ingress of hot gas. Provided it is returned to the gas path with low radial velocity the impact on turbine aerodynamic efficiency is negligible.
- Bearing chamber sealing: approximately 0.02 kg/s is required per chamber.
- Leakage from high to low pressure air system flow paths: in complex air systems up to 2% may leak between neighbouring flow paths.
- Thrust balance pistons: it is not possible to generalise here as if additional or increased air system flows are required they are highly specific to an engine design.
- Engine auxiliary cooling: the amount of flow required for aero-engines varies significantly depending upon engine and installation configuration.
- Handling bleeds: typically there will be approximately 5% per bleed valve. Up to around four bleed valves may be arranged downstream of each compressor.
- Customer bleed extraction: for industrial engines this will usually be less than 1%. For aircraft engines around 0.01 kg/s per passenger is required. It is such a large flow that it often warrants the complexity of having two source points, for low and high altitude. For marine engines, up to 10% of intake flow is required.

5.15.3 *Magnitudes of turbine and NGV blade cooling flows*

Chapter 1 gives some guidance regarding which engine applications warrant the complexity of turbine NGV or blade cooling. Introducing cooling has a significant impact upon cost, this is evident from the complex internal blade cooling passages shown in Reference 38.

Furthermore, the benefit of approximately the first 50° of increase in SOT achieved by cooling is lost due to increased flow bypassing the turbines and not doing work, and by spoiling of the turbine efficiency as it returns. The magnitude of the former results from an engine design point performance calculation, typical magnitudes of the latter are provided in section 5.9. Hence for turbine cooling to be worthwhile a significant increase in SOT must be achieved.

Chart 5.16 presents typical NGV and blade cooling air flows versus SOT suitable as first-order estimates for preliminary engine design point performance calculations. Accurately evaluating the amount of cooling air required for a given set of NGVs and blades is complex as it depends upon a multitude of parameters such as:

- Life required
- Technology level: both materials and cooling
- Combustor OTDF for NGVs and RDTF for blades (see section 5.9)
- Cooling air temperature
- Corrosive environment: fuel type and any presence of salt in the atmosphere
- Reaction: low reaction reduces blade metal temperature for a given SOT
- Centrifugal stress due to rotational speed causing creep – blades only
- Blade configuration: shrouded versus unshrouded

5.15.4 *Air system flows in design point calculations*

The air system flow percentages can be defined as either a fraction of engine inlet flow, or as a fraction of the flow entering the component where they are extracted, the former being used herein. The following calculations are performed at the source station:

- The air system flow percentage is converted into a physical mass flow and deducted from the gas path mass flow.
- The gas path total pressure and temperature are unchanged.
- If the position is part way along the compressor then compressor input power is calculated from Formula F5.15.1.

When the air system flow is returned the following calculations are performed:

- The physical mass flow is added to the main gas path flow.
- The main gas path total pressure is unchanged, i.e. the air system flow is considered to have lost the difference between the source and sink pressures during its journey.
- The mixed total temperature is calculated by Formula F5.15.2. The iteration loop shown is required in that the CP of the mixed gas must be guessed initially. Usually the air system flow is considered not to have been heated along its flow path and hence is returned with the source temperature. It is only for highly sophisticated engine performance models, or where the air system flow passes through a heat exchanger, that any heat pick up is modelled.

Special consideration must be given to air system flows which are returned to the turbines with respect to which do work, and which do not. Industry standard practice is illustrated in Fig. 5.38 and summarised below:

- Disc cooling or sealing air entering the gas path at the front or rear of the rotor blade row does *not* do work in that stage, but does in downstream stages. Hence in performance calculations it is mixed in after the turbine stage.
- NGV aerofoil film or platform cooling introduced upstream of the nozzle throat, and trailing edge cooling ejected with high velocity, are considered to achieve NGV exit momentum and so *do* work in that stage. Hence the aerofoil and platform cooling is mixed in at the throat station 405, and the trailing edge ejection is then mixed in at the SOT station 41 upstream of the rotor. The NGV capacity is calculated at station 405.

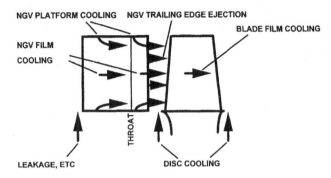

(a)   Real turbine

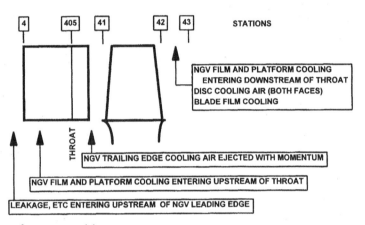

(b)   Engine performance models

Notes:
Turbine capacity for comparison to map is calculated at station 405.
Turbine efficiency, expansion ratio and power are calculated between stations 41 (SOT) and 42.
'Rotor inlet capacity' is sometimes calculated, based on all flows except rotor blade and rear disc.
Work extracted from blade overtip leakage by tip fences is accounted via efficiency level.

Fig. 5.38   Cooled turbine – performance modelling.

- NGV film or platform cooling entering downstream of the throat does not achieve nozzle exit velocity and hence is *not* considered to do work. This is mixed in after the rotor blade row and hence only does work in any downstream stages.
- Rotor blade film cooling is *not* considered to do work in that blade row. Hence it is also mixed in downstream of the blades, and only does work in any downstream stages.
- The spoiling effect of cooling flows on turbine efficiency is discussed in section 5.9.3.

If a multi-stage turbine is modelled as one turbine then the following apply.

- The fractions of total work done by each stage must be estimated.
- For mechanical design consideration SOT station 41 must be calculated as above.
- A further *pseudo SOT* station 415 must be evaluated, it is from this station that work output is calculated. A fraction of the cooling air flows entering downstream are mixed in between station 41 and 415 such that the overall work output is the same as that calculated if the air was considered to enter stage by stage and the rules above applied.
- The remaining mass flow is mixed in downstream of the last stage at station 51 if no other turbine is present, such that it does no work at all.

Apart from a small quantity of film or/and platform cooling that enters downstream of the throat, and any spoiling effects, *cooling the first stage NGV has no fundamental effect on engine performance*. However cooling later stage NGVs does have an effect as that air has then bypassed the first turbine stage. For recuperated engines it is advantageous to cool the first NGV with *recuperator air side delivery air*, as opposed to compressor delivery air. This is because while the flow is increased because of its higher temperature it has negligible impact upon engine performance, but it has been able to exchange heat from the exhaust that otherwise would have been lost to the cycle.

5.15.5 *Estimating air system flow magnitudes*

Example calculation C5.7 shows how air system flow magnitudes for a given engine design may be estimated from the above guidelines. The sample calculations in Chapter 6 illustrate the air offtake and return calculations, including those for a cooled turbine.

5.16   Air systems – off design performance

5.16.1 *Modulation of flows*

For engine off design performance models the most common practice is to maintain a fixed percentage for all air system flows at all off design conditions. A further small gain in accuracy is achieved by maintaining a fixed capacity $W\sqrt{T}/P$ at the extraction point for flows where the path is choked. This is particularly true for handling bleeds. For highly unchoked handling bleeds, with multiple valves discharging into a common manifold, sophisticated modelling or at least representative overall graphs are required.

For extreme accuracy the parallel flow path network calculations described in References 37 and 38, as well as calculations to evaluate main gas path source and sink static pressures, must be merged with the engine off design calculations. This is cumbersome and is rarely attempted.

5.17   Mechanical losses – design point performance and basic sizing

For all engine configurations except ramjets there are a number of components and mechanisms that lead to power loss from an engine shaft. The total power loss can be up to 5% of that being transmitted along the spool and it is important to include this in performance calculations. In addition the power extracted to drive engine auxiliaries must be considered.

5.17.1 *Bearings – configuration, power loss and basic sizing guidelines*

Journal bearings support the shaft radially, and in the special case of *thrust bearings* they react the net axial thrust load on the spool. Power is lost due to friction in the bearing *race* and manifests itself as *heat to oil*. *Ball* and *roller* bearings employ an inner and outer race with balls or rollers between which are also free to rotate. The former react both radial and axial loads, whereas the latter only reacts radially. *Hydrodynamic bearings* do not employ balls or rollers between the inner and outer race. The choice of which bearing system to employ is usually dictated by mechanical design issues as opposed to performance. Ball and roller bearings have the following advantages.

- The required oil flow is between 5 and 10% of that for hydrodynamic bearings.
- They can tolerate greater shaft misalignment.
- Power loss is approximately 10% of that for hydrodynamic bearings.

Conversely hydrodynamic bearings have the following advantages.

- They generally have a higher life for a given duty.
- They have a simpler oil supply system as no jets onto the bearing race are required.
- A single thrust bearing is capable of withstanding a far higher load, the highest thrust load that a single ball bearing is capable of is around 125 kN.

The net result of the above is that ball and roller bearings are most commonly employed in gas turbine engines. However, large industrial engines usually utilise hydrodynamic bearings due to life considerations, and to balance very high thrust loads. References 39 and 40 provide further details regarding bearing selection and design.

Bearings may be lubricated by either *synthetic* or *mineral oil*, Chapter 13 presents basic properties for both. Synthetic oil is used for the majority of applications, and exclusively in hot areas due to its higher auto-ignition temperature. Sometimes mineral oil is employed for driven equipment and power turbines in industrial applications due to its lower cost. This is particularly true if hydrodynamic bearings are utilised.

Formulae F5.17.1–F5.17.4 allow power loss to be calculated at the design point. This loss is then combined with disc windage described below to yield a *mechanical efficiency* (section 5.17.3) before being applied in engine design point performance models.

To first-order accuracy, the maximum bearing race pitch line diameters for acceptable life may be estimated by keeping the *DN number* less than 2.5E06 mm rpm; the DN number being the product of rotational speed and bearing race diameter as defined in Formula F5.17.5. The other key factor in bearing selection is that critical speeds must be acceptable to avoid shaft whirl. On small engines this can have a significant impact upon engine layout, and hence the performance cycle design. If a large bearing diameter is required for shaft stiffness, then a lower rotational speed must be selected to maintain an acceptable DN number for bearing life. Hence this may impact the achievable pressure ratio within a specified engine outer diameter.

5.17.2   Windage – mechanism and power loss

Shaft power is also lost due to *windage*, the frictional work done on air between a rotating compressor or turbine disc and a static structural member. This applies whether or not there is a nett flow through the chamber.

Formula F5.17.6 enables disc windage to be calculated. Before accounting it as part of a mechanical efficiency term it must be checked that it is not already included in the turbine or compressor efficiency. If the latter have come from a rig test then it is likely that it is already included, whereas the converse is usually true if the component efficiencies have come from a computer prediction.

5.17.3   Mechanical efficiency

Mechanical efficiency, as defined by Formula F5.17.7, combines the individually calculated bearing and windage losses into one term. This may then be applied to the power balance on the shaft as per Formula F5.17.8. Equally the parasitic losses may be subtracted from the turbine output power without the intermediate step of calculating mechanical efficiency.

In the early phases of a project, mechanical efficiency may be estimated from previous experience as opposed to deriving it from the formulae provided in sections 5.17.1 and 5.17.2. If ball and roller bearings are utilised mechanical efficiency may range from 99 to 99.9%, increasing with engine size. If, alternatively, some hydrodynamic bearings are utilised then mechanical efficiency may be as low as 96% for small industrial and automotive engines.

5.17.4 *Engine auxiliaries – power extraction and basic sizing guidelines*

In addition to the losses accounted via the mechanical efficiency term power will also be extracted to drive 'engine auxiliaries', such as the oil and fuel pumps. This power is invariably extracted from the HP spool and is quite different from 'customer power extraction' which is part of any installed losses as described in Chapter 6. It is good practice, and less prone to error, to account engine auxiliaries separately from mechanical efficiency. Typically, at the design point, 0.5% of shaft power will be required for a small engine, and less than 0.1% for a large engine. However if natural gas fuel is used this value may be higher if it must be pumped from a low pressure main to that required by the fuel injection system. Natural gas pumping power requirements may be calculated using the data provided in Chapter 13.

The total volume of engine auxiliaries is less than 5% for a large engine , but up to 20% for a small RPV engine.

5.17.5 *Gearboxes*

Engine auxiliaries will usually be driven via a gearbox. Any loss here is included with the engine auxiliary requirements given above. However shaft power engines may also drive the load via a speed reducing gearbox. This enables the selection of the optimum rotational speed for power turbine efficiency, independent of that required by a generator, propeller or natural gas pipeline compressor. The cost, weight and volume of an output gearbox is undesirable and so in the engine concept design phase every effort is made to avoid it.

The maximum practical power output for which a gearbox is viable is around 80 MW. Typically design point gearbox efficiency (Formula F5.17.9) is between 97.5 and 99%.

5.17.6 *Applying basic efficiency and sizing guidelines*

Sample calculation C9.1 demonstrates the use of the formulae provided herein for deriving mechanical losses. It is for the starting regime, however the calculation process is similar when above idle.

5.18 Mechanical losses – off design performance

5.18.1 *Mechanical efficiency*

For off design conditions bearing and windage losses may be calculated using Formulae F5.17.1–F5.17.4 and F5.17.6. These are then combined to derive mechanical efficiency which is applied to the shaft power balance via Formula F5.17.8. (Equally the powers could be simply added to the compressor drive power without also calculating mechanical efficiency, but seeing a value for it is informative.)

5.18.2 *Engine auxiliaries*

For off design operation engine auxiliary losses are small, and also they often do not change dramatically as, for instance, an electric liquid fuel pump will pump excess fuel beyond that required by the combustor, with the balance spilled back to the fuel tank. Hence it is often acceptable to keep the power extraction constant throughout off design operation. A mechanically driven pump would operate on a cube law versus speed.

The one exception is that for small engines during starting, particularly the dry crank phase, where engine auxiliaries may be a significant power extraction from the shaft. In this instance the formulae presented in Chapter 9 may be employed to model these losses. Where extreme accuracy is required in above idle modelling then they may also be used.

5.18.3 *Gearboxes*

Formula F5.18.1 should be used to modulate gearbox losses during off design operation. At idle the gearbox loss will be around 65% of the full load value, in MW.

5.19 Mixers – design point performance and basic sizing

For turbofans a *mixer* may be employed to combine the hot and cold streams prior to exhausting through a *common propelling nozzle*. References 41 and 42 provide an excellent introduction to the fundamental theory and practical design of these devices. *Mixed*, as opposed to *separate jets* turbofans are considered for a variety of reasons.

- If an afterburner (see section 5.21) is to be employed then mixing the cold and hot streams upstream of it will offer a far greater afterburning thrust boost.
- At cruise a small specific thrust and SFC improvement may be achieved if the cycle is *designed specifically for a mixer*.
- The optimum fan pressure ratio for specific thrust and SFC is significantly lower than for the separate jets configuration. This leads to lower weight and cost for both the fan and the fan turbine.
- The reverse thrust increases when a bypass duct blanking style thrust reverser (typical of the design on high bypass ratio turbofans) is deployed. This is because the forward thrust still being produced by the core stream is diminished due to the large dump pressure loss in the mixer chamber.
- For military applications where avoiding heat seeking missiles is vital, the IR signature is reduced by the lower temperatures in the common propelling nozzle plane.
- Jet noise is proportional to jet velocity to the power of 8. With a mixer jet velocities are far lower than in the core stream of a separate jets engine.

In deciding whether to adopt a mixer or not these considerations must be balanced against the disadvantages of the additional cost and weight. Furthermore if the bypass duct style of thrust reverser is employed then complex sealing arrangements are required when it is not deployed to minimise overboard leakage. The impact on cowl drag is heavily dependent on the installation design.

The net result of the above is that all turbofans employing an afterburner are mixed. This is also generally true for subsonic RPV turbofans due to stealth considerations. Until recent years medium to high bypass turbofans for subsonic civil transport aircraft employed separate jets. However due to ever increasing bypass ratios leading to a worthwhile thrust and SFC gain, coupled with the increased emphasis on low noise, many modern engines are mixed.

5.19.1 *Configurations*

Figure 5.39 shows the configuration of the three mixer types which in order of increasing length requirement are:

- Injection mixer
- Lobed annular mixer
- Plain annular mixer

Owing to the high pressure loss of the injection mixer the lobed or plain annular configurations are most common. They comprise hot and cold *mixer chutes* followed by a *mixing chamber*. When lobes, as opposed to a circular wall, are used at the end of the chutes where mixing is initiated then the perimeter is increased by up to three times. This has the effect of significantly reducing the required mixing chamber length.

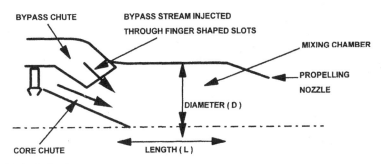

(a)   Forced injection mixer

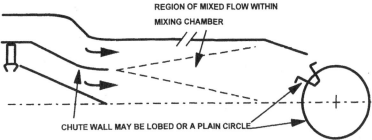

(b)   Lobed annular and plain annular mixers

Fig. 5.39   Mixer configurations.

5.19.2   *Scaling an existing mixer design*

To a first order, if an existing mixer is linearly scaled then its performance will be unchanged, provided that the same hot to cold total temperature and pressure ratios are maintained. The referred inlet mass flow for each stream will be increased by the ratio of the linear scale factor squared.

5.19.3   *Gross thrust, net thrust and SFC improvement*

The derivation of the theoretical thrust and SFC gain of a mixed relative to separate jets turbofan is described in References 41 and 42. The expressions are complex and hence for early concept design studies a simplified method is presented here to evaluate mixer performance at the design point. The engine design point performance is analysed as for separate jets, and then a theoretical gross thrust gain evaluated as well as factors to account for real effects. Hence the gross thrust for the mixed engine can be estimated. Finally the required fan pressure ratio for the mixed cycle is derived from charts presented in section 5.19.4.

Chart 5.17 shows the theoretical gross thrust gain for a mixed engine versus bypass ratio and hot to cold stream total temperature ratio. The following comments apply:

- The gross thrust for both mixed and separate jets engines is for each configuration *at its respective optimum fan pressure ratio*. As described in section 5.19.4, the optimum fan pressure ratio for a mixed turbofan is significantly lower than that for separate jets with the same core.
- The downstream propelling nozzle expansion ratio is greater than 2.5:1, hence jet velocities are high enough such that the mixer is a significant benefit.
- No pressure loss is accounted for the chutes or mixing chamber.

- The mixer is designed for the optimum length to diameter ratio such that there is zero temperature spread at the mixer exit.
- The mixed engine has equal hot and cold stream total pressures at the mixer chute exit plane.

Propelling nozzle expansion ratio is usually greater than 2.5 for most turbofans at cruise and hence the charts apply. At expansion ratios much below 2 the gross thrust gain becomes insignificant.

Chart 5.18 shows how much of the resultant theoretical gross thrust gain will be attained versus mixing chamber length to diameter ratio. This figure accounts for the mixing chamber pressure loss, the mixer chute pressure losses and for the degree of temperature spread at the mixer exit. Chart 5.19 shows a further debit which must be applied to the gross thrust gain resulting from Chart 5.17 if the hot and cold stream total pressures are not equal.

From the above it is apparent that for a 5:1 bypass ratio turbofan with a total temperature ratio of 3, equal total pressures and a lobed mixer with a length to diameter ratio of 2, the actual gain in gross thrust at 0.8 Mach number cruise is around 2%. Initially this may not seem like a very worthwhile return, however because momentum drag remains unchanged the gain in net thrust will be around 4%. For low bypass ratio turbofans operating at a Mach number of 2 the increase in net thrust will be approximately 3% for each 1% gain in gross thrust. The improvement in SFC is as per net thrust in that the fuel burnt for same core is unchanged (since bypass ratio, overall pressure ratio and SOT are held constant).

Charts 5.17, 5.18 and 5.19 may be used in conjunction with the separate jets cycle diagrams presented in Chapter 6 to predict the impact of a mixer on SFC and specific thrust for a given SOT, overall pressure ratio and bypass ratio. However it must be remembered that the values derived apply to a mixed turbofan of different fan pressure ratio from that of the corresponding separate jets engine.

5.19.4   *Optimum fan pressure ratio for mixed turbofans*

It is evident from the above that the engine cycle *must be designed for a mixer from the outset* such that ideally equal *total* pressures at the mixer chute exit plane are achieved. Chapter 6 presents design point diagrams for separate jets turbofans showing the optimum fan pressure ratio for each combination of SOT, overall pressure ratio and bypass ratio. Charts 5.20 and 5.21 show the impact of a mixer on optimum fan pressure ratio versus overall pressure ratio, for a selection of SOTs and bypass ratios. Values at other SOTs and bypass ratios may be found by interpolation. The level of fan pressure ratio presented for mixed turbofans will ensure that total pressures in the hot and cold mixer chute exit plane are equal.

It is apparent from Charts 5.20 and 5.21 that optimum fan pressure ratio is significantly lower for a mixed engine at all flight Mach numbers, and combinations of other cycle parameters. The magnitude of this reduction increases as bypass ratio is reduced.

5.19.5   *Guide to basic sizing parameters*

Reference 42 provides a comprehensive design data base and methodology for mixer design. For early concept design mixer geometry may be sketched using the following guidelines.

Chute exit Mach number and static pressure
The chutes should be designed for an exit Mach number of between 0.35 and 0.55. Gross thrust gain is insensitive to the Mach number levels and their ratio. The static pressure in the exit plane of the mixer chutes must be equal for the hot and cold streams. Q curves apply here and hence area may be found for given flow conditions.

Mixing chamber diameter
The mean exit Mach number should be between 0.35 and 0.5 to allow satisfactory mixing and pressure loss. Again area and hence diameter may be found using Q curves and the known flow conditions.

Mixing chamber length
Mixing chamber length should be set to achieve a good percentage of theoretical thrust gain via Chart 5.18. In practice for many installations engine length restrictions may place an upper limit of around 1.25 on the length to diameter ratio. It is only in a minority of occasions, such as if the engine is mounted in the fuselage, that higher ratios are allowed.

5.19.6 *Applying basic efficiency and sizing guidelines*

Sample calculation C5.8 shows how the guidance provided in this section may be applied to a mixer design for a given engine application.

5.20 Mixers – off design performance

5.20.1 *Off design operation*

As a mixed engine is throttled back from its design point at high rating the propelling nozzle expansion ratio falls. Once it is below 2.5 mixer gross thrust gain falls rapidly. This will occur at higher percentage of thrust when static, than when at high flight Mach number. Furthermore, as described above, net thrust increase for a given gross thrust gain decreases as flight Mach number is reduced.

Mixer operation must be modelled at all corner points of the operational envelope to ensure satisfactory engine off design performance.

5.20.2 *Off design performance modelling*

In off design operation all the parameters affecting mixer performance vary, such as cold to hot stream temperature and pressure ratios, and propelling nozzle expansion ratio. Hence to model a mixer the complete methodology presented in Reference 42 must be used such that station data are calculated through the mixer and gross thrust calculated in the conventional manner using the resulting propelling nozzle conditions.

The matching scheme for a separate jets turbofan is described in Chapter 7. For a mixed engine this matching scheme must be modified in that there is now only one, as opposed to two, propelling nozzle capacities to use as matching constraints. The matching constraint used instead for a mixed engine is that the *static pressure in the mixer chute exit plane must be equal for both the hot and cold chutes.*

5.21 Afterburners – design point performance and basic sizing

Afterburning, sometimes called *reheat,* is a mechanism for augmenting thrust for supersonic aircraft engines. An additional combustor is introduced between the last turbine and the propelling nozzle. The dramatic increase in nozzle temperature increases nozzle gas velocity, and hence thrust. Owing to the accompanying reduction in propulsive efficiency (see Chapter 6) SFC deteriorates significantly.

The engine would only be used *wet,* i.e. with the afterburner lit, at certain key points in the operational envelope. For instance, high Mach number military fighter engines use their afterburners for takeoff and for high supersonic flight speeds. The engines for the supersonic civil transport aircraft Concorde only operate wet at takeoff and on accelerating through the sound barrier. At all other conditions the engines for both aircraft types operate dry with the afterburner unlit.

References 14 and 43 provide further details regarding afterburner design. Furthermore the guidelines presented here with respect to efficiency and basic sizing may also be applied to ramjet combustors. This is because the ramjet combustor is faced with similar inlet conditions and diameter constraints as an afterburner.

5.21.1   *Configuration*

Figure 5.40 shows the most common afterburner configuration for a turbojet. Gas leaving the last turbine stage must be diffused to provide a velocity low enough for satisfactory combustion. Radial aerofoil struts support circumferential V gutters which provide a turbulent mixing regime to sustain a flame, this is analogous to the primary zone with a double torroid as described for conventional combustors in section 5.7. Fuel is sprayed behind the V gutters via a manifold housed within the struts. To achieve satisfactory afterburner loading (see section 5.7) then its volume must be significantly greater than that of the main engine combustor due to the relatively lower pressure. Hence its diameter is usually equal to that of the main engine, and it is long relative to a conventional combustor. Owing to the high flame temperatures a cooled afterburner liner must be employed.

Where an afterburner is applied to a turbofan an upstream mixer, as described in section 5.19, is highly beneficial to combine the hot and cold streams. As described in section 5.22 for handling purposes the propelling nozzle downstream of an afterburner must be of variable area. Also, as described in section 5.13, for aircraft capable of Mach numbers much greater than 1 it is also usually con–di.

5.21.2   *Scaling an existing afterburner design and non-dimensional performance*

All of the comments in section 5.7.2 regarding linearly scaling an existing combustor design are equally applicable to an afterburner.

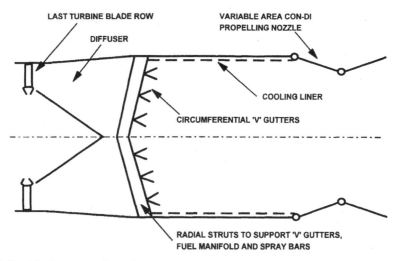

Fig. 5.40   Afterburner configuration.

5.21.3 *Efficiency*

Afterburner efficiency reflects both chemical combustion efficiency as per Formula F5.21.1, and the effect of exit temperature profiles on the ability to produce thrust. The chemical efficiency is low because loading is usually high, due both to the low pressure relative to the main combustor, and geometric constraints limiting the available volume. The temperature profile reduces thrust because energy is wasted in overly hot streams, which at the same driving pressure ratio aquire higher velocity. Afterburner efficiency is typically around 90% at the high altitude and supersonic flight Mach number design point. While, strictly, Chart 5.22 is applicable only to conventional combustor configurations, it may be used for first-order estimates of afterburner efficiency with around 7% points deducted (F5.21.1).

As for a combustor, it is essential to rig test an afterburner prior to engine testing in a development programme. It is only at this point that the efficiency characteristics of the afterburner will be accurately determined.

5.21.4   *Temperature rise*

The propelling nozzle exit total temperature is usually of the order of 1850–2000 K, this is the highest attainable due to the following restrictions:

- *Dissociation*: this is the endothermic reaction where combustion products revert in composition to reactants (e.g. $CO_2 \rightarrow CO + O$). Dissociation should not be confused with combustion efficiency and occurs at high temperatures and low pressures.
- Around 10% of turbine delivery air will be required for afterburner wall cooling. This air will not usually participate in the combustion process and will mix in downstream of the afterburner lowering the average propelling nozzle temperature.
- Temperature rise in the afterburner may be limited by *vitiation* of air by oxygen usage in the main combustor. This will normally only be the case for low turbine exit temperatures, where a high afterburner fuel flow would be desired.
- *Reheat buzz* described in section 5.22.

For preliminary design work Formulae F3.37–F3.41 and Chart 3.15 may be used to evaluate afterburner temperature rise as a function of fuel air ratio. While these are rigorous for main combustors they only provide first-order accuracy for afterburners as dissociation may occur above 1900 K. When dissociation is present then pressure is an additional variable that must be introduced into Chart 3.15. Reference 44 facilitates rigorous temperature rise computation with dissociation.

5.21.5   *Pressure loss*

The afterburner 'cold loss' comprises that in the turbine exit diffuser and that due to the struts, V gutters, etc. The diffuser must reduce the turbine exit Mach number to around 0.25 in the afterburner. To minimise engine frontal area turbine exit Mach number will generally be at the higher end of the guidelines provided in section 5.9. The resulting design point total pressure loss will be between 5 and 7%.

As described in section 5.7 for conventional combustors, there is also an afterburner 'fundamental' or 'hot loss' in the combustion section of the flame tube. Owing to the greater temperature rise this will be between 5 and 10% of total pressure at the design point.

5.21.6   *Thrust gain and SFC deterioration*

As stated above, the objective of an afterburner is to augment thrust. Chart 5.22 shows the ratio of wet to dry net thrust versus the ratio of wet to dry propelling nozzle temperatures, for lines of constant flight Mach number. This figure may be used for both turbojets and turbofans to make first-order estimates of the available thrust augmentation. Each point is for an unchanged gas generator operating point. Hence once the dry engine propelling nozzle temperature is known, thrust augmentation can be determined for a given afterburner temperature.

Ignoring the additional fuel flow and pressure loss of the afterburner, Formula 5.21.2 shows that the ratio of wet to dry gross thrust is equal to the square root of the wet to dry propelling nozzle temperature ratio. The line for zero Mach number, where net and gross thrust are equal, on Chart 5.22 is in fact this ratio but does have some allowance for the effects of afterburner pressure loss and fuel flow. The pressure loss effect outweighs that of the fuel flow, meaning that the thrust ratio is in fact less than that predicted by the square root of the temperature ratio.

Chart 5.22 shows that at higher flight Mach numbers the net thrust gain is considerably greater than at static conditions for a given jet pipe temperature ratio. This is because the afterburner increases gross thrust by approximately the square root of the temperature ratio, but momentum drag is unchanged. Since, when operating, dry net thrust is the relatively small difference between large values of gross thrust and momentum drag, the increased wet gross thrust has a bigger impact.

Some typical gas generator cycles are also shown on Chart 5.22, all for a wet propelling nozzle temperature of 1900 K. The best net thrust augmentation for a turbojet is 28% and 95% at Mach numbers of zero and 2 respectively. This is for a low SOT cycle of 1500 K, where turbine exit temperature is low. However a 1.5 bypass ratio mixed turbofan with the same SOT and pressure ratio has net thrust augmentation of 330% at Mach 2. This is due to the lower turbine exit temperature allowing a higher temperature ratio, and net thrust being an even smaller proportion of gross thrust.

Chart 5.23 shows wet to dry SFC ratio versus wet to dry propelling nozzle temperature ratio. Again for both the wet and dry case the gas generator operating point is unchanged. Formula F5.21.3 shows how, to a first order, the increased fuel flow may be estimated, and Formula F5.21.4 the SFC ratio. Again once the dry engine performance is available then the SFC change for a given afterburner temperature may be evaluated. These formulae show that SFC ratio is highly dependent upon gas generator cycle parameters such as compressor delivery temperature and SOT. Chart 5.23 shows that SFC is always worse with the afterburner operative. This is due to the addition of fuel at low pressure, pressure loss and a combustion efficiency of circa 90% outweighing the thrust gain. SFC deteriorates by around 20% for a turbojet at Mach 2 and a temperature ratio of 1.2, but by up to 60% for a 1.5 bypass ratio turbofan at Mach 2 and a temperature ratio of 3.

5.21.7 Basic sizing parameters

Guidelines for designing first pass scantlings for an afterburner are presented below.

Axial Mach numbers

The gutter entry Mach number should be set to between 0.2 and 0.3 for satisfactory combustion stability and light off capability. While this is higher than for a main combustor it is usually not practical to go any lower due to engine frontal area considerations. Local Mach in the recirculating zone downstream of the V gutters is far lower to sustain combustion. Around 10% of air will be used to cool the outer wall, the outer annulus Mach number should be kept to around 0.1. If the resulting diameter is greater than that set by the airframe manufacturer, or of other engine components, then higher Mach numbers may be inevitable. This will have a detrimental impact upon performance.

Loading

Ideally this should be less than 100 kg/s atm m^3 to achieve an efficiency of around 90%.

Length

This should be set to give the required loading in conjunction with the area derived for the required axial Mach numbers. In practical designs airframe restrictions will normally limit it to less than 2.5 times the diameter.

5.21.8 *Applying basic efficiency and sizing guidelines*

Sample calculation C5.3 illustrates the application of the basic sizing guidelines for a conventional combustor. The process is similar for an afterburner, but using the guidelines provided in this section. Sample calculation C5.9 shows how approximate changes in thrust and SFC for afterburning an engine at a given operating point may be quickly derived as per section 5.21.6.

5.22 Afterburners – off design performance

5.22.1 *Operation*

Figure 5.41 shows both engine net thrust and aircraft drag versus flight Mach number for typical military fighter operation. The afterburner is operative for takeoff, but to maintain good SFC it is not used or indeed required for subsonic flight, except in combat. However a flight Mach number of 0.9 is the highest attainable dry as here aircraft drag exceeds engine

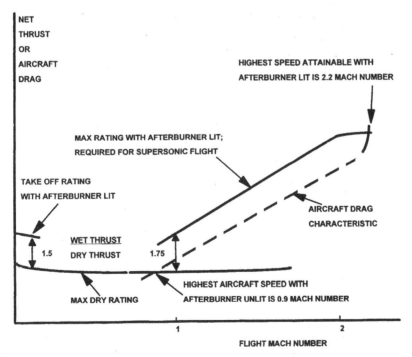

Notes:
Thrust gains shown are indicative, actual values depend upon engine cycle.
With afterburner lit SOT is held at the same value as at Mach 0.9 unlit.
Limiting attainable aircraft Mach numbers shown are for this illustration only.

Fig. 5.41 Typical afterburner operation for a low bypass ratio military mixed turbofan.

maximum dry thrust. Hence at this Mach number the afterburner is lit and the additional thrust enables fast aircraft acceleration through the sound barrier. The maximum flight Mach number is now 2.2 where aircraft drag again exceeds engine thrust.

The afterburner will usually have a *maximum wet*, and a *minimum wet* rating. These ratings relate to the degree of afterburning and for both the gas generator may be at different ratings. The following comments apply to the extreme combinations.

Minimum gas generator
- Mininum wet: this is used on approach where in emergency the pilot may need to slam the throttle to go around. This gives approximately 25% of the maximum wet rating with the gas generator at full throttle.
- Maximum wet: this is not commonly used.

Maximum gas generator
- Mininum wet: this is used in combat situations, or for low supersonic Mach number flight. This gives approximately 90% of the maximum wet rating.
- Maximum wet: this is used for takeoff and at the highest flight Mach numbers attainable.

5.22.2   *Variable area propelling nozzle*

To avoid compressor surge problems it *is essential to have a variable area propelling nozzle* downstream of an afterburner. This is because when the afterburner is lit the dramatic increase in propelling nozzle temperature would rematch the gas generator pushing compressors to surge. To maintain the same gas generator operating point the variable nozzle area must be increased with the square root of afterburner exit temperature. The most common control system strategies monitor referred parameter relationships dry, and then modulate propelling nozzle area to maintain these relationships when wet. Turbine expansion ratios and compressor pressure ratios are the parameters most commonly used as they are highly sensitive to changes in nozzle temperature enabling the control swiftly to change nozzle area.

At high nozzle pressure ratios an overlarge nozzle is of benefit in increasing thrust, via increased engine air flow; this is *over restoring*. At low nozzle pressure ratio a smaller nozzle area is better, to raise it, known as *under restoring*. Both routes are limited by compressor surge margins.

When operating dry the propelling nozzle area must be a few per cent larger than if the same engine were not fitted with an afterburner. This is to maintain the same gas generator operating point since the additional afterburner pressure loss and fuel flow increase the referred flow of the jet.

5.22.3   *Temperature rise, efficiency, pressure losses and wall cooling*

As at the design point, Formulae F3.37–F3.41 may be used to first-order accuracy at off design conditions to determine temperature rise once inlet temperature and fuel air ratio are known. Should the afterburner exit temperature be less than 1900 K then dissociation is unlikely and these calculations are rigorous. Reference 44 shows how to calculate temperature rise rigorously with dissociation. The lowest temperature rise of around 200 °C occurs at the minimum wet rating with the gas generator throttled back.

Again chemical efficiency may be correlated at off design conditions versus afterburner loading. For first-order calculations, around 7% should be deducted from the levels given on Chart 5.5. If an engine programme is committed then this characteristic must be determined from rig testing as per section 5.8. Efficiency may fall to as low as 30% at the minimum wet rating with the gas generator throttled back.

The cold and hot pressure loss coefficients, and hence pressure losses, are determined as for a conventional combustor using Formulae F5.7.9 and F5.7.10. The percentage of air used for afterburner cooling remains constant at off design conditions.

5.22.4 *Stability*

In practical operation an afterburner will never encounter rich extinction. However at rich mixtures approaching stoichiometric an audible instability called *afterburner buzz* may occur. Buzz is noise generated by the combustion process and is more prevalent at higher afterburner pressures and lower afterburner Mach numbers. If the afterburner is continuously operated with buzz present then mechanical damage is likely. As stated in section 5.21, it is one of the practical design phenomena that limit the achievable afterburner exit temperature. Weak extinction must be designed against for low afterburner fuel air ratios. The lower limit to the minimum wet rating with the gas generator throttled back is typical of the restrictions imposed by weak extinction.

5.23 Heat exchangers – design point performance and basic sizing

Recuperators, *regenerators* and *intercoolers* are the three basic types of *on engine* heat exchangers used for industrial, automotive and marine gas turbines. While they have been considered for aero-turboshaft applications, particularly long range helicopters or turboprops, there is currently no engine in production where they have been employed due to weight, volume and reliability considerations.

Recuperators or regenerators transfer waste heat from the engine exhaust to compressor delivery. The difference is that whereas heat transfer occurs *through* the passage walls of a recuperator, a regenerator *physically transports* heat between the streams. The heat transfer increases combustor inlet temperature and hence reduces the fuel required to achieve a given SOT. Intercoolers are used to remove heat from between compressors, reducing inlet temperature to the second compressor. Work input required to raise a given pressure ratio for the second compressor is reduced proportionally to the inlet temperature as per Formula F5.1.4. As described in Chapter 6, recuperators or regenerators improve thermal efficiency while providing a small loss in specific power due to the additional pressure losses. Conversely intercoolers improve specific power but, except at the highest pressure ratios, deteriorate thermal efficiency. A heat exchanged and intercooled cycle improves both thermal efficiency and specific power.

Other types of heat exchangers are employed with gas turbines, a full description of their performance and sizing is beyond the scope of this book. In combined cycle or CHP heat recovery steam generators (HRSGs) are used to raise steam from the engine exhaust heat. These are usually of *shell and tube* design where the hot gas turbine exhaust gas flows through a shell over a bank of tubes. The pressurised water and steam flow in a counterflow direction through the tubes and are heated accordingly. Other novel uses of heat exchangers include a bank of tubes located in the bypass duct of some military turbofans. Here the cold side is the bypass air flowing over the tubes and the hot side is a small percentage of compressor delivery air flowing through them. The latter is cooled prior to it being used to cool the turbine blades and NGVs reducing the amount of air required. Advanced cycle industrial gas turbines also may use *cooled cooling air* to reduce the amount required. Here the heat is exchanged into the natural gas fuel, and hence in turn is returned to the cycle. This does pose some challenges for the fuel injector at off design conditions.

5.23.1 *Configurations*

Figure 5.42 presents the basic recuperator configurations. The *primary surface* recuperator comprises corrugated metal sheets of around 0.15–0.2 mm thick stacked together with air and gas counter flowing through alternate layers. Heat is transferred from the gas (hot) stream to the air (cold) stream directly through the sheets. The *secondary surface* recuperator is more robust with the corrugated sheets being brazed to circa 0.3–0.5 mm thick support sheets. The term secondary surface is used in that the bulk of the heat must conduct along the

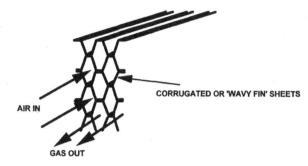

(a)   Primary surface recuperator

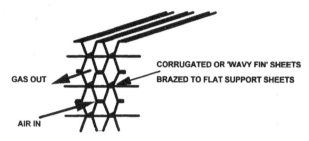

(b)   Secondary surface recuperator

Note:

For both configurations gas passage area must exceed that of air side to minimise pressure loss in
fixed volume.

Fig. 5.42   Recuperator configurations.

secondary corrugated sheets before being transferred to the cold side through the support
sheets. For both configurations the inlet and outlet manifolds and headers are a complex
fabrication.

The regenerator shown in Fig. 5.43 is a radically different heat exchanger concept. Here a
rotating ceramic disc with axial passages is employed, driven at between 20 and 30 rpm by an
electric motor or an engine shaft via reduction gearing. The cold air and hot gas are ducted to
flow through the matrix at opposite sides of the regenerator disc. The disc passages are
alternately heated and cooled as they rotate between the hot and cold streams, thereby
transferring heat. The area open to the gas side is significantly greater than that for the air side
due to its lower density so that low pressure drop can be maintained. The passages are of
around 0.5 mm hydraulic diameter (Formula F5.23.3), and typically triangular with a wall
thickness of around 0.2 mm.

As shown in Fig. 5.43, a seal is employed on both sides of the disc to minimise air in
(compressor delivery pressure) leaking to gas out (exhaust diffuser inlet pressure), or air out
leaking to the gas in. This is usually a carbon or brush seal since the disc must rotate against it.
In addition to any underseal leakage there is a small amount of *carry over* leakage of air that
was in the cold side passages being transported into the gas side by the disc rotation. Because
of the hot walls being rotated into the cold side the flow length of the passages needed is far
less than for a recuperator with typically discs thicknesses of only 60 mm.

The intercooler usually uses a liquid cooling medium, with brazed primary or secondary
heat exchangers as described for recuperators. If the cold sink is sea water, then it is usual to
avoid it reaching the engine. Either air is ducted out and back via scrolls, or an intermediate
freshwater/glycol loop is used. Condensation at HP compressor inlet is prevented by partially
bypassing the cold sink.

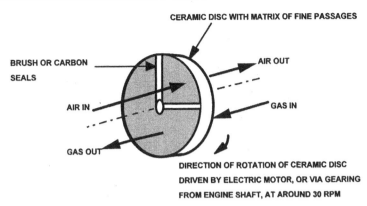

Fig. 5.43   Regenerator configuration.

5.23.2   *Scaling a heat exchanger*

If an existing heat exchanger design is linearly scaled then, to a first-order, the inlet flow area per unit mass flow is preserved, and hence the flow velocity. Manufacturing and integrity considerations normally require that the physical fin form is retained unscaled. In this case achieving the same temperature drop requires an unchanged residence time and hence unchanged (unscaled) flow length. The manufacturer should be consulted regarding any other 'scaling' scenarios.

5.23.3   *Recuperator – thermal effectiveness, pressure loss and basic sizing*

As defined by Formula F5.23.1, thermal *effectiveness* or *thermal ratio* for a recuperator is the ratio of the air temperature rise to the ideal value, the latter being the difference between the gas and air inlet temperatures. Chart 5.24 shows effectiveness versus mass flow rate divided by volume for both primary and secondary surface recuperators at the design point. This chart may be employed to size the recuperator volume for a given flow conditions and target effectiveness. The volume is that of the heat exchanger matrix excluding the manifolds and headers. Effectiveness improves as volume is increased for given flow conditions due to the higher surface area for heat transfer. Chart 5.24 is for a uniform inlet flow profile, a 20% velocity ratio peak to mean decreases effectiveness by around 1% point.

   Chart 5.24 also shows the heat exchanger matrix percentage pressure losses versus mass flow rate divided by volume. These decrease as volume is increased for a given mass flow as velocities are reduced. The gas side pressure loss is significantly higher than that for the air side due to its lower density dictating higher velocity in the matrix to conserve overall volume. Low air side velocities require little extra volume. In addition, further total pressure loss occurs in the inlet and outlet ducting, which is significant in terms of overall performance and must not be ignored. As per section 5.13, the actual levels depend upon the complexity of the geometry and the duct inlet Mach number, typical design point levels are as follows.

Compressor delivery to air inlet	3–6% (Lambda 0.7–1.5)
Air outlet to combustor inlet	1–2.5% (Lambda 2.5–6)
Turbine outlet to gas inlet	2–6% (Lambda 0.4–1.2)
Gas outlet	This is normally included with the exhaust duct for which guidance is given in section 5.13.

5.23.4   *Regenerators – effectiveness, pressure losses, leakage and basic sizing*

Chart 5.25 shows effectiveness versus mass flow per unit area for a regenerator with 60 mm disc thickness. Effectiveness improves as disc area is increased for a given mass flow, again due to the higher area available for heat transfer.

Chart 5.25 also shows the regenerator matrix pressure losses and under seal leakage versus mass flow divided by area. Again due to lower velocities pressure loss reduces as disc area is increased for a given mass flow. Inlet and outlet duct pressure losses are comparable to those provided for a recuperator.

Under seal leakage increases as disc area is increased for a given mass flow as the seal perimeter must increase. Carry over leakage is between 0.25 and 0.5%, and is primarily a function of disc area.

For given flow conditions and target effectiveness, disc area can be derived from Chart 5.25. For manufacturing and strength considerations the largest practical diameter is around 600 mm. Engine layouts with up to two discs have been employed.

5.23.5   *Intercooler – effectiveness and pressure loss*

Intercooler effectiveness is defined by Formula F5.23.2. Total pressure losses including ducting are 5–7%, or up to 10% if the air is ducted a significant distance to an off engine intercooler.

5.23.6   *Intercooler sizing*

Owing to the wide variation in heat exchanger performance levels these guidelines provide first-order estimates only. Detailed data should be sought from a manufacturer as soon as possible.

An intercooler heat exchanger passing the air should be sized to give a Mach number of at most 0.04–0.05, based on the unblocked area. The metal and the liquid passages will increase the actual Mach number beyond this. At these levels the flow length required is around 0.4–0.6 m, depending on the fin form, but may be reduced if smaller flow velocities can be achieved.

An off engine liquid to liquid heat exchanger would be around 0.4–0.5 m^3/MW at a temperature difference of 100 °C, and should give internal flow velocities of around 0.3 m/s. For smaller temperature differences the size would increase in inverse proportion for the same number of MW exchanged.

5.23.7   *Recuperators versus regenerators*

Historically, recuperators have suffered from low cycle fatigue problems resulting from thermal cycling, and *fouling* (gas side passage blockage) due to carbon build up from the combustion process. Indeed it was for these very reasons that regenerators were first conceived in the 1960s. Ceramic discs have good low cycle fatigue strength, and are self-cleaning as air and gas alternately flow in opposite directions through the same passages, blowing off any carbon deposits. However, in recent years recuperators have been developed to be more resilient to thermal cycling. Also cleaning cycles have been designed to burn off carbon by bypassing the cold stream for a brief duration. Fouling has also been improved by the advent of low emissions combustors which necessarily must be carbon free. Hence the traditional concerns regarding recuperators have diminished.

Owing to the limitations on regenerator disc diameter, and the number of discs practical for engine layout discussed above, they are only practical for engine mass flows of less than 2 kg/s or around 500 kW power output. Hence they have been employed for many automotive engine development projects, but all heat exchanged engines of higher output power use recuperators.

Whether the recuperator or regenerator is better with respect to engine performance depends on the engine cycle and space restrictions. The guidelines presented above enable the reader to estimate design point performance, and basic sizes, for both devices. In the majority of cases the recuperator provides better overall engine performance but requires a higher volume.

Finally, regenerators are of lower first cost than most recuperator designs. This is particularly true if nickel based alloys, as opposed to stainless steel must be used to provide sufficient recuperator low cycle fatigue strength. The maintence required to replace discs continually may offset this, however. Both devices have a limit on inlet temperature of around 900–1000 K, depending on the materials used. For recuperators the lower levels are achieved by stainless steels, and the higher by nickel alloys.

5.23.8 *Applying basic efficiency and sizing guidelines*

Sample calculation C5.10 illustrates the application of the basic thermal effectiveness, pressure loss and sizing guidelines presented herein.

5.24 Heat exchangers – off design performance

5.24.1 *Effectiveness*

Recuperators and regenerators
Charts 5.24 and 5.25 show that recuperator or regenerator effectiveness increases at part power, as physical mass flow reduces while the volume or area remain fixed. Indeed at idle effectiveness may be up to 10% points higher than at full power. As described in Chapter 7, recuperators and regenerators are almost invariably employed together with variable area power turbine nozzle guide vanes (VANs). These enable recuperator inlet temperature to be kept high at part power such that the maximum heat may be recovered. This contributes to a far flatter part load SFC versus power curve than for a simple cycle engine. Care must be taken to ensure that the mechanical integrity limit temperature is not exceeded at part power.

For initial off design engine performance modelling a curve from Chart 5.24 or 5.25 may be loaded either as a polynomial curve fit or as an array of values for interpolation. Alternatively Formula 5.24.1 is good to first-order accuracy for both recuperators and regenerators, and relates effectiveness simply to physical flow level. This simple relationship results because a recuperator operates on the air side with a high temperature and downstream capacity essentially fixed by that of the HP turbine. Later, in a detailed design and engine development phase, the proprietary codes of the heat exchanger supplier must be incorporated into the off design engine performance model.

Intercoolers
Effectiveness also increases at part power, but the air inlet conditions have a greater influence than for a recuperator and downstream capacity of the HP compressor varies more. Formula F5.24.2 relates effectiveness to the parameter group for inlet flow, and is good to first-order accuracy.

5.24.2 *Pressure losses*

Recuperators and regenerators
For the inlet and outlet ducts where there is no work or heat transfer then the methodology described in section 5.16 is employed to model pressure loss variation. For the heat exchanger matrix the air side percentage total pressure loss may actually increase at part power, due to increased heat transfer, and the gas side inlet capacity and hence percentage pressure loss decreases significantly. A curve fit may be applied to curves taken from Charts 5.24 or 5.25,

or Formulae F5.24.3 and F5.24.4 used with the constants being calculated at the design point. In the later stages of a project the supplier's proprietary code should be employed.

Intercoolers
Percentage total pressure loss reduces at part power, due to lower referred air inlet mass flow and increased heat extraction. Formula F5.24.5 relates percentage pressure loss to the parameter group for inlet flow, and is good to first-order accuracy.

5.24.3 Regenerator leakage

Both under seal and carry over leakage can be considered as fixed percentages at all off design conditions.

5.25 Alternators – design point performance

The function of an *alternator* is to convert gas turbine shaft power output into AC (alternating current) electrical power. A *dynamo* produces direct current (DC). The generic term *generator* may be applied to either.

5.25.1 Configuration

A typical alternator comprises a wound rotor rotating inside wound coils, connected as *pole pairs*. The rotor is *energised* by an *excitation* current supplied via slip rings. Its rotation induces alternating current in the stationary coils.

5.25.2 Scaling an existing alternator design

To a first order, achievable power is proportional to volume. Length changes should be valid, but care is needed in changing diameter, as in synchronous use rotational speed must be fixed. In this event the manufacturer should be consulted.

5.25.3 Frequency and voltage

As shown by Formula F5.25.1, alternator output frequency is proportional to the number of pairs of magnetic poles and the rotational speed. For the vast majority of installations the electrical power is fed to a grid or used locally in a 'mains' environment. Owing to the requirements of electrical equipment the frequency must be maintained within a tight tolerance at either 50 Hz or 60 Hz dependent upon the country in question. Hence alternator speed must be as follows.

3600 rpm: two pole, 60 Hz	1800 rpm: four pole, 60 Hz
3000 rpm: two pole, 50 Hz	1500 rpm: four pole, 50 Hz

Usually two pole alternators are employed with gas turbines since the output shaft speed for optimum turbine design is not as low as 1500 rpm for even the biggest engines. For small engines where the output speed may be far in excess of 3000/3600 rpm then a speed reduction gearbox must be employed.

There are niche applications where frequency need not be maintained at a constant value, such as high speed alternators for gas turbine propelled hybrid vehicles (see Chapter 1). In this instance the engine output speed, and alternator frequency, can vary dramatically throughout the operational envelope. Power electronics are used to smooth and rectify it before delivering DC (direct current) to the battery or wheel motors.

Formula F5.25.2 shows that once the rotational speed has been set the peak output voltage is a function of the magnetic field flux density combined with the number and area of windings. If the electricity is to be used locally then typically it will be 240, 220 or 110 V depending upon the country to suit the electrical equipment that it supplies. If it is being exported to a grid system then alternator output voltage will be far higher as described in Chapter 1.

5.25.4 *Power output, current and efficiency*

To a first order, power output is proportional to alternator volume. Once output voltage is set as above then current depends on the load. For a purely resistive load, Ohm's law (Formula F5.25.3) states that current is the ratio of voltage to resistance. Power is then the product of voltage and current. For inductive or capacitive loads, *power factor* is the cosine of the *phase angle* between the alternating voltage and current waveforms. Pure inductive loads cause current to lag voltage, and pure capacitive loads cause current to lead voltage, by 90° in both cases. When loads of these types are combined the resistive and lead/lag impedances are separately added arithmetically, as per Formula F5.25.4. An overall impedance magnitude is found by a root sum square of the two values. Reference 45 discusses AC circuit theory.

Alternator efficiency is defined as the electrical power output divided by the shaft power input. Generally at the design operating point efficiency would be between 97.5 and 98.5%. The loss mechanisms include friction at bearings and slip rings, windage, heating (I^2R) losses in stator windings, and eddy currents in the metal frame. Low power factors increase the impact of heating and eddy current losses.

5.25.5 *Polar moment of inertia*

For a single shaft configuration, for engine transient and start performance analysis it is essential to know the alternator polar moment of inertia. If the alternator is directly driven then it is added to that of the engine shaft. However if it is driven via a gearbox then it is *referred* to, and combined with, the engine shaft inertia via Formula F9.6.

5.26 Alternators – off design performance

5.26.1 *Frequency, power output and current*

As described above, in most applications constant frequency and peak voltage are maintained by ensuring that the gas turbine output and alternator speeds, are constant at all power output levels where the alternator is loaded. The power varies primarily due to the current which decreases at part load. Power factor determines output power for given voltage and current levels. Efficiency falls slightly with falling power factor, as stated.

5.26.2 *Efficiency*

A typical characteristic for efficiency versus percentage load and power factor is shown in Chart 5.26. For early off design engine performance modelling this may be loaded as an array and interpolated. However the efficiency characteristic from the alternator manufacturer should be incorporated at the earliest opportunity.

Formulae

In general the formulae provided below utilise constant values of CP and gamma, based on the mean temperature through the process. Formulae for gamma and CP are provided in Chapter 3, along with the iteration process which must often be followed. The accuracy gain if the fully rigorous enthalpy and entropy method is utilised is also described in Chapter 3, again along with the calculation process.

F5.1.1   Compressor input power (kW) = fn(mass flow (kg/s), blade speed (m/s), change in whirl velocity (m/s))

$$PW2 = W2 * U * (Vwhirl\ into\ rotor - Vwhirl\ out\ of\ rotor)/1000$$

(i)   Whirl velocities are the vector components of absolute or relative gas velocity perpendicular to the axial direction.
(ii)   The form shown is for no change in radius.

F5.1.2   Compressor input power (kW) = fn(mass flow (kg/s), temperature rise (K), CP (kJ/kg K))

$$PW2 = W2 * CP23 * (T3 - T2)$$

F5.1.3   Compressor isentropic efficiency = fn(specific enthalpy rise (kJ/kg K), temperature rise (K))

$$E2 = (H3isentropic - H2)/(H3 - H2)$$

or approximating that CP is constant at the mean temperature:

$$E2 = (T3isentropic - T2)/(T3 - T2)$$

where, from rearranging F3.21:

$$T3isentropic = T2 * (P3/P2\ )^{\wedge}((\gamma - 1)/\gamma)$$

F5.1.4   Compressor temperature rise = fn(inlet temperature (K), pressure ratio, isentropic efficiency)

$$T3 - T2 = T2 * (P3Q2^{\wedge}((\gamma - 1)/\gamma) - 1)/ETA2$$

(i)   Derived by combining F5.1.1 with F3.21; $T3isentropic/T2 = (P3/P2)^{\wedge}(\gamma/(\gamma - 1))$

F5.1.5   Compressor loading = fn(specific enthalpy rise (J/kg), blade speed (m/s))

$$LOADING = CP * (T3 - T2)/U^{\wedge}2$$
or
$$LOADING = CP * (T2 * (P3Q2^{\wedge}((\gamma - 1)/\gamma) - 1)/ETA2)/U^{\wedge}2$$

F5.1.6   Mean stage loading for multi-stage compressor = fn(mean specific heat (J/kg), exit temperature (K), inlet temperature (K), mean blade speed (m/s), number of stages)

$$LOADINGmean = CP * (T3 - T2)/(Umean^{\wedge}2 * Nstages)$$

F5.1.7   Velocity ratio = fn(axial velocity (m/s), blade speed (m/s))

$$VRATIO = Vaxial/U$$

F5.1.8 Aspect ratio = fn(blade height, blade chord)

AR = Height/Chord

(i) Axial chord or true chord may be used.

F5.1.9 DeHaller number = fn(inlet velocity (m/s), exit velocity (m/s))

DeH = V2/V1

(i) Limiting minimum value is 0.72.

F5.1.10 Diffusion factor = fn(inlet velocity (m/s), exit velocity (m/s), change in whirl velocity (m/s), pitch chord ratio)

DF = 1 − (V2/V1) + DVwhirl ∗ (S/C)/(2 ∗ V1)

(i) Limiting maximum value is 0.6, or 0.4 for rotor tip sections.
(ii) This is used to select blade pitch chord ratio, hence helping select blade numbers.

F5.2.1 Axial compressor Reynolds number correction to efficiency (%pt) = fn(map efficiency (%), Reynolds number, critical Reynolds number)

RE = W2 ∗ C2/(A2 ∗ VIS2)

RE.crit = 0.63 ∗ C2/K.cla

If RE < RE.crit

DE2 = 100 − (100 − ETA2.map) ∗ (RE/RE.crit)$^{\wedge}$0.13

If RE >= RE.crit

DE2 = 0.

(i) C2 is blade average chord, A2 is inlet annulus area.
(ii) K.cla is blade surface roughness, centreline average. Typical values (in 10^{-3} mm) are:

Precision cast surface,	2–3
Typical polished forging	0.75–1
Highly polished	0.25–0.5

F5.2.2 Axial compressor Reynolds number correction to flow = fn(efficiency correction (%pt), pressure ratio)

DE2 = value from Formula F5.2.1

P3Q2.RE = value from Formula F5.1.4 using ETA2 = ETA2.map − DE2 and unchanged temperature rise

W2.RE = W2.map ∗ SQRT(P3Q2.RE/P3Q2.map)

F5.2.3 DC60 (fraction) = fn(total pressures in intake (kPa), inlet dynamic head (kPa))

(Paverage lowest 60° sector − Paverage for 360°)/(P − PS) average 360°

(i) Note that for modern compressors a 90° sector is also considered, hence a DC90 value.

F5.2.4 Axial compressor rms tip clearance (mm) = fn(individual stage tip clearance (mm))

TC.RMS = $\sqrt{\left(\sum (\text{TCstage}^{\wedge}2)\right)}$

F5.2.5   Applying factors and deltas to a compressor map

WRTP2 = FACTOR1 * WRTP2map + DELTA1
ETA2 = FACTOR2 * ETA2map + DELTA2
P3Q2 = ((P3Q2map − 1) * FACTOR3 + DELTA3) + 1
NRT2 = NRTmap * FACTOR4 + DELTA4

(i)   FACTOR2 is usually set to 1 and DELTA4 to 0.

F5.3.1   Centrifugal compressor input power (kW) = fn(mass flow (kg/s), blade speeds (m/s), whirl velocities (m/s))

PW2 = W2 * (Uex * Vwhirl out of rotor − Uin * Vwhirl into rotor)/1000

(i)   Whirl velocities are the vector components of absolute gas velocity perpendicular to the axial and radial directions.

F5.3.2   Slip factor (definition) = fn(whirl component of exducer absolute velocity (m/s), exducer blade speed (m/s))

Fslip = Vwhirl/Uex

F5.3.3   Slip factor (value) = fn(number of impeller vanes)

Fslip = 1 − 0.63 * π/Nvanes

(i)   This is the Stanitz correlation.
(ii)   Typically the number of vanes is between 20 and 30 due to manufacturing limitations, hence slip factor is between 0.9 and 0.935.

F5.3.4   Specific speed = fn(rotational speed (rpm), mass flow (kg/s), inlet total temperature and pressure (K, Pa), CP (J/kg K), actual temperature rise (K))

NS = N * VOLUMETRICFLOW^0.5/TRISE.ideal^0.75
NS = N * 0.1047 * (W2 * T2 * 10131.2/P2)^0.5/(CP * 10.718
      * (T3 − T2) * ETA2)^0.75

(i)   This is the Balje non-dimensional definition.
(ii)   Volumetric flow rate is at inlet, and is in m^3/s.
(iii) This term is frequently used in imperial units of rpm/ft$^{0.75}$ s$^{0.5}$, to arrive at this multiply the above non-dimensional definition by 129.

F5.3.5   Pressure ratio = fn(efficiency, power input factor, slip factor, exducer tip speed (m/s), CP (J/kg K), inlet temperature (K))

P3Q2 = (1 + (ETA2 * Fpower input * Fslip * Uex^2)/(CP * T2))^(γ/(γ − 1))

(i)   Power input factor is the power lost to back plate and shroud windage, it is typically 1.02–1.05.
(ii)   Slip factor is defined by F5.3.2 and F5.3.3.
(iii) This is for axial inlet flow, if this is not the case then Fslip * U2 is replaced by (Vwhirl3 − Vwhirl2) * U.
(iv)   This is valid for straight radial vanes only. For backswept vanes use efficiency and Formula F5.3.6.

F5.3.6   Impeller exit blade speed (m/s) = fn(slip factor, backsweep angle (deg), CP (J/kg K), temperature rise (K), inlet rms whirl velocity (m/s), inlet rms blade speed (m/s), exit relative velocity (m/s))

Uex = sqrt (C + sqrt (C^2 − 4 * A * D))/(2 * A))

where:

$A = 1 + (Fslip/tan(beta.ex))^2$

$B = CP * Trise + Vwhirl.in.rms * U.in.rms$

$C = 2 * B * (1 + Fslip/tan(beta.ex)^2) + Vex.rel^2$

$D = B^2 * (1 + 1/tan(beta.ex)^2)$

For straight radial vanes, i.e. zero backsweep, use instead:

$Uex = SQRT(CP * Trise + Vwhirl.in.rms * U.in.rms)/Fslip$

F5.3.7 Length parameter = fn(impeller length (m), exducer tip radius (m), inducer tip radius (m), inducer hub radius (m))

$LP = L/(Rex\ tip - (Rind\ tip + Rind\ hub)/2)$

F5.3.8 Bend parameter = fn(axial straightener inner wall radius (m), diffuser vane outer radius (m), diffuser vane height (m))

$BP = (Raxstraightener - Rdiffuser\ vane)/Hdiffuser\ vane$

F5.4.1 Efficiency reduction due to impeller tip clearance (%) = fn(exducer fraction tip clearance, exducer/inducer shroud radius ratio)

If F.clnce > 0.02

$D.ETA = (0.48 * F.clnce + 0.02) * Rex.tip/Rind.tip.$

If F.clnce <= 0.02

$D.ETA = (2.48 * F.clnce - 50 * F.clnce^2) * Rex.tip/Rind.tip.$

F5.5.1 Pod drag (N) = fn(air density (kg/m³), true air speed (m/s), nacelle drag factor, nacelle surface area (m²))

$Pod\ Drag = 0.5 * RHO * VTAS^2 * C * A$

(i) The drag factor C varies between 0.002 and 0.003.

F5.5.2 Nacelle surface area (m²) = fn(length (m), diameters (m))

$Nacelle\ Area = PI * L * (D.ENGINE + D.INTAKE + D.NOZZLE)/3$

F5.7.1 Combustion efficiency = fn(fuel input (kg/s), fuel burnt (kg/s))

$ETA3 = WFburnt/WF$

F5.7.2 Combustor loading (kg/s atm$^{1.8}$ m³) = fn(mass flow (kg/s), inlet pressure (atm), inlet temperature (K), combustor volume (m³))

$LOADING = W/(VOL * P31^1.8 * 10^(0.00145 * (T31 - 400)))$

(i) For efficiency correlations, volume is of the whole can, but does not include the outer annuli.

(ii) For stability correlations, volume and flow are for the primary zone only.

(iii) Some companies use the inverse of the above.

F5.7.3   Combustion intensity $(kW/atm\,m^3)$ = fn(fuel flow (kg/s), combustion efficiency, fuel lower calorific value (kJ/kg), inlet pressure (atm), combustor volume (m^3))

INTENSITY = Wf * ETA3 * LHV/(P31 * VOL)

(i)   Volume is of the whole can, but does not include the outer annuli.

F5.7.4   Residence time (s) = fn(length (m), combustor average Mach number, gas constant (J/kg K), temperature (K))

V = M * SQRT(γ * R * T)
Time = L/V

(i)   M is from Q curves, Formula F3.33.

F5.7.5   Equivalence ratio = fn(fuel air ratio, stoichiometric fuel air ratio)

PHI = FAR/FARstoichiometric

F5.7.6   Stoichiometric air fuel ratio = fn(carbon hydrogen ratio by weight)

AFRstoichiometric = 11.494 * (CHratio + 3)/(CHratio + 1)

F5.7.7   OTDF or RTDF = fn(maximum exit temperature, mean exit temperature, inlet temperature (K))

OTDF or RTDF = (T4max − T4mean)/(T4mean − T31)

(i)   For OTDF T4max is that occurring anywhere in the combustor exit plane.
(ii)   For RTDF T4max is the maximum of the circumferential mean temperature, from root to tip in the combustor exit plane.

F5.7.8   Approximate combustion efficiency (%) = fn(loading $(kg/s\,atm^{1.8}\,m^3)$)

ETA3 = −5.46974E − 11 * LOAD^5 + 3.97923E − 08
       * LOAD^4 − 8.73718E − 06 * LOAD^3 + 0.000300007
       * LOAD^2 − 0.004568246 * LOAD + 99.7

Above is for the unconstrained design from Chart 5.5.

F5.7.9   Combustor cold loss (kPa) = fn(cold loss factor, inlet air flow (kg/s), inlet temperature (K), inlet pressure (kPa))

DPcold = Kcold * P31 * (W31 * SQRT(T31)/P31)^2

F5.7.10   Combustor hot loss (kPa) = fn(hot loss factor, inlet air flow (kg/s), inlet temperature (K), exit temperature (K), inlet pressure (kPa))

DPhot = Khot * P31 * (T4/T31 − 1) * (W31 * SQRT(T31)/P31)^2

F5.9.1   Turbine output power (kW) = fn(mass flow (kg/s), blade speed (m/s), change in whirl velocity (m/s))

PW4 = W4 * U * (Vwhirl out of rotor − Vwhirl into rotor)/1000

(i)   Whirl velocities are the vector components of absolute or relative gas velocity perpendicular to the axial direction.
(ii)   The form shown is for no change in radius.

F5.9.2   Turbine output power (kW) = fn(mass flow (kg/s), temperature drop (K), CP (kJ/kg K))

$PW4 = W4 * CP45 * (T4 - T5)$

F5.9.3   Turbine isentropic efficiency = fn(specific enthalpy drop (kJ/kg K), temperature drop (K))

$ETA4 = (H4 - H5)/(H4 - H5isentropic)$

or approximating that CP is constant at the mean temperature:

$ETA4 = (T4 - T5)/(T4 - T5isentropic)$

F5.9.4   Turbine temperature drop = fn(inlet temperature (K), isentropic efficiency, expansion ratio)

$T4 - T5 = T4 * ETA4 * (1 - 1/P4Q5^{\wedge}((\gamma - 1)/\gamma))$

(i)   Derived by combining F5.9.3 with F3.21.

F5.9.5   Turbine loading = fn(specific enthalpy drop (J/kg), blade speed (m/s)

$LOADING = CP45 * (T4 - T5)/U^{\wedge}2$

F5.9.6   Velocity ratio = fn(axial velocity (m/s), blade speed (m/s))

$VRATIO = Vaxial/U$

F5.9.7   Turbine reaction = fn(static temperature drop rotor (K), static temperature drop stage (K))

$REACTION = dTSrotor/dTSstage$

(i)   Reaction may also be defined based on pressure drops.

F5.10.1–F5.10.3   Turbine Reynolds corrections

F5.10.1 $ETA4 = 1 - (1 - ETA4map)/(RE/REcritical)^{\wedge}K1$
F5.10.2 $WRTP4 = WRTP4map * (1 - K2 * (ETA4map - ETA4)/ETA4map)$
F5.10.3 $RE = W4 * C5/(A4 * VIS5)$

 (i)   K1 and K2 lie between 0.05 and 0.25, and 0.4 and 0.6, respectively.
 (ii)   C5 is the exit annulus height.
(iii)   The critical Reynolds number is around 1×10^5.

F5.10.4   Applying factors and deltas to a turbine map

$WRTP4 = FACTOR1 * WRTP4map + DELTA1$
$ETA4 = FACTOR2 * ETA4map + DELTA2$
$NRT2 = NRTmap * FACTOR4 + DELTA4$

(i)   FACTOR2 is usually set to 1 and DELTA4 to 0.

F5.11.1   Radial turbine output power (kW) = fn(mass flow (kg/s), blade speeds (m/s), whirl velocities (m/s))

$PW4 = W4 * (Uex * Vwhirl \ out \ of \ rotor - Uin * Vwhirl \ into \ rotor)/1000$

(i)   Whirl velocities are the vector components of absolute gas velocity perpendicular to the axial and radial directions.

F5.11.2   Radial turbine specific speed = fn(rotational speed (rpm), mass flow (kg/s), inlet total temperature and pressure (K, Pa), CP (J/kg K), actual temperature drop (K))

$$NS = N * VOLUMETRIC\ FLOW^{\wedge}0.5/TDROP.ideal^{\wedge}0.75$$
$$NS = N * 0.1047 * (W5 * T5 * 10131.2/P5)^{\wedge}0.5/(CP * 10.718$$
$$* (T4 - T5)/ETA4)^{\wedge}0.75$$

(i)   This is the Balje non-dimensional definition.

(ii)   Volumetric flow rate is at exit, and is in m^3/s.

(iii)   This term is frequently used in imperial units of rpm/ft$^{0.75}$ s$^{0.5}$, to arrive at this multiply the non-dimensional value by 129.

F5.11.3   Radial turbine blade tip speed (m/s) = fn(CP (J/kg K), gamma, inlet total temperature (K), expansion ratio)

$$U^{\wedge}2 = CP45 * T4 * ETA4 * (1 - 1/P4QP5^{\wedge}((\gamma - 1)/\gamma))$$

(i)   This is the combination of Formulae F5.11.1 and F5.9.4 and hence is for zero degrees wheel inlet blading such that relative velocity is radial axial exit flow, and also for zero exit swirl. Hence the change in whirl velocity equals blade speed.

(ii)   If inlet or exit flow angles are different then velocity triangles must be used to derive the change in whirl velocity as a function of wheel tip speed.

(iii)   The above does not include disc windage.

(iv)   Where back plate scalloping is employed wheel rim speed will be less than the blade tip speed obtained here.

F5.13.1   Duct total pressure loss (%) = fn(duct pressure loss (kPa), duct inlet pressure (kPa))

$$DPQP = 100 * (DP)/Pin$$

F5.13.2   Duct inlet dynamic head (kPa) = fn(inlet total and static pressures (kPa))

$$DHEAD = Pin - PSin$$

(i)   At Mach numbers less than 0.25 dynamic head may be approximated to the Bernoulli expression which is only valid for incompressible flow.

$$DHEAD = 0.5 * RHO * Vin^{\wedge}2$$

F5.13.3   Pressure loss coefficient = fn(inlet pressure (kPa), exit pressure (kPa), inlet static pressure (kPa))

$$LAMBDA = (Pin - Pout)/(Pin - PSin)$$

F5.13.4   Duct total pressure loss (kPa) = fn(lambda, inlet dynamic head (kPa))

$$Pin - Pout = LAMBDA * (Pin - PSin)$$

F5.13.5   Duct total pressure loss equivalent to one dynamic head (%) = fn(inlet total and static pressure (kPa))

$$(Pin - PSin)/Pin = (1 - (1/(Pin/PSin))) * 100$$

F5.13.6   Duct total pressure loss (%) = fn(lambda, inlet total and static pressures (kPa))

$$(Pin - Pout)/Pin = LAMBDA * (1 - (1/(Pin/PSin))) * 100$$

F5.13.7   Total pressure loss coefficient at sudden expansion = fn(exit area, inlet area)

LAMBDA $= (1 - A1/A2)^\wedge 2$

F5.13.8   Total pressure loss coefficient due to friction in a pipe of constant cross-sectional area = fn(friction factor, length (m), hydraulic diameter (m))

LAMBDA $= F * L/Dh$

(i)   A typical value for friction factor is 0.04. Other details may be found from the 'Moody chart' presented in Reference 30.

(ii) Hydraulic diameter Dh is defined in Formula F5.23.3.

F5.13.9   Ram recovery factor = fn(ram total pressure (kPa), ambient pressure (kPa), intake exit pressure (kPa))

RRF $= (P1 - Pamb)/(P0 - Pamb)$

F5.13.10   Inlet total pressure loss (%) = fn(ram recovery factor, ram total pressure (kPa), ambient pressure (kPa))

$(P1 - P0)/P0 = 100 * (1 - RRF) * (P0 - Pamb)/P0$

F5.13.11   Intake efficiency (%) = fn(ram total pressure (kPa), ambient pressure (kPa), intake exit pressure (kPa), gamma)

ETA1 $= (P1/Pamb)^\wedge((\gamma - 1)/\gamma) - 1)/(P0/Pamb)^\wedge((\gamma - 1)/\gamma) - 1)$

F5.13.12 and F5.13.13   Propelling nozzle thrust (kN) = fn(propelling nozzle and free stream velocities, propelling nozzle and inlet mass flows (kg/s), propelling nozzle and ambient static pressures (kPa), propelling nozzle geometric area (m^2))

F5.13.12   FG $= (Wn * Vn * CX + (PS9 - Pamb) * A9)$
FD $= W1 * V0$
FN $= FG - FD$

Notes: If the propelling nozzle is choked then PS9 is obtained via Formula F5.13.14. If the propelling nozzle is unchoked then PS9 = Pamb and:

F5.13.13   FG $= Wn * Vn * CX$

F5.13.14   Propelling nozzle velocity (m/s) and static pressure (kPa) = fn(inlet total pressure (kPa), ambient pressure (kPa), inlet total temperature (K), effective area (m^2))

Evaluate nozzle pressure ratio:

P9Qlamb $= P9/lamb$

Calculate MN from above using Formula F3.32.

If MN would exceeds 1, nozzle is choked, so evaluate the throat static pressure from the total to static pressure ratio given by Formula F3.32 with MN = 1:

PS9 $= P9/(P9QPS9)$

Calculate nozzle static temperature from the total to static temperature ratio given by Formula F3.31. If nozzle is choked use $MN = 1$ otherwise use $MN =$ calculated value:

TS9 = T9/(T9QTS9)

Calculate nozzle velocity from Formula F3.35.

F5.13.15   Coefficient of discharge = fn(effective and geometric areas (m^2))

CD = A9effective/A9geometric

F5.13.16   Coefficient of thrust = fn(actual gross thrust (kN), ideal gross thrust (kN))

CX = FG/FGideal

(i)   Formulae F5.13.12 and F5.13.13 show the application of CX.

F5.13.17   Coefficient of velocity = fn(actual jet velocity (m/s), ideal jet velocity (m/s))

CV = V9/V9ideal

F5.13.18   Ratio gross thrust con–di nozzle to that for convergent nozzle for same nozzle inlet conditions = fn(propelling nozzle and free stream velocities for convergent nozzle and for con–di nozzle designed to run full (m/s), propelling nozzle inlet mass flow (kg/s), propelling nozzle and ambient static pressures (kPa), propelling nozzle geometric area (m^2))

FG = W9 * V9full * CX/(W9 * V9 * CX + (PS9 − Pamb) * A9)

F5.13.19   Swan neck duct parameter = fn(inlet mean diameter (m), outlet mean diameter (m), length (m))

SNDP = (Dout − Din)/L

F5.14.1   Duct flow capacity squared is approximately proportional to dynamic head

(Win * SQRT(Tin)/Pin)$^\wedge$2 ∝ (Pin − PSin)/Pin

F5.14.2   Duct total pressure loss (%) = fn(alpha, inlet capacity (kg $\sqrt{K}$/s kPa))

DP/P = ALPHA * (Win * SQRT(Tin)/Pin)$^\wedge$2

F5.14.3   Pseudo loss coefficient alpha = fn(lambda, total pressure (kPa), static pressure (kPa), mass flow (kg/s), total temperature (K))

ALPHA = (LAMBDA * (1 − (1/(Pin/PSin))) * 100)/(Win * SQRT(Tin)/Pin)$^\wedge$2

F5.14.4   Pressure ratio for supersonic intake shock system = fn(free stream total pressure (kPa), flight Mach number)

P1 = P0 * (1 − 0.075 * (M − 1)$^\wedge$1.35)

(i)   This is purely for the shock system. Other duct pressure losses will occur within the intake.

F5.15.1 Compressor power factor for interstage bleed extraction = fn(bleed flow fractions (of inlet flow), bleed work fractions (of work done on compressor discharge air))

$$\text{PWfac} = 1 - \sum(\text{WBQ} * \text{DHBQ})$$

F5.15.2 Gas path total temperature after return of air system flow (K) = fn(gas path and air system mass flows (kg/s), gas path total temperature before return (K), air system flow total temperature (K), CP of gas path and air system flows (kJ/kg K))

Approximation for concept design:

$$\text{Tmixed} = \left(\sum(\text{WB} * \text{TB}) + \text{Wgas} * \text{Tgas}\right) \Big/ \left(\sum\text{WB} + \text{Wgas}\right)$$

Improved accuracy:

$$\text{Tmixed} = \left(\sum(\text{CPB} * \text{WB} * \text{TB}) + \text{CPgas} * \text{Wgas} * \text{Tgas}\right)$$
$$\Big/ \left(\sum(\text{WB} + \text{Wgas}) * \text{CPgas}\right)$$

(i)   The CP values should be obtained from Formulae F3.23 and F3.24 based on the mean temperature between $0\,^\circ$C and the respective actual value.
(ii) This form may also be used with the following approximate standardised constant specific heats:
      Compressors: CP = 1.005 kJ/kg, $\gamma = 1.4$, $\gamma/(\gamma - 1) = 3.5$
      Turbines: CP = 1.150 kJ/kg, $\gamma = 1.333$, $\gamma/(\gamma - 1) = 4.0$
(iii) The above standardised CP values are close to those based on the mean temperature between $0\,^\circ$C and the respective actual value. In practice the CP variation in compressors is not excessive, especially for low pressure ratios. In turbines the CP level is often actually higher than the value show, especially at high SOT levels.

Rigorous enthalpy basis:

$$\text{Hmixed} = \left(\sum(\text{WB} * \text{HB}) + \text{Wgas} * \text{Hgas}\right) \Big/ \left(\sum\text{WB} + \text{Wgas}\right)$$

(i)   Enthalpy values are from Formula F3.27.
(ii) Temperature is found by iteration using Formula F3.27.

F5.17.1 Ball bearing power loss (kW) = fn(pitch circle diameter (mm), speed (rpm), oil viscosity (kg/m s), oil flow (l/h))

$$\text{PWPAR} = (4.87\text{E-}15 * \text{D}^{\wedge}3.95 * \text{N}^{\wedge}1.75 * \text{VISoil}^{\wedge}0.4$$
$$+ 3.19\text{E-}10 * \text{N} * \text{D} * \text{Qoil})$$

F5.17.2 Roller bearing power loss (kW) = fn(pitch circle diameter (mm), speed (rpm), oil viscosity (kg/m s), oil flow (l/h))

$$\text{PWPAR} = (2.07\text{E-}15 * \text{D}^{\wedge}3.95 * \text{N}^{\wedge}1.75 * \text{VISoil}^{\wedge}0.4$$
$$+ 1.56\text{E-}10 * \text{N} * \text{D} * \text{Qoil})$$

F5.17.3 Hydrodynamic radial bearing power loss = fn(diameter (mm), speed (rpm), oil viscosity (kg/m s), oil flow (l/h))

$$\text{PWPAR} = (2.48\text{E-}14 * \text{D}^{\wedge}3.95 * \text{N}^{\wedge}1.75 * \text{VISoil}^{\wedge}0.4$$
$$+ 1.87\text{E-}09 * \text{N} * \text{D} * \text{Qoil})$$

F5.17.4   Hydrodynamic thrust bearing power loss = fn(diameter (mm), speed (rpm), oil viscosity (kg/m s), oil flow (l/h))

$$PWPAR = (5.84E\text{-}14 * D^\wedge 3.95 * N^\wedge 1.75 * VISoil^\wedge 0.4$$
$$+ 3.83E\text{-}09 * N * D * Qoil)$$

F5.17.5   Bearing DN number (rpm mm) = fn(pitch circle diameter (mm), speed (rpm))

$$DN = D * N$$

F5.17.6   Disc windage power per face (kW) = fn(air density (kg/m^3), diameter (m), rim speed (m/s))

$$PWPAR = 3.75E\text{-}03 * RHO * D^\wedge 2 * Urim^\wedge 3$$

F5.17.7   Shaft mechanical efficiency (%) = fn(turbine power (kW), power losses (kW))

$$EMech = 100 * \left(PW4 - \left(\sum PWPAR\right)\right) / PW4$$

(i)   PWPAR is the summation of values calculated from Formula F5.17.6 and the appropriate choice from Formulae F5.17.1–F5.17.4.

F5.17.8   Use of shaft mechanical efficiency, output power = fn(turbine power (kW), mechanical efficiency (%))

$$PW = PW4 * EM$$

(i)   PW is then the power available to drive a compressor or the output load.
(ii)   In the latter case it is termed *PWSD*.

F5.17.9   Definition of gearbox efficiency (%) = fn(output power (kW), input power (kW))

$$E = 100 * (PWout//PWin)$$

F5.18.1   Off design gearbox efficiency (%) = fn(design point input power (kW), design point output power (kW), input power (kW))

$$PWpar.des = PWin.des - PWout.des$$
$$PWpar = PWpar.des * (0.6 + 0.4 * (PWin/PWindes))$$
$$E = 100 * ((PWin - PWpar)/PWin)$$

F5.19.1   Mixer gross thrust gain = fn(theoretical gain (%), factor)

$$FGmixed/FGunmixed = (Chart5.17.1 - 1) * Chart5.18/100 * Chart5.19 + 1$$

(i)   Chart 5.17 provides the theoretical gross thrust gain and Charts 5.18 and 5.19 the reductions due to 'real' effects.

F5.21.1   Approximate afterburner efficiency (%) = fn(loading (kg/s atm$^{1.8}$ m^3))

$$ETA3 = -5.46974E\text{-}11 * LOAD^\wedge 5 + 3.97923E\text{-}08 * LOAD^\wedge 4 - 8.73718E\text{-}06$$
$$* LOAD^\wedge 3 + 0.000300007 * LOAD^\wedge 2 - 0.004568246$$
$$* LOAD + 92.7$$

F5.21.2 Afterburner gross thrust ratio = fn(exit temperature (K), inlet temperature (K))

FGwet/FGdry = SQRT(T7/T6)

(i) T7 is wet, T6 is dry.

(ii) Ratio is for an unchanged gas generator operating point.

(iii) This is approximate as it ignores the additional fuel mass flow and the afterburner pressure loss.

F5.21.3 Approximate afterburner wet to dry fuel flow ratio = fn(afterburner inlet and exit temperature (K), main combustor inlet and exit temperature (K), bypass ratio)

Turbojet:

WFwet/WFdry = ((T7 − T6) + (T41 − T3))/(T41 − T3)

Turbofan:

WFwet/WFdry = ((1 + BPR) * (T7 − T6) + (T41 − T3))/(T41 − T3)

F5.21.4 Approximate SFC ratio = fn(fuel flow ratio, gross thrust ratio)

SFCwet/SFCdry = (WFwet/WFdry)/(FGwet/FGdry)

(i) The wet to dry fuel flow and gross thrust ratios are derived from F5.21.2 and F5.21.3.

F5.23.1 Definition of recuperator or regenerator effectiveness on thermal ratio = fn(air inlet temperature (K), air exit temperature (K), gas inlet temperature (K))

EF307Q6 = (T308 − T307)/(T6 − T307)

F5.23.2 Definition of intercooler effectiveness = fn(air inlet temperature (K), air exit temperature (K), cold sink temperature (K))

EF23 = (T23 − T25)/(T23 − Tsink)

F5.23.3 Hydraulic diameter (m) = fn(flow area (m^2), wetted perimeter (m))

Dh = 4 * A/L

F5.24.1 Approximate off design recuperator or regenerator effectiveness = fn(air mass flow (kg/s), design air mass flow (kg/s), design effectiveness)

EF307 = 1 − (W307/W307des) * (1 − EF307.des)

F5.24.2 Approximate off design intercooler effectiveness = fn(inlet air flow (kg/s), inlet air temperature (K), inlet air pressure (kPa), design inlet air flow (kg/s), design inlet air temperature (K), design inlet air pressure (kPa), design effectiveness)

EF23 = 1 − (W23 * SQRT(T23)/P23)/(W23 * SQRT(T23)/P23).des)
       * (1 − EF23.des)

F5.24.3   Approximate off design recuperator air side pressure loss (%DP/P) = fn(design pressure loss (%DP/P), air flow (kg/s), inlet temperature (K), inlet pressure (kPa), exit temperature (K), design air flow (kg/s), design inlet temperature (K), design inlet pressure (kPa), design exit temperature (K))

$$P307D308Q307 = P307D308Q307des * (W307/P307)^{\wedge}2$$
$$* T308^{\wedge}1.55/T307^{\wedge}0.55/(W307des/P307des)^{\wedge}2$$
$$* T308des^{\wedge}1.55/T307des^{\wedge}0.55$$

F5.24.4   Approximate off design recuperator gas side pressure loss (%DP/P) = fn(design pressure loss (%DP/P), gas flow (kg/s), inlet temperature (K), design gas flow (kg/s), design inlet temperature (K), design inlet pressure (kPa))

$$P6D601Q6 = P6D601Q6.des * W6^{\wedge}2 * T6/(W6.des^{\wedge}2 * T6.des)$$

F5.24.5   Approximate off design intercooler pressure loss = fn(design pressure loss (%DP/P), inlet air flow (kg/s), inlet air temperature (K), inlet air pressure (kPa), design inlet air flow (kg/s), design inlet air temperature (K), design inlet air pressure (kPa))

$$P23D25Q23 = P23D25Q23.des * W23^{\wedge}2 * T23$$
$$* P23.des^{\wedge}2/(W23.des^{\wedge}2 * T23.des * P23^{\wedge}2)$$

F5.25.1   Alternator frequency (Hz) = fn(number of pole pairs, rotational speed (rpm))

$$F = Polepairs * N/60$$

F5.25.2   Alternator peak voltage (V) = fn(flux density (Ta), number of coil turns, area (m^2))

$$V = B * A * N$$

F5.25.3   Current (A) = fn(voltage (V), resistance (ohms))

$$I = V/R$$

F5.25.4   Combined impedance (ohms) = fn(resistive impedance (ohms), inductive impedance (ohms), capacitive impedance (ohms))

$$R = SQRT\left(\sum (Resistance)^{\wedge}2 + \sum (Capacitance)^{\wedge}2 + \sum (Inductance)^{\wedge}2 \right)$$

(i)   Inductive and capacitive impedances cause current waveform to lag or lead voltage by 90°, respectively. Reference 45 discusses AC circuit theory.

Sample calculations

For all sample calculations presented herein fixed CP and γ have been used as below:

Compressors: $CP = 1.005\,kJ/kg$, $\gamma = 1.4$, $\gamma/(\gamma - 1) = 3.5$
Turbines: $CP = 1.150\,kJ/kg$, $\gamma = 1.333$, $\gamma/(\gamma - 1) = 4.0$

For engine design purposes this leads to unacceptable levels of error, and CP at mean T, or ideally the fully rigorous enthalpy/entropy polynomials, presented in Chapter 3, should be employed. Sample calculation C3.2 shows how these may be incorporated into the component performance calculations presented below. This involves a lot of iteration and hence is suited to a computer program or spreadsheet.

C5.1 **Evaluate first pass scantlings and efficiency for an axial flow compressor for a 60 Hz single spool power generation engine, with ISO base load design point requirements of 15:1 pressure ratio and 450 kg/s mass flow. The engine drives the alternator directly without any intermediate gearbox. The compressor will eventually be designed using high technology methodology.**

Number of stages, efficiency and exit conditions

From Chart 5.2 select 14 stages: this is from the 'high technology' band, and is the larger number of stages to attain best efficiency at the expense of engine weight.

Select a stage loading of 0.3: this at the bottom of the guidelines presented in section 5.1.4, again to strive for the highest efficiency.

From Chart 5.1, polytropic efficiency $= 91.0\%$. Utilising F3.42 and F5.1.4 calculate isentropic efficiency, T3 and P3:

$\text{ETA2} = (15^{\wedge}(1/3.5) - 1)/(15^{\wedge}(1/(3.5*0.91)) - 1)$
$\text{ETA2} = 87.1\%$

$T3 - 288.15 = 288.15 * (15^{\wedge}(1/3.5) - 1)/0.871$
$T3 = 674.5\,\text{K}$

$P3 = 101.325 * 15$
$P3 = 1520\,\text{kPa}$

Calculate pitch line blade speed and radius

For the first pass consider the inner and outer hade angles to be equal, hence the pitch line radius is constant through the compressor. Derive pitch line blade speed from F5.1.6, and then radius, remembering that rotational speed must be 3600 rpm for synchronous operation:

$0.3 = 1005 * (674.5 - 288.15)/(\text{Upitch}^{\wedge}2 * 14)$
$\text{Upitch} = 304\,\text{m/s}$

$304 = \text{Rpitch} * 3600 * 2 * \text{PI}/60$
$\text{Rpitch} = 0.806\,\text{m}$

Consider stage 1

Since an industrial engine is under consideration and frontal area is less of a concern, then set inlet Mach number $= 0.4$, at the lower end of the guidelines. From Q curves tabulated in Chart 3.8 $Q = 25.4137\,\text{kg}\,\sqrt{\text{K}}/\text{m}^2\,\text{kPa s}$. Hence:

$25.4137 = 450 * 288.15^{\wedge}0.5/(A2 * 101.325)$
$A2 = 3.0\,\text{m}^2$

$3.0 = \text{PI} * (\text{Rtip}^{\wedge}2 - \text{Rhub}^{\wedge}2)$
$0.806 = (\text{Rtip} + \text{Rhub})/2$

and from the tip velocity triangle with axial inlet flow:

Tip relative Mach no. $= (0.4^{\wedge}2 + (\text{Utip}/(1.4 * 287.05 * 288.15)^{\wedge}0.5)^{\wedge}2)^{\wedge}0.5$

and at the pitch line

$\text{Va}/\text{Upitch} = 0.4 * (1.4 * 287.05 * 288.15)^{\wedge}0.5/304$

and at the hub

$\text{Loading} = 0.3 * (\text{Upitch}/\text{Uhub})^{\wedge}2$

Solving the above gives:

Rtip $= 1.11\,\text{m}$, Rhub $= 0.502\,\text{m}$, Blade height $= 0.61\,\text{m}$, Hub tip ratio $= 0.45$,
Tip speed $= 418\,\text{m/s}$, Hub speed $= 189\,\text{m/s}$, Tip relative Mach number $= 1.29$,
Va/Upitch $= 0.45$, Hub loading $= 0.78$, Tip rel angle $= 71.9\,\text{deg}$

Relative to the guidelines in section 5.1 the tip and hub speeds and hub loading are acceptable, while the tip relative Mach number is on the upper limit. However hub tip ratio is lower than the guideline of 0.65 and Va/U is lower than the guideline of 0.5. Hence set inlet Mach number = 0.5 and repeat, this yields:

> Rtip = 1.052 m, Rhub = 0.560, Blade height = 0.492 m, Hub tip ratio = 0.53,
> Tip speed = 397 m/s, Hub speed = 211 m/s, Tip relative Mach number = 1.27,
> Va/Upitch = 0.56, Hub loading = 0.62, Tip rel angle = 66.8 deg

Hence hub tip ratio is still outside the guidelines, however at this point continue on to see how the rest of the scantlings compare to the guidelines. This will better define the design changes required for the next major iteration.

Stage 14 exit
Set exit Mach number to 0.30 with zero exit swirl to minimise downstream duct pressure loss. Hence from, Chart 3.8, $Q = 19.8575\,kg\,\sqrt{K}/s\,m^2\,kPa$

> $19.8575 = 450 * 674.5^{\wedge}0.5/(A3 * 1520)$
> $A3 = 0.387\,m^2$

> $0.387 = PI * (\ Rtip^{\wedge}2 - Rhub^{\wedge}2)$
> $0.806 = (Rtip + Rhub)/2$

and at the pitch line:

> $Va/Upitch = 0.3 * (1.4 * 287.05 * 674.5)^{\wedge}0.5/304$

Solving the above gives:

> Rtip = 0.844 m, Rhub = 0.768 m, Blade height = 0.076 m, Hub tip ratio = 0.91,
> Tip speed = 318 m/s, Hub speed = 289 m/s, Va/Upitch = 0.51,
> Hub loading = 0.78

Hence all of the above parameters are acceptable relative to the guidelines.

Annulus:
From basic geometry:

> Mean blade height = (0.492 + 0.076)/2
> Mean blade height = 0.284 m

Set mean aspect ratio to 2.5. This is the median value from the guidelines reflecting that for a high pressure ratio single spool compressor the front stages will be at the top of the band, and the rear stages at the bottom for mechanical design reasons. Also set all 27 axial gaps to 20% of the upstream axial chord as per the guidelines. Hence:

> Length = 28 * 0.284/2.5 + 27 * 0.284/2.5 * 0.2
> Length = 3.794 m

Calculate inner and outer hade angle from basic geometric relationships:

> tan(alpha) = (0.284 − 0.076)/(2 * 3.794)
> alpha = 1.57°

Hence this is acceptable relative to the guidelines.

Further steps
The above is only the start of the basic sizing of the compressor, providing confidence that the design is in the correct 'ball park'. Further steps include the following:

- Repeat all of the above with a lower loading, say 0.25 for the next pass. This will push the annulus out in radius improving the first stage hub tip ratio. This parameter could be aided by further increasing the inlet Mach number, however the highest end of the guidelines is for aero-engines and does bring some loss in efficiency.
- Pass through the compressor stage by stage calculating all of the above, at this point the loading distribution may be varied from the constant value for each stage used above. The aspect ratio should be allowed to fall from front to back.
- One extra and one less stage should be tried to gain an understanding of the exchange rates.
- A constant hub radius for all stages should be tried. As stated in the guidelines if all other parameter groups can be maintained at acceptable levels then this is the lowest cost option for an industrial engine. Here the number of stages should be reconsidered as the mean blade speed will be lower on the rear stages, which increases their loading.
- The diffusion factor and DeHaller numbers should be evaluated and maintained within the guidelines.
- The annulus should be sketched out in unison with other engine components to ensure that the gas path is compatible.

C5.2 Evaluate first pass scantlings and efficiency for a centrifugal compressor for a remotely piloted vehicle turbojet. Choose ISA SLS takeoff as the compressor design point; this requires a pressure ratio 4.5 : 1 and mass flow of 1.25 kg/s. The turbine requires a rotational speed of 53 000 rpm. The compressor will eventually be designed using low technology methodology

Calculate specific speed, efficiency and hence exit conditions
Guess ETA2 = 0.8, hence via F5.1.4:

$$T3 - 288.15 = 288.15 * (4.5^\wedge(1/3.5) - 1)/0.8$$
$$T3 = 481.5\,K$$

Apply F5.3.4 guessing ETA2 = 0.80

$$NS = 53\,000 * 0.1047 * (1.25 * 288.15 * 10131.2/101325)^\wedge 0.5/(1005 * 10.718$$
$$* (481.5 - 288.15) * 0.8)^\wedge 0.75$$
$$NS = 0.718$$

From Chart 5.3, polytropic efficiency = 83.2%. The lowest line has been used for the following reasons.

- The compressor is of small size.
- Because it is an aero application a low diffuser radius ratio will be chosen, and the first pass will be done without impeller backsweep.
- There is only low design technology available.

Apply F3.42 to calculate isentropic efficiency:

$$ETA2 = (4.5^\wedge(1/3.5) - 1)/(4.5^\wedge(1/(3.5 * 832)) - 1)$$
$$ETA2 = 79.4\%$$

Specific speed should now be re-calculated using this level of isentropic efficiency and the above then repeated. However the resulting error will be small and this step is omitted here for brevity.

Exducer
Apply F5.3.5 to calculate impeller tip speed: inlet flow is axial and the exducer vanes have no backsweep:

$$4.5 = (1 + (0.794 * 1.035 * 0.90 * Utip^2)/(1005 * 288.15))^\wedge(3.5)$$
$$4.5^\wedge(1/3.5) = 1 + Utip^2/391545$$
$$Utip = 458\,m/s$$

As per the guidelines in section 5.3 this is just acceptable for an aluminium impeller, which is desirable for weight and cost. Calculate impeller tip (rim) radius:

Rtip = 458 * 60/(53 000 * 2 * PI)
Rtip = 0.083 m

Inducer

Set inlet Mach number to 0.5; the middle of the range given in the guidelines. Hence from Chart 3.8, Q = 30.1613 kg $\sqrt{K}$/s kPa m^2 and:

30.1613 = 1.25 * 288.15^0.5/(A2 * 101.325)
A2 = 0.0069 m^2

Guess Rtip = 0.05 m, hence Rhub = 0.017 m, Hub tip ratio = 0.34.

Re-guess Rtip = 0.052 to bring hub tip ratio into the guideline band, hence Rhub = 0.023, hub tip ratio = 0.44, Utip = 289 m/s.

Calculate tip relative Mach number from basic trigonometry:

Tip rel Mach number = (0.5^2 + (289/(1.4 * 287.05 * 288.15)^0.5)^2)^0.5
Tip rel Mach number = 0.985

cos(Tip rel angle) = 0.5/0.985
Tip rel angle = 59°

Hence the inducer parameters are all within the design guidelines.

Impeller length

Set *impeller length parameter* to 1.1, the bottom of the guidelines due to the aero application, hence applying F5.3.7:

1.1 = L/(0.083 − (0.052 − 0.023)/2)
L = 0.068 m

Vaneless space

Using the *radius ratio* of 1.05 given in the guidelines Rdiffuser in = 0.087 m.

Diffuser

Set diffuser exit to impeller tip radius ratio to 1.4, the middle of the guidelines for turbojets, hence Rdiffuser out = 0.116 m.

Also consider the radial component of exit Mach number to be 0.25, hence Q = 16.8191 kg $\sqrt{K}$/s kPa m^2 and diffuser height may be derived:

16.8191 = 1.25 * 481.5^0.5/(Adiff * 4.5 * 101.325)
Adiff = 0.0036 m^2

Hdiff = 0.0036/0.116
Hdiff = 0.031 m

Diffuser radial to axial bend

Set the bend parameter to 0.4, at the lower end of the guideline range due to the aero application, hence:

0.4 = (Raxstraightener − 0.116)/0.031
Raxstraightener = 0.128 m

Hence the inner radius of the axial straightener may be derived via Q curves and an exit Mach number of 0.2.

Further steps

- Sketch out the velocity triangles and ensure they are sensible.
- Try another design at 40° of backsweep and determine the increase in compressor diameter. The application requirements will determine whether this is acceptable or not.

C5.3 **Conduct basic sizing for a conventional, as opposed to DLE, industrial engine combustor. The configuration is a single pipe combustor which has the following requirements at ISO base load:**

mass flow $= 7\,\text{kg/s}$ $T4 = 1400\,\text{K}$

$P3 = 900\,\text{kPa}$ ($8.88\,\text{atm}$) $WF = 0.146\,\text{kg/s}$

$T3 = 610\,\text{K}$ $LHV = 43124\,\text{kJ/kg}$ – kerosene

Combustor volume

As per the design guidelines (in sections 5.7 and 5.8) set combustor loading to $1\,\text{kg/s}\,\text{atm}^{1.8}\,\text{m}^3$ (based upon total mass flow and combustor volume) for 99.9% efficiency. Hence from F5.7.2:

$1 = 7/(\text{VOL} * 8.88^{\wedge}1.8 * 10^{\wedge}(0.00145 * (610 - 400)))$

$\text{VOL} = 0.068\,\text{m}^3$

Loading at idle should also be checked as per the guidelines provided, however for an industrial engine which does not have altitude operation or altitude relight then usually setting this level at ISO base load is sufficient.

Check that combustor intensity is satisfactory via F5.7.3:

$\text{INTENSITY} = 0.146 * 0.999 * 43\,124/(8.88 * 0.068)$

$\text{INTENSITY} = 10.41\,\text{MW/atm}\,\text{m}^3$

This is considerably less than the design guideline maximum level of $60\,\text{MW/atm}\,\text{m}^3$.

Primary zone air flow and can area

From the guidelines design the primary zone for an equivalence ratio of 1.02, hence:

$\text{FAR} = 1.02 * 0.067$

$\text{FAR} = 0.0683$

$\text{Wprimary} = 0.146/0.0683$

$\text{Wprimary} = 2.14\,\text{kg/s}$

Set primary zone exit Mach number $= 0.02$, the lower end of the design guidelines. Hence from Chart 3.8, $Q = 1.3609$ and taking primary zone exit temperature to be $2300\,\text{K}$:

$1.3609 = 2.14 * 2300^{\wedge}0.5/(\text{Acan} * 900)$

$\text{Acan} = 0.084\,\text{m}^2$

Combustor radii

Can radius is derived from area:

$0.084 = \text{PI} * \text{Rcan}^{\wedge}2$

$\text{Rcan} = 0.164\,\text{m}$

Set 0.1 Mach number for outer annuli, as per design guidelines, hence $Q = 6.9414$ $\text{kg}\,\sqrt{\text{K}}/\text{s}\,\text{kPa}\,\text{m}^2$ and:

$6.9414 = 7 * 610^{\wedge}0.5/(\text{Aouter} * 900)$

$\text{Aouter} = 0.028\,\text{m}^2$

$0.028 = \text{PI} * (\text{Router}^{\wedge}2 - 0.164^{\wedge}2)$

$\text{Router} = 0.189\,\text{m}$

Combustor length

From volume and area:

$$L = 0.068/0.084$$
$$L = 0.81 \, m$$

Now check residence time using F5.7.4:

$$V = 0.02 * (1.333 * 287.05 * 2300)^{\wedge}0.5$$
$$V = 18.76 \, m/s$$

$$Time = 0.81/18.76$$
$$Time = 43 \, ms$$

This is significantly longer than the minimum value of 3 ms given in the guidelines. However, only primary zone Mach number has been used. While it is acceptable it shows that there is some scope to reduce combustor area, length and volume.

Secondary and tertiary air flows

From the design guidelines the secondary zone should be set up for an equivalence ratio of 0.6, hence:

$$FAR = 0.6 * 0.067$$
$$FAR = 0.0402$$

$$Wsecondary = 0.146/0.0423$$
$$Wsecondary = 3.45 \, kg/s$$

$$Wtertiary = 7 - 2.14 - 3.45$$
$$Wtertiary = 1.41 \, kg/s$$

C5.4 **Define first pass scantlings for the last stage turbine of a single spool power generation engine. The engine is around 10 MW and there is a gearbox between the output shaft and the generator. The turbine will be designed using the highest technology methods. The overall turbine requirements are:**

Mass flow = 45.0 kg/s	P41 = 1700 kPa
T41 = 1500 K	Expansion ratio = 16 : 1

Calculate requirements for last stage

Initially guess efficiency = 90% for whole turbine, hence applying F5.9.4 and F5.9.2 to the whole turbine:

$$1500 - T5 = 1500 * 0.9 * (1 - 1/16^{\wedge}(1/3.5))$$
$$T5 = 761 \, K$$

$$PW4 = 45 * 1.150 * (1500 - 761)$$
$$PW4 = 38\,240 \, kW$$

For the first pass select three stages; this gives an average stage expansion ratio of 2.5 : 1, which is in the middle of the guidelines provided in section 5.9.4. Also for the first pass consider an equal expansion ratio for all three stages. Hence for the last stage with guessed efficiency = 90%:

$$Tin - 761 = Tin * 0.9 * (1 - 1/2.5^{\wedge}(1/3.5))$$
$$Tin = 960 \, K$$

$$Pin = 1700/2.5/2.5$$
$$Pin = 272 \, kPa$$

Pout = 272/2.5
Pout = 108.8 kPa

PWst3 = 45 * 1.150 * (960 − 761)
PWst3 = 10 298 kW

Hence stage 3 is doing 27% of the work, generally the last stage is designed to do less work than the earlier ones due to its lower inlet temperature requiring a higher expansion ratio for the same work.

Exit plane
Set exit Mach number = 0.3, the bottom end of the range provided in the guidelines to minimise downstream duct pressure loss. Hence from Chart 3.8, Q = 19.3834:

19.3834 = 45 * 761^0.5/(Aout * 108.8)
Aout = 0.588 m^2

Set AN^2 = 50E06 rpm^2 m^2; as per the guidelines this is the highest allowable value for last stage stress considerations. For brevity here AN^2 will be based upon exit area as opposed to mean blade annulus area:

50E06 = N^2 * 0.588
N = 9221 rpm

Set hub tip ratio = 0.6; this is the bottom of the range provided in the guidelines. This has been chosen to minimise disc stress since the AN^2 is on the upper limit:

0.588 = PI * (Rtip^2 − Rhub^2)
0.6 = Rhub/Rtip

Solving the above gives Rtip = 0.541 m, Rhub = 0.324 m, Rpitch = 0.433 m, Utip = 522 m/s, Uhub = 313 m/s, Upitch = 418 m/s.
As per the guidelines the hub speed is acceptable with respect to stress.

Loading and Va/U
Apply F5.9.5 and F5.9.6:

LOADING = 1150 * (960 − 761)/418^2
LOADING = 1.31

Va/U = 0.3 * (1.333 * 287.05 * 761)^0.5/418
Va/U = 0.39

From Chart 5.8, the optimum value of Va/U for the loading of 1.3 is 0.72. The value of 0.39 loses 2 points of efficiency. Hence the exit Mach number should be increased and the calculation process should be repeated.

The efficiency derived from Chart 5.8 from the above pass is 92.1%. The following debits should be made as per the guidelines:

Inlet capacity:	1.0% − inlet capacity is 5.1 kg $\sqrt{K}$/s kPa
Technology level:	nil − highest level is available
Cooling:	nil − see Chart 5.16 and Tin = 960 K
Other parameters:	nil − consider no compromises initally though this may later have to be updated

Hence *predicted efficiency = 91.1%*.

Further steps
Once the above has been repeated to give a satisfactory combination of Va/U for loading and speed the following steps are required.

- Check blade inlet hub relative Mach number as per the guidelines by constructing the hub velocity triangle.
- Check that pitch line exit swirl angle is acceptable by constructing the pitch line velocity triangles and Formula F5.9.1.
- Check reaction is around 0.5.
- Set the blade aspect ratio to around 6 as per the guidelines. Hence calculate the hade angles and ensure they are within the guidelines.
- Repeat all of the above for the first two stages. To keep parameters within bounds for all stages may require some change in the work split. Also the radii of the last stage may have to be modified to suit the geometry of the first two.
- Sketch out the annulus.

C5.5   Derive basic scantlings for a radial gas generator turbine for a truck automotive engine, maximum power conditions are as below. The turbine will be designed using low technology methodology.

Inlet temperature = 1300 K   Expansion ratio = 1.7
Inlet pressure = 450 kPa   Mass flow = 1.5 kg/s

Calculate exit conditions
Apply F5.9.4 guessing efficiency = 88.0%:

$$1300 - \text{Tout} = 1300 * 0.88 * (1 - 1/1.7^\wedge(1/3.5))$$
$$\text{Tout} = 1139\,\text{K}$$

$$\text{Pout} = 450/1.7$$
$$\text{Pout} = 265\,\text{kPa}$$

Set rotational speed and derive efficiency
Set specific speed to 0.6; the optimum for total to static efficiency as per Chart 5.9. Apply F5.11.2:

$$0.6 = N * 0.1047 * (1.5 * 1139 * 10131.2/265000)^\wedge 0.5/(1150 * 10.718$$
$$* (1300 - 1139)/0.88)^\wedge 0.75$$
$$0.6 = N * 0.846/58\,192$$
$$N = 41\,270\,\text{rpm}$$

Hence from Chart 5.9, isentropic efficiency (total to total) = 90.5%. As per the guidelines 2 points should be deducted due to the moderate size and low technology level available. Hence predicted efficiency is 88.5%.

The above two steps should be repeated with this efficiency level. However this is omitted here for brevity.

Turbine wheel geometry
Consider radial vanes and zero exit swirl angle, hence applying F5.11.3:

$$\text{Utip in} = 1150 * 1300 * 0.885 * (1 - 1/1.7^\wedge(1/3.5))$$
$$\text{Utip in} = 431\,\text{m/s}$$

$$\text{Rtip in} = 431/(41\,270 * 2 * \text{PI}/60)$$
$$\text{Rtip in} = 0.1\,\text{m}$$

The tip speed is comfortably within the guidelines for mechanical integrity.

Turbine wheel exit geometry
From Chart 5.10, exit tip to inlet tip diameter ratio = 0.7 for 0.6 specific speed, hence:

$$\text{Rtip exit} = 0.7 * 0.1$$
$$\text{Rtip exit} = 0.07\,\text{m}$$

Set exit Mach number to 0.3, the lower end of the range given in the guidelines, hence Q = 19.3834:

$$19.3834 = 1.5 * 1300^\wedge 0.5/(A * 265)$$
$$A = 0.016\,m^2$$

$$0.016 = PI * (0.07^\wedge 2 - Rhub\ exit^\wedge 2)$$
$$Rhub\ exit = negative$$

Hence 0.3 Mach number is not achievable so try 0.4. This gives Rhub exit = 0.021 m and hub tip ratio = 0.3. The latter is within the guidelines and hence acceptable for the first pass.

NGV

From Chart 5.9 NGV exit angle is circa 72°. From Chart 5.10, NGV height to wheel tip diameter ratio is 0.091 for 0.6 specific speed. Hence:

$$NGV\ height = 0.091 * 2 * 0.1$$
$$NGV\ height = 0.018\,m$$

Set vaneless space radius ratio to 1.10 and NGV radius ratio to 1.45, as per the guidelines, the higher value of the latter has been used as frontal area is not a concern.

$$Rexit = 1.1 * 0.1$$
$$Rexit = 0.11\,m$$

$$Rin = 1.45 * 0.11$$
$$Rin = 0.16\,m$$

Further steps

- Draw out velocity triangles and ensure that they are sensible.
- Sketch the radial turbine.
- Ensure that the exit annulus is commensurate with the geometry of the downstream power turbine.

C5.6 Evaluate the loss coefficient, psuedo loss coefficient and design point pressure loss for a conical exhaust diffuser downstream of the turbine exhaust bullet. The diameter of the duct at this plane is 0.5 m, and the allowable length is restricted to 1.5 m by the installation. At ISO base load swirl angle into the duct is zero and inlet conditions are: mass flow = 10 kg/s, inlet temperature = 760 K

Duct geometry

Set duct exit Mach number to 0.05 Mach number, as per the guidelines in section 5.13.7. From Chart 3.8, Q = 3.3982, hence:

$$1.0017 = Pexit/101.325$$
$$Pexit = 101.5\,kPa$$

$$3.3982 = 10 * 760^\wedge 0.5/(Aexit * 101.5)$$
$$Aexit = 0.80\,m^2$$

$$Rexit = 0.50\,m$$

$$Ain = PI * 0.25^\wedge 2$$
$$Ain = 0.196\,m^2$$

$$Aratio = 0.8/0.196$$
$$Aratio = 4.1$$

$$tan(Wall\ angle) = (0.5 - 0.25)/1.50$$
$$Wall\ angle = 9.5°$$

$$Cone\ angle = 19.0°$$

Loss coefficient and pressure loss

From Chart 5.11, lambda = 0.40. It should be noted that the loss coefficients quoted in Fig. 5.37 also include the diffusing section from turbine exit to the end of the exhaust bullet, which is the inlet plane here. Also the Mach number on Fig. 5.37 is at turbine exit.

Now calculate inlet Q guessing Pin = 106 kPa:

$$Q = 10 * 760^0.5/(0.196 * 106)$$
$$Q = 13.27$$

From Chart 3.8 inlet Mach number = 0.20, and one dynamic head = 2.6164% DP/P. Hence applying F5.13.6:

$$DP/P = 0.40 * 2.6164$$
$$DP/P = 1.05\%$$

$$Pin = 101.5/(1 - 0.0105)$$
$$Pin = 102.6 \, kPa$$

The above should be repeated calculating Q with this level of inlet pressure. This step has been omitted here for brevity.

Pseudo loss coefficient

Applying F5.14.3:

$$ALPHA = (0.40 * 0.026164)/(10 * 760^0.5/(0.196 * 102.6))^2$$
$$ALPHA = 0.0000988$$

This may now be used for all off design calculations together with any modulation for swirl angle as per Chart 5.15.

C5.7   **Estimate the uninstalled air system flow requirements at ISO base load for a single spool industrial engine which has an SOT of 1250 K. The inlet mass flow is 10 kg/s, the engine has two turbine stages with 60% of work derived from the first stage and 40% from the second, and two bearings. Thrust balance is attained by a large hydrodynamic thrust bearing, and the NGV/blade cooling technology available is low.**

Actual air system

Using the guidelines provided in section 5.15.2:

- 0.5% (0.05 kg/s) for each of the four disc faces
- 0.2% (0.02 kg/s) for each bearing chamber for sealing to overboard
- 0.5% (0.1 kg/s) leakage – moderate complexity air system
- No air for thrust balance is required
- No handling bleeds at base load power
- The engine is uninstalled so there is no customer bleed extraction

Actual turbine cooling requirements

Taking the low technology lines from Chart 5.16:

- NGV stage 1 = 2.8% (0.28 kg/s)
- Blades stage 1 = nil
- NGV stage 2 = nil (inlet temperature will be below 1100 K)
- Blades stage 2 = nil

Hence total air flow extractions by compressor delivery = 5.5%.

Return points of air system flows for the performance model if only one turbine map is used

- Consider all of the 2.8% NGV cooling flow to enter upstream of the throat or from trailing edge ejection with momentum. Hence this is all included in SOT station 41.
- The cooling for the two faces of the stage 1 disc do work in the second disc, hence 40% (0.4% or 0.04 kg/s) of this flow is mixed in to evaluate T415, the pseudo SOT from which work output is calculated.
- The remaining 0.6% of disc 1 cooling, and the 1.0% for disc 2 cooling is mixed in downstream of stage 2 as this does not do any work.
- Consider the 0.5% leakage to mix in downstream of stage 2 also.

C5.8   **Find the approximate gain in gross thrust and improvement in SFC for a turbofan at cruise at 0.8 Mach number, 11 000 m and ISA if the engine is designed with a mixer as opposed to designed unmixed. Also what is the change in the required fan pressure ratio? Pod length restrictions limit the mixer L/D to 1.25. Performance parameters for the unmixed engine are:**

BPR = 6 : 1	SOT = 1470 K
Thot = 700 K	FPR = 1.8
Tcold = 300 K	OPR = 37 : 1

Mixing gain

Calculate temperature ratio:

Tratio = 700/300
Tratio = 2.33

From Charts 5.17–19 considering a lobed mixer with an L/D of 1.25:

- Theoretical gross thrust, mixed/unmixed = 1.021
- Percentage of theoretical gross thrust gain achieved – pressure losses and imperfect mixing = 57%
- Fraction of theoretical thrust gain achieved – mixing effect = 1.0 (will design for equal hot and cold stream pressures)

Apply F5.19.1 to calculate actual gross thrust gain:

FGmixed/FGunmixed = (1.021 − 1) ∗ 57/100 ∗ 1 + 1
FGmixed/FGunmixed = 1.012

Optimum fan pressure ratio

From Chart 5.20 the optimum FPR for 1470 K at 6 : 1 BPR and 37 : 1 OPR is approximately 0.2 less than that for the unmixed engine.

C5.9   **Evaluate the approximate changes in gross thrust and gross SFC for a turbojet operating at the below gas generator operating point when its afterburner is lit:**

T6dry = T6wet = 1110 K	T41wet = T41dry = 1500 K
T9wet = 1900 K	T3wet = T3dry = 681 K

Thrust gain

Apply F5.21.2:

FGwet/FGdry = SQRT(1900/1110)
FGwet/FGdry = 1.31

SFC change

Apply F5.21.3 and F5.21.4 to derive approximate fuel flow and SFC changes:

WFwet/WFdry = ((1900 − 1110) + (1500 − 681))/(1500 − 681)
WFwet/WFdry = 1.96

SFCwet/SFCdry = 1.96/1.31
SFCwet/SFCdry = 1.5

Comparison with Charts 5.22 and 5.23

This sample calculation is for a turbojet with an SOT of 1500 K, PR of 15:1 at ISA SLS. The thrust ratio calculated is similar to the point plotted for this engine on Chart 5.22 (temperature ratio = 1.7). However the chart does account for the additional mass of the fuel flow and the afterburner pressure loss, which the above approximate calculations do not.

The calculated SFC ratio is consistent with Chart 5.23. It is towards the top of the band provided, the ratio being lowered by its 'zero bypass ratio', but being increased due to the static flight condition.

C5.10 **Determine the volume required and the associated pressure losses for a recuperator for an automotive engine to provide a thermal effectiveness of 86.0%. There is a 10% velocity ratio, peak to mean, of the gas flow entering it. The recuperator is of secondary surface construction and will be designed using the highest technology level. Engine pressure ratio is 4.5:1 and recuperator mass flow is 1.5 kg/s**

Recuperator volume

As per the exchange rate provided in section 5.23.3 there will be a 0.5% loss in effectiveness due to the inlet velocity profile. To achieve 86.0% effectiveness the value for a uniform inlet velocity profile needs to be 86.5%. Hence from Chart 5.24, using the high technology line for a secondary surface recuperator, W/VOL = 10 kg/s m^3. Hence:

VOL = 1.5/10
VOL = 0.15 m^3

Pressure losses

Use the median lines for a secondary surface recuperator on Chart 5.24, high technology level, but low pressure ratio. Hence:

DP/Pgas side = 7.0%
DP/Pair side = 1.4%

The pressure losses for the inlet and exit ducting must also be included in any performance calculations as provided in section 5.23.3.

Charts

Chart 5.1 Axial compressor polytropic efficiency versus stage loading.

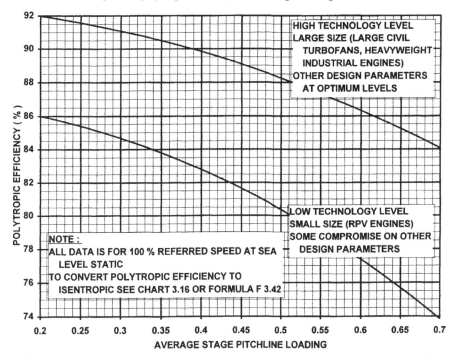

Chart 5.2 Axial compressors: pressure ratio versus number of stages.

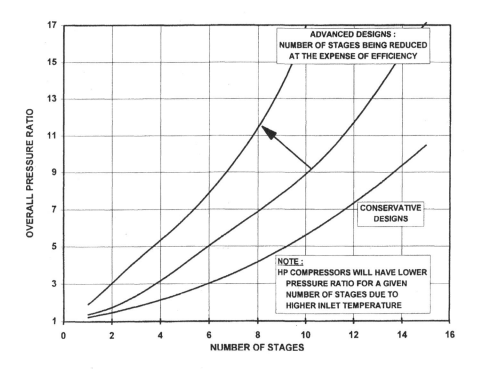

Chart 5.3 Centrifugal compressor: polytropic efficiency versus specific speed.

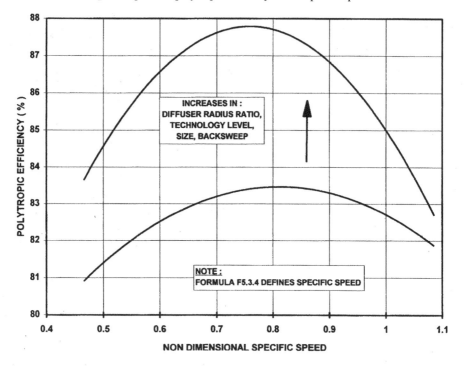

Chart 5.4 Single stage axial fan polytropic efficiency versus average pitchline loading.

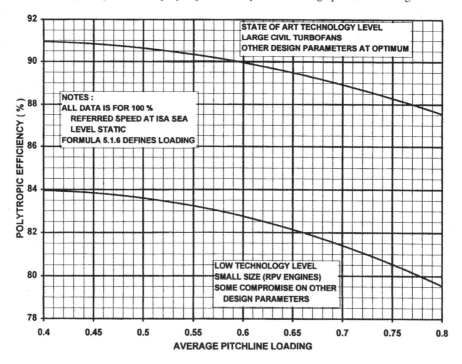

Chart 5.5   Combustion efficiency versus loading.

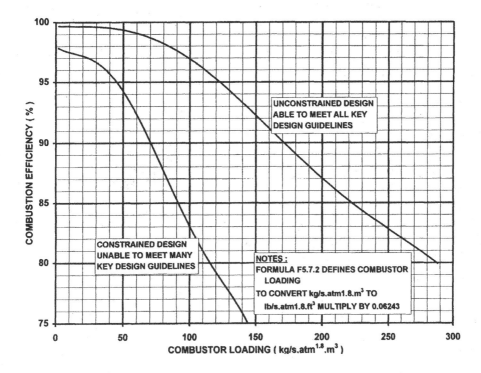

Chart 5.6   Emission levels of NO_x and CO versus combustion temperature.

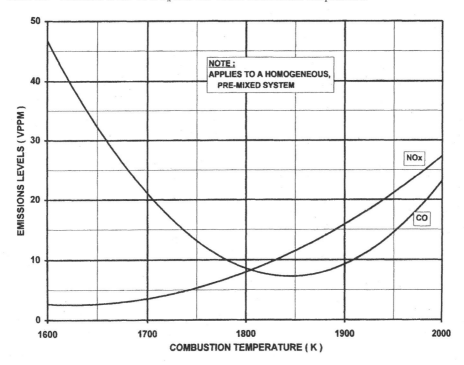

Chart 5.7   Combustion stability versus primary zone loading and equivalence ratio.

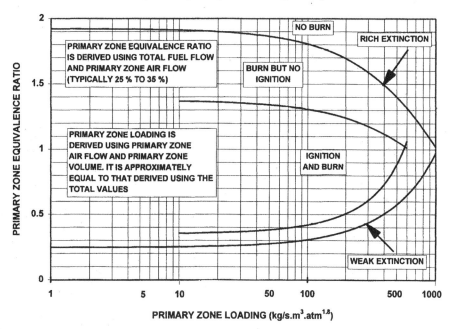

(a)   Burn and ignition stability loops

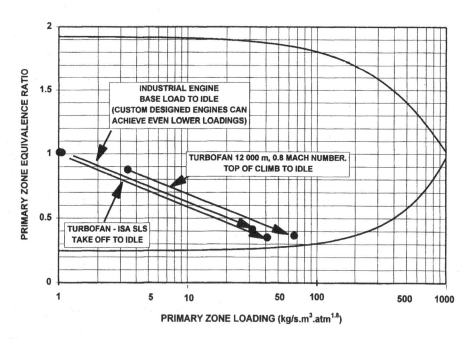

(b)   Locus of engine operating points

Chart 5.8   Axial turbine efficiency versus loading and axial velocity ratio.

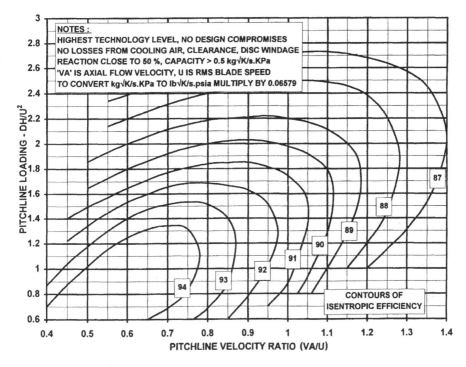

Chart 5.9   Radial turbine efficiency versus specific speed and nozzle exit angle.

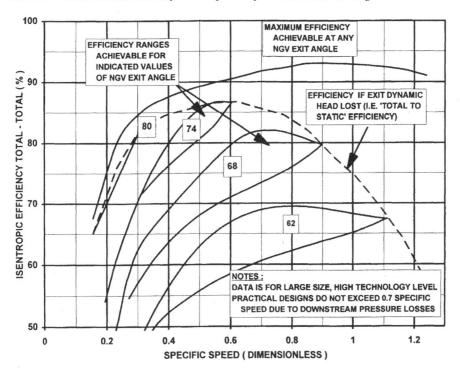

Chart 5.10   Radial turbine optimum diameter ratios versus specific speed.

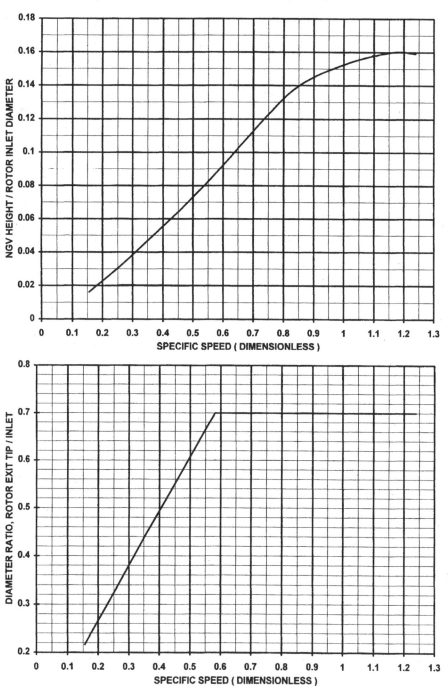

Chart 5.11   Conical diffuser loss coefficient versus cone angle and area ratio.

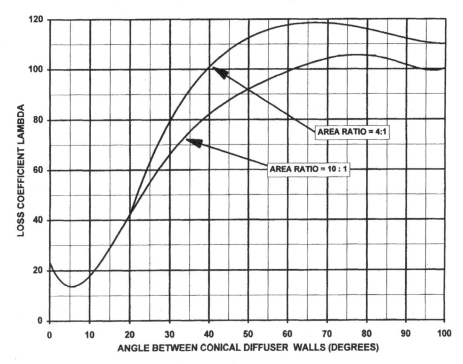

Chart 5.12   Comparison of convergent and con–di propelling nozzles.

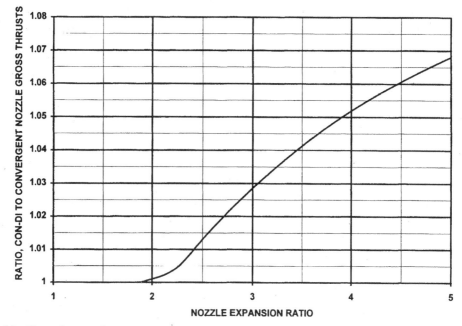

(a)   Gross thrust ratio versus expansion ratio

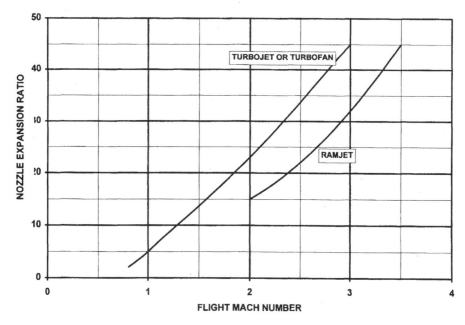

(b)   Typical nozzle expansion ratios versus flight Mach number

Chart 5.13 Propelling nozzle discharge coefficients.

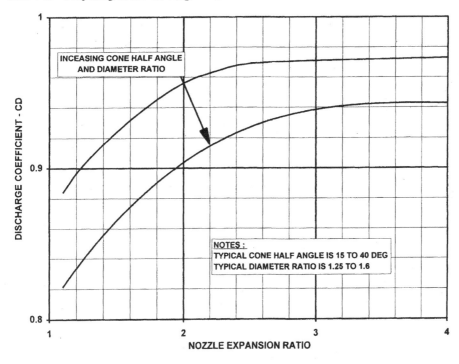

Chart 5.14 Propelling nozzle thrust coefficient.

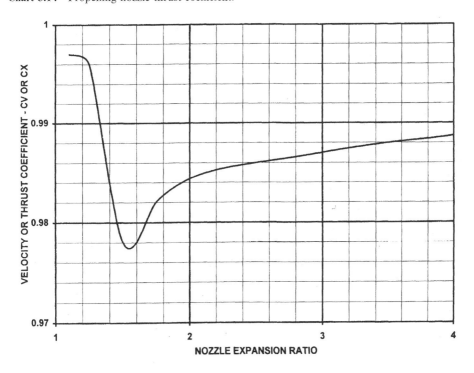

Chart 5.15   Effect of duct inlet swirl on loss coefficient.

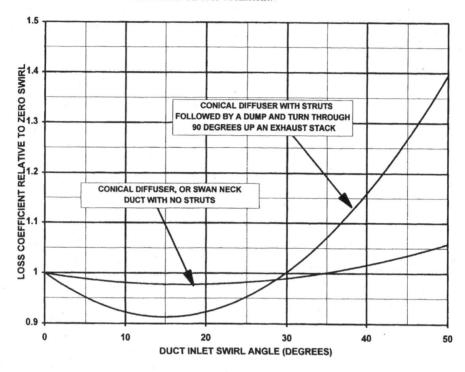

Chart 5.16   Turbine NGV and blade cooling flow requirements versus SOT.

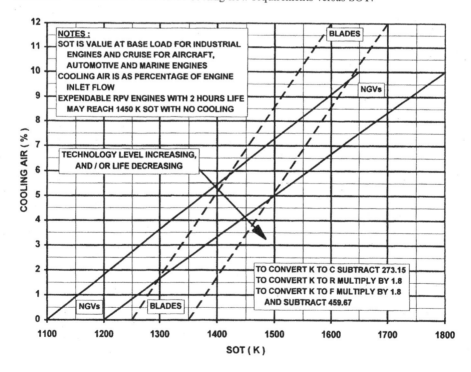

Chart 5.17   Theoretical gross thrust mixing gain versus bypass ratio and temperature ratio.

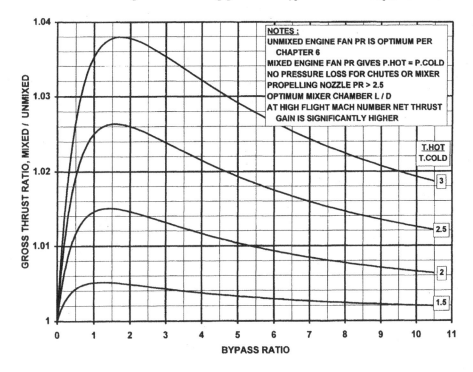

Chart 5.18   Effect of mixer geometry on gross thrust gain.

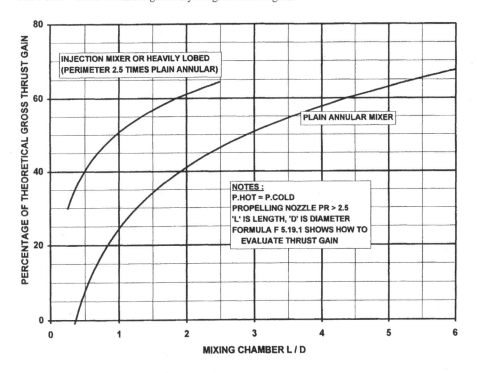

Chart 5.19   Effect of unequal stream pressures on mixer gross thrust gain.

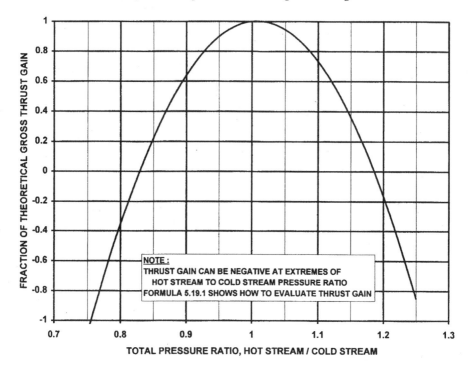

Chart 5.20 Mixed turbofan cycles: optimum fan PR at 11 000 m, ISA, 0.8M.

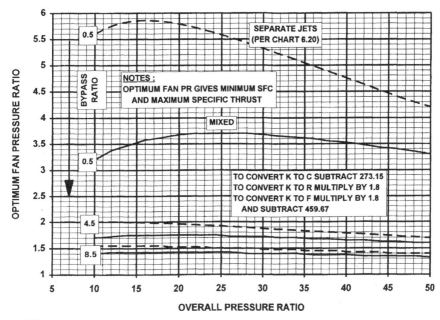

(a) SOT = 1400 K

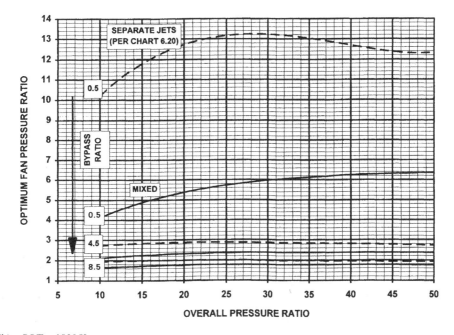

(b) SOT = 1800 K

Chart 5.21   Mixed turbofan cycles: optimum fan PR at 11 000 m, ISA, 2.2M.

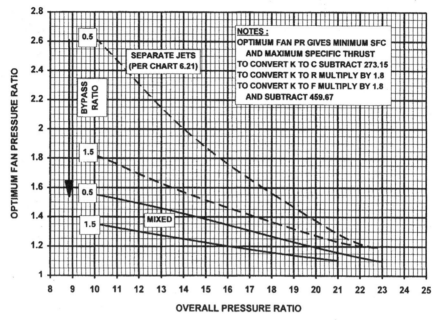

(a)   SOT = 1400 K

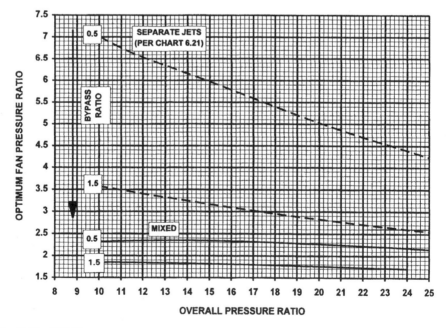

(b)   SOT = 1800 K

Chart 5.22   Afterburner net thrust gain versus jet pipe temperature ratio and flight Mach number.

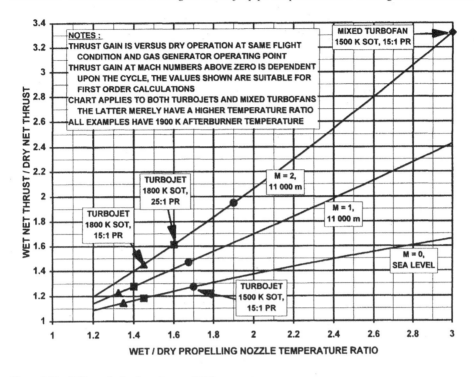

Chart 5.23   Effect of afterburning on SFC.

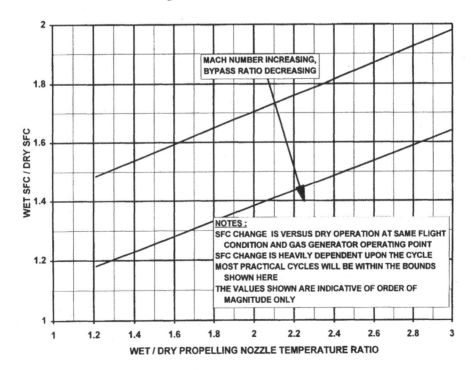

Chart 5.24   Recuperator design point performance versus mass flow per unit volume.

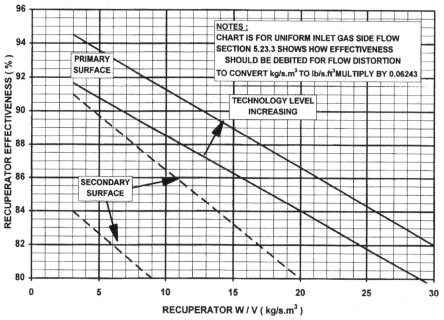

(a)   Effectiveness

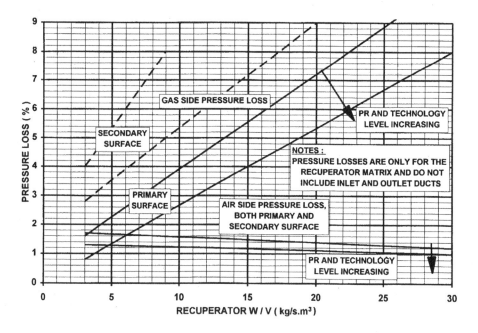

(b)   Pressure losses

Chart 5.25   Regenerator design point performance versus flow per unit area.

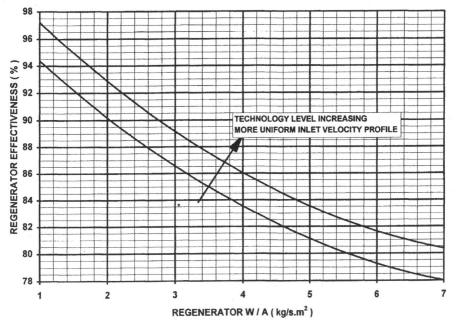

(a)   Effectiveness

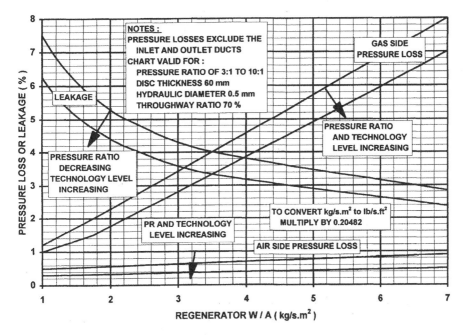

(b)   Pressure losses and leakage

Chart 5.26   Alternator efficiency versus load and power factor.

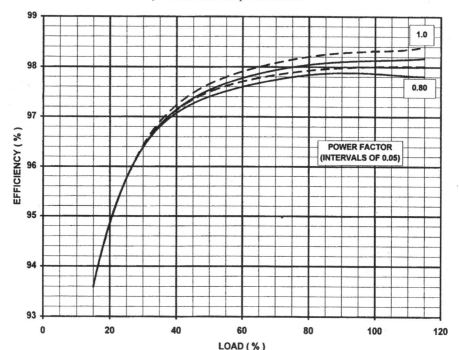

References

1. H. Cohen, G. F. C. Rogers and H. I. H. Saravanamuttoo (1995) *Gas Turbine Theory*, 4th edn, Longman, Harlow.
2. J. H. Horlock (1958) *Axial Flow Compressors*, Butterworth, London.
3. N. A. Cumpsty (1989) *Compressor Aerodynamics*, Longman, London.
4. S. L. Dixon (1969) *Thermodynamics of Turbomachinery*, Pergamon Press, London.
5. R. C. Pampreen (1969) *Small Turbomachinery Compressor and Fan Aerodynamics*, ASME 73-GT-69, ASME, New York.
6. J. P. Smed, F. A. Pisz, J. A. Kain, N. Yamaguchi and S. Umeura (1991) *501F Compressor Development Program*, ASME 91-GT-226, ASME, New York.
7. F. Carchedi and G. R. Wood (1982) Design and development of a 12:1 PR compressor for the Ruston 6 MW gas turbine. *ASME Journal of Engineering for Power*, **104**, 823–31.
8. A. R. Howell and W. J. Calvert (1981) *Axial Compressor Performance Prediction by Stage Stacking*, HMSO, London.
9. J. A. Raw and G. C. Weir (1980) *The Prediction of Off-Design Characteristics of Axial and Axial/Centrifugal Machines*, Society of Automotive Engineers, Pennsylvania.
10. P. M. Came (1978) The development, application and experimental evaluation of a design procedure for centrifugal compressors. *Proceedings of the Institution of Mechanical Engineers*, **192** (5), 49–67.
11. M. V. Herbert (1980) *A Method of Performance Prediction for Centrifugal Compressors*, Aeronautical Research Council, HMSO, London.
12. C. Rodgers and R. A. Langworthy (1974) *Design and Test of a Small Two Stage High Pressure Ratio Centrifugal Compressor*, ASME 74-GT-137, ASME, New York.
13. R. W. Chevis and R. J. Varley (1980) *Centrifugal Compressors for Small Aero and Automotive Gas Turbine Engines*, AGARD 55th Specialists' Meeting, Paper 21, May 1980, Brussels.
14. A. H. Lefebvre (1983) *Gas Turbine Combustion*, Hemisphere, New York.
15. F. A. Williams (1965) *Combustion Theory*, Addison Wesley, Reading MA.
16. G. Leonard and J. Stegmaier (1989) *Development of an Aeroderivative Gas Turbine Dry Low Emissions System*. ASME 89-GT-255, ASME, New York.

17. T. Sattelmeyer, M. P. Felchlin, J. Haumann and D. Styner (1992) Second generation low emission combustors for ABB gas turbines: burner development and tests at atmospheric pressure. *Transactions of ASME*, **114**, 118–124.
18. A. H. Summerfield, D. Pritchard, D. W. Tuson and D. A. Owen (1993) *Mechanical Design and Development of the RB211 Dry Low Emissions Engine*, ASME 93-GT-245, ASME, New York.
19. N. C. Corbett and N. P. Lines (1993) *Control Requirements for the RB211 Low Emission Combustion System*, ASME 93-GT-12, ASME, New York.
20. J. H. Horlock (1966) *Axial Flow Turbines*, Butterworth, London.
21. S. C. Kacker and U. Okapuu (1982) A mean line prediction method for axial flow turbine efficiency. *ASME Journal of Engineering for Power*, **104**, 111–119.
22. D. G. Ainley and G. C. R. Mathieson (1951) *A Method of Performance Estimation for Axial Flow Turbines*, Aeronautical Research Council, HMSO, London.
23. J. Dunham and P. M. Came (1970) Improvements to the Ainley–Mathieson method of turbine performance prediction. *ASME Journal of Engineering for Power*, **92**, 252–256.
24. H. E. Rohlik (1968) *Analytical Determination of Radial Inflow Turbine Design Geometry for Maximum Efficiency*, NASA Technical Note TN D 4384, Washington DC.
25. G. F. Hiett and I. H. Johnston (1964) Experiments concerning the aerodynamic performance of inward flow radial flow turbines. *Proceedings of the Institution of Mechanical Engineers*, **178**, Part 3I(ii).
26. C. Rodgers (1966) *Efficiency and Performance Characteristics of Radial Turbines*, Society of Automotive Engineers, Pennsylvania.
27. J. Mowill and S. Strom (1983) An advanced radial-component industrial gas turbine. *ASME Journal of Engineering for Power*, **105**, 947–952.
28. Jane's Information Group (1997) *Janes All The World's Aircraft 1997–8*, Jane's Information Group, Coulsdon, Surrey.
29. *The Diesel and Gas Turbine Worldwide Catalog*, Diesel and Gas Turbine Publications, Brookfield, Wisconsin.
30. B. S. Massey (1975) *Mechanics of Fluids*, Van Nostrand Reinhold, London.
31. R. E. Grey and H. D. Wilstead (1950) *Performance of Conical Jet Nozzles in Terms of Flow and Velocity Coefficients*, NACA 933, Washington DC.
32. J. Fabri (ed.), (1956) Air intake problems in supersonic propulsion. *AGARD 11th Combustion and Propulsion Panel Meeting*, Paris, December 1956, Pergamon Press, Elmsford, New York.
33. W. Swan (1974) Performance problems related to installation of future engines in both subsonic and supersonic transport aircraft. *2nd International Symposium on Air Breathing Engines*, Sheffield, UK, March 1974.
34. M. Summerfield, C. R. Foster and W. C. Swan (1954) Flow separation in overexpanded supersonic exhaust nozzles. *Jet Propulsion*, **24**, 319–321.
35. R. C. Adkins and J. O. Yost (1983) *A Compact Diffuser System for Annular Combustors*, ASME 83-GT-43, ASME, New York.
36. ESDU (1975) *Performance of Circular Annular Ducts in Incompressible Flow*, Fluid Mechanics Internal Flow Vol. 4, ESDU, London.
37. D. S. Miller (1978) *Internal Flow Systems*, British Hydrodynamic Research Association, London.
38. K. P. L. Fullager (1973) The design of air cooled turbine rotor blades. In: *Design and Calculation of Constructions Subject to High Temperature*, Royal Institute of Engineers in the Netherlands, Delft.
39. ESDU (1965) *General Guide to the Choice of Journal Bearing Type*, Tribiology Vol. 1, ESDU 65007, ESDU, London.
40. ESDU (1967) *General Guide to the Choice of Thrust Bearing Type*, ESDU 67003, ESDU, London.
41. H. Pearson (1962) Mixing of exhaust and bypass flow in a bypass engine. *Journal of the Royal Aeronautical Society*, **66**, 528.
42. T. H. Frost (1966) Practical bypass mixing systems for fan jet aero engines. *The Aeronautical Quarterly*, **May**, 141–160.
43. R. L. Marshal, G. E. Canuel and D. J. Sullivan (1967) Augmentation systems for turbofan engines. *Combustion in Advanced Gas Turbine Systems*, Cranfield International Symposium Series, Vol 10, Pergamon Press, Elmsford, NY.
44. H. M. Spiers (1961) *Technical Data on Fuel*, The British National Committee, London.
45. E. Hughes (1973) *Electrical Technology*, Longman, Harlow.

Chapter 6
Design Point Performance and Engine Concept Design

6.0 Introduction

Design point performance is central to the engine concept design process. The engine configuration, cycle parameters, component performance levels and sizes are selected to meet a given specification. This chapter describes this performance input, which cannot be divorced from component design. Design point performance must be defined before analysis of any other operating conditions is possible. The resulting overall performance of the final engine will be crucial to its commercial success or otherwise.

Generic design point diagrams and sample design point calculations are presented for all major gas turbine types. The diagrams are referred to extensively in Chapter 1, which describes gas turbine engine applications. Section 3.6.5 includes schematic T–S diagrams for the major engine cycles. Configurations of all gas turbine types discussed are presented just before Chapter 1.

6.1 Design point and off design performance calculations

6.1.1 *The engine design point and design point performance calculations*

For initial definition work, the operating condition where an engine will spend most time has been traditionally chosen as the *engine design point*. For an industrial unit this would normally be ISO base load, or for an aero-engine cruise at altitude on an ISA day. Alternatively some important high power condition may be chosen. Either way, at the design point the engine configuration, component design and cycle parameters are optimised. The method used is the *design point performance calculation*. Each time input parameters are changed and this calculation procedure is repeated, *the resulting change to the engine design requires a different engine geometry, at the fixed operating condition*.

For the concept design phase described here the *component design points* are usually at the same operating condition as the engine design point. In a detailed design phase however, this may not be true. For example in the detailed design phase an aero-engine fan may be designed at the top of climb, the highest referred speed and flow, whereas the engine design point would be cruise. In this chapter the term *design point* refers to the engine design point in the concept design phase, which is taken to be coincident with the component design points.

6.1.2 *Off design performance calculations*

With the engine geometry fixed by the design point calculation, the performance at other key operating conditions can be evaluated, such as ISA SLS takeoff for an aero-engine. In this instance the calculation procedure is the *off design performance calculation*, which is covered in Chapter 7. Here *geometry is fixed and operating conditions are changing*. In the concept design phase design point and off design performance calculations must be used iteratively, as described in section 6.12.

6.2 Design point performance parameters

6.2.1 *Engine performance parameters*

A number of key parameters that define overall engine performance are utilised to assess the suitability of a given engine design to the application, or compare several possible engine designs. These *engine performance parameters* are described below.

- Output power or net thrust (PW, FN – Formulae F6.1–F6.4)
  The required output power or net thrust is almost always the fundamental goal for the engine design. It is evaluated via the overall cycle calculation. The term *effective* or *equivalent power* is used for turboprops and turboshafts, where any residual thrust in the exhaust is converted to power via Formula F1.18 and added to the shaft output power.
- Exhaust gas power
  For a turboshaft engine core this is the output power that would be produced by a power turbine of 100% efficiency. It is of interest when engine cores are tested or supplied without their free power turbine, which may remain with the installation or be supplied by another collaborating company.
- Specific power or thrust (SPW, SFN – Formulae F6.5 and F6.6)
  This is the amount of output power or thrust per unit of mass flow entering the engine. It provides a good first-order indication of the engine weight, frontal area and volume. It is particularly important to maximise specific power or thrust in applications where engine weight or volume are crucial, or for aircraft which fly at high Mach numbers where the drag per unit frontal area is high. For turboprops and turboshafts an *effective specific power* may be evaluated, based on the effective power from Formula F1.18.
- Specific fuel consumption (SFC – Formulae F6.7 and F6.8)
  This is the mass of fuel burnt per unit time per unit of output power or thrust. It is important to minimise SFC for applications where the weight and/or cost of the fuel is significant versus the penalties of doing so. When quoting SFC values it is imperative to state the calorific value of the fuel, and whether it is the higher or lower heating value (see Chapter 13), to ensure valid 'back to back' comparisons. Again, for turboprops and turboshafts an *effective SFC* may be evaluated.
- Thermal efficiency for shaft power engines (ETATH – Formula F6.9)
  This is the engine power output divided by the rate of fuel energy input, usually expressed as a percentage. It is effectively the reciprocal of SFC, but is independent of fuel calorific value. However when quoting thermal efficiencies it is still important to state whether the values are based upon the higher or lower fuel heating value (see Chapter 13). For combined cycle applications the terms *gross* and *net* thermal efficiency are used. Gross thermal efficiency does not deduct the power required to drive the steam plant auxiliaries, whereas net values do.

  Thermal efficiency is usually quoted for industrial gas turbines, and SFC for aircraft turboshafts and turboprops. For marine and automotive gas turbines both terms are commonly used. This difference reflects the higher importance of fuel weight in aero applications, and the lower likely variation of fuel calorific values.
- Heat rate for shaft power cycles (HRATE – Formula F6.10)
  Heat rate is a parameter used only in the power generation industry, and is the rate of fuel energy input divided by the useful power output. Hence it is comparable to SFC but is independent of fuel calorific value. Again it is important to state whether higher or lower calorific value has been assumed in calculating fuel energy input, and for combined cycle applications whether the values are gross or net.

  Formulae F6.11–F6.13 show the interrelationships between shaft power engine SFC, thermal efficiency and heat rate.

- Exhaust temperature (T6)
  For engines used in combined cycle for industrial power generation, high exhaust temperature is vital in maximising overall efficiency. For combined heat and power the optimum
  value depends on the relative demand of heat versus power. In both cases there is a limit
  to the allowable exhaust temperature due to mechanical integrity considerations in the
  steam plant.
      For military aircraft applications low exhaust gas temperature is important to reduce the
  infrared *signature* presented to heat seeking missiles.
- Exhaust mass flow (W6)
  For engines used in combined cycle or combined heat and power applications the exhaust
  mass flow is important in indicating the heat available in the gas turbine exhaust, and
  hence the overall plant thermal efficiency.

For aircraft thrust engines there are a number of secondary performance parameters. These do
not in themselves describe overall engine performance, but do help the engine designer understand the variation of the primary performance parameters across a number of designs. These
parameters are as follows:

- Thermal efficiency (ETATH – Formula F6.14)
  Thermal efficiency for aircraft thrust engines is defined as the rate of addition of kinetic
  energy to the air divided by the rate of fuel energy supplied, usually expressed as a
  percentage. The energy in the jet is proportional to the difference in the squares of jet and
  flight velocities. Generally thermal efficiency increases as pressure ratio and SOT increase
  together (see section 6.2.2), as this results in a higher jet velocity for a given energy input.
- Propulsive efficiency (ETAPROP – Formulae F6.15 and F6.16)
  Propulsive efficiency for aircraft thrust engines is defined as the useful propulsive power
  produced by the engine divided by the rate of kinetic energy addition to the air, again
  usually expressed as a percentage. The net thrust is proportional to the difference in the jet
  and flight velocities. Since power is force times velocity, propulsive power is proportional
  to the flight speed times the difference in the jet and flight velocities.
      Formula F6.16 shows that propulsive efficiency is improved by low jet velocities, due to
  lower energy wastage as jet kinetic energy. This requires high pressure ratio and low SOT.
  However low jet velocities produce lower thrust output, hence to achieve high propulsive
  efficiency, as well as a required thrust, high engine mass flow must be coupled with low jet
  velocities. This leads to engines of low specific thrust, which are large and heavy. Turbofan
  engines are based upon this principal.

As shown by Formula F6.17, thrust SFC is directly proportional to flight speed, and inversely
proportional to both thermal and propulsive efficiencies. The dependency on the efficiencies is
intuitively obvious. The choice of pressure ratio and SOT for minimum SFC is a compromise
between maximising propulsive and thermal efficiency. An optimum pressure ratio and SOT
occur, as indicated in the generic design point diagrams presented in sections 6.7–6.11.
    The dependency of thrust SFC on flight speed arises because fuel flow relates to power with
the thermal and propulsive efficiencies fixed, and propulsive power is directly proportional
to flight speed for a given thrust.

6.2.2   *Cycle design parameters*

Section 3.6.5 discussed the fundamental thermodynamics of gas turbine cycles in relation
to temperature entropy diagrams. These immediately show that the changes in pressure and
temperature that the working fluid experiences strongly affect the engine performance parameters. The degree of change of pressure and temperature are reflected via the following *cycle
design parameters.*

Overall pressure ratio

This is total pressure at compressor delivery divided by that at the engine inlet.

Stator outlet temperature (SOT)

This is the temperature of the gas able to do work at entry to the first turbine rotor, and is shown diagramatically in Chapter 5. Other terms are also used to reflect maximum temperatures in a cycle:

- Rotor inlet temperature (RIT): this term is sometimes used in North America, and means the same as SOT
- Combustor outlet temperature (COT): this is the temperature at the first turbine nozzle guide vane leading edge
- Turbine entry temperature (TET): this can have either of the above meanings

The standard definition for SOT, used herein, is:

> *The fully mixed out temperature resulting from combustion delivery gas mixing with all cooling air which enters upstream of the first turbine rotor, and is able to do work due to having momentum comparable to the nozzle guide vane flow.*

Hence, in this definition, nozzle guide vane or platform cooling air entering upstream of the throat, or trailing edge cooling air ejected with the same momentum and direction as the main flow, would be included. However front disc face cooling air flow would not be considered in evaluating SOT since it will not do work in the turbine rotor. For most gas turbine engine types it is desirable to raise SOT to as high a level as possible within mechanical design constraints.

In addition for turbofan engines two further cycle design parameters occur due to the parallel gas paths.

Fan pressure ratio

This is the ratio of fan delivery total pressure to that at fan inlet, and is usually lower for the core stream than the bypass due to lower blade speed.

Bypass ratio (BPR – Formulae F6.18–F6.20)

This is the ratio of mass flow rate for the cold stream to that for the hot stream. Practical limitations on bypass ratio are discussed in section 5.5.

6.2.3 Component performance parameters

The plethora of parameters that define component performance in terms of efficiency, flow capacity, pressure loss, etc. are described in Chapter 5. As the level of *component performance parameters* improves, at fixed values of *cycle design parameters*, then the *design point engine performance parameters* (section 6.2.1) also improve.

Changes in component performance parameters have a secondary effect on the optimum values of engine cycle design parameters.

6.2.4 Mechanical design parameters

For a given performance design point to be practical the *mechanical design parameters* must be kept within the limits of the materials, manufacturing and production technology available. Chapter 5 and Reference 2 discuss mechanical design constraints in relation to gas turbine engines. Coverage beyond that is available in classical stress texts such as Reference 3. As a brief guide the major items to be considered are:

- Creep as a function of material type, metal temperatures, stress level or AN^2
- Oxidation as a function of material and coating type, and metal temperatures
- Cyclic life (low cycle fatigue) as a function of material type and metal temperatures
- Disc and blade tensile stress as a function of rim speed or AN^2
- Casing rupture as a function of compressor delivery pressure
- Choke or stall flutter as a function of fan or compressor referred speed
- Vibration (high cycle fatigue) of rotating components as a function of rotational speed and excitation parameters such as upstream blade numbers and pressure levels
- Shaft critical speeds

6.2.5 *Life parameters*

The two major life parameters are:

(1) Time between overhauls (TBO)
(2) Cyclic life (also called low cycle fatigue life): this is the number of times the engine is started, accelerated to full power, and eventually shut down, between overhauls.

The TBO is governed mainly by creep and oxidation life, while cyclic life is dictated by thermal stess levels. Typical life requirements for the major gas turbine applications are as follows.

Application	TBO (hours)	Cycles
Power generation – Base load	25 000–50 000	3000
Power generation – Standby/Peak lopping	25 000	10 000
Gas and oil pumping	25 000–100 000	5000–10 000
Automotive – Family saloon	5000	10 000
Automotive – Truck	10 000	5000–10 000
Marine – Military	5000–20 000	2000–3000
Marine – Fast ferry	5000–10 000	3000
Aero-engine – Civil	15 000	3000
Aero-engine – Military fighter*	25–3000	25–3000

*There is a large difference between past achievements and future targets, as the emphasis has moved from performance to cost of ownership.

6.2.6 *Fuel type*

Kerosene is the standard aviation fuel while marine engines burn diesel and most industrial applications use natural gas. The highly distilled forms of diesel used make little difference to performance compared with kerosene, but natural gas gives performance improvements because of the higher resulting specific heat of the combustion products. The impact of fuel types on thermal efficiency and specific power is described in Chapter 13.

6.3 Design point calculation and diagram

6.3.1 *Calculations*

Design point calculations for various gas turbine types are presented in sample calculations C6.1–C6.5, which also illustrate the increases in both accuracy and complexity as simplifying assumptions are discarded. The calculation routines are ideally suited to digital computer programs and hence the examples have been laid out in this fashion. Though rotational speed is not involved directly in the design point calculation it is inferred in that the levels of component efficiencies employed depend heavily upon it, especially if engine diameter is fixed.

6.3.2   *Design point diagrams*

A *design point diagram* is created by plotting engine performance parameters versus the cycle parameters. To produce the diagram the design point calculations are repeated varying each cycle parameter in turn through the range of interest; every point is a different engine geometry. Generic design point diagrams for all gas turbine engine configurations are presented in sections 6.7–6.11. These diagrams are extremely useful in the engine design process for selecting the optimum cycle parameters, or for comparing the performance of different gas turbine types. Initially design point diagrams are prepared using constant component performance parameter levels for all combinations of cycle parameters. Later in the engine design process different individual component performance levels are used for each combination of cycle parameters, based on aerothermal component design work.

6.3.3   *Referred parameters*

The design point charts may be applied to any altitude, if the referred form of specific power or thrust, SFC, etc. are used, as per Chapter 4. In this way the datum values for the charts are adjusted. A change in flight Mach number does change engine matching however, unless all nozzles are choked, and also changes ram drag. The charts are not directly adaptable to other flight Mach numbers.

6.4   Linearly scaling components and engines

During the concept design process the scaling of existing components, or occasionally a complete engine, will be considered wherever possible. If viable this will significantly reduce programme cost and development risk.

The scaling parameters, presented in Chapter 4, along with rules for their use, describe component and engine performance parameter changes for a given linear scale factor. To a first order SFC, thermal efficiency, specific power and specific thrust do not change with scaling, hence the generic design point diagrams are largely independent of scale factor.

6.5   Design point exchange rates

Design point exchange rates show the impact of a small change, typically 1%, in leading component performance levels on engine design point performance parameters. All other component efficiencies and cycle parameters are held constant as the parameter in question is varied. In practice this would require many other components to be redesigned, to change flow sizes. Tables of design point exchange rates are extremely useful in the engine design process in highlighting the most sensitive component design areas, and should always be produced.

Synthesis exchange rates are quite different, and are utilised for off design analysis where as the component in question is modified no other components are redesigned. All components move or *rematch* to a different non-dimensional operating point. In consequence component performance parameter levels change, as well as cycle parameters.

6.6   Ground rules for generic design point diagrams

Sections 6.7–6.11 provide and discuss generic design point diagrams for all major gas turbine types. The ground rules used in creating these diagrams are as follows:

- Design point computer programs have been used, which employ the calculation procedures illustrated in sample calculations C6.1–C6.5.
- The levels of component efficiencies used are included in Charts 6.1–6.4 along with design point exchange rates.
- Cycle parameters encompass the full range likely to be used for a given configuration.
- For combustor temperature rise, Formula F3.38A is used; for all other components, gas properties are based on CP at mean temperature.
- Constant component performance levels have been used, e.g. efficiency, pressure loss, effectiveness. This means constant polytropic rather than isentropic efficiencies are used for fans, compressors and turbines, reflecting a constant technology level over the range of pressure or expansion ratios, as explained in Chapter 5.
- Constant percentage cooling flows are used.
- Liquid fuel is used, i.e. kerosene or diesel.
- For combined heat and power diagrams the exhaust stack temperature is 150 8C. With diesel fuel this level prevents condensation, and hence sulphidation of the stack. With natural gas the stack temperature is more typically 100 8C, due to the lower sulphur content.
- For thrust engines convergent nozzles have been used for subsonic flight Mach numbers, and convergent–divergent nozzles at 2.2 Mach number.
- For thrust engines all diagrams are without afterburning. The impact of afterburning is discussed in Chapter 5.
- For turbofans the diagrams are for separate core and bypass jets. The impact of an exhaust mixer is discussed in Chapter 5.
- Graph scales are normalised relative to the levels of a typical design point for that engine configuration. This is to ease comparison of various design points, showing at a glance whether a parameter is 'high' or 'low'.
- Key temperature limits that would restrict the practical combination of cycle parameters, via mechanical integrity, are provided for each cycle type.

For real engine projects component performance levels and trends will differ from the constant values used for the generic diagrams herein. To overcome this each section provides design point exchange rates and the component performance assumptions, in addition to the design point diagrams. Moreover, the effects of variable component efficiencies and cooling flows tend to mutually cancel, in that usually engines with high SOT and cooling flows reflect a higher technology level, and therefore have higher component efficiencies.

Section 6.12 describes the engine concept design process. In this, generic design point diagrams are useful for:

- Providing an indication of the optimum combination of cycle parameters for a given engine type
- Comparing the performance of different engine configurations which may be considered for a given engine requirement
- Providing exchange rates for the effect of changing the cycle parameters at constant component efficiencies during the engine design process
- Choosing a narrow range of cycle parameters at which to conduct initial component design to obtain applicable levels of component performance

6.7 Open shaft power cycles: generic design point diagrams and exchange rates

With one exception all gas turbine engine configurations are *'open cycle'*. Air which enters the front of the engine is not recirculated, but is exhausted back into the atmosphere. The following generic design point diagrams relate to open shaft power cycles:

- Simple cycle :
    - Chart 6.5 – thermal efficiency versus pressure ratio and SOT
    - Chart 6.6 – specific power versus pressure ratio and SOT
    - Chart 6.1 – design point exchange rates, and component efficiencies used
    - Chart 6.7 – key temperatures versus pressure ratio and SOT
- Recuperated cycle
    - Chart 6.5 – thermal efficiency versus pressure ratio and SOT
    - Chart 6.6 – specific power versus pressure ratio and SOT
    - Chart 6.1 – design point exchange rates, and component efficiencies used
    - Chart 6.7 – key temperatures versus pressure ratio and SOT
    - Chart 6.8 – effect of recuperator effectiveness on optimum pressure ratio for thermal efficiency
- Intercooled cycle
    - Chart 6.5 – thermal efficiency versus pressure ratio and SOT
    - Chart 6.6 – specific power versus pressure ratio and SOT
    - Chart 6.1 – design point exchange rates, and component efficiencies used
    - Chart 6.7 – key temperatures versus pressure ratio and SOT
    - Chart 6.9 – effect of intercooler position on thermal efficiency
    - Chart 6.10 – effect of intercooler position on specific power
- Intercooled and recuperated cycle
    - Chart 6.5 – thermal efficiency versus pressure ratio and SOT
    - Chart 6.6 – specific power versus pressure ratio and SOT
    - Chart 6.1 – design point exchange rates, and component efficiencies used
    - Chart 6.7 – key temperatures versus pressure ratio and SOT
    - Chart 6.8 – effect of recuperator effectiveness on optimum pressure ratio for thermal efficiency
    - Chart 6.9 – effect of intercooler position on thermal efficiency
    - Chart 6.10 – effect of intercooler position on specific power
- Combined cycle
    - Chart 6.5 – thermal efficiency versus pressure ratio and SOT
    - Chart 6.6 – specific power versus pressure ratio and SOT
    - Chart 6.1 – design point exchange rates, and component efficiencies used

6.7.1   *Simple cycle*

This is the basic shaft power cycle. First air is compressed, and then it is heated by burning fuel in the combustor. Next expansion through one or more turbines produces power in excess of that required to drive the compressors, which is available as output. Whether the engine is single spool or free power turbine is only reflected in a design point diagram by any small change in assessed overall turbine efficiency. The following are apparent from Charts 6.5–6.8:

- Thermal efficiency and specific power generally increase with SOT. Though less convenient, combustor temperature rise is a better gauge of the cycle's ability to deliver power. When this is low excess power is low and the losses due to component inefficiencies dominate, reducing both thermal efficiency and specific power.
- The optimum pressure ratio for thermal efficiency increases with SOT, being 12:1 at 1100 K and rising to over 40:1 at 1800 K. The fact that there is an optimum is because as well as high combustion temperature rise, high thermal efficiency also requires low exhaust temperature to minimise energy wastage. Maximum thermal efficiency occurs at the minimum value of the ratio of combustor temperature rise to exhaust temperature, reflecting the ratio of heat input to heat wastage. Achieving this mainly requires low exhaust temperature, hence the optimum pressure ratios are relatively high. Chart 6.7 shows that at too high a pressure ratio the low combustor temperature rise offsets the low exhaust temperature, which reduces thermal efficiency.

- The optimum pressure ratio for specific power also increases with SOT but is only around half that for thermal efficiency, being only 7:1 at 1100 K rising to 20:1 at 1800 K. Maximum specific power occurs at the maximum difference between the combustor temperature rise and the exhaust temperature, which reflects work output. Achieving this is less dominated by exhaust temperature, and more by the need to reduce compressor work. Again, the trends may be deduced from Chart 6.7.
- For mechanical integrity the combustor entry temperature must be limited via Chart 6.7 to between 850 and 950 K, depending on the technology level.
- If the engine is to be used in combined cycle or combined heat and power then, via Chart 6.7, exhaust temperature must be limited to between 800 and 900 K, depending on the technology level of the heat recovery system.

6.7.2   *Recuperated cycle*

This is as per the simple cycle except that a heat exchanger transfers some of the heat in the exhaust to the compressor delivery air. If the gas and air streams do not mix the heat exchanger is known as a *recuperator*; if they do mix, as with the automotive ceramic rotating matrix variety, it is known as a *regenerator*. The design point diagrams herein reflect a recuperator, the impact of a regenerator being discussed in Chapter 5. The following are apparent from Charts 6.5–6.9:

- Thermal efficiency and specific power generally increase with SOT.
- At optimum pressure ratios the thermal efficiency is around 10% better than that for the simple cycle, due to the heat recovery reducing the fuel requirement. This difference reduces as SOT increases as the corresponding simple cycle becomes more efficient.
- The optimum pressure ratio for thermal efficiency is comparatively low since the difference between the exhaust and compressor delivery temperatures is high, hence more heat may be recovered. This is the dominant effect of pressure ratio. For the 0.88 effectiveness used for Charts 6.5 and 6.6 the optimum pressure ratio for thermal efficiency is 5:1 at 1100 K rising to 11.5:1 at 1800 K.
- The optimum pressure ratio for specific power is 7.5:1 at 1100 K and 23:1 at 1800 K, and is independent of effectiveness. It is very similar to that for simple cycle, being only slightly reduced by the additional pressure losses in the recuperator.
- Effectiveness is one of the few component design parameters that has a strong effect upon the optimum pressure ratio for thermal efficiency. As it is increased the optimum pressure ratio decreases, since increased heat recovery more than offsets a poor simple cycle efficiency.
- Combustor inlet temperature rises with falling pressure ratio for a given SOT. This is because of the reduced power requirement to drive the compressor, which reduces the temperature drop in the turbines.
- For mechanical integrity the recuperator and combustor entry temperatures must be limited via Chart 6.7 to between 850 and 950 K, depending on the technology level.

6.7.3   *Intercooled cycle*

In this configuration the temperature of the air is reduced part way through the compression process using a heat exchanger and an external medium such as water. This increases power output via reduced compressor work, which is directly proportional to compressor inlet temperature. Generally thermal efficiency reduces as more fuel is required to reach a given SOT. The following are apparent from Charts 6.5–6.10:

- Thermal efficiency and specific power generally increase with SOT.
- Thermal efficiency is marginally worse than for the simple cycle below the optimum simple cycle pressure ratio and significantly better above it. This is driven by the relative

magnitudes of the extra power output and the extra fuel flow required to raise the lower compressor delivery temperature to the given SOT.

- The optimum pressure ratio for thermal efficiency is 18:1 at 1100 K and over 40:1 at 1800 K. This is higher than for simple cycle, as the intercooler gives most benefit with a comparatively high pressure ratio.
- Chart 6.9 shows that as the intercooler position is moved thermal efficiency peaks with it 30% through the compression ratio, such that the first compressor pressure ratio would be the overall value raised to an exponent of 0.30. (The design point diagrams are drawn for the intercooler placed 50% through the compression such that the compressors have equal pressure ratios.)
- Specific power is approximately 20–40% higher than for simple cycle due to the reduction in the power to drive the compressor stages after the intercooler. This drive power is directly proportional to compressor inlet temperature for a given pressure ratio.
- The optimum pressure ratio for specific power is 12:1 at 1100 K and over 40:1 at 1800 K. Again this is higher than for simple cycle, and for the same reason.
- Chart 6.10 shows that as the intercooler position is moved specific power peaks with the intercooler 50% through the compression ratio, such that the compressors would have equal pressure ratios. This corresponds to rejecting more heat than for the case of optimum thermal efficiency, and clearly requires more fuel energy input.

6.7.4 *Intercooled and recuperated cycle*

This is the combination of intercooling and recuperation. The thermal efficiency loss due to intercooling is offset by increased exhaust heat recovery due to the lower recuperator air side entry temperature. The following are apparent from Charts 6.5–6.10:

- Thermal efficiency and specific power generally increase with SOT.
- Thermal efficiency is around 20% higher than for simple cycle for a given SOT at the respective optimum pressure ratios, as in concert both the intercooler and recuperator reduce fuel consumption.
- The optimum pressure ratio for thermal efficiency is 7:1 at 1100 K and 20:1 at 1800 K. However the cycle curves are flatter versus pressure ratio than for the other configurations above.
- Again, increasing recuperator effectiveness significantly reduces the optimum pressure ratio for thermal efficiency (Chart 6.8).
- Specific power is around 10% lower than for an intercooled only cycle due to the recuperator pressure losses, but still significantly higher than for simple cycle.
- The optimum pressure ratio for specific power is 13:1 at 1100 K and over 30:1 at 1800 K. Again, the cycle curves are flatter versus pressure ratio than for the other configurations above.
- For mechanical integrity the recuperator and combustor entry temperatures must be limited via chart 6.7 to between 850 and 950 K, depending on the technology level.
- Charts 6.9 and 6.10 show that, as the intercooler position is moved, thermal efficiency peaks with it 40% through the pressure rise, and specific power peaks with it at 50%.

6.7.5 *Combined cycle*

This is where a steam (Rankine) cycle also generates power, using exhaust heat from a simple cycle gas turbine. In addition, *supplementary firing* may be employed whereby a separate boiler supplements the gas turbine heat output when necessary. The curves shown are for a triple pressure reheated steam plant. The following are apparent from Charts 6.5 and 6.6:

- Thermal efficiency and specific power generally increase with SOT.
- Thermal efficiency exceeds that of all other configurations by around 20–30%.

- The optimum pressure ratio for thermal efficiency is 4:1 at 1100 K and around 21:1 at 1800 K. These values are lower than for simple cycle, as here increasing pressure ratio at a given SOT reduces the exhaust heat available for the steam cycle, reducing its efficiency. At the highest SOT levels the thermal efficiency curves are flatter versus pressure ratio than for all other configurations above; the steam cycle trends offset those of the gas turbine, benefiting from the higher exhaust temperature when gas turbine efficiency falls.
- As a rule of thumb, the pressure ratio which gives the best combined cycle thermal efficiency, while retaining an acceptable gas turbine exit temperature, is the optimum value for simple cycle specific power.
- The optimum pressure ratio for specific power is less than 4:1 at 1100 K and 7:1 at 1800 K. These values are even lower than those for simple cycle due to the effect of high gas turbine exit temperature on steam plant power.
- The exhaust temperature limits of 800–900 K discussed for simple cycle engines must be considered here.

6.8 Combined heat and power: generic design point diagrams and exchange rates

This configuration differs from those above in that the engine exhaust heat is also utilised in the application, either directly in an industrial process or to raise steam for space heating. A varying amount of exhaust heat may be utilised or wasted, hence *heat to power ratio* is an important parameter in cycle comparisons. Again, supplementary firing may be employed. Relevant chart numbers are:

> Chart 6.1 – design point exchange rates, and component efficiencies used
> Chart 6.11 – thermal efficiency versus heat to power ratio and pressure ratio

The following are apparent from these charts.

- Thermal efficiency increases with SOT.
- Thermal efficiency increases almost linearly with heat to power ratio, as would be expected.
- Increasing pressure ratio increases thermal efficiency, though with diminishing returns above a value of 12:1. This reflects the corresponding simple cycle trends.
- Relatively high exhaust temperature, as given by a high specific power gas turbine, is often beneficial as *high grade* heat is more useful for a variety of processes. The exhaust temperature limits of 800–900 K discussed for simple cycle engines must be considered here, however.

6.9 Closed cycles: generic design point diagrams and exchange rates

Here the working fluid is recirculated through the cycle. Heat exchangers are employed in place of the intake/exhaust and combustion processes, passing heat to and from external media. Maximum temperatures are limited by heat exchanger mechanical integrity, and unlike for open cycles compressor entry pressure may be controlled to vary power level. Applications involve energy sources unsuited to direct combustion within a gas turbine engine, such as nuclear reactors or alternative fuels, as discussed in Chapter 1. The working fluid is normally helium, due to its high specific heat. Relevant chart numbers are:

> Chart 6.12 – thermal efficiency versus pressure ratio, SOT and inlet temperature
> Chart 6.13 – power ratio versus pressure ratio and SOT

The following are apparent from these charts:

- Thermal efficiency and 'power ratio' generally increase with SOT.
- The 'power ratio' as presented is specific power multiplied by the ratio of the density of helium at the inlet temperature and pressure to that of air at ISO conditions. This reflects the impact on the required engine size.
- The optimum pressure ratio for thermal efficiency is around 4:1.
- Reducing inlet temperature raises thermal efficiency and power ratio, via increased non-dimensional SOT, i.e. SOT/T1, as per Chapter 4.
- Raising inlet pressure increases power ratio, via increased fluid density.
- The optimum pressure ratio for power ratio is around 6:1.
- Practical SOT levels are low, due to heat exchanger integrity limits. The values shown are 1100 K and 1200 K.

6.10 Aircraft engine shaft power cycles: generic design point diagrams and exchange rates

Here a simple cycle configuration produces shaft power, along with a residual exhaust thrust. The charts at ISA SLS are essentially as per those for a simple cycle ground based application, but with SFC plotted rather than thermal efficiency. The relevant charts are:

Chart 6.14 – SFC and specific power versus pressure ratio and SOT, ISA SLS
Chart 6.15 – SFC and specific power versus pressure ratio and SOT, ISA, 6000 m, 0.5 M
Chart 6.2 – design point exchange rates, and component efficiencies used

The following are apparent from these charts:

- SFC and specific power generally improve with SOT.
- At ISA SLS the optimum pressure ratio for SFC is 12:1 at 1100 K SOT and over 30:1 at 1700 K. At the typical turboprop cruise flight condition of 6000 m 0.5 M it is 17:1 at 1100 K SOT, and over 30:1 at 1700 K. The small increase in optimum pressure ratio is due to the increased cycle temperature ratio at the cooler inlet conditions at altitude, despite the change in matching produced by the flight Mach number.
- At both conditions the optimum pressure ratio for specific power is around 7:1 at 1100 K SOT and 19:1 at 1700 K.

6.11 Aircraft engine thrust cycles: generic design point diagrams and exchange rates

The following generic design point diagrams are presented:

- Turbojets
    Chart 6.16 – SFC and specific thrust versus pressure ratio and SOT, ISA, 11 000 m, 0.8 M
    Chart 6.17 – SFC and specific thrust versus pressure ratio and SOT, ISA, 11 000 m, 2.2 M
    Chart 6.18 – compressor delivery pressure and temperature versus pressure ratio
    Chart 6.3 – design point exchange rates, and component efficiencies used ISA, 11 000 m, 0.8 M
    Chart 6.4 – design point exchange rates, and component efficiencies used ISA, 11 000 m, 2.2 M
- Turbofans
    Chart 6.18 – compressor delivery pressure and temperature versus pressure ratio
    Chart 6.19 – SFC and specific thrust versus fan PR and bypass ratio, ISA, 11 000 m, 0.8 M

- Ramjets

6.11.1 *Turbojets*

Here a simple cycle configuration produces thrust via high exhaust gas velocities. The charts presented in this section are for engines without afterburning. The effect of an afterburner is described in Chapter 5. The following are apparent from Charts 6.16–6.18.

- At ISA, 11 000 m, 0.8 Mach number, SFC deteriorates (i.e. increases) as SOT increases, but specific thrust improves. This is due to the higher resulting jet velocities, and hence worse propulsive efficiency as per section 6.2.1.
- At ISA, 11 000 m, 0.8 Mach number, SFC improves with pressure ratio, reflecting increased thermal efficiency as per section 6.2.1. For specific thrust the optimum pressure ratio is approximately 7:1 at 1100 K SOT, rising to 14:1 at 1800 K.
- At 2.2 Mach number there are similar trends, except that at the lowest SOT levels SFC rises again at the highest pressure ratios, due to reducing specific thrust and hence dominance of cycle losses. For the highest SOT levels SFC is better than at 0.8 Mach number, due to increased ram compression and propulsive efficiency per section 6.2.1. Overall, specific thrust is worse due to the higher intake ram drag.
- For a given pressure ratio, increasing Mach number from 0.8 to 2.2 increases compressor delivery temperature by 300–500 K, and pressure by a factor of around 7. For mechanical integrity compressor delivery temperature must be kept below approximately 850–950 K, via Chart 6.18.

6.11.2 *Turbofans*

Here air from the first compressor of what is otherwise a turbojet engine bypasses the remaining compressor and turbine stages. This air produces thrust directly, additional to that from the hot exhaust gases. The charts presented in this section are for separate jets and without afterburning. The impact of mixing the jets, or employing an afterburner, is covered in Chapter 5. The following are apparent from Charts 6.18–6.25:

- Chart 6.19 shows that for each combination of bypass ratio, overall pressure ratio and SOT there is an *optimum fan pressure ratio*, giving both maximum specific thrust and minimum SFC. Optimum specific thrust and SFC occur at the same fan pressure ratio since for a fixed core engine combustion mass flow is fixed, as are combustion entry and exit temperature; fuel flow must therefore be fixed also. The only remaining influence on SFC is specific thrust.
- At the optimum fan pressure ratio the overall energy conversion to thrust is maximised. Thrust is proportional to jet velocity (momentum), and jet kinetic energy to the square of

it. At low fan pressure ratio excessive core jet velocity wastes energy, and so does the bypass jet velocity at high fan pressure ratio. There is also wastage in the inefficiency of energy transfer from the core to bypass stream, as the LP turbine and fan bypass efficiencies are less than 100%. The optimum ratio of bypass jet velocity to that of the core is simply the energy transfer efficiency, which is the product of the LP turbine and fan bypass efficiencies. The optimum *core* jet velocity is therefore always the higher, by a factor of around 1.2; the optimum would be equal velocities if the energy transfer efficiency were 100%. (The optimum velocity ratio can be proved by algebraic differentiation to find the maximum thrust from a fixed energy level.)

- Increasing bypass ratio at fixed SOT and overall pressure ratio reduces the optimum fan pressure ratio sharply. This is due to reduced core nozzle pressure ratio and hence jet velocity, resulting from an increased LP turbine power requirement to compress the bypass stream.
- At 0.8 Mach number optimum fan pressure ratio is mostly almost constant versus overall pressure ratio for a fixed SOT and bypass ratio (see Chart 6.20). Exceptions are the lowest SOT and bypass ratio. Low SOT decreases the core nozzle pressure ratio, increasing sensitivity to changes in fan drive power. Conversely, low bypass ratio reduces the effect of fan pressure ratio on fan drive power.
- At 2.2 Mach number optimum fan pressure ratio falls as overall pressure ratio increases, particularly at low bypass ratios (see Chart 6.21). This is because at 2.2 Mach number excessive core nozzle jet velocities result from the intake ram compression, which raises all engine pressures. The highest core nozzle jet velocities occur at low bypass ratio (which requires low LP turbine power), and/or at low overall pressure ratio with all else fixed (which requires low HP turbine power). In either case the optimum fan pressure ratio to achieve optimum jet velocities is higher at 2.2 Mach number.
- Increasing SOT at a fixed overall pressure ratio and bypass ratio increases optimum fan pressure ratio at all Mach numbers. This is due to increased ability to drive a fan whilst maintaining core nozzle pressure ratio.

Charts 6.22–6.25 show every point at optimum fan pressure ratio. The main effects are as follows.:

- Increasing bypass ratio at a fixed SOT and overall pressure ratio

    At 0.8 Mach number increasing bypass ratio generally makes SFC significantly better, and specific thrust significantly worse. Above bypass ratios of around 5 there are diminishing returns for SFC, however, and at low SOT and high pressure ratio increasing bypass ratio actually makes SFC worse due to loss of core stream thrust.

    At 2.2 Mach number only low bypass ratios are practical, and even at the highest SOT there are only small benefits for SFC beyond a bypass ratio of 2.5, which is the highest value shown. At the lowest SOT increasing bypass ratio again actually makes SFC worse.
- Increasing overall pressure ratio at a fixed SOT and bypass ratio

    At 0.8 Mach number increasing pressure ratio mostly improves SFC, due to improved thermal efficiency. At low SOTs specific thrust falls with increasing overall pressure ratio, corresponding to low core jet velocities as discussed above, and at high SOTs specific thrust remains almost constant.

    At 2.2 Mach number optimum overall pressure ratio is lower, as significant compression happens anyway due to intake ram recovery. For all SOTs the rate of fall of specific thrust with increasing pressure ratio is steeper than at 0.8 Mach number, due to nett thrust being a smaller proportion of gross thrust and hence more sensitive to cycle changes.
- Increasing SOT at a fixed bypass ratio and overall pressure ratio

    At 0.8 Mach number increasing SOT at the highest bypass ratio of 10.5:1 improves SFC due to increased core jet velocity. The opposite is true at a bypass ratio of 0.5, where core jet velocities are already high and the engine is more like a turbojet. For intermediate bypass ratios the optimum value of SOT for SFC increases with bypass ratio. Hence SOT and bypass ratio are usually increased together to achieve the best SFC. Increasing SOT always improves specific thrust. Optimum levels of overall pressure ratio and bypass ratio increase with SOT, as more pressure is available downstream of the turbines to produce core thrust.

This scenario is also apparent at 2.2 Mach number, though increased intake ram drag means low SOT levels are not practical.

- Mach number

   In contrast to a turbojet, SFC is always better at 0.8 than 2.2 Mach number, even at 0.5 bypass ratio. Like a turbojet, turbofan specific thrust is always better at 0.8 than 2.2 Mach number due to intake ram drag.

   Only low bypass ratios and higher SOTs are practical at 2.2 Mach number, and SFC levels for turbofans and turbojets are more similar. Chart 6.18 shows that for a fixed pressure ratio, increasing Mach number from 0.8 to 2.2 increases compressor delivery temperature by 300–500 K (against the practical limit of 850–950 K), and pressure by a factor of around 7.

6.11.3 *Ramjets*

Here fuel is burnt in air compressed solely by the intake ram effect; there is no turbo-machinery. Engine performance parameters are therefore a function only of SOT and flight Mach number. The following are apparent from Chart 6.26.

- Generally increasing SOT worsens SFC, but improves specific thrust. The maximum temperature range extends up to the stoichiometric limit, in the absence of blading life concerns. Accordingly specific thrusts far higher than for a turbojet may be achieved.
- SFC improves with Mach number up to a value of approximately 3, then stays flat as compression and ram effects (including temperature increase) roughly cancel. Specific thrust peaks at a Mach number of approximately 2.6.
- Operation is not possible at very low Mach number. Combustion will not be practical, and there is insufficient ram compression to provide an exhaust pressure ratio for a jet velocity.
- At high flight Mach number, minimising supersonic wave drag dictates that the ramjet be of constant diameter. This constrains the nozzle exit diameter to be less than the ideal value for full supersonic acceleration of the exhaust gas. The charts presented are therefore an 'upper bound' of the performance achievable.

6.12   The engine concept design process

This section presents a simplified overview of the complex concept design process. To lead an engine concept design effectively requires a full understanding of gas turbine performance *and* all the associated design processes and their interrelationship. Equally, others involved in the process must have some knowledge of performance if they are to contribute to it beyond their own subject areas.

6.12.1 *Statement of requirements*

A *statement of requirements* or *specification* for a new engine may be presented by a customer organisation, or may be specified within the cycle designer's company to meet a perceived market need. Realism is required in terms of technical challenge as well as potential development and unit production costs. The document must not just reflect what 'Marketing' or 'the Customer' desires but also what 'Engineering' can achieve, hence it is usually written jointly.

With respect to performance the statement of requirements must contain a minimum of:

- Performance targets both at design point and other key operating conditions. The relationship to *average* or *minimum* engine and in service deterioration must also be specified, as well as future *growth potential*
- Full definition of operating conditions for the above, and also the entire operational envelope, including ambient temperature, pressure or pressure altitude, humidity, flight Mach number and installation losses
- The starting envelope, as well as starting and above idle transient response times
- Fuel type and emissions requirements
- Engine diameter, length and weight
- Time between overhauls and cyclic life
- Design and development programme duration and cost, as well as unit production cost
- Any derivative engines which should be considered

6.12.2 'First cut' design points

Generic design point diagrams such as those presented in sections 6.7–6.11 will show what engine configurations, together with the levels of cycle parameters, are likely to meet the statement of requirements. Next, initial component performance targets are set; the data presented in Chapter 5 are sufficient at this point. For the narrow range of cycle parameters of interest, refined design point diagrams are then produced. Usually component efficiency variation with engine cycle parameters is unknown at this stage. As far as possible the cycle designer takes account of mechanical design limits such as temperature levels, and considers the use of existing or scaled components.

From these design point diagrams a first cut design point for each engine configuration under consideration is formally issued to component designers. An example of multiple configurations is that both simple cycle and recuperated engines may be competitive for a shaft power application.

For clarity all design points should have an identifying version number, as many more updates will follow in what is a *highly iterative design process*

6.12.3 'First cut' aero-thermal component design

The component specialists now reassess the component performance inputs assumed in the design points, and perform initial component sizing. Their analysis uses component design computer codes and empirical data. Mechanical design parameters are monitored and where there is doubt regarding limits mechanical technologists must be consulted. Chapter 5 provides a basic introduction to component design.

Proper choice of rotational speed is crucial. It is key to achieving the target compressor and turbine efficiency levels as well as number of stages and diameter, and the compressor and turbine designers must agree upon it. The practicality or otherwise of using the number of spools that the cycle designer had in mind when setting the initial design points will soon be apparent, and weight and cost implications will become clear.

Component performance levels are now *bid* back to the performance engineer for each design point.

6.12.4 *Design point calculations and aero-thermal component design iterations*

By repeating the design point calculations using the new bid component performance levels the concept design team can decide which engine configuration is the most suitable. *If a bid is changed even for only one component, it will change the design point requirements for all others.*

The process is repeated a number of times for the chosen configuration so that a number of cycle parameter combinations, each with specific component performance levels, can be compared.

6.12.5 *Engine layout*

It is important to commence engine layout drawings as early as possible to highlight potential difficulties. For example, for a turbofan there would normally be a rigid engine diameter constraint. The core diameter required by the compressor and turbine designs, combined with the bypass duct annulus area to keep Mach number to a level where pressure loss is acceptable, may exceed this limit. In this instance the cycle designer may have to reduce bypass ratio, or the component designers increase rotational speed.

6.12.6 *Off design performance*

Once a good candidate design point is emerging from the above iterations the cycle designer must then *freeze* the engine geometry and issue off design performance. This should cover other key operating points as well as all corners of the operational envelope. The off design modelling methods are discussed in Chapter 7. Often at this juncture the component designers will not have created off design maps for the 'frozen' component designs so maps from existing components of similar design must be used, with *factors* and *deltas* applied to align them to the design point, as described in Chapter 5. The performance engineer must decide how the engine will be controlled for key ratings (see Chapter 7) before running the off design cases.

It is good practice at this stage for the performance engineer to highlight the operational cases where the maximum and minimum value of all key parameters occur, and any margins which need to be applied for say operating temperatures for a minimum engine. Appropriate margins are discussed in section 6.13.

6.12.7 *Performance, aero-thermal and mechanical design investigation*

The off design performance data must be examined in detail by all parties. The cycle designer must check whether the engine design meets other performance requirements besides that at the design point. The aero-thermal and mechanical component designers must ensure that the components are satisfactory throughout the operational envelope with respect to component performance levels, stress and vibration.

6.12.8 *Basic starting and above idle transient performance assessment*

Engine operability is assessed by ensuring that surge and weak extinction margins are commensurate with the required accel, decel and start times (see Chapters 8 and 9). A judgement is made with respect to the need for variable stator vanes or blow off valves (BOVs); these also impact compressor design, unless the BOVs are at compressor delivery, and may be inherited anyway if an existing compressor design is utilised. Unless there are unusual or severe requirements, modelling of transient performance or starting would not normally be conducted in the concept design phase. However when such requirements are present or if a novel engine configuration is employed, then transient performance modelling is *essential* as transient considerations can significantly impact the fundamental engine design in terms of cycle, number of spools, etc.

6.12.9 *Iterations*

A number of iterations through the whole process concludes the concept design phase. The resulting engine design is then compared with the statement of requirements and a judgement made as to whether to proceed into a detailed design and development programme.

6.13   Margins required when specifying target performance levels

It is important that no ambiguities are present when performance targets for an engine design are specified, and that certain less obvious issues are not forgotten; some of the latter are outlined below.

6.13.1   *Minimum engine*

Owing to manufacture and build tolerances there will be significant engine to engine variation in production. Usually the engine designers deal with *average engine* data, that is the average performance of all production engines. However if the customer has been guaranteed a given performance level then even the *minimum engine* must achieve this standard. Hence a design margin must be built into the average engine performance targets. The shortfall in power or thrust and SFC of a minimum engine in a production run will be 1–3% versus the average engine if run at constant temperature, depending on build quality and engine complexity. If the customer has been guaranteed levels of power or thrust then this minimum engine will run up to 20 K hotter than an average engine to achieve power or thrust. This must be allowed for in the mechanical design.

6.13.2   *Design and development programme shortfall*

At the outset of a programme the risk of component performance levels falling short of the values used in any preliminary design point analysis must be assessed. Ideally some engine performance risk analysis should be employed to quantify the potential shortfall versus confidence level. Suitable margins should then be applied to the performance target levels.

However, at the outset of a programme this level of risk analysis may not be commensurate with time or resource constraints and the available component performance data base. In this instance judgement must be used. The level of the margins should reflect the confidence of component performance levels employed in any preliminary analysis. Each case must be considered individually, however typical margins are presented below for three confidence levels that the components will meet their design targets.

	High	Medium	Low
Power or thrust	+2.5%	+5%	+7.5%
SFC	−2%	−4%	−6%

When in competitive tender for an engine application the luxury of SFC margins may not be allowed due to commercial constraints. However for shaft power engines building in power margin is often practical by sizing the engine larger. In this instance, should the component performance levels employed in the preliminary analysis be achieved then almost invariably the customer will accept the extra power. In aircraft applications it may be more difficult to provide margin for thrust due to tight restrictions on weight and frontal area.

6.13.3   *Growth potential*

Almost any engine must be capable of adaptation to higher power or thrust levels. In aircraft applications the airframe usually requires it, intentionally or otherwise. In addition, addressing a wider market with development costs spread over a family of engines is attractive economically. Examples for turbofans include the Rolls-Royce RB211/Trent family (thrust growth of 180 to 400+ kN in 25 years), the Rolls-Royce civil and military Spey family, and the General Electric CF6 range. Margin is required on levels of temperature and speed, and there must be a practical route to increasing engine mass flow.

6.13.4 *Engine deterioration*

At the concept design stage it is unusual to specify performance requirements after deterioration over a number of hours in service. However if this is the case then a margin must be allowed for from the outset. The amount required depends on whether in service the engine is governed to fixed levels of temperature, speed or thrust/power. Chapter 7 describes the various rating methods. Depending on engine complexity typical margins at 10 000 hours, expressed at constant SOT, are: +3 to +6% SFC; −5 to −15% power.

Performance degradation is usually exponential with time, and after 10 000 hours there is minimal further deterioration. Practices vary regarding the accounting or not of *running in* effects (Chapter 11) caused by initial seal cutting as part of deterioration. The aim of running in is to improve performance levels by ensuring a gentle first contact on key seals, which removes less material than would a first in-service rub.

One further effect is compressor *fouling* where accretion of airborne particles on aerofoil surfaces reduces flow and efficiency, raising engine temperature levels. This applies mainly to ground based engines and is recoverable via compressor *washing*, for which appropriate fluid injection nozzles must be provided. The temperature increase depends entirely on site location, filter effectiveness and washing frequency.

6.13.5 *Installation losses*

Usually performance targets are stipulated as *uninstalled*, with the engine performance quoted from the engine intake flange to the engine exhaust or propelling nozzle flange. If the performance targets are *installed* then the magnitude of all installation effects must be stated, as follows:

- Plant or airframe intake pressure loss
- Plant exhaust or airframe jet pipe pressure loss
- Customer auxiliary power offtake (gas turbine accessory power requirements should be accounted even for uninstalled performance)
- Customer bleed offtake
- Whether shaft power output and thermal efficiency are at turbine, gearbox or alternator output
- Whether the thrust and SFC include any pod drag

Formulae

F6.1 and F6.2   Shaft power (kW) = fn(power turbine mass flow (kg/s), delta enthalpy or CP (kJ/kg K) and delta temperature (K))

For rigorous calculations change in enthalpy must be used:

F6.1 $PW = W * DH$

For calculations to within 1% accuracy CP at the mean temperature may be utilised:

F6.2 $PW = W * CP * DT$

F6.3 and F6.4   Thrust = fn(mass flow (kg/s), air speed (m/s), jet velocity (m/s), nozzle area (m²), nozzle static pressure (kPa), ambient pressure (kPa))

F6.3 $FG = (W9 * V9) + A9 * 1000 * (PS9 - PAMB)$

$FD = W0 * V0$
$FN = FG - FD$

When the propelling nozzle is unchoked PS9 = PAMB and:

F6.4   FG = (W9 * V9)

FN = FG − FD

(i)   Stations 0 and 9 are free stream and propelling nozzle respectively.
(ii) For a bypass engine with separate jets the thrust of both streams must be evaluated using the above and then summed.

F6.5 and F6.6   Specific power (kW s/kg), specific thrust (N s/kg) = fn(power (kW), thrust (N), inlet mass flow (kg/s))

F6.5   SPW = PW/W0   Specific power
F6.6   SFN = FN/W0   Specific thrust

F6.7 and F6.8   Specific fuel consumption (kg/kW h, kg/N h) = fn(fuel flow (kg/s), shaft power (kW), thrust (N))

F6.7   SFC = 3600 * WF/PW   Shaft power engines
F6.8   SFC = 3600 * WF/FN   Thrust engines

F6.9   Shaft power engine thermal efficiency (%) = fn(power (kW), fuel flow (kg/s), fuel calorific value (kJ/kg))

ETATH = 100 * PW/(WF * HV)

(i)   Whether the fuel lower (LHV), or the higher calorific value (HHV), has been used should always be stated, it is usual to use LHV.

F6.10   Heat rate (kJ/kW h) = fn(power (kW), fuel flow (kg/s), fuel calorific value (kJ/kg))

HRATE = 3600 * WF * HV/PW

(i)   Whether the fuel lower (LHV), or the higher calorific value (HHV), has been used should always be stated, it is usual to use LHV.

F6.11–F6.13   SFC (kg/kW h), thermal efficiency (%) and heat rate (kJ/kW h) inter-relationships

F6.11   HRATE = SFC * HV
F6.12   HRATE = 100 * 3600/ETATH
F6.13   ETATH = 100 * 3600/(SFC * HV)

F6.14   Thrust engine thermal efficiency (%) = fn(mass flow (kg/s), air speed (m/s), jet velocity (m/s), fuel flow (kg/s), fuel lower calorific value (kJ/kg))

ETATH = 100 * 0.5 * (W9 * V9^2 − W0 * V0^2))/(WFE * LHV * 1000)

(i)   Stations 0 and 9 are free stream and propelling nozzle respectively.

F6.15 and F6.16   Propulsive efficiency (%) = fn(thrust (N), mass flow (kg/s), air speed (m/s), jet velocity (m/s))

F6.15   ETAPROP = 100 * FN * V0/(0.5 * (W9 * V9^2 − W0 * V0^2))

If the propelling nozzle is unchoked, and approximating that inlet and propelling nozzle mass flows are equal then:

F6.16   ETAPROP $= 100 * 2/(1 + V9/V0)$

(i)   Stations 0 and 9 are free stream and propelling nozzle respectively.

F6.17   Thrust engine SFC (kg/N h) = fn(air speed (m/s), thermal efficiency (fraction), propulsive efficiency (fraction), fuel lower calorific value (kJ/kg K))

SFC $= 3600 * V1/($ETATH$ *$ ETAPROP $*$ LHV $* 1000)$

F6.18   Bypass ratio = fn(cold and hot stream mass flow (kg/s))

BPR $=$ Wcold/Whot

F6.19 and F6.20   Cold and hot stream mass flow (kg/s) = fn(bypass ratio, inlet mass flow (kg/s))

F6.19   Wcold $=$ W0$/(1 + 1/$BPR$)$
F6.20   Whot $=$ W0$/(1 +$ BPR$)$

Sample calculations

The sample design point calculations in this section show the reader the basic steps through each component, and also how to combine them such that gas path parameters at each station and overall performance parameters may be calculated. The following calculations are provided

C6.1	Single spool turbojet
C6.2	Single spool turboshaft
C6.3	Intercooled and recuperated single spool turboshaft
C6.4	Two spool separate jets turbofan
C6.5	Combined heat and power (CHP) using engine from C6.2

Sample calculation C6.1 is highly simplified, as it ignores the air system and uses F3.40 (fixed CP) for combustor temperature rise. C6.2–C6.4 include the air system and F3.38A (rigorous fit to enthalpy curves) for combustor temperature rise, illustrating the increases in both accuracy and complexity. C6.5 utilises simple charts for combined heat and power performance. The calculation process is laid out as for a computer program or spreadsheet, which is appropriate given the length.

For all the sample design point calculations, fixed values of CP and γ have been used as below:

Compressors: CP $= 1.005$ kJ/kg, $\gamma = 1.4$, $\gamma/(\gamma - 1) = 3.5$
Turbines: CP $= 1.150$ kJ/kg, $\gamma = 1.333$, $\gamma/(\gamma - 1) = 4.0$

For engine design purposes this is an unacceptable error. Here CP at mean T should be used, or ideally the fully rigorous enthalpy/entropy polynomials presented in Chapter 3. Sample calculation C3.2 shows how these more accurate methods may be superimposed onto the design point calculations presented below. This involves a great deal of iteration and hence is best suited to a computer program or spreadsheet.

These sample design point calculations may form the starting point for an engine design:

- Having done the first pass design point calculation using the component efficiency levels provided, the reader may use the data base provided in Chapter 5 to derive more appropriate efficiency levels. These would reflect the size of engine being considered and the component design technology level available. The design point calculation may then be repeated.
- The reader may then conduct some basic component sizing using the data provided in Chapter 5 and sketch out the physical layout of the engine resulting from the design point calculation. If the resulting dimensions are not suitable then the reader may change the cycle and component performance parameters, and repeat the design point calculation.

C6.1   Conduct a design point calculation for a single spool turbojet at ISA, 11 000 m, 0.8 Mach number with the following cycle parameters:

Mass flow $= 100$ kg/s
Pressure ratio $= 25:1$
SOT $= 1400$ K
Compressor polytropic efficiency $= 88\%$
Turbine polytropic efficiency $= 89\%$
Combustion efficiency $= 99.9\%$
Kerosene fuel with LHV $= 43\,100$ kJ/kg
Mechanical efficiency $= 99.5\%$
Intake pressure loss $= 0.5\%$
Compressor exit diffuser pressure loss $= 2.0\%$
Combustor pressure loss $= 3.0\%$
Jet pipe pressure loss $= 1.0\%$
Propelling nozzle CD, CX $= 0.97, 0.99$
Air system extractions $=$ none

Station numbering is as per C6.2.

Ambient conditions
From Chart 2.1 for ISA conditions TAMB $= 216.7$ K, PAMB $= 22.628$ kPa

Free stream conditions
Use Q curve Formulae F3.31 and F3.32 to derive free stream total temperature and pressure:

$T1/216.7 = (1 + (1.4 - 1)/2 * 0.8^2)$
$T1 = 244.4$ K

$P1/22.628 = (244.4/216.7)^{3.5}$
$P1 = 34.47$ kPa

Intake
Total temperature is unchanged along duct, and total pressure loss is applied as a percentage:

$T2 = T1 = 244.4$ K

$P2 = 34.47 * (1 - 0.005)$
$P2 = 34.30$ kPa

$W2 = 100$ kg/s

Compressor

Apply F3.42 to derive isentropic efficiency:

$$E2 = (25^\wedge(1/3.5) - 1)/(25^\wedge(1/(3.5 * 0.88)) - 1)$$
$$E2 = 0.818$$

Apply F5.1.4 to derive T3:

$$T3 - 244.4 = 244.4/0.818 * (25^\wedge(1/3.5) - 1)$$
$$T3 = 695.1 \, K$$

$$P3 = 25 * 34.3$$
$$P3 = 857.5 \, kPa$$

Deduct air offtakes to derive W3:

$$W3 = 100 \, kg/s (\text{no air offtakes})$$

$$PW2 = 100 * 1.005 * (695.1 - 244.4)$$
$$PW2 = 45\,295 \, kW$$

Hence compressor design parameters (see C7.4) are:

$$W\sqrt{T}/P = 100 * 244.47^\wedge 0.5/34.3$$
$$W\sqrt{T}/P = 45.58 \, kg\,\sqrt{K}/s\,kPa$$

$$PR = 25$$

Isentropic efficiency $= 81.8\%$

Compressor exit diffuser

Total temperature is unchanged, derive exit pressure by applying percentage total pressure loss:

$$T31 = T3 = 695.1 \, K$$

$$P31 = 857.5 * (1 - 0.02)$$
$$P31 = 840.35 \, kPa$$

$$W31 = 100 \, kg/s$$

Combustor and SOT station

Calculate FAR to station 41, i.e. after the cooling air that will do work has mixed in before the turbine, using F3.40. Apply percentage pressure loss to derive P41:

$$T4 = 1400 \, K$$

$$P4 = 840.35 * (1 - 0.03)$$
$$P4 = 815.1 \, kPa$$

$$FAR = 1.15 * (1400 - 695.1)/0.999/43100$$
$$FAR = 0.0188$$

$$WF = 0.0188 * 100$$
$$WF = 1.88 \, kg/s$$

$$W4 = 100 + 1.88$$
$$W4 = 101.9 \, kg/s$$

Turbine

This engine cycle has no cooling air hence parameters at the SOT station 41 are equal to those at station 4. Equate power to that of compressor and use F5.9.2 to derive exit temperature:

PW41 = 45295/0.995
PW41 = 45523 kW

45523 = 101.9 ∗ 1.15 ∗ (1400 − T5)
T5 = 1010.5 K

Derive isentropic efficiency E41 via F3.44:

Guess expansion ratio of 4 to derive isentropic efficiency from polytropic efficiency:

E41 = (1 − 4^(−1/4 ∗ 0.89))/(1 − 4^(−1/4))
E41 = 0.906

Derive expansion ratio P41Q5 via F5.9.4:

1400 − 1010.5 = 0.906 ∗ 1400 ∗ (1 − P41Q5^(−1/4))
P41Q5 = 4.34

One should now go back and recalculate turbine isentropic efficiency using this expansion ratio. However for brevity this step is omitted.

Calculate P5 from expansion ratio:

P5 = 815.1/4.34
P5 = 187.8 kPa

W5 = W4 = 101.9 kg/s

Hence turbine design parameters (see C7.4) are:

W$\sqrt{T}$/P = 101.9 ∗ 1400^0.5/815.1
W$\sqrt{T}$/P = 4.68 kg $\sqrt{K}$/s kPa

CPdT/T = 1.15 ∗ (1400 − 1010.5)/1400
CPdT/T = 0.32 kJ/kg K

Isentropic efficiency = 90.6%

Jet pipe

Total temperature is unchanged, derive exit pressure by applying percentage total pressure loss:

T7 = T5 = 1010.5 K

P7 = 187.8 ∗ (1 − 0.01)
P7 = 185.9 kPa

W7 = W5 = 101.9 kg/s

Propelling nozzle

P7QN = 185.9/22.628
P7QN = 8.215

Evaluate nozzle flow parameters. Via the Q curves presented in Chart 3.8 or 3.10 the nozzle is choked. From F3.32, or Chart 3.8 P/PS at M = 1 is 1.8509 for γ = 1.33. Use Q curve formulae to derive other nozzle parameters.

Use the choking pressure ratio to calculate the throat static pressure:

> 185.9/PSN = 1.8506
> PSN = 100.45 kPa

F3.31, T/TS at M = 1:

> 1010.5/TSN = (1 + (1.333 − 1)/2 * 1^2)
> TSN = 866.3 K

F3.35, V/SQRT(T):

> VN = 1 * SQRT(1.333 * 287.05 * 866.3)
> VN = 575.7 m/s

F3.33, Q (choking value):

> 39.6971 = 101.9 * SQRT(1010.5)/(AN * 185.9)
> ANeffective = 0.439 m^2

F5.13.15, definition of CD:

> ANgeometric = 0.439/0.97
> ANgeometric = 0.453 m^2

F5.13.13, gross thrust:

> FG = 101.9 * 575.7 + 0.438 * (100.45 − 22.628) * 1000
> FG = 92776 N

Final calculations

Calculate free stream velocity and hence momentum drag. Then calculate net thrust applying CX to the gross thrust:

F2.6, with M = 1:

> V1 = 0.8 * SQRT(1.4 * 287.05 * 216.7)
> V1 = 236.1 m/s

F5.13.13, momentum drag and net thrust:

> FD = 100 * 236.1
> FD = 23610 N
>
> FN = 92 776 * 0.99 − 23 610
> FN = 68 238 N

Specific nett thrust:

> SFN = 68 238/100
> SFN = 682.4 N s/kg

F6.8 SFC:

> SFC = 1.88 * 3600/68 238
> SFC = 0.0992 kg/N h

Notes: The above performance level is better than that from Chart 6.16. The difference is due to not considering any air system, the approximation of constant gas properties and the approximate temperature rise method employed.

C6.2   Conduct a design point calculation for a simple cycle, single spool turboshaft of:

Mass flow $= 100\,\text{kg/s}$
Pressure ratio $= 20:1$
SOT $= 1400\,\text{K}$
Compressor polytropic efficiency $= 88\%$
Turbine polytropic efficiency $= 89\%$
Combustion efficiency $= 99.9\%$
Kerosene fuel with LHV $= 43\,100\,\text{kJ/kg}$
Mechanical efficiency $= 99.5\%$
Intake pressure loss $= 0.5\%$
Compressor exit diffuser pressure loss $= 3.0\%$
Combustor pressure loss $= 3.0\%$
Exhaust pressure loss $= 5\%$

Air system is as per Chart 6.1 which is summarised in the diagram below together with station numbering.

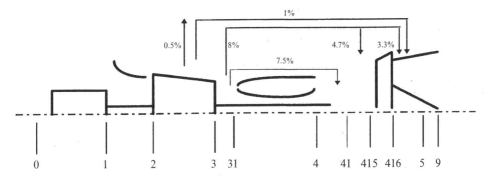

Station 41 is after the NGV cooling has mixed in and hence is the SOT station to which turbine is physically exposed.

Station 415 is the 'pseudo SOT' station, as described in section 5.15.4. A proportion of the 8% cooling air which enters downstream of station 41 is mixed in between 41 and 415 in proportion to the amount of work it achieves in the turbine.

Ambient conditions
From Chart 2.1 for ISA conditions TAMB $= 288.15\,\text{K}$, PAMB $= 101.325\,\text{kPa}$.

Intake
Total temperature is unchanged along duct, and total pressure loss is applied as a percentage:

$\text{T2} = \text{T1} = \text{TAMB} = 288.15\,\text{K}$

$\text{P1} = \text{PAMB} = 101.235\,\text{kPa}$
$\text{P2} = \text{P1} * (1 - 0.005)$
$\text{P2} = 100.82\,\text{kPa}$

$\text{W2} = \text{W1}$

Compressor
Apply F3.42 to derive isentropic efficiency:

$\text{E2} = (20^{\wedge}(1/3.5) - 1)/(20^{\wedge}(1/(3.5 * 0.88)) - 1)$
$\text{E2} = 0.823$

Apply F5.1.4 to derive T3:

$T3 - 288.15 = 288.15/0.823 * (20^\wedge(1/3.5) - 1)$
$T3 = 762.1\,K$

Calculate P3 from given pressure ratio:

$P3 = 20 * 100.82$
$P3 = 2016.4\,kPa$

Deduct air offtakes to derive W3:

$W3 = 100 - 100 * (0.005 + 0.01)$
$W3 = 98.5\,kg/s$

Calculate input power required based on F5.15.1. Considering interstage bleed to be taken at the mean temperature through the compressor:

$Tbleed = (762.1 + 288.15)/2$
$Tbleed = 525.1\,K$

$PW2 = 98.5 * 1.005 * (762.1 - 288.15) + 1.5 * 1.005 * (525.1 - 288.15)$
$PW2 = 47\,275\,kW$

Compressor exit diffuser

Total temperature is unchanged, derive exit pressure by applying percentage pressure loss, derive exit mass flow by deducting air system flows:

$T31 = T3 = 762.1\,K$

$P31 = 2016.4 * (1 - 0.03)$
$P31 = 1955.9\,kPa$

$W31 = 98.5 - 100 * 0.08 - 100 * 0.075$
$W31 = 83\,kg/s$

Combustor and SOT station

Calculate FAR to station 41, i.e. after the cooling air that will do work has mixed in before the turbine, using F3.38A. Apply percentage pressure loss to derive P41:

$T41 = 1400\,K$

$P41 = 1955.9 * (1 - 0.03)$
$P41 = 1897.2\,kPa$

$FAR1 = 0.10118 + 2.00376E\text{-}05 * (700 - 762.1) = 0.09994$
$FAR2 = 3.7078E\text{-}03 - 5.2368E\text{-}06 * (700 - 762.1)$
$\qquad - 5.2632E\text{-}06 * 1400 = -0.003335$
$FAR3 = 8.889E\text{-}08 * ABS(1400 - 950) = 0.00004$
$FAR = (0.09994 - SQRT(0.09994^\wedge2 + 0.003335) - 0.00004)/0.999$
$FAR = 0.0184$

$WF = 0.0184 * (83.0 + 100 * 0.075)$
$WF = 1.665\,kg/s$

$W41 = 83.0 + 100 * 0.075 + 1.665$
$W41 = 92.165\,kg/s$

Now calculate T4 using mixing enthalpy balance for constant specific heats, F5.15.2:

W4 = 83.0 + 1.665
W4 = 84.665 kg/s

7.5 * 1.005 * 762.1 + 84.665 * 1.15 * T4 = 92.165 * 1.15 * 1400
T4 = 1465.0 K

P4 = P41 = 1897.2 kPa

Calculate pseudo SOT station 415 adding in cooling air proportion that mixes in downstream but does work as per section 5.15.4:

W415 = 92.165 + 100 * 0.047
W415 = 96.865 kg/s

F5.15.2 Mixing enthalpy balance for constant specific heats:

92.165 * 1.15 * 1400 + 4.7 * 1.005 * 762.1 = 96.865 * 1.15 * T415
T415 = 1364.4 K
P415 = P41 = 1897.2 kPa

Turbine

Derive P5 via ambient pressure and the exhaust pressure loss:

P5 * (1 − 0.05) = 101.325
P5 = 106.66 kPa
P416 = P5 = 106.66 kPa
P415Q416 = 1897.2/106.66
P415Q416 = 17.79

Derive E415 via F3.44:

E415 = (1 − 17.79^(−1/4 * 0.89))/(1 − 17.79^(−1/4))
E415 = 0.922

Evaluate T416 via F5.9.4:

T415 − T416 = E415 * T415 * (1 − P415Q416^(−1/4))
1364.4 − T416 = 0.922 * 1364.4 * (1 − 17.79^(−1/4))
T416 = 718.9 K

Calculate power via F5.9.2:

PW415 = W415 * CP * (T415 − T416)
PW415 = 96.865 * 1.15 * (1364.4 − 718.9)
PW415 = 71905 kW

Finally calculate capacity at stations 4 and 41:

WRTP4 = 84.752 * SQRT(1464.9)/1897.2
WRTP4 = 1.710 kg√K/s kPa

WRTP41 = 92.165 * SQRT(1400)/1897.2
WRTP41 = 1.818 kg√K/s kPa

Cooling air return downstream of turbine

W5 = 96.865 + 100 * 0.01 + 100 * 0.033
W5 = 101.165 kg/s

Mixing enthalpy balance for constant specific heats, F5.15.2:

$$101.165 * 1.15 * T5 = 96.865 * 1.15 * 718.9$$
$$+ 100 * 0.01 * 1.005 * 525.1 + 100 * 0.033 * 1.005 * 762.1$$
$$T5 = 714.6\,K$$

Exhaust

Total temperature is unchanged along the duct. Use Q curve Formulae F3.32 and F3.33 to determine exhaust plane pressure and area:

$$T9 = T5 = 714.6\,K$$

$$P9/PS9 = (1 + (1.333 - 1)/2 * 0.05^2)^{(4)}$$
$$P9/PS9 = 1.00167$$
Since $PS9 = PAMB$ then $P9 = 101.49\,kPa$

$$W9 = W51 = 101.165\,kg/s$$

$$Q = 1000 * SQRT(2 * 4/287.05) * 1.00167^{(-2/1.333)}$$
$$* (1 - 1.00167^{(-1/4)})$$
$$Q = 3.405\,kg\sqrt{K}/s\,m^2\,kPa$$
$$3.405 = 101.165 * SQRT(714.6)/(A9 * 101.49)$$
$$A9 = 7.826\,m^2$$

Final calculations

Use F6.5, F6.7 and F6.9:

$$PW = 71\,905 - 47\,275/0.995$$
$$PW = 24\,392\,kW$$

$$SPW = 24\,392/100$$
$$SPW = 243.92\,kW\,s/kg$$

$$SFC = 1.752 * 3600/24\,392$$
$$SFC = 0.2586\,kg/kW\,h$$

$$ETATH = 3600/(0.2586 * 43\,100) * 100$$
$$ETATH = 32.30\%$$

Notes: The above values are similar to the datum for Charts 6.5 and 6.6. Any difference is due to the charts being prepared using CP at mean T instead of constant CP.

The calculation process is similar for a free power turbine engine except that:

- Gas generator turbine work is equated to compressor work, hence T45 is found.
- P45 is then derived from work and isentropic efficiency using Formula F5.9.4.
- P5 is derived in the same way.
- Hence for the free power turbine T5 may be found knowing P45, T45, P5 and F5.9.4.
- Hence power output may be calculated from F5.9.2.

C6.3   Conduct a design point calculation for an intercooled and recuperated single spool turboshaft of:

Mass flow = 100 kg/s
Overall pressure ratio = 11:1
Each compressor PR after allowing for duct losses = 3.42
SOT = 1400 K
Intercooler effectiveness = 81%
Recuperator effectiveness = 86%

Compressor polytropic efficiency = 88%
Turbine efficiency = 89%
Combustion efficiency = 99.9%
Kerosene fuel with LHV = 43 100 kJ/kg
Mechanical efficiency = 99.5%
Intake pressure loss = 0.5%
Intercooler inlet duct pressure loss = 2.5%
Intercooler pressure loss = 2.5%
Intercooler exit duct pressure loss = 1.0%
Recuperator air side inlet duct pressure loss = 3.0%
Recuperator air side pressure loss = 3.0%
Recuperator exit duct pressure loss = 1.0%;
Combustor pressure loss = 3.0%
Recuperator gas side inlet duct pressure loss = 4.0%
Recuperator gas side pressure loss = 4.0%;
Exhaust pressure loss = 1.0%
Intercooler sink temperature = ambient

Air system is as per Chart 6.1 which is summarised in the diagram with C6.2, but:

NGV cooling air is taken from recuperator delivery and blade and disc cooling air has been taken from recuperator inlet.

Station numbering is as per C6.1 but with additional stations for the additional components:

compressor 1 exit = 21	compressor 2 inlet = 26	combustor in = 31
intercooler inlet = 23	recup air in = 307	recup gas in = 6
intercooler exit = 25	recup air out = 308	recup gas out = 601

Intake
Total temperature is unchanged along duct, and total pressure loss is applied as a percentage:

$$T2 = T1 = TAMB = 288.15\,K$$

$$P1 = PAMB = 101.325\,kPa$$
$$P2 = P1 * (1 - 0.005)$$
$$P2 = 100.82\,kPa$$

$$W2 = W1$$

Compressor 1
Apply F3.42 to derive isentropic efficiency:

$$E2 = (3.42^\wedge(1/3.5) - 1)/(3.42^\wedge(1/(3.5 * 0.88)) - 1)$$
$$E2 - 0.858$$

Apply F5.1.4 to derive T21:

$$T21 - 288.15 = 288.15/0.858 * (3.42^\wedge(1/3.5) - 1)$$
$$T21 = 428.8\,K$$

Calculate P21 from given pressure ratio:

$$P21 = 3.42 * 100.82$$
$$P21 = 344.80\,kPa$$

Deduct any cooling flow offtakes:

$$W21 = W2 = 100\,kg/s$$

Calculate input power required based on F5.15.1:

$$PW2 = 100 * 1.005 * (428.8 - 288.15)$$
$$PW2 = 14\,135\,kW$$

Intercooler inlet duct

Apply percentage total pressure loss. Total temperature and mass flow are unchanged:

$$P23 = 344.80 * (1 - 0.025)$$
$$P23 = 336.18\,kPa$$

$$T23 = T21 = 428.8\,K$$

$$W23 = W21 = 100\,kg/s$$

Intercooler

Apply F5.23.2 to derive T25, and apply percentage total pressure loss:

$$0.81 = (428.8 - T25)/(428.8 - 288.15)$$
$$T25 = 314.9\,K$$

$$P25 = 336.18 * (1 - 0.025)$$
$$P25 = 327.8\,K$$

$$W25 = W23 = 100\,kg/s$$

Intercooler exit duct

$$T26 = T25 = 314.9\,K$$

$$P26 = 327.8 * (1 - 0.01)$$
$$P26 = 324.5\,kPa$$

$$W26 = W25 = 100\,kg/s$$

Compressor 2

Apply F3.42 to derive isentropic efficiency:

$$E26 = (3.42^{\wedge}(1/3.5) - 1)/(3.42^{\wedge}(1/(3.5 * 0.88)) - 1)$$
$$E26 = 0.858$$

Apply F5.1.4 to derive T3:

$$T3 - 314.9 = 314.9/0.858 * (3.42^{\wedge}(1/3.5) - 1)$$
$$T3 = 469.4\,K$$

Calculate P3 from given pressure ratio:

$$P3 = 3.42 * 324.5$$
$$P3 = 1109.79\,kPa$$

Deduct any cooling flow offtakes:

$$W3 = 100 - 100 * (0.005 + 0.01)$$
$$W3 = 98.5\,kg/s$$

Calculate input power required based on F5.15.1:

$$PW26 = 100 * 1.005 * (469.4 - 314.9)$$
$$PW26 = 15\,527\,kW$$

Recuperator inlet duct

T307 = T3 = 469.4 K

P307 = 1109.79 * (1 − 0.03)
P307 = 1076.5 kPa

Deduct cooling flow offtakes:

W307 = 98.5 − 100 * 0.08
W307 = 90.5 kg/s

Recuperator

First guess recuperator gas side inlet temperature T6 = 840 K. Then apply F5.23.1, the definition of recuperator effectiveness:

0.86 = (T308 − 469.4)/(840 − 469.4)
T308 = 788.1 K

Apply recuperator total pressure loss:

P308 = 1076.5 * (1 − 0.03)
P308 = 1044.2 kPa

W308 = W307 = 90.5 kg/s

Recuperator outlet duct

T31 = T308 = 788.1 K

Apply duct total pressure loss:

P31 = 1044.2 * (1 − 0.01)
P31 = 1033.76 kPa

Deduct cooling flow offtakes:

W31 = 90.5 − 100 * 0.075
W31 = 83.0 kg/s

Combustor

Calculate FAR to station 41, i.e. after the cooling air that will do work has mixed in before the turbine, using F3.38A. Apply percentage pressure loss to derive P41:

T41 = 1400 K

P41 = 1033.76 * (1 − 0.03)
P41 = 1002.75 kPa

FAR1 = 0.10118 + 2.00376E-05 * (700 − 788.1) = 0.09941
FAR2 = 3.7078E-03 − 5.2368E-06 * (700 − 788.1)
        −5.2632E-06 * 1400 = −0.003199
FAR3 = 8.889E-08 * ABS(1400 − 950) = 0.00004
FAR = (0.09941 − SQRT(0.09941^2 − 0.003199) − 0.00004)/0.999
FAR = 0.0177

WF = 0.0177 * (83.0 + 100 * 0.08)
WF = 1.611 kg/s

W41 = 83.0 + 100 * 0.075 + 1.611
W41 = 92.1 kg/s

Now calculate T4 using mixing enthalpy balance for constant specific heats, F5.15.2:

W4 = 83.0 + 1.609
W4 = 84.61 kg/s

7.5 * 1.005 * 788.1 + 84.61 * 1.15 * T4 = 92.1 * 1.15 * 1400
T4 = 1462.9 K

P4 = P41 = 1002.75 kPa

Calculate pseudo SOT station 415 by adding in cooling air proportion that mixes in downstream but does work as per section 5.15.4:

W415 = 92.1 + 100 * 0.047
W415 = 96.8 kg/s

92.1 * 1.15 * 1400 + 4.7 * 1.005 * 467.9 = 96.8 * 1.15 * T415
T415 = 1351.9 K

P415 = P41 = 1002.75 kPa

Turbine

Derive P5 via ambient pressure and the exhaust pressure loss:

101.325 = P5 * (1 − 0.04) * (1 − 0.04) * (1 − 0.01)
P416 = P5 = 111.06 kPa

P415Q416 = 1002.75/111.06
P415Q416 = 9.029

Derive E415 via F3.44:

E415 = $(1 − 9.029^{(−1/4 * 0.89)})/(1 − 9.029^{(−1/4)})$
E415 = 0.915

Evaluate T416 via F5.9.4:

T415 − T416 = E415 * T415 * $(1 − P415Q416^{(−1/4)})$
1351.9 − T416 = 0.915 * 1351.9 * $(1 − 9.029^{(−1/4)})$
T416 = 828.5 K

Calculate power via F5.9.2:

W415 = W41 = 96.8 kg/s
PW415 = 96.8 * 1.15 * (1351.9 − 828.59)
PW415 = 58 255 kW

Finally calculate capacity at stations 4 and 41:

WRTP4 = $84.61 * 1462.9^{0.5}/1002.75$
WRTP4 = 3.227 kg $\sqrt{K}$/s kPa

WRTP41 = $92.1 * 1400^{0.5}/1002.75$
WRTP41 = 3.437 kg $\sqrt{K}$/s kPa

Cooling air downstream of turbine

W5 = 96.8 + 100 * 0.01 + 100 * 0.033
W5 = 101.1 kg/s

Mixing enthalpy balance for constant specific heats, F5.15.2:

$$101.1 * 1.15 * T5 = 96.8 * 1.15 * 828.5$$
$$+100 * 0.01 * 1.005 * 428.8 + 100 * 0.033 * 1.005 * 467.9$$

$$T5 = 810.3 \,\text{K}$$

Recuperator inlet duct

$$T6 = T5 = 810.3 \,\text{K}$$

Iterate until T6guess is within 0.1 deg of T6calc

$$T6error = 810.3 - 840$$
$$T6error = -29.7 \,\text{K}$$

$$T6guess2 = 810.3 \,\text{K}$$
GOTO RECUPERATOR INLET

To keep the length of this sample calculation manageable this iteration will not be conducted here, but instead will proceed with the values resulting from the first T6 guess of 840 K. This will result in some error in the calculated performance, because fuel flow will be too high and the gas path parameters downstream of the recuperator air side will be in error.

$$W6 = W5 = 101.1 \,\text{kg/s}$$

$$P6 = 111.06 * (1 - 0.04)$$
$$P6 = 106.62 \,\text{kPa}$$

Recuperator gas side

Heat transferred from the gas side must equal heat received by the air side:

$$101.1 * 1.15 * (810.3 - T601) = 90.5 * 1.005 * (788.1 - 469.4)$$
$$T601 = 561.0 \,\text{K}$$

$$P601 = 106.62 * (1 - 0.04)$$
$$P601 = 102.36 \,\text{kPa}$$

$$W601 = W6 = 101.1 \,\text{kg/s}$$

Exhaust

Total temperature is unchanged along the duct. Use Q curve Formulae F3.32 and F3.33 to determine exhaust plane pressure and area with the design Mach number of 0.05:

$$T9 = T601 = 561.0 \,\text{K}$$

$$P9/PS9 = (1 + (1.333 - 1)/2 * 0.05^\wedge 2)^\wedge (4)$$
$$P9/PS9 = 1.00167$$

Since PS9 = PAMB then P9 = 101.49 kPa

$$W9 = W6 = 101.1 \,\text{kg/s}$$
$$Q = 1000 * \text{SQRT}(2 * 4/287.05) * 1.00167^\wedge(-2/1.333) * (1 - 1.00167^\wedge(-1/4)))$$
$$Q = 3.405 \,\text{kg}\sqrt{\text{K}}/\text{s m}^2 \,\text{kPa}$$
$$3.405 = 101.252 * \text{SQRT}(561.0)/(A9 * 101.49)$$
$$A9 = 6.940 \,\text{m}^2$$

Final calculations

Use F6.5, F6.7 and F6.9:

$$PW = 58\,255 - (15\,527 + 14\,135)/0.995$$
$$PW = 28\,444 \,\text{kW}$$

SPW = 28 444/100
SPW = 284.44 kW s/kg

SFC = 1.602 * 3600/28 444
SFC = 0.2028 kg/kW h

ETATH = 3600/(0.2028 * 43 100) * 100
ETATH = 41.2%

Notes: The above values are similar to the datum for Charts 6.5 and 6.6. Any difference is due to the charts being prepared using CP at mean T and because in the sample calculation the iterations were omitted which would have ensured T6guess was within 0.1 K of T6calculated.

C6.4 Conduct a design point calculation for a two spool separate jets turbofan at ISA, 11 000 m, 0.8 Mach number:

Mass flow = 100 kg/s
SOT = 1400 K
Bypass ratio = 4.5
Pressure ratio = 25 : 1
   From chart 6.20 optimum fan PR = 1.9 hence:
   fan tip PR = fan root PR = 1.9
   compressor PR = 14.0
Fan and compressor polytropic efficiency = 88%
Turbine polytropic efficiency = 89%
Combustion efficiency = 99.9%
Kerosene fuel with LHV = 43 100 kJ/kg
Mechanical efficiency = 99.5%
Intake pressure loss (installation + engine duct) = 0.5%
Bypass duct pressure loss = 3.0%
Inter compressor duct pressure loss = 2.0%
Compressor exit diffuser pressure loss = 2.0%
Combustor pressure loss = 3.0%
Inter turbine duct pressure loss = 1.0%
Jet pipe pressure loss = 1.0%
Propelling nozzle CD, CX = 0.97, 0.99

Air system as per C6.2 but the flows are a fraction of W26. Station numbering is as per C6.1 but with additional stations as necessary:

fan root exit = 24	fan tip exit = 3	bypass duct exit = 17
compressor inlet = 26	HP turbine exit = 44	LP turbine inlet = 46
LP turbine exit = 48	core nozzle exit = 7	

Ambient conditions
From Chart 2.1 for ISA conditions TAMB = 216.7 K, PAMB = 22.628 kPa.

Free stream conditions
Use Q curve Formulae F3.31 and F3.32 to derive free stream total temperature and pressure:

T0/216.7 = (1 + (1.4 − 1)/2 * 0.8^2)
T0 = 244.4 K

P0/22.628 = (244.4/216.7)^3.5
P0 = 34.47 kPa

Intake

Total temperature is unchanged along duct, and pressure loss is applied as a percentage:

$$T2 = T0 = 244.4\,\text{K}$$

$$P2 = 34.47 * (1 - 0.005)$$
$$P2 = 34.30\,\text{kPa}$$

$$W2 = 100\text{kg/s}$$

Fan

Apply F3.42 to derive isentropic efficiency:

$$E2 = E12 = (1.9^\wedge(1/3.5) - 1)/(1.9^\wedge(1/(3.5*0.88)) - 1)$$
$$E2 = E12 = 0.869$$

Apply F5.1.4 to derive T13 and T24:

$$T24 - 244.4 = 244.4/0.869 * (1.9^\wedge(1/3.5) - 1)$$
$$T24 = T13 = 301.0\,\text{K}$$

Calculate P24 and P13 from given pressure ratio:

$$P24 = P13 = 1.9 * 34.3$$
$$P24 = P13 = 65.17\,\text{kPa}$$

Calculate core and bypass flows:

$$W24 = 100/(1 + 4.5)$$
$$W24 = 18.18\,\text{kg/s}$$

$$W13 = 100 - 18.18$$
$$W13 = 81.82\,\text{kg/s}$$

Calculate input power required based on F5.15.1:

$$PW2 = 100 * 1.005 * (301.0 - 244.4)$$
$$PW2 = 5688\,\text{kW}$$

Bypass duct

Total temperature is unchanged, derive exit pressure by applying percentage pressure loss, derive exit mass flow by deducting air system flows:

$$T17 = T13 = 301.0\,\text{K}$$

$$P17 = 65.17 * (1 - 0.03)$$
$$P17 = 63.21\,\text{kPa}$$

$$W17 = W13 = 81.82\,\text{kg/s}$$

Inter compressor duct

Total temperature is unchanged, derive exit pressure by applying percentage pressure loss, derive exit mass flow by deducting air system flows:

$$T26 = T24 = 301.0\,\text{K}$$

$$P26 = 65.17 * (1 - 0.02)$$
$$P26 = 63.87\,\text{kPa}$$

$$W26 = W24 = 18.18\,\text{kg/s}$$

Compressor

Apply F3.42 to derive isentropic efficiency:

$E26 = (14^\wedge(1/3.5) - 1)/(14^\wedge(1/(3.5*0.88))) - 1$
$E26 = 0.830$

Apply F5.1.4 to derive T3:

$T3 - 301.0 = 301.0/0.830 * (14^\wedge(1/3.5) - 1)$
$T3 = 709.2\,K$

Calculate P3 from given pressure ratio:

$P3 = 14 * 63.87$
$P3 = 894.2\,kPa$

Deduct any cooling flow offtakes:

$W3 = 18.18 - 18.18 * (0.005 + 0.01)$
$W3 = 17.91\,kg/s$

Calculate input power required considering interstage bleed to be taken at the mean temperature through the compressor, using F5.15.1:

$Tbleed = (709.2 + 301.0)/2$
$Tbleed = 505.1\,K$

$PW26 = 17.91 * 1.005 * (709.2 - 301.0) + 0.273 * 1.005 * (505.1 - 301.0)$
$PW26 = 7403.4\,kW$

Compressor exit diffuser

Total temperature is unchanged, derive exit pressure by applying percentage total pressure loss, derive exit mass flow by deducting air system flows:

$T31 = T3 = 709.2\,K$

$P31 = 894.2 * (1 - 0.02)$
$P31 = 876.3\,kPa$

$W31 = 17.91 - 18.18 * 0.08 - 18.18 * 0.075$
$W31 = 15.09\,kg/s$

Combustor and SOT station

Calculate FAR to station 41, i.e. after the cooling air that will do work has mixed in before the turbine, using F3.38A. Apply percentage pressure loss to derive P41:

$T41 = 1400\,K$

$P41 = 876.3 * (1 - 0.03)$
$P41 = 850.0\,kPa$

$FAR1 = 0.10118 + 2.00376E\text{-}05 * (700 - 709.2)$
$\quad\quad\quad = 0.100996$
$FAR2 = 3.7078E\text{-}03 - 5.2368E\text{-}06 * (700 - 709.2)$
$\quad\quad\quad\quad -5.2632E\text{-}06 * 1400$
$\quad\quad\quad = -0.0036125$
$FAR3 = 8.889E\text{-}08 * ABS(1400 - 950)$
$\quad\quad\quad = 0.00004$
$FAR = (0.100996 - SQRT(0.100996^\wedge 2 - 0.0036125) - 0.00004)/0.999$
$FAR = 0.0198$

$WF = 0.0198 * (15.09 + 100 * 0.075)$
$WF = 0.326\,kg/s$

W41 = 15.09 + 18.18 ∗ 0.075 + 0.326
W41 = 16.78 kg/s

Now calculate T4 using using mixing enthalpy balance for constant specific heats, F5.15.2:

W4 = 15.09 + 0.326
W4 = 15.42 kg/s

1.36 ∗ 1.005 ∗ 709.2 + 15.42 ∗ 1.15 ∗ T4 = 16.78 ∗ 1.15 ∗ 1400
T4 = 1468.8 K

P4 = P41 = 850.0 kPa

Calculate pseudo SOT station 415 adding in cooling air proportion that mixes in downstream but does work as per section 5.15.4:

W415 = 16.78 + 18.18 ∗ 0.047
W415 = 17.63 kg/s

16.78 ∗ 1.15 ∗ 1400 + 0.85 ∗ 1.005 ∗ 709.2 = 17.63 ∗ 1.15 ∗ T415
T415 = 1362.4 K

P415 = P41 = 850.0 kPa

HP turbine

Equate power to that of compressor and use F5.9.2 to derive exit temperature:

PW415 = 7403.4/0.995
PW415 = 7440.6 kW

7440.6 = 17.63 ∗ 1.15 ∗ (1362.4 − T416)
T416 = 995.4 K

Derive isentropic efficiency E415 via F3.44:

Guess expansion ratio of 4 to derive isentropic efficiency from polytropic efficiency:

E415 = (1 − 4^(−1/4 ∗ 0.89))/(1 − 4^(−1/4))
E415 = 0.906

Derive expansion ratio P415Q416 via F5.9.4:

1362.4 − 995.4 = 0.906 ∗ 1362.4 ∗ (1 − P415Q416^(−1/4))
P415Q416 = 4.10

One should now go back and recalculate turbine isentropic efficiency using this expansion ratio. However for brevity this step is omitted.
   Calculate P416 from expansion ratio:

P416 = 850.0/4.1
P416 = 207.31 kPa

W416 = W415 = 17.63 kg/s
P44 = P416 = 207.31 kPa

No other cooling flows return, so:

T44 = T416 = 995.4 K

W44 = W416 = 17.63 kg/s

Inter turbine duct

Total temperature is unchanged. Derive exit pressure by applying percentage pressure loss, derive exit mass flow by returning any air system flows:

$T46 = T44 = 995.4\,K$

$P46 = 207.3 * (1 - 0.01)$
$P46 = 205.22\,kPa$

$W46 = W44 = 17.63\,kg/s$

LP turbine

Equate power to that of fan and use F5.9.2 to derive exit temperature:

$PW46 = 5688/0.995$
$PW46 = 5717\,kW$

$5717 = 17.63 * 1.15 * (995.4 - T5)$
$T48 = 713.4\,K$

Derive isentropic efficiency E46 via F3.44:

Guess expansion ratio of 4 to derive isentropic efficiency:

$E46 = (1 - 4^{\wedge}(-1/4 * 0.89))/(1 - 4^{\wedge}(-1/4))$
$E46 = 0.894$

Derive expansion ratio P46Q48 via F5.9.4:

$995.4 - 713.4 = 0.906 * 995.4 * (1 - P46Q48^{\wedge}(-1/4))$
$P46Q48 = 4.481$

One should now go back and recalculate turbine isentropic efficiency using this expansion ratio. However for brevity this step is omitted.

$P48 = 205.22/4.481$
$P48 = 45.80\,kPa$

$W48 = W46 = 17.63\,kg/s$

Cooling air downstream of turbine and jet pipe pressure loss

$W5 = 17.63 + 18.18 * 0.01 + 18.18 * 0.033$
$W5 = 18.41\,k/s$

Mixing enthalpy balance for constant specific heats, F5.15.2:

$18.41 * 1.15 * T5 = 17.63 * 1.15 * 713.4 + 18.18 * 0.01 * 1.005 * 505.1$
$$+18.18 * 0.033 * 1.005 * 709.2$$
$T5 = 707.7\,K$

$P7 = 45.8 * (1 - 0.01)$
$P7 = 45.34\,kPa$

Cold propelling nozzle

$P17QN = 63.21/22.628$
$P17QN = 2.793$

From Chart 3.8 the choking pressure ratio for $\gamma = 1.4$ is 1.8929, hence the cold nozzle is choked, throat Mach number $= 1$ and the throat static pressure is higher than ambient. Use curve formulae to derive other nozzle parameters:

Use the choking pressure ratio to calculate the throat static pressure:

63.21/PSN = 1.8506
PSN = 34.16 kPa

F3.31, T/TS at M = 1:

301.0/TSN = (1 + (1.333 − 1)/2 * 1^2)
TSN = 258.0 K

F3.35, V/SQRT(T):

VN = 1 * SQRT(1.333 * 287.05 * 258.0)
VN = 314.2 m/s

F3.33, Q (choking value):

39.6971 = 81.82 * SQRT(301)/(AN * 63.21)
ANeffective = 0.566 m²

F5.13.15, definition of CD:

ANgeometric = 0.566/0.97
ANgeometric = 0.584 m²

F5.13.13, gross thrust:

FG = 81.82 * 314.2 + 0.566 * (34.16 − 22.628) * 1000
FG = 32 235 N

Hot propelling nozzle

P7QN = 45.34/22.628
P7QN = 2.004

Again the nozzle is choked and the calculation process is as per cold nozzle:

45.34/PSN = 1.8506
PSN = 24.50 kPa

707.7/TSN = (1 + (1.333 − 1)/2 * 1^2)
TSN = 607.5 K

VN = 1 * SQRT(1.333 * 287.05 * 607.5)
VN = 482.1 m/s

39.6971 = 18.41 * SQRT(707.7)/(AN * 45.34)
ANeffective = 0.272 m²

ANgeometric = 0.272/0.97
ANgeometric = 0.281 m²

FG = 18.41 * 482.1 + 0.272 * (24.5 − 22.628) * 1000
FG = 9385 N

Final calculations

Calculate free stream velocity and hence momentum drag. Then calculate net thrust by applying CX to the gross thrust:

F2.6, with M = 1:

V1/SQRT(244.4) = 0.8 * SQRT(1.4 * 287.05)/SQRT(244.4/216.7)
V1 = 236.1 m/s

F5.13.13, momentum drag and net thrust:

FD = 100 ∗ 236.1
FD = 23 610 N

FN = (32 235 + 9385) ∗ 0.99 − 23 610
FN = 17 594 N

SFN = 17 594/100
SFN = 175.9 N s/kg

SFC = 0.326 ∗ 3600/17 594
SFC = 0.0667 kg/N h

Notes: The above values are similar to those from Chart 6.19. Any difference is due to the approximation of constant gas properties.

C6.5   Calculate combined heat and power performance with no supplementary firing for the simple cycle gas turbine from C6.2, ignoring plant intake and exhaust pressure losses. As per the guidance in section 6.8 take the stack exit temperature to be 150 8C and the steam plant efficiency to be 90%.

Calculate heat transferred to steam plant, then apply the steam plant efficiency to derive that usefully used. Hence calculate CHP thermal efficiency, specific power and the heat to power ratio:

Heat transferred = W ∗ CP ∗ (Tin − Tstack)
PWtransfer = 101.252 ∗ 1.15 ∗ (714.6 − 423.15)
PWtransfer = 33 936 kW

PWsteam = 33 936 ∗ 0.9
PWsteam = 30 543 kW

PWchp = 24 457 + 30 543
PWchp = 55 000 kW

SPWchp = 55 000/100
SPWchp = 550 kW s/kg

SFCchp = 1.752 ∗ 3600/55 000
SFCchp = 0.1147 kg/kW h

ETATHchp = 3600/(0.1147 ∗ 43 100) ∗ 100
ETATHchp = 72.82%

HEATPWratio = 55 000/24 457
HEATPWratio = 2.25

Notes: These values are similar to Chart 6.11. Any difference is due to the gas properties and because plant intake and exhaust pressure losses of 1 and 2%, respectively, were included in the gas turbine design point calculations for Chart 6.11.

Charts

Chart 6.1   Design point exchange rates for shaft power cycles at ISO.

(a)   Simple cycle at 20 : 1 pressure ratio and 1400 K SOT

	Datum value (%)	Change in value (% Points)	Resulting changes	
			Thermal efficiency (%)	Specific power (%)
Compressor polytropic efficiency	88	+1	1.93	3.02
Combustion efficiency	99.9	−1	−0.92	0.08
Turbine polytropic efficiency	89	+1	2.22	2.22
Turbine cooling air percentage	7	+1	−1.01	−2.09
Intake pressure loss	0.5	+1	−0.70	−0.70
Combustor pressure loss	3	+1	−0.72	−0.72
Exhaust pressure loss	5	+1	−0.73	−0.73

(b)   Recuperated cycle at 7 : 1 pressure ratio and 1400 K SOT

	Datum value (%)	Change in value (% Points)	Resulting changes	
			Thermal efficiency (%)	Specific power (%)
Compressor polytropic efficiency	88	+1	1.51	1.64
Combustion efficiency	99.9	−1	−0.94	0.04
Turbine polytropic efficiency	89	+1	1.38	1.98
Recuperator effectiveness	86	+1	0.68	−0.03
Turbine cooling air percentage	7	+1	−1.40	−1.71
Intake pressure loss	0.5	+1	−0.75	−1.06
Recup. air side/combustor pressure loss	3/3	+1	−0.77	−1.09
Recup. gas side/exhaust pressure loss	5/1	+1	−0.78	−1.11

(c)   Intercooled cycle at 20 : 1 pressure ratio and 1400 K SOT

	Datum value (%)	Change in value (% Points)	Resulting changes	
			Thermal efficiency (%)	Specific power (%)
Compressor polytropic efficiency	88	+1	1.33	1.72
Combustion efficiency	99.9	−1	−0.92	0.08
Turbine polytropic efficiency	89	+1	1.72	1.72
Intercooler effectiveness	81	+1	0.04	0.35
Turbine cooling air percentage	7	+1	−0.72	−1.81
Intake pressure loss	0.5	+1	−0.54	−0.54
Combustor pressure loss	3	+1	−0.41	−0.52
Intercooler pressure loss	2.5	+1	−0.56	−0.56
Exhaust pressure loss	5	+1	−0.57	−0.57

Note: Intercooler positioned so that compressors have equal pressure ratios

contd.

Chart 6.1   *contd.*

(d)   Intercooled recuperated cycle at 11:1 pressure ratio and 1400 K SOT

	Datum value (%)	Change in value (% Points)	Resulting changes	
			Thermal efficiency (%)	Specific power (%)
Compressor polytropic efficiency	88	+1	1.33	1.40
Combustion efficiency	99.9	−1	−0.93	0.05
Turbine polytropic efficiency	89	+1	1.18	1.78
Intercooler effectiveness	81	+1	0.17	0.22
Recuperator effectiveness	86	+1	0.54	−0.03
Turbine cooling air percentage	7	+1	−1.23	−1.69
Intake pressure loss	0.5	+1	−0.50	−0.73
Intercooler pressure loss	2.5	+1	−0.50	−0.53
Recup. air side/combustor pressure loss	3/3	+1	−0.51	−0.75
Recup. gas side/exhaust pressure loss	4/1	+1	−0.50	−0.74

Note: Intercooler positioned so that compressors have equal pressure ratios

(e)   Combined cycle at 13:1 pressure ratio and 1400 K SOT

	Datum value (%)	Change in value (% Points)	Resulting changes	
			Thermal efficiency (%)	Specific power (%)
Compressor polytropic efficiency	88	+1	0.70	1.43
Combustion efficiency	99.9	−1	−0.92	0.08
Turbine polytropic efficiency	89	+1	0.86	0.86
Turbine cooling air percentage	7	+1	−0.45	−1.54
Intake pressure loss	0.5	+1	−0.32	−0.32
Combustor pressure loss	3	+1	−0.33	−0.33
Exhaust/steam plant pressure loss	5/2	+1	−0.34	−0.34
Steam cycle efficiency	34.7	+1	1.00	1.00

contd.

Chart 6.1 *contd.*

(f) Combined heat and power cycle at 13:1 PR and 1400 K SOT

	Datum value (%)	Change in value (% Points)	Resulting changes			
			HTP ratio = 0		HTP ratio = max	
			Thermal efficiency (%)	Specific power (%)	Thermal efficiency (%)	Specific power (%)
Compressor polytropic efficiency	88	+1	1.40	2.08	0.18	0.84
Combustion efficiency	99.9	−1	−0.93	0.07	−0.91	0.08
Turbine polytropic efficiency	89	+1	1.97	1.97	0.09	0.09
Turbine cooling air percentage	7	+1	−0.73	−1.81	−0.23	−1.33
Intake pressure loss	0.5	+1	−0.76	−0.76	−0.03	−0.03
Combustor pressure loss	3	+1	−0.78	−0.78	−0.03	−0.03
Exhaust/steam plant pressure loss	5/2	+1	−0.80	−0.80	−0.03	−0.03
Steam cycle efficiency	0.9	+1	0.00	0.00	0.65	0.65

Air system assumptions for all design point cases:
0.5% to overboard from 50% through the compression process for bearing sealing
7% for turbine cooling from compressor delivery
    60% returned upstream of the turbines (assumed to do useful work in the turbines)
    40% returned downstream of the turbines (assumed to do no work)
1% for disc cooling from 50% through the compression process to turbine exit (does no work)
1% for disc cooling from compressor delivery
    50% returned upstream of the turbines (does work)
    50% returned downstream of the turbines (does no work)

Chart 6.2   Turboshaft and turboprop design point exchange rates.

	Datum value (%)	Change in value (% Points)	ISA SLS		ISA, 6000 m, 0.5 M	
			SFC (%)	Specific power (%)	SFC (%)	Specific power (%)
Compressor polytropic efficiency	88	+1	−1.74	2.70	−1.16	1.97
Combustion efficiency	99.9	−1	0.92	0.08	0.93	0.07
Turbine polytropic efficiency	89	+1	−2.12	2.12	−1.72	1.72
Turbine cooling air percentage	7	+1	0.91	−1.99	0.62	−1.71
Intake pressure loss	2	+1	0.71	−0.71	0.54	−0.54
Combustor pressure loss	3	+1	0.72	−0.72	0.55	−0.55
Exhaust pressure loss	3	+1	0.72	−0.72	0.55	−0.55

Cycle parameters:
SOT = 1400 K, Overall PR = 17.5 : 1

Air system assumptions for all design point cases:
0.5% to overboard from 50% through the compression process for bearing sealing
7% for turbine cooling from compressor delivery
   60% returned upstream of the turbines (assumed to do useful work in the turbines)
   40% returned downstream of the turbines (assumed to do no work)
1% for disc cooling from 50% through the compression process to turbine exit (does no work)
1% for disc cooling from compressor delivery
   50% returned upstream of the turbines (does work)
   50% returned downstream of the turbines (does no work)

Chart 6.3 Turbojet and turbofan design point exchange rates at 11 000 m, ISA, 0.8 M.

	Datum value (%)	Change in value (% Points)	Resulting changes								
			Turbojet		0.5 BPR		4.5 BPR		10.5 BPR		
			SFC (%)	SFN (%)	SFC (%)	SFN (%)	SFC (%)	SFN (%)	SFC (%)	SFN (%)	
Fan tip polytropic efficiency	88	+1			-0.14	0.14	-0.48	0.48	-0.66	0.67	
Fan root polytropic efficiency	88	+1			-0.16	0.69	-0.15	0.33	-0.09	0.18	
Compressor polytropic efficiency	88	+1	-0.01	0.99	-0.16	0.65	-0.53	1.37	-0.78	1.72	
Combustion efficiency	99.9	-1	0.96	0.05	0.90	0.12	0.93	0.08	0.93	0.08	
Turbine polytropic efficiency	89	+1	-0.28	0.28	-0.52	0.53	-0.99	1.01	-1.20	1.23	
Mechanical efficiency	99.5	-1	0.61	-0.60	0.83	-0.82	1.66	-1.61	2.09	-2.01	
Turbine cooling air percentage	7	+1	-0.34	-0.77	-0.28	-0.83	0.20	-1.29	0.47	-1.56	
Intake pressure loss	0.5	+1	0.17	-0.17	0.05	-0.34	0.93	-1.19	2.42	-2.58	
Bypass duct pressure loss	3	+1			-0.02	0.02	0.93	-0.72	2.15	-2.06	
Fan to compressor pressure loss	1.5	+1			0.00	-0.29	0.19	-0.48	0.28	-0.56	
Diffuser/combustor pressure loss	2/3	+1	0.17	-0.17	0.17	-0.17	0.33	-0.33	0.39	-0.39	
Hot exhaust pressure loss	1.5	+1	0.17	-0.17	0.16	-0.16	0.32	-0.32	0.38	-0.38	
Hot nozzle velocity coefficient	99	-1	0.89	-0.88	0.70	-0.69	0.49	-0.49	0.42	-0.41	
Cold nozzle velocity coefficient	99	-1			0.20	-0.20	1.47	-2.86	3.18	-2.99	

Note: Exchange rates based on SOT 1400 K, OPR 25:1, Optimum fan PR for bypass cycles
Performance and assumptions as for Chart 6.4.

Chart 6.4   Turbojet and turbofan design point exchange rates at 11 000 m, ISA, 2.2 M.

	Datum value (%)	Change in value (% Points)	Turbojet SFC (%)	Turbojet SFN (%)	0.5 BPR SFC (%)	0.5 BPR SFN (%)	4.5 BPR SFC (%)	4.5 BPR SFN (%)
Fan tip polytropic efficiency	88	+1			−0.23	0.23	−0.38	0.38
Fan root polytropic efficiency	88	+1			−0.45	1.10	−0.54	0.81
Compressor polytropic efficiency	88	+1	−1.03	3.40	−0.55	1.50	−1.70	3.09
Combustion efficiency	99.9	−1	0.91	0.10	0.70	0.31	0.63	0.38
Turbine polytropic efficiency	89	+1	−0.57	0.58	−0.64	0.65	−1.19	1.21
Mechanical efficiency	99.5	−1	2.35	−2.24	1.47	−1.43	2.91	−2.75
Turbine cooling air percentage	7	+1	0.44	−1.53	0.06	−1.16	0.96	−2.02
Intake pressure loss	0.5	+1	0.23	−0.23	0.05	−0.56	0.79	−1.27
Bypass duct pressure loss	3	+1			−0.16	0.16	0.13	−0.13
Fan to compressor pressure loss	1.5	+1			0.02	−0.53	0.46	−0.95
Diffuser/combustor pressure loss	2/3	+1	0.24	−0.23	−0.02	0.02	0.13	−0.13
Hot exhaust pressure loss	1.5	+1	0.23	−0.23	−0.02	0.02	0.12	−0.12
Hot nozzle velocity coefficient	99	−1	2.64	−2.51	1.58	−1.53	2.01	−1.93
Cold nozzle velocity coefficient	99	−1			0.52	−0.51	4.25	−3.92

Note: SFN is specific net thrust
Exchange rates based on SOT 1600 K, OPR 25:1, optimum fan PR for bypass cycles

Fan root performance:
Fan root pressure ratio = fan tip pressure ratio × 0.95
Fan root polytropic efficiency = 88% (as for fan tip)

Air system assumptions for all design point cases:
0.5% to overboard from 50% through the compression process for bearing sealing
7% for turbine cooling from compressor delivery
   60% returned upstream of the turbines (assumed to do useful work in the turbines)
   40% returned downstream of the turbines (assumed to do no work)
1% for disc cooling from 50% through the compression process to turbine exit (does no work)
1% for disc cooling from compressor delivery
   50% returned upstream of the turbines (does work)
   50% returned downstream of the turbines (does no work)

Chart 6.5 Shaft power cycles: thermal efficiency versus pressure ratio and SOT at ISO (design point diagrams).

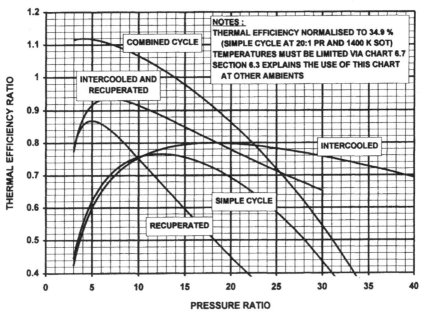

(a)   SOT = 1100 K

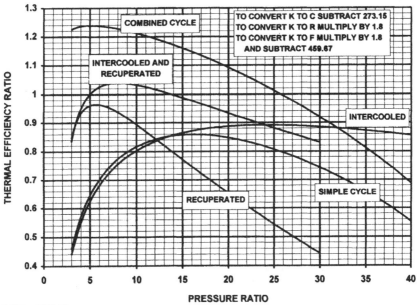

(b)   SOT = 1200 K

Chart 6.5   *contd.*

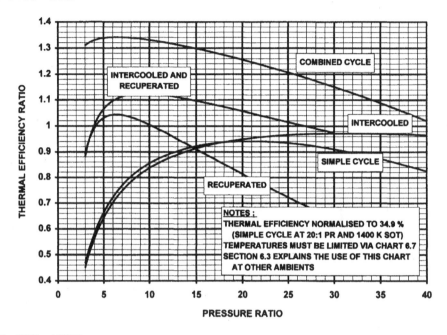

(c)   SOT = 1300 K

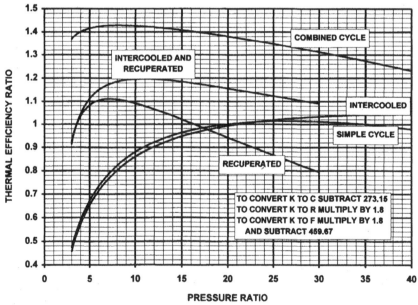

(d)   SOT = 1400 K

Chart 6.5 *contd.*

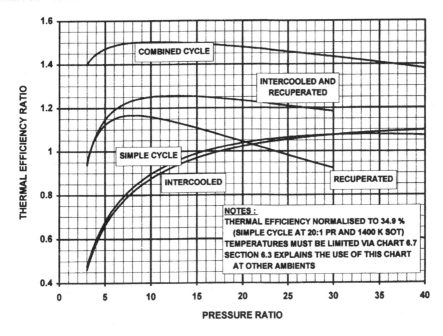

NOTES :
THERMAL EFFICIENCY NORMALISED TO 34.9 %
(SIMPLE CYCLE AT 20:1 PR AND 1400 K SOT)
TEMPERATURES MUST BE LIMITED VIA CHART 6.7
SECTION 6.3 EXPLAINS THE USE OF THIS CHART
AT OTHER AMBIENTS

(e)   SOT = 1500 K

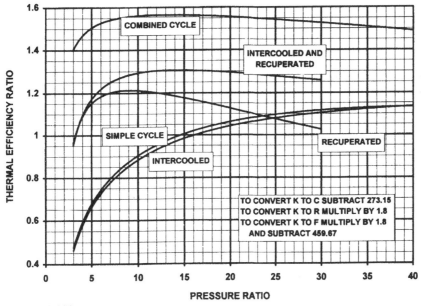

TO CONVERT K TO C SUBTRACT 273.15
TO CONVERT K TO R MULTIPLY BY 1.8
TO CONVERT K TO F MULTIPLY BY 1.8
AND SUBTRACT 459.67

(f)   SOT = 1600 K

Chart 6.5   *contd.*

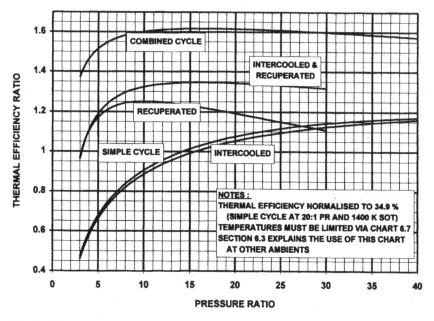

(g)   SOT = 1700 K

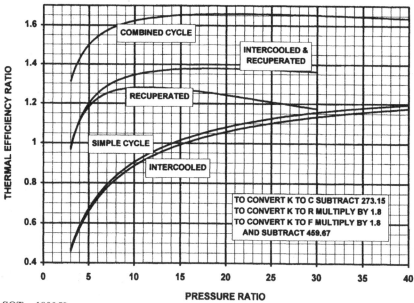

(h)   SOT = 1800 K

Chart 6.6 Shaft power cycles: specific power versus pressure ratio and SOT at ISO (design point diagrams).

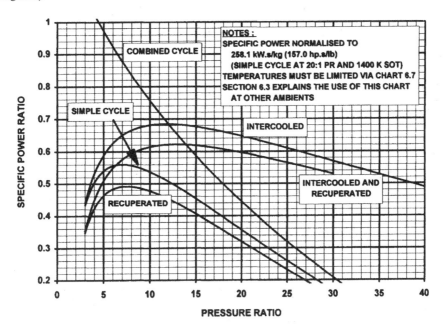

(a) SOT = 1100 K

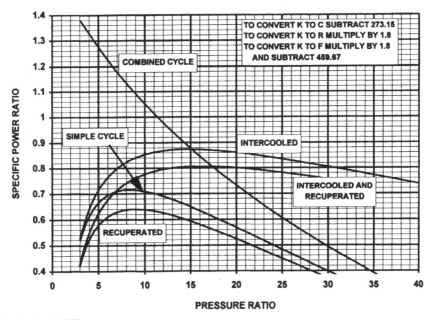

(b) SOT = 1200 K

Chart 6.6   *contd.*

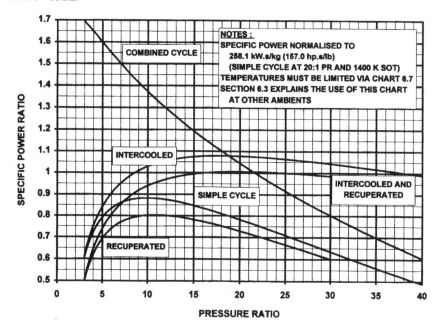

(c)   SOT = 1300 K

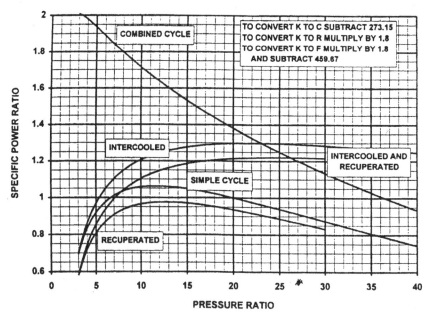

(d)   SOT = 1400 K

Chart 6.6 *contd.*

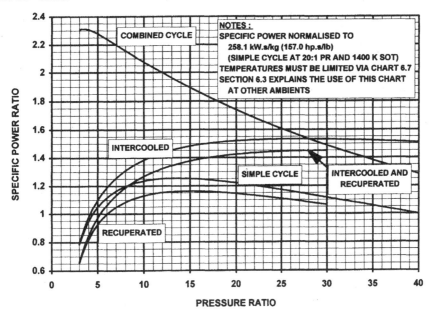

(e) SOT = 1500 K

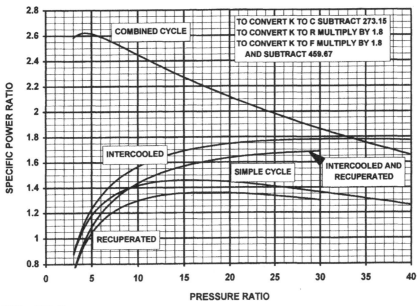

(f) SOT = 1600 K

Chart 6.6   *contd.*

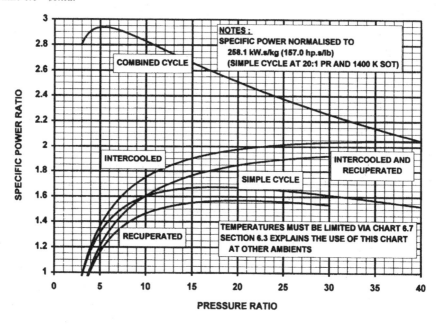

(g)   SOT = 1700 K

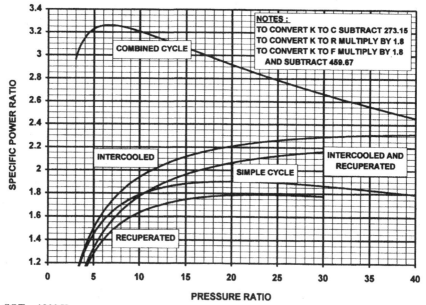

(h)   SOT = 1800 K

Chart 6.7   Shaft power cycles: key temperatures versus pressure ratio and SOT at ISO (design point diagrams).

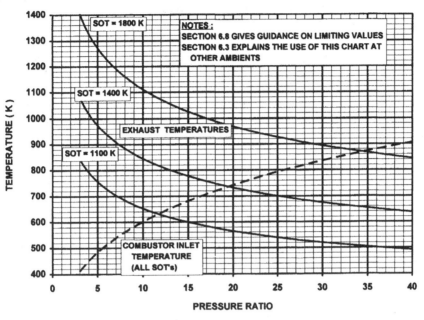

(a)   Simple cycle

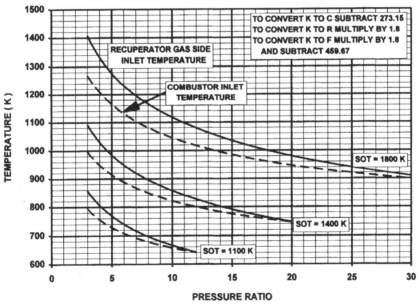

(b)   Recuperated cycle

Chart 6.7 *contd.*

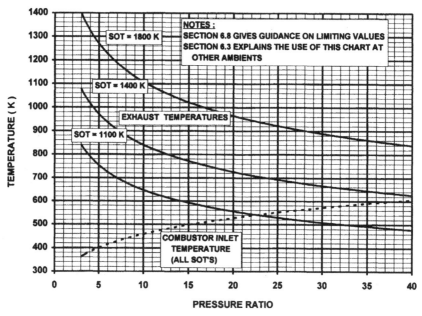

(c)   Intercooled cycle

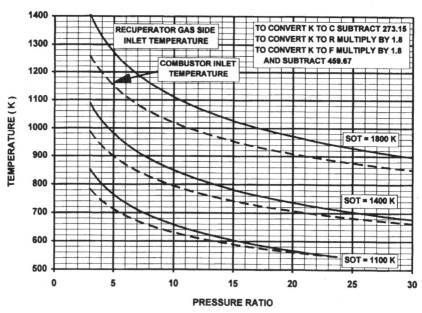

(d)   Intercooled and recuperated cycle

Chart 6.8   Heat exchanged cycles: effect of recuperator effectiveness on optimum pressure ratio for thermal efficiency (design point diagrams).

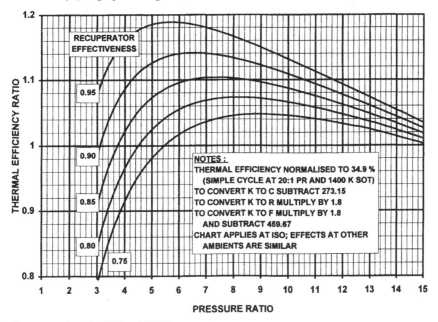

(a)   Recuperated cycle, SOT = 1400 K

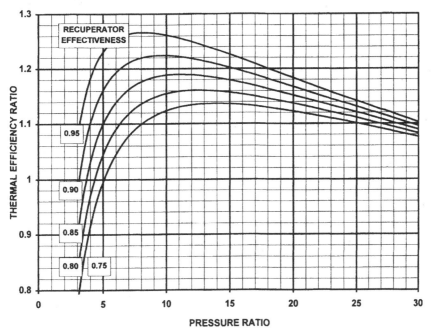

(b)   Intercooled and recuperated cycle, SOT = 1400 K

Chart 6.9   Heat exchanged cycles: effect of intercooler position on thermal efficiency (design point diagrams).

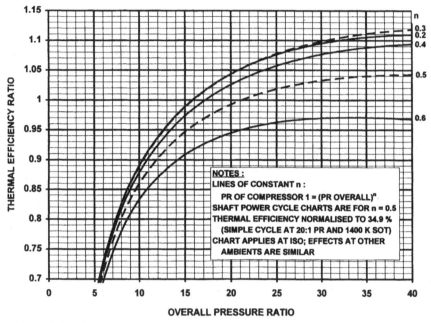

(a)   Intercooled cycle, SOT = 1400 K

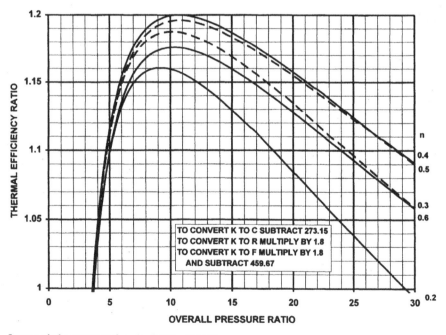

(b)   Intercooled recuperated cycle, SOT = 1400 K

Chart 6.10 Heat exchanged cycles: effect of intercooler position on specific power (design point diagrams).

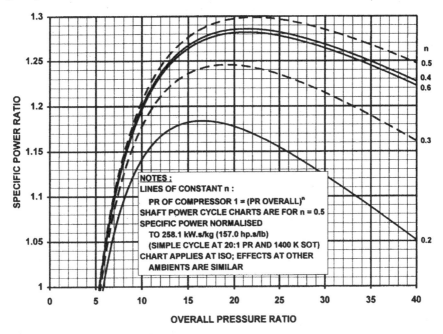

NOTES :
LINES OF CONSTANT n :
PR OF COMPRESSOR 1 = (PR OVERALL)^n
SHAFT POWER CYCLE CHARTS ARE FOR n = 0.5
SPECIFIC POWER NORMALISED
TO 258.1 kW.s/kg (157.0 hp.s/lb)
(SIMPLE CYCLE AT 20:1 PR AND 1400 K SOT)
CHART APPLIES AT ISO; EFFECTS AT OTHER
AMBIENTS ARE SIMILAR

(a) Intercooled cycle, SOT = 1400 K

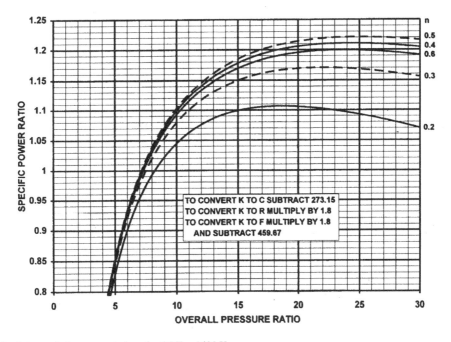

TO CONVERT K TO C SUBTRACT 273.15
TO CONVERT K TO R MULTIPLY BY 1.8
TO CONVERT K TO F MULTIPLY BY 1.8
AND SUBTRACT 459.67

(b) Intercooled recuperated cycle, SOT = 1400 K

Chart 6.11 Combined heat and power: thermal efficiency versus heat to power ratio and pressure ratio at ISO (design point diagrams).

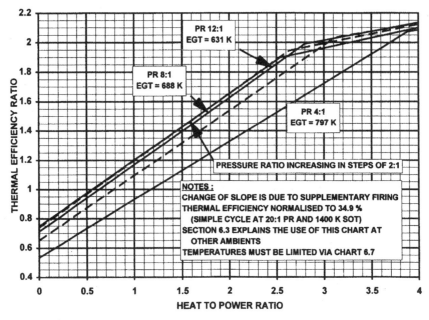

(a)   SOT = 1100 K

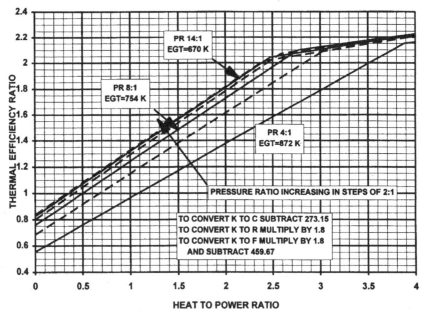

(b)   SOT = 1200 K

Chart 6.11 *contd.*

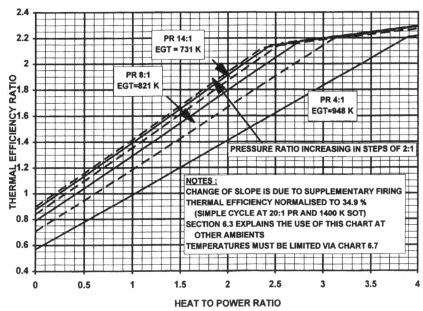

(c) SOT = 1300 K

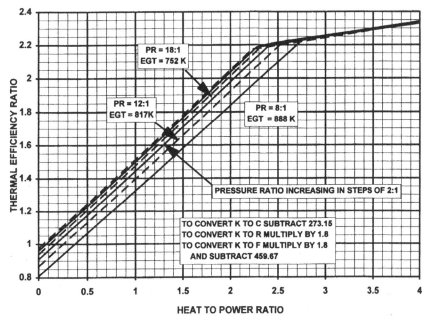

(d) SOT = 1400 K

Chart 6.12 Closed cycles: thermal efficiency versus pressure ratio, SOT and inlet temperature (design point diagrams).

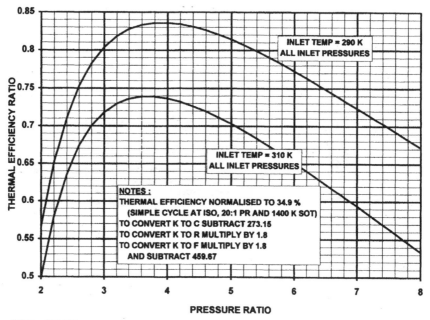

(a) SOT = 1100 K

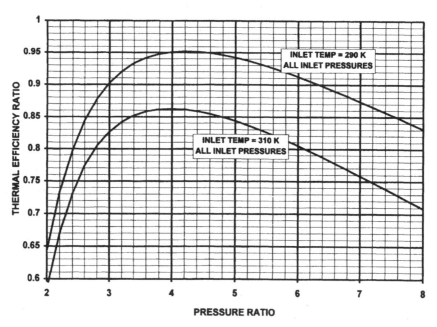

(b) SOT = 1200 K

Chart 6.13 Closed cycles: Power ratio versus pressure ratio, SOT and inlet temperature (design point diagrams).

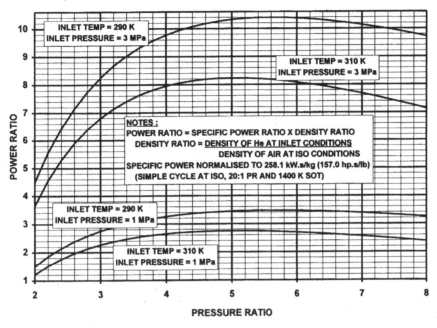

(a) SOT = 1100 K

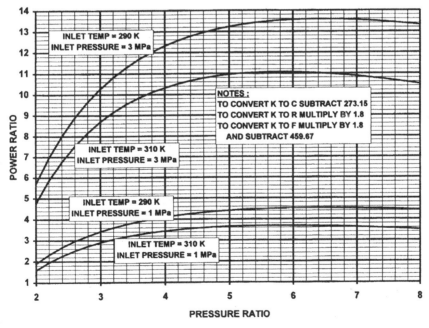

(b) SOT = 1200 K

Chart 6.14 Turboshaft and turboprop cycles: SFC and specific power versus overall pressure ratio and SOT at ISA SLS (design point diagrams).

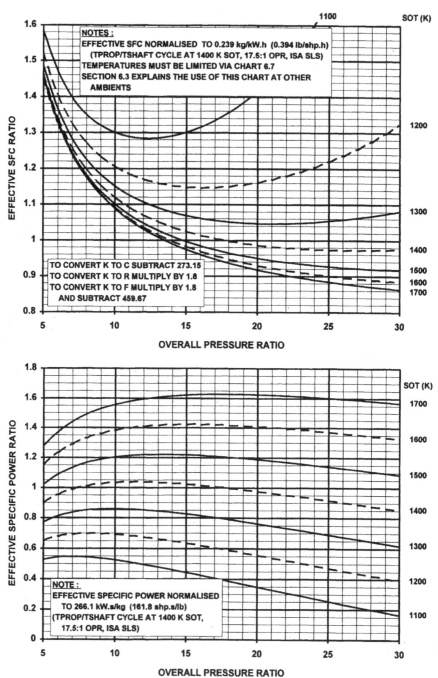

Chart 6.15 Turboshaft and turboprop cycles: SFC and specific power versus overall pressure ratio and SOT at ISA, 6000 m, 0.5 M (design point diagrams).

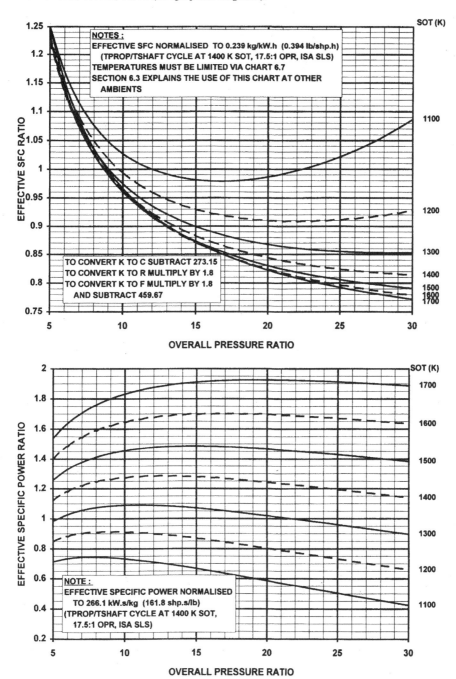

Chart 6.16   Turbojet cycles: SFC and specific thrust versus overall pressure ratio and SOT, at 11 000 m, ISA, 0.8 M (design point diagrams).

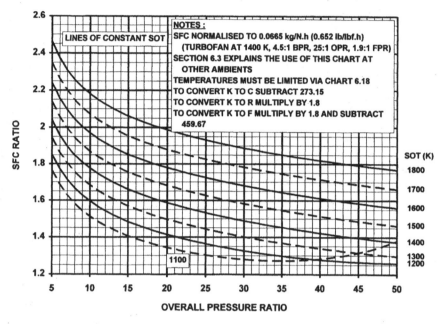

(a)   Uninstalled SFC

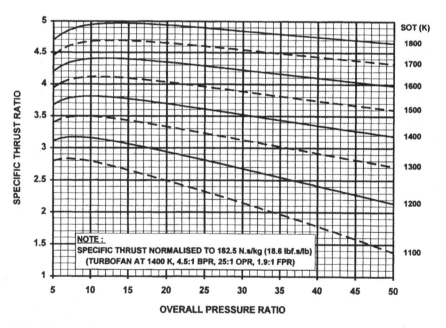

(b)   Uninstalled specific thrust

Chart 6.17 Turbojet cycles: SFC and specific thrust versus overall pressure ratio and SOT, at 11 000 m, ISA, 2.2 M (design point diagrams).

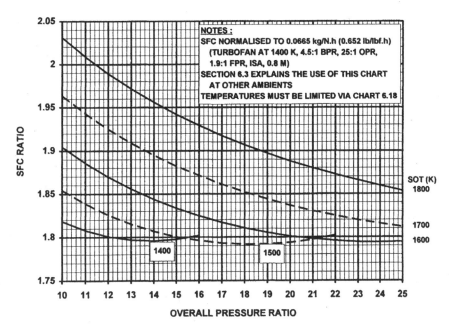

(a) Uninstalled SFC

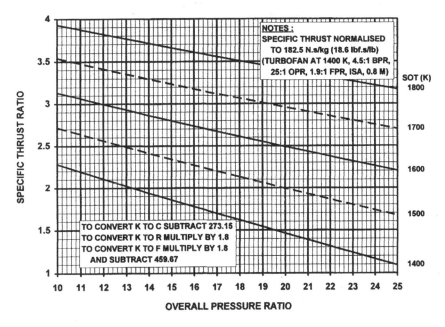

(b) Uninstalled specific thrust

Chart 6.18   Turbojet and turbofan cycles: compressor delivery temperature and pressure versus overall pressure ratio at 11 000 m, ISA, 0.8 M (design point diagrams).

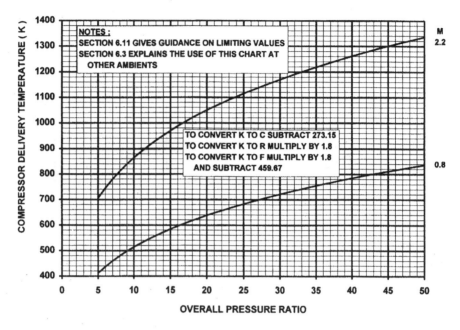

(a)   Compressor delivery temperature

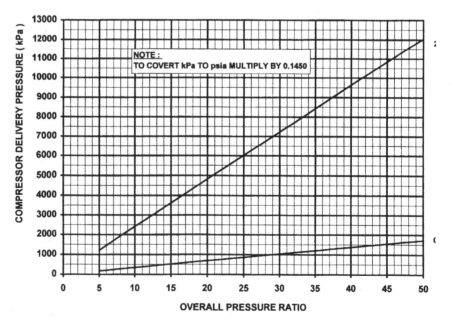

(b)   Compressor delivery pressure

Chart 6.19 Turbofan cycles: SFC and specific thrust versus fan pressure ratio and bypass ratio at 11 000 m, ISA, 0.8 M (design point diagrams).

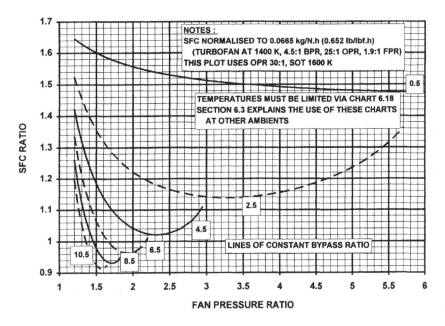

(a) Uninstalled SFC, at constant SOT and OPR

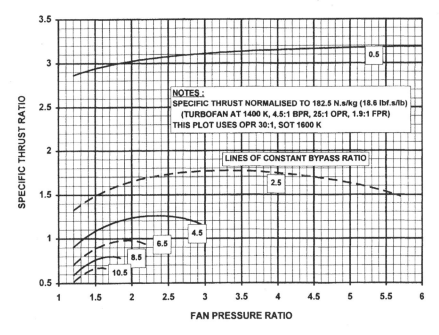

(b) Uninstalled specific thrust, at constant SOT and OPR

Chart 6.20 Turbofan cycles: optimum fan PR versus overall PR and bypass ratio at 11 000 m, ISA, 0.8 M (design point diagrams).

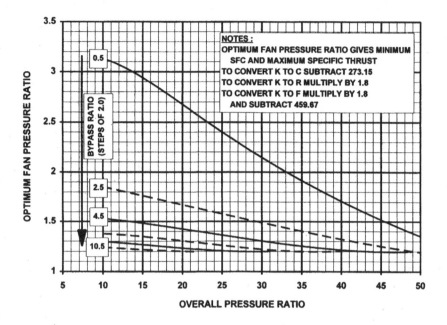

(a)   SOT = 1100 K

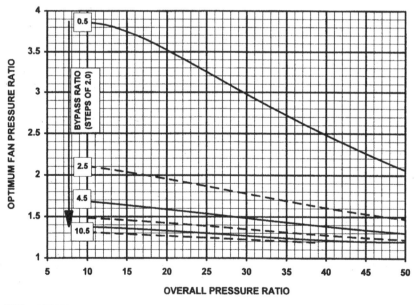

(b)   SOT = 1200 K

Chart 6.20 *contd.*

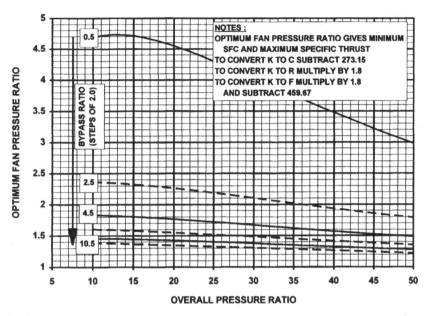

(c) SOT = 1300 K

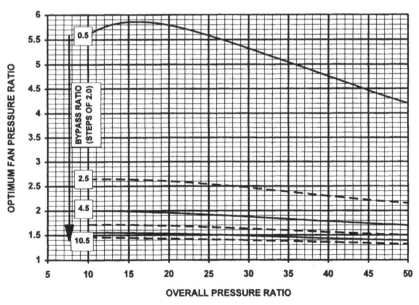

(d) SOT = 1400 K

Chart 6.20 *contd.*

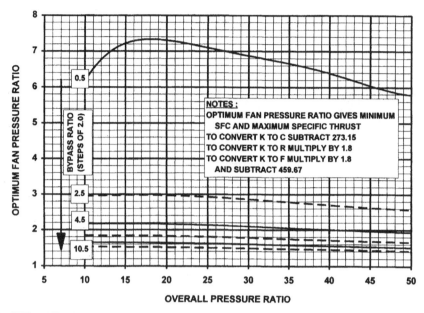

(e)   SOT = 1500 K

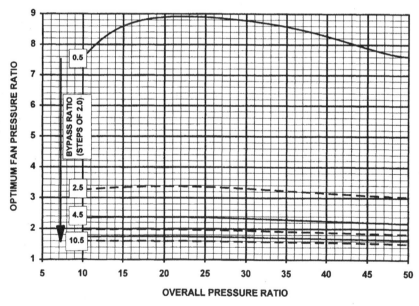

(f)   SOT = 1600 K

Chart 6.20   *contd.*

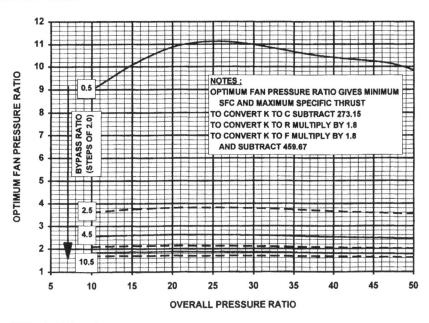

(g)   SOT = 1700 K

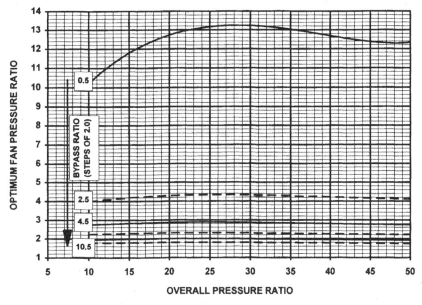

(h)   SOT = 1800 K

Chart 6.21   Turbofan cycles: optimum fan PR versus overall PR and bypass ratio at 11 000 m, ISA, 2.2 M (design point diagrams).

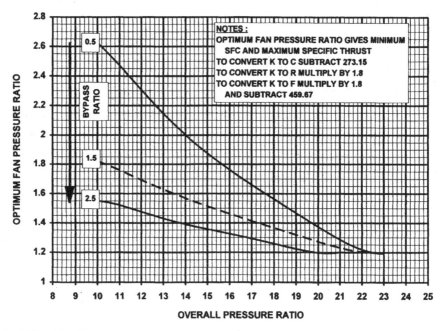

(a)   SOT = 1400 K

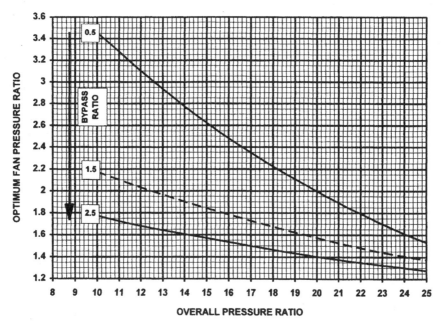

(b)   SOT = 1500 K

Chart 6.21 *contd.*

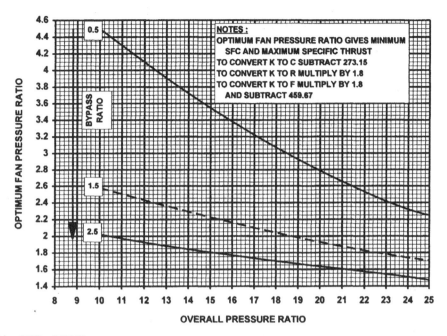

(c)   SOT = 1600 K

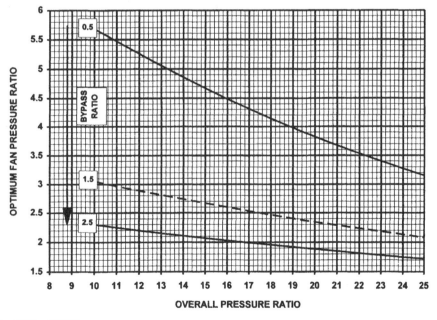

(d)   SOT = 1700 K

Chart 6.21   *contd.*

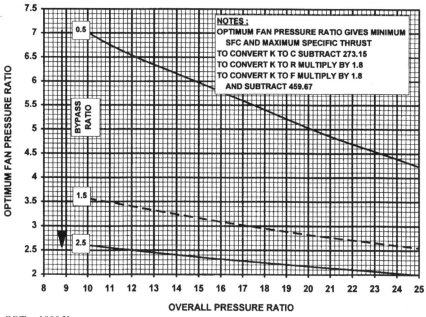

(e)   SOT = 1800 K

Chart 6.22   Turbofan cycles: SFC versus overall PR and bypass ratio at optimum fan PR at 11 000 m, ISA, 0.8 M (design point diagrams).

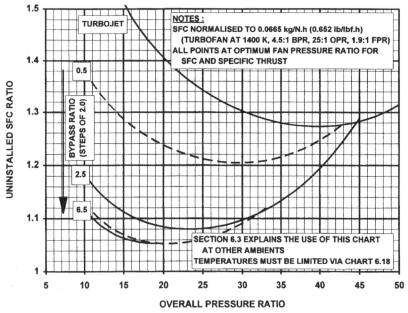

(a)   SOT = 1100 K

Chart 6.22 *contd.*

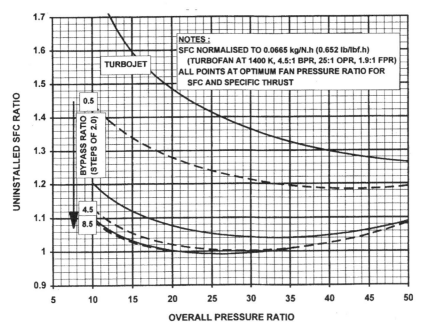

NOTES :
SFC NORMALISED TO 0.0665 kg/N.h (0.652 lb/lbf.h)
(TURBOFAN AT 1400 K, 4.5:1 BPR, 25:1 OPR, 1.9:1 FPR)
ALL POINTS AT OPTIMUM FAN PRESSURE RATIO FOR
SFC AND SPECIFIC THRUST

(b)   SOT = 1200 K

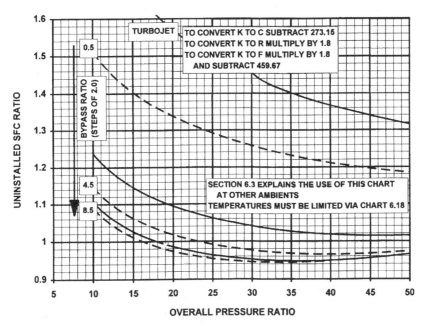

TO CONVERT K TO C SUBTRACT 273.15
TO CONVERT K TO R MULTIPLY BY 1.8
TO CONVERT K TO F MULTIPLY BY 1.8
AND SUBTRACT 459.67

SECTION 6.3 EXPLAINS THE USE OF THIS CHART
AT OTHER AMBIENTS
TEMPERATURES MUST BE LIMITED VIA CHART 6.18

(c)   SOT = 1300 K

Chart 6.22   *contd.*

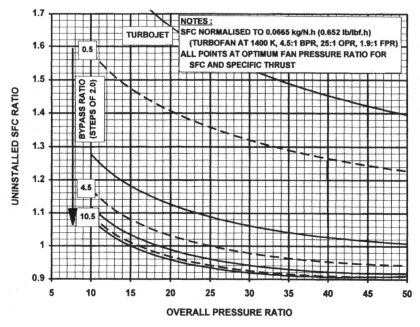

(d)   SOT = 1400 K

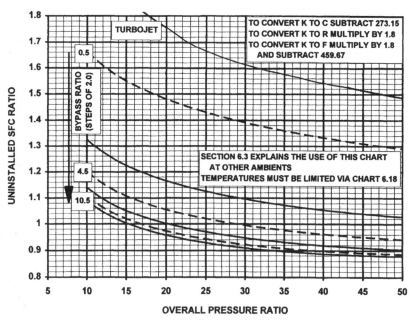

(e)   SOT = 1500 K

Chart 6.22   *contd.*

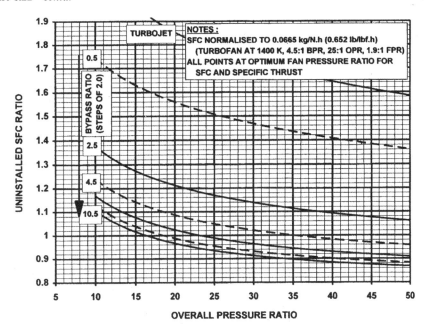

NOTES :
SFC NORMALISED TO 0.0665 kg/N.h (0.652 lb/lbf.h)
(TURBOFAN AT 1400 K, 4.5:1 BPR, 25:1 OPR, 1.9:1 FPR)
ALL POINTS AT OPTIMUM FAN PRESSURE RATIO FOR
SFC AND SPECIFIC THRUST

(f)   SOT = 1600 K

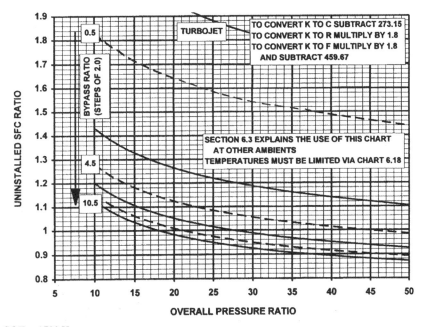

TO CONVERT K TO C SUBTRACT 273.15
TO CONVERT K TO R MULTIPLY BY 1.8
TO CONVERT K TO F MULTIPLY BY 1.8
AND SUBTRACT 459.67

SECTION 6.3 EXPLAINS THE USE OF THIS CHART
AT OTHER AMBIENTS
TEMPERATURES MUST BE LIMITED VIA CHART 6.18

(g)   SOT = 1700 K

Chart 6.22   *contd.*

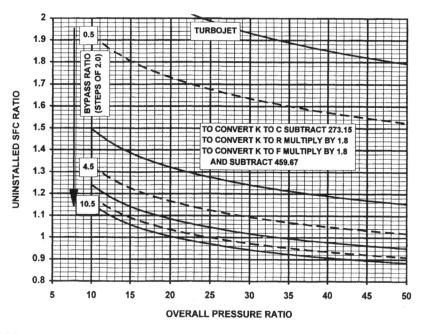

(h)   SOT = 1800 K (Notes as per part (a))

Chart 6.23   Turbofan cycles: SFC versus overall PR and bypass ratio at optimum fan PR at 11 000 m, ISA, 2.2 M (design point diagrams).

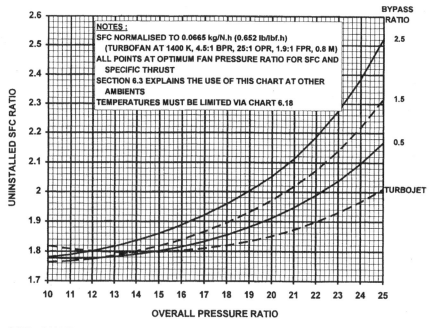

(a)   SOT = 1400 K

Chart 6.23 *contd.*

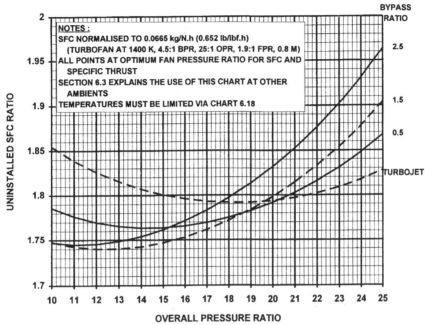

(b) SOT = 1500 K

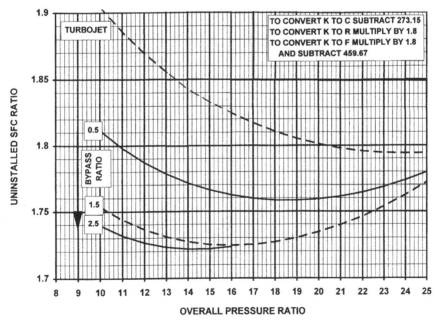

(c) SOT = 1600 K

Chart 6.23 *contd.*

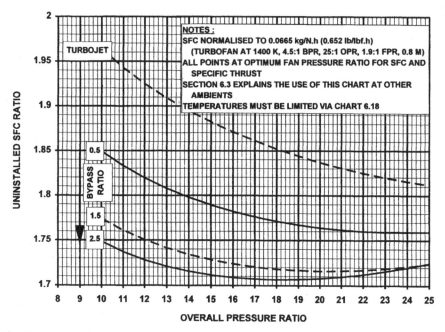

(d)   SOT = 1700 K

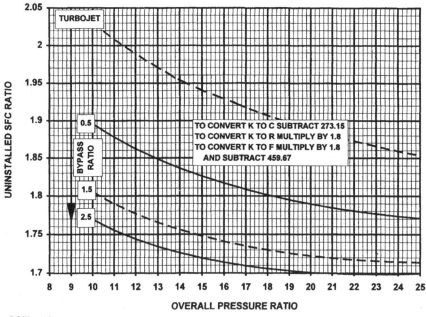

(e)   SOT = 1800 K

Chart 6.24 Turbofan cycles: specific thrust versus overall PR and bypass ratio at optimum fan PR at 11 000 m, ISA, 0.8 M (design point diagrams).

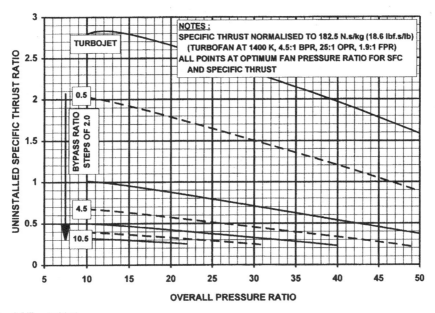

(a) SOT = 1100 K

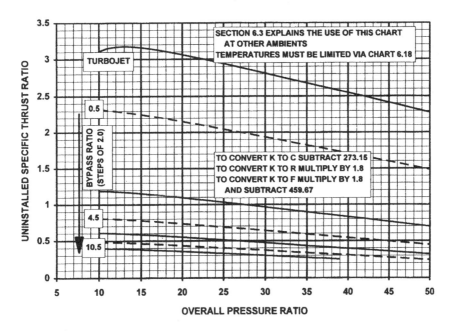

(b) SOT = 1200 K

Chart 6.24 *contd.*

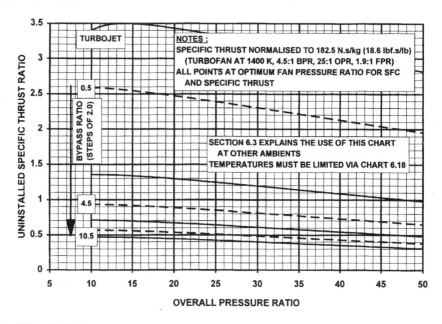

(c) SOT = 1300 K

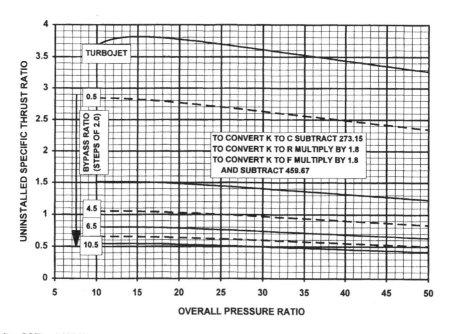

(d) SOT = 1400 K

Chart 6.24 *contd.*

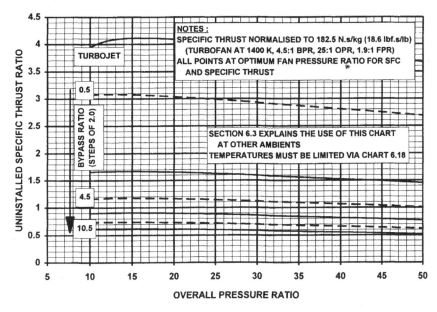

(e) SOT = 1500 K

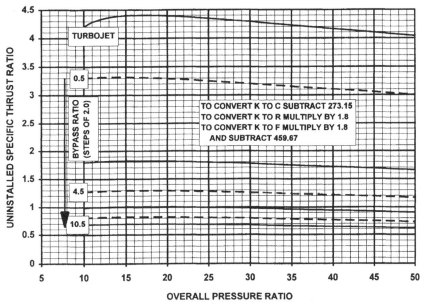

(f) SOT = 1600 K

Chart 6.24 *contd.*

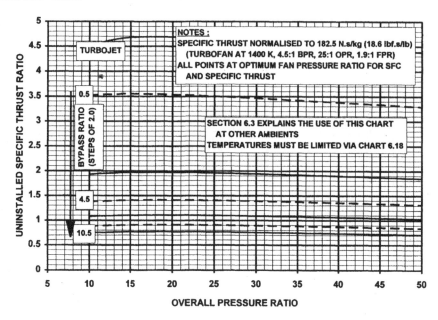

(g)   SOT = 1700 K

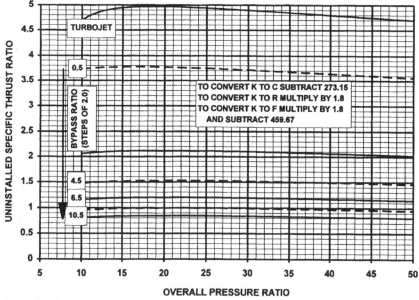

(h)   SOT = 1800 K

Chart 6.25 Turbofan cycles: specific thrust versus overall PR and bypass ratio at optimum fan PR at 11 000 m, ISA, 2.2 M (design point diagrams).

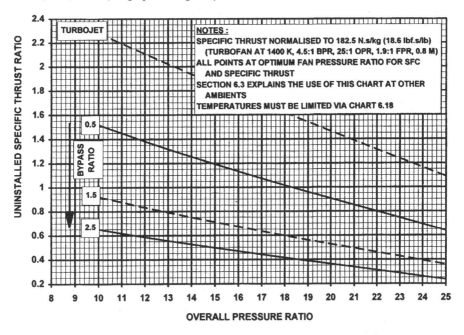

(a) SOT = 1400 K

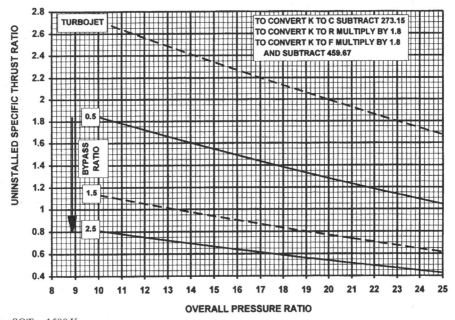

(b) SOT = 1500 K

Chart 6.25   *contd.*

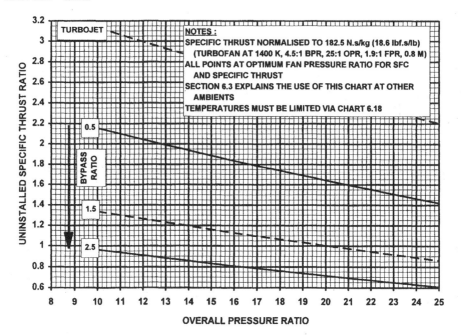

(c)   SOT = 1600 K

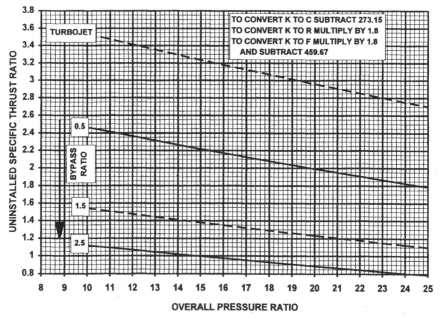

(d)   SOT = 1700 K

Chart 6.25 *contd.*

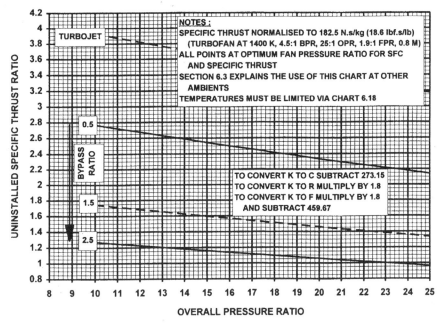

(e)   SOT = 1800 K

Chart 6.26   Ramjet cycles : SFC and specific thrust versus T4 and M at 15 240 m, ISA (design point diagrams).

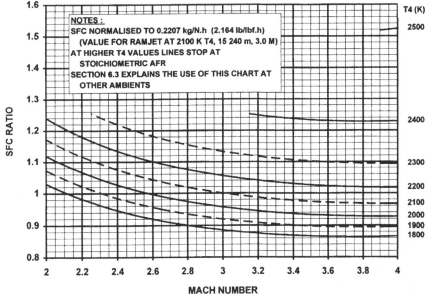

(a)   SFC

Chart 6.26   *contd.*

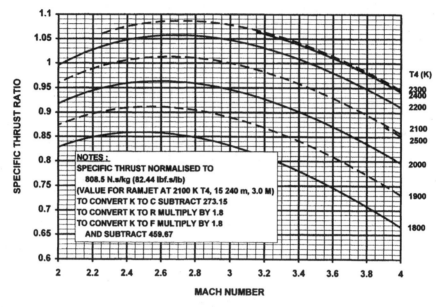

(b)   Specific thrust

References

1. H. Cohen, G. F. C. Rogers and H. I. H. Saravanamuttoo (1996) *Gas Turbine Theory*, 4th edn, Longmans, Harlow.
2. R. T. C. Harman (1987) *Gas Turbine Engineering*, Macmillan, Basingstoke.
3. J. P. Den Hartog (1961) *Strength of Materials*, Dover, New York.

Chapter 7
Off Design Performance

7.0 Introduction

This chapter describes the steady state performance variation of a fixed engine design as the *operational condition* is changed. This is in contrast to Chapter 6 which described the performance variation at a fixed operational condition as the *engine design* was modified. The operational condition is defined by given values of:

- The power or thrust level, and
- The point within the operational envelope

Major off design effects are presented, along with the underlying physical mechanisms and modelling methods. Off design performance is inseparable from many other aspects of performance: to appreciate it fully familiarity with the subject matter in the four preceding chapters is essential:

- Chapter 2: The operational envelope
  The environmental and flight conditions which an engine may encounter have a significant effect on engine performance.
- Chapter 4: Dimensionless, quasi-dimensionless, referred and scaling parameter groups
  Turbomachinery responds to, and hence is modelled via, groups of parameters, e.g. Mach number and pressure ratio.
- Chapter 5: Gas turbine engine components
  Off design engine performance is the *synthesis* of component performance, and is strongly affected by major component performance trends.
- Chapter 6: Design point performance and engine concept design
  Definition of a candidate engine design precedes evaluation of off design performance. During the engine concept design process this sequence is repeated in an iterative process, hence off design performance influences the final choice of engine design.

7.1 Generic off design characteristics

This section presents generic off design characteristics for each engine type, plotted in terms of referred parameters.

As discussed in Chapter 4, engines behave non-dimensionally, such that fixing the value of one or more referred parameter groups also fixes that of the others. As the power or thrust level is varied, the referred parameter groups therefore follow a unique running line or families of running lines which are independent of ambient conditions. To a first order the off design performance may be defined via charts showing the interrelationship of the referred parameter groups. For a given operational condition, knowledge of the absolute values of inlet pressure and temperature allows *actual* performance parameter values to be easily calculated. The value of these charts cannot be over emphasised. They enable 'on the spot' judgements such as during engine testing, or discussing in a meeting the impact of an extreme operating point.

Accuracy is increased if additional charts or scale factors are also used for the real second-order effects of P1, T1 and combustion efficiency, as described in Chapter 4. Such effects are often highly specific to each individual engine design. For generating fully rigorous data the thermodynamic matching models described in section 7.3 must be employed.

7.1.1 Turboshaft, single spool

Here a turbine drives both the engine compressor and the output load. Chart 7.1 shows the referred performance charts and Fig. 7.1 the compressor working line. As described in Chapter 4, when power is extracted from a shaft two referred parameter groups, rather than just one, must be fixed in order to fix all others. Referred power is used as a base parameter for the charts for this configuration alone, because it is almost solely employed for power generation where the shaft must rotate at synchronous speed irrespective of power level. The second base parameter for Chart 7.1 is referred speed, with lines plotted corresponding to synchronous operation at −40 8C, 15 8C and 40 8C. In this way Chart 7.1 maps out the full off design operating range of this engine type. Sample calculation C7.1 demonstrates the use of these curves.

Referred compressor delivery pressure and temperature increase significantly as day temperature is reduced and referred speed increases. For a given day temperature, at part power referred turbine temperatures fall sharply as fuel flow is reduced. Referred compressor delivery pressure and temperature fall only very slowly, due to the fixed speed. SFC becomes worse as the cycle efficiency reduces at the low firing temperature, and also as compressor efficiency falls. Chart 7.1 shows that at 20% power on an ISO day SFC is almost twice that at 100% power.

Figure 7.1 shows that for a given day temperature as fuel flow, hence SOT and output power, are increased the compressor operating point moves up the constant referred speed line. Equally if day temperature increases referred speed falls. If the engine is *flat rated* to hold constant power then on hot days surge margin will reduce, as referred SOT must increase.

7.1.2 Turboshaft, single spool gas generator and free power turbine

In this configuration the engine compressor is driven by one turbine, and a second turbine drives the output load. Chart 7.2 shows the referred performance charts and Fig. 7.2 the compressor working line. Charts 7.2a show leading gas generator parameters versus referred

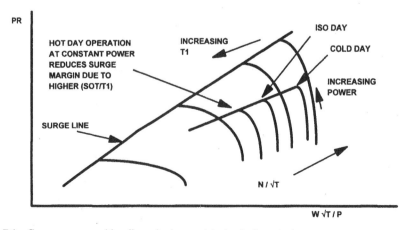

Fig. 7.1 Compressor working line, single spool turboshaft or turboprop.

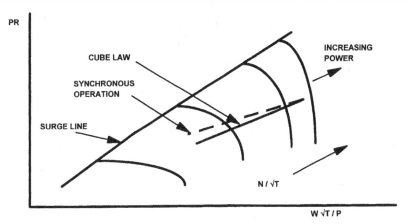

Fig. 7.2 Compressor working line, single spool gas generator with free power turbine.

speed, and SFC versus referred power. Unlike the single spool engine the gas generator speed is not tied to that of the load, hence it reduces as power falls and is an appropriate base parameter for the charts. At part load cycle pressures and temperatures again fall, but compressor efficiency remains higher than for a single spool turboshaft. At 20% power SFC is around 1.6 times that at 100% power.

For a given power or gas generator speed the power turbine speed may vary over a wide range, depending on the absorption characteristics of the driven load. Chart 7.2b shows the shaft output power and SFC versus power turbine speed, for lines of constant gas generator speed. To a first order each line of referred gas generator speed is a fixed gas generator operating point, and hence the curves of SFC and output power versus power turbine speed are primarily a function of power turbine efficiency. The variation in power turbine capacity produces a further second-order effect on gas generator operation.

The engine may operate at any point on the map shown in Chart 7.2b. For power generation the running line is vertically up and down a constant synchronous speed line. For cube law operation the running line coincides roughly with the peaks of the referred gas generator speed lines (maximum shaft power at a gas generator speed) and troughs of the SFC contours (lowest/best SFC). This is because power turbine efficiency and capacity are higher on a cube law. Sample calculation C7.2 shows the evaluation of referred parameters at two different levels of power turbine speed.

Chart 7.2c shows referred torque versus referred power turbine speed, for lines of constant referred gas generator speed. Classically *stall torque* (that at zero output speed) has been estimated as around 2 times that at full power and 100% speed. Also shown on Chart 7.2c is the maximum available torque versus speed for the single shaft configuration covered in section 7.1.1. At low output speed very little torque is available, indeed at zero output speed the engine cannot operate. Hence it is apparent that *for mechanical drive applications operating on a cube law a free power turbine configuration must be used.* For example, for vehicle acceleration an automotive application requires high torque and power at low output speed, which only a free power turbine engine can deliver.

An alternative description of engine behaviour is to consider power turbine inlet conditions. As defined in Chapter 6, the engine core produces 'exhaust gas power'. The power turbine converts this to shaft power, its isentropic efficiency being simply the ratio of shaft power to exhaust gas power. The lines of constant referred gas generator speed can be considered as lines of constant referred exhaust gas power, if the second-order effects of power turbine capacity variation are ignored. The shape of the lines on Chart 7.2b is then seen to reflect power turbine efficiency variation.

Figure 7.2 shows that the compressor referred speed changes with power level. The power turbine speed law sets the part load trend; the reductions in capacity and efficiency produced by synchronous operation reduce surge margin at low power.

7.1.3   *Recuperated turboshaft, single spool gas generator and free power turbine*

Here two turbines drive the engine compressor and output load respectively, as per the preceding configuration, but a recuperator recovers exhaust heat to the combustor inlet. Chart 7.3 shows the referred performance charts, and Fig. 7.3 shows the compressor working line. Both fixed and variable power turbine geometry are shown. There is actually one set of referred parameter plots for each VAN (variable area nozzle) setting, as each gives a unique engine geometry. Each set of plots is akin to the fixed geometry operation shown as dotted lines. The solid lines shown for variable geometry, which is normally used in practice, represent 'slices' through the fixed geometry plots.

The VAN is controlled to initially close at part power to maintain constant power turbine entry temperature, T46, to provide high heat recovery and hence improve part power SFC. Once it reaches its closure stop at around 40% power it remains closed down to almost idle. In fact SFC does not show the usual simple cycle increase at part power until around 50% power, as the temperature difference across the recuperator rises substantially at part power maintaining near constant combustor inlet temperature (T31). At idle the VAN is fully open again to reduce temperature levels to ease the thermal transients, especially for the recuperator, during start and shut down.

With a VAN, power falls versus compressor speed faster than for the unrecuperated turboshaft above, due to the reduced gas generator turbine expansion ratio. For any given level of fuel flow a closed VAN will result in a lower compressor speed, hence lower inlet air flow, with a higher fuel air ratio and hence higher turbine temperatures. Accordingly the T46 plot shows a flat region, the T41 plot a retarded fall versus power, and T5 actually rises at part power in this region. The difference in these trends is due to the reducing temperature drop across each turbine as power falls, and hence expansion ratios.

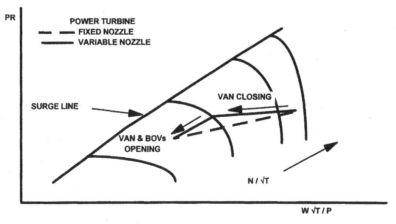

Notes:
Above is for power turbine operating on a cube law.
Synchronous operation raises part load working line for fixed nozzle or results in modified van schedule for variable nozzle.
BOVs are 'blow off valves', or handling bleeds.

Fig. 7.3   Compressor working line, recuperated turboshaft, single spool gas generator with free power turbine.

The effects of power turbine speed on engine performance are similar to Charts 7.2b and c, though with two exceptions. SFC will vary less due to recuperator heat recovery offsetting power turbine efficiency variation, and above 50% power the effects on gas generator performance will be offset by the VAN running more open to maintain constant T46. (In addition, this may increase power turbine efficiency and actually improve SFC versus the cube law.)

Figure 7.3 shows that closing the VAN raises the compressor working line, due to reduced gas generator turbine expansion ratio requiring higher SOT at each speed level.

7.1.4 *Turboshaft, two spool gas generator and free power turbine*

In this configuration two engine compressors and their driving turbines comprise the gas generator, and a free power turbine drives the output load. Chart 7.4 shows the referred performance charts and Fig. 7.4 the compressor working lines. Here LP compressor referred speed is the base parameter for the charts. The scale is longer than for engines with a compressor on one spool: LP speed reduces drastically at very low power due to the unchoked power turbine reducing the LP turbine expansion ratio and hence power output. HP turbine expansion ratio remains much closer to its design value, and hence so does HP speed. The trends in the main parameters are basically similar to those for a single spool free power turbine turboshaft; the main impact of the number of spools is on compressor design.

The effects of variable power turbine speed are broadly as per Charts 7.2b and c, except that the resultant variation in LP speed in the low power region will be greater. As the engine is throttled back, Fig. 7.4 shows that the LP compressor working line moves significantly towards surge, while that of the HP compressor remains more away from it.

7.1.5 *Turboshaft, intercooled and recuperated*

Here an intercooler is added between two engine compressors, and a recuperator recovers exhaust heat. Chart 7.5 shows the referred performance charts and Fig. 7.5 shows the compressor working lines. Again, lines are shown for both fixed and variable power turbine geometry, and the solid lines shown for the latter represent 'slices' through a family of fixed

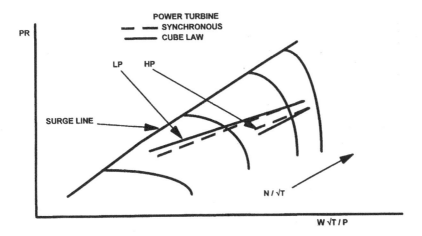

Note:
Blow off valves (BOVs) produce downward steps on opening (not shown).

Fig. 7.4 Compressor working lines, turboshaft or turboprop, twin spool gas generator with free power turbine.

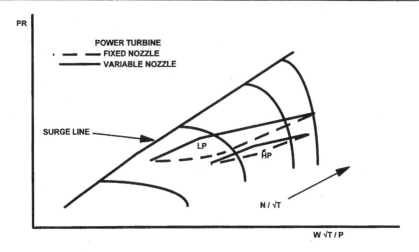

Notes:
Blow off valves (BOVs) produce downward steps on opening (not shown).
Intercooler inoperative operation raises LP compressor working line.
Recuperator bypassed operation lowers HP compressor working line.
Synchronous power turbine operation results in modified VAN schedule.

Fig. 7.5   Compressor working lines, intercooled, recuperated turboshaft, twin spool gas generator with free power turbine.

geometry plots. An additional variable to map this configuration fully in terms of referred parameters is the level of intercooler heat extraction, there being one set of curves for each ratio of sink temperature to ambient. For practicality, Chart 7.5 is for a single ratio of heat sink temperature to ambient; plotting all possible curves would be prohibitive.

Intercooling alone increases power via reduced HP compressor work; with a recuperator the reduced air side inlet temperature increases the exhaust heat recovery, resulting in a corresponding SFC improvement. This benefit is particularly strong below 30% power, where the HP compressor work would be an increasing fraction of the total turbines' work.

Again the VAN is controlled to close at part power to maintain constant power turbine entry temperature, T46, to improve part power SFC. As for the recuperated cycle, SFC does not show the simple cycle increase at part power until around 50% power, as the temperature difference across the recuperator rises substantially, maintaining near constant combustor inlet temperature (T31). The other effects of the VAN on turbine temperatures are as per the recuperated turboshaft. One difference is that here with two compressors and a VAN the LP speed scale is expanded by the reduced LP turbine expansion ratio which reduces LP speed at a power when the VAN is closed.

The effects of power turbine speed on engine performance are similar to Charts 7.2b and c, though with the same two exceptions as in section 7.1.3.

Figure 7.5 shows that closing the VAN raises both compressor working lines, due to reduced gas generator turbine expansion ratios. The main effect on the LP compressor is reduced speed however, and if the LP turbine were choked (which at part power it is not) then VAN closure would actually drop the LP compressor working line.

7.1.6   *Turbojet, single spool*

Here a compressor is driven by a turbine, which exhausts into a propelling nozzle. The pressure ratio across this nozzle results in high exhaust gas velocities, and hence jet thrust.

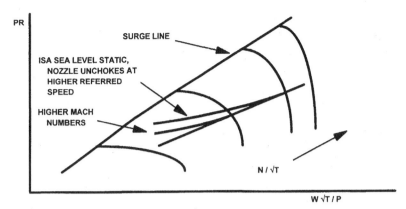

Fig. 7.6   Compressor working lines, subsonic turbojet, single spool.

Subsonic operation

Chart 7.6 shows the referred performance charts and Fig. 7.6 the compressor working line. An additional parameter relative to the land based engines is flight Mach number. As described in Chapter 4, the engine has a unique referred running line for each level of flight Mach number. Only once the propelling nozzle chokes do these running lines become coincident. The thrust level at which choking occurs depends on flight Mach number, which produces different levels of nozzle pressure ratio.

Sample calculation C7.3 demonstrates the use of Chart 7.6. Referred fuel flow, air flow and turbine temperatures show a strong variation versus Mach number at lower referred speeds, via variation of turbine expansion ratio into the unchoked propelling nozzle. As Mach number increases at a referred speed level referred mass flow increases, referred fuel flow reduces and the referred turbine temperatures are therefore lower. The exact variation of referred P3 depends on the compressor map shape, the small amount shown corresponds to relatively flat speed lines.

Unlike the landbased engines, SFC improves significantly down to around 50% thrust due to increasing propulsive efficiency outweighing falling thermal efficiency of the core engine cycle. This is due to lower exhaust velocities and temperatures, and hence less energy used for any level of exhaust momentum (i.e. thrust). At lower thrust levels SFC worsens again, due to rapidly deteriorating thermal efficiency.

The largest effect of flight Mach number is via inlet momentum drag, which reduces nett thrust and therefore worsens SFC. The 'ram' compression partly offsets this, and the available physical (rather than referred) thrust is high at cruise altitude. For early civil airliners with turbojets rather than turbofans, achieving takeoff thrust rather than cruise sized the engines. Section 7.1.7 includes a comparison of the effect of Mach number on both engine types.

Figure 7.6 shows that the compressor working line falls with increasing flight Mach number when the propelling nozzle is unchoked.

7.1.7   Turbofan, two spool, subsonic operation

Here air from a first *LP compressor* or *fan* partially bypasses the main engine core, to reach a separate bypass nozzle via a bypass duct. The *bypass ratio* is the ratio of bypass to core flow. The core has an HP compressor, and two turbines. Chart 7.7 shows the referred performance charts and Fig. 7.7 the compressor working lines. The main advantage of a bypass engine is improved SFC via increased flow and reduced exhaust temperatures and velocities, which improves propulsive efficiency and hence leave less wasted energy in the jet plume. As for a

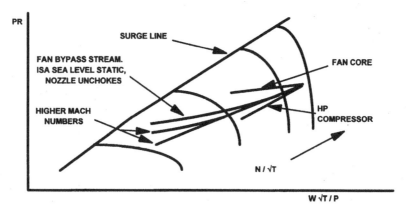

Fig. 7.7   Compressor working lines, subsonic turbofan, twin spool.

turbojet, SFC improves down to low thrusts due to improved propulsive efficiency, before rising again due to low thermal efficiency. Unlike for turbojets, the cruise rather than takeoff condition sizes the engine, and hence excess thrust is available at takeoff.

As with a pure turbojet, flight Mach number affects the degree of choking of the propelling nozzles. The effect is particularly apparent on the fan bypass working line, and the bypass ratio increases at low thrust, especially at high Mach number; here the core compressor capacity falls more quickly due to reducing speed than that of the bypass propelling nozzle, giving a very 'flat' fan core working line.

A major consequence of the increased inlet mass flow is increased inlet momentum drag, which significantly reduces thrust at high Mach number. Figure 7.8 shows *actual* net thrust versus Mach number at maximum rating for both a turbojet and a high bypass ratio turbofan.

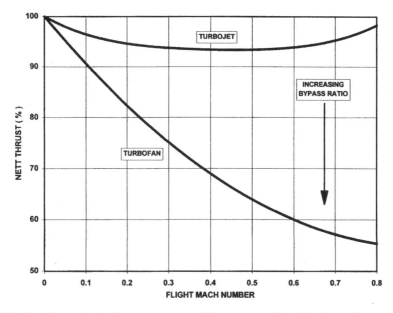

Fig. 7.8   Turbojet and turbofan maximum rated thrust versus flight Mach number.

For a turbojet, initially thrust falls as Mach number increases, it recovers to the static value at around 0.8 Mach number, and above this it increases rapidly. This shape is due to the combination of referred net thrust falling with Mach number, inlet pressure increasing and raising actual thrust for a given referred value, and momentum drag and propulsive efficiency both increasing with Mach number. For a turbofan however thrust continues to fall versus Mach number, because the increasing ram drag due to the high mass flow outweighs all the other effects. For this reason, as well as pod drag, high flight Mach numbers require either a low bypass ratio turbofan with reheat, or a turbojet.

7.1.8 *Supersonic turbojet or turbofan*

Such operation is complex and depends heavily on the chosen cycle parameters, hence no sufficiently generic charts can be provided. Section 7.3.8 describes the interaction of the key effects on off design performance. The high pressure rise produced by the flight Mach number interacts strongly with the basic cycle pressure ratio.

Supersonic operation produces a choked intake, which fixes engine mass flow. Variable intake geometry is provided for efficient operation down to zero Mach number: otherwise too high an engine flow demand simply results in excessive pressure losses before the engine, the high $W\sqrt{T}/P$ being achieved by reduced P. Variable areas are provided for blow in, spillage and throat.

Owing both to an afterburner being normally fitted, and to address varying flight Mach number, a variable nozzle area is also provided. Afterburning, as described in Chapter 5, is used for supersonic operation or at least transonic acceleration; this requires an increased nozzle area to prevent engine surge due to reduced final turbine expansion ratio. In addition, varying flight Mach number changes the nozzle entry pressure, so a major aim in setting the nozzle area is to ensure the best combination of nozzle pressure ratio and con–di area ratio are achieved. Not all the criteria for nozzle area can be met simultaneously, so some compromise is inevitable.

7.1.9 *Ramjet*

Here the cycle pressure rise comes only via diffusion of free stream air from high flight Mach number, with no rotating turbomachinery. Each engine design must be closely tailored to its specific application in terms of intake and nozzle performance to address the required flight Mach number range. Accordingly no sufficiently generic charts can be prepared. Section 7.3.9 describes the major mechanisms affecting off design performance.

7.2 Off design performance modelling – methodology

This chapter concentrates on the almost universal form of steady state off design model, which is a *thermodynamic matching model*. Any model is also commonly referred to as a *deck*, from the early days of computers when a model was loaded as a deck of hole-punched cards. Other model types, such as look up tables, are rarely used and hence are not covered.

Chapters 4 and 5 explain how component behaviour may be represented by maps based on parameter groups. Once a component geometry is fixed its map is unique, representing performance at all off design conditions. For a given engine operating condition, such as ISO and 100% power, the operating point on each component map is also unique, being dependent on the maps of the components that it is either connected or *matched* to. There is then a unique set of component operating points at each engine operating condition. A simple illustration is the compressor test rig described in Chapter 5, in which there is a different operating point on the compressor map for each downstream throttle setting (which is the only component to match to), and speed (which is analogous to engine operating condition).

Off design engine performance modelling determines each component operating point as it is matched to the others; it is often termed *matching,* and is highly iterative. It requires successive 'guesses' of the operating point on some component maps. These are updated as iteration continues until other known conditions are satisfied. Once this iteration is complete, overall cycle parameters such as thrust or shaft power can easily be derived. This process is best illustrated via a detailed description of the modelling required, which is presented in the following sections for the various major engine configurations.

Many essential model capabilities are recognised in the relevant international standards for customer decks, as discussed in Appendix A. These include the ability to solve to specified values of key engine parameters, via *power codes,* and to restrict any solutions that would exceed input limiting values. Standard software tools are increasingly available commercially.

7.2.1 *Serial nested loops versus matrix iteration*

Off design matching calculations are almost invariably done on a computer. They utilise *stations* representing the engine flowpath in proper sequence. Appendix A presents the internationally recognised engine nomenclature, which includes station numbering. At each station thermodynamic parameters are calculated based on upstream component performance.

For a thermodynamic steady state model the iteration described above is achieved either via *serial nested loops,* or via a *matrix solution.* The former is often easier to understand physically, and to implement via personal computer programs. Accordingly it is this method that is illustrated by the sample calculations presented herein. However this technique becomes computationally inefficient for more than five nested loops, and hence in larger more elaborate systems the matrix solution method is now more common. Either way for each engine there are several *matching guesses,* and an equal number of *matching constraints.* During iteration the matching guesses are continually updated until the matching constraints are satisfied. Compressor beta values are typically matching guesses, and calculated turbine capacities being equal to those from the maps are typically matching constraints.

For serial nested loops the matching guesses and matching constraints are paired and solved in a *nested* sequence, whereby for each pass though an 'outer' iteration loop each iteration loop within it is repeated until convergence.

In matrix iteration the overall interaction is recognised and the equations are solved simultaneously. This requires a numerical method utilising *partial derivatives,* which are the effect of changing each matching guess individually on the errors in all the matching constraints. The basic steps in this methodology are as follows.

- Choose initial values of matching guesses, e.g. as fixed input values or by reading from supplied graphs.
- Complete one pass through the engine, almost as for a design point calculation.
- Calculate the error between calculated values of the matching constraints and values looked up from the component maps.
- Make a small change in each matching guess in turn and repeat the preceding two steps.
- From the error values obtained evaluate the partial derivatives of the errors in each matching constraint with respect to each matching guess. When complete, this step produces the *matrix of partial derivatives.*
- Invert the matrix of partial derivatives, using a standard numerical technique such as *Gauss–Jordan inversion* or *LU decomposition.*
- Simultaneously change all matching guesses by the amounts given by multiplying the inverted partial derivatives matrix by the errors matrix, per Formula F7.1. This step may be enhanced by using a standard numerical technique such as the *damped Newton method.*
- Repeat the above processes until the errors between calculated values of the matching constraints and values looked up from the component maps are within an allowable tolerance band, e.g. 0.1%.

Chapter 11 describes how matrix iteration may be extended to engine test data analysis. Here, due to complexity, nested loops become impractical.

7.2.2 *Gas properties*

All modelling must use valid thermodynamic gas properties. Sample calculations presented herein use the gas properties evaluated at mean temperature through a component. The use of CP and gamma, and the more rigorous method of using entropy and enthalpy, are described in Chapter 3.

7.3   Off design performance modelling – flow diagrams and sample calculations

Flow diagrams for off design performance calculations using serial nested loops are presented for each major engine configuration. The model is shown run to a fixed LP speed, apart from the single spool turboshaft engine which must be run to both speed and power. For all configurations, to run to other parameters generally requires an additional outermost loop with LP speed as the matching guess, and the error between the calculated and required values of

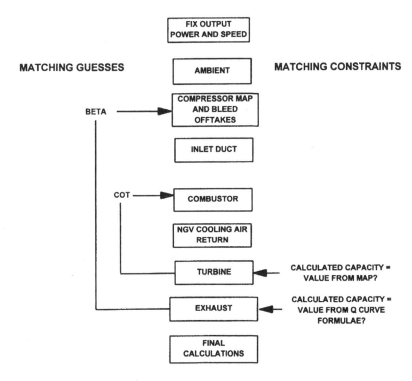

Notes:

Matching guesses shown require initial values, updated by each pass through the relevant loop.

Output speed is set by the load, e.g. constant synchronous value.

For simplicity turbine cooling flow returns downstream of the first NGV are not shown.

Recirculating bleeds require additional guesses and matches at return station for flow quantity and temperature.

Fig. 7.9 Matching diagram, turboshaft or turboprop, single spool.

the specified parameter as the matching constraint. Occasionally a different choice of matching guesses may avoid the need for this additional loop.

Section 7.3.10 describes the detailed calculation process for each individual component. Sample calculation C7.4 follows the methods shown for a turbojet. These use Formulae F7.2 and F7.3 for updating matching guesses at the end of each iteration. The allowable convergence tolerance for the errors betweeen calculated and map values of matching constraints should be less than 0.1%. For reference, Charts 7.8–7.14 illustrate more elaborate matching schemes which suit the matrix method.

7.3.1 *Turboshaft, single spool*

Figure 7.9 shows the matching diagram, with the engine run to specified power and speed. The latter is the synchronous value required for power generation. There are two iteration loops, to satisfy the flow capacities of the gas generator turbine and of the exhaust discharge area. These require adjustment of the guessed values of COT and compressor beta respectively.

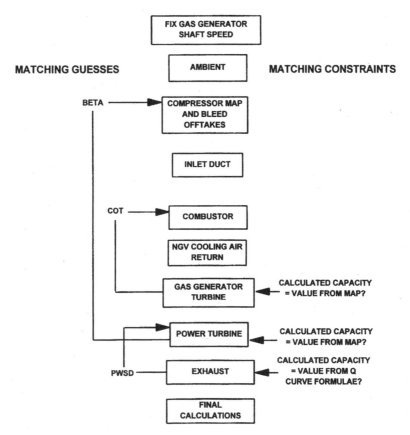

Notes:

Matching guesses shown require initial values, updated by each pass through the relevant loop.

Power turbine speed is calculated from the load characteristic, e.g. ship's propeller cube law.

For simplicity turbine cooling flow returns downstream of the first NGV are not shown.

Recirculating bleeds require additional guesses and matches at return station for flow quantity and temperature.

Fig. 7.10 Matching diagram, turboshaft or turboprop, single spool gas generator with free power turbine.

7.3.2 *Turboshaft, single spool gas generator and free power turbine*

Figure 7.10 shows the matching diagram, with the model run to fixed gas generator speed. There are three iteration loops, to satisfy the flow capacities of the gas generator turbine, the power turbine and the exhaust discharge area. These require adjustment of the guessed values of COT, compressor beta and output power respectively. For each guessed value of output power the power turbine speed is set according to the load characteristic, e.g. a cube law for a ship's propeller or synchronous speed for a generator.

The loops for output power and compressor beta overlap. If the calculated exhaust capacity does not match the value from Q curves then the output power guess is changed, the power turbine speed updated, and the the power turbine calculation repeated as per the loop shown.

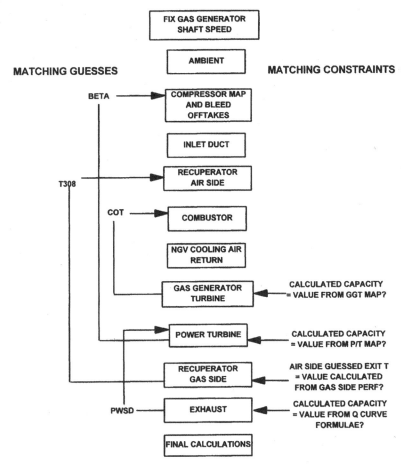

Notes:

Matching guesses shown require initial values, updated by each pass through the relevant loop.

Power turbine speed is calculated from the load characteristic, e.g. ship's propeller cube law.

For simplicity turbine cooling flow returns downstream of the first NGV are not shown.

Recirculating bleeds require additional guesses and matches at return station for flow quantity and temperature.

A variable area power turbine nozzle would allow additional iteration, e.g. to fixed power turbine inlet temperature.

Fig. 7.11 Matching diagram, recuperated turboshaft, single spool gas generator with free power turbine.

The new speed and power values now result in a different level of capacity from the map. This will now differ from the power turbine capacity calculated from inlet conditions. A new guess must now be made for beta, and the compressor calculations repeated. This whole process is repeated until convergence.

7.3.3   *Recuperated turboshaft, single spool gas generator and free power turbine*

Figure 7.11 shows the matching diagram, with the engine run to fixed gas generator speed. Again there are three engine iterations, for the turbine and exhaust flow capacities, requiring adjustment of the guessed values of COT, compressor beta, and power turbine output power.

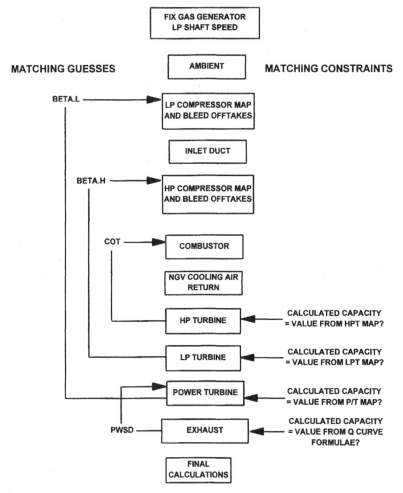

Notes:

Matching guesses shown require initial values, updated by each pass through the relevant loop.

Power turbine speed is calculated from the load characteristic, e.g. ship's propeller cube law.

For simplicity turbine cooling flow returns downstream of the first NGV are not shown.

Recirculating bleeds require additional guesses and matches at return station for flow quantity and temperature.

Fig. 7.12   Matching diagram, turboshaft or turboprop, twin spool gas generator with free power turbine.

The recuperator requires an additional iteration for the air side exit temperature, as it depends on the gas side parameters which have not yet been evaluated. (Guessing a value normally allows direct calculation of air side pressure loss, though in the most elaborate models that also requires iteration.) The guessed value of air side exit temperature is compared with calculations using the gas side parameters; if they do not match the iteration loop is repeated. The recuperator loop is reached before that of the load; repeating it will result in re-iteration of the beta and COT loops.

If the power turbine incorporates a variable area nozzle, its setting may either be specified directly or used in an additional outer iteration to achieve some other specified parameter. This is usually power turbine entry temperature (T46), which provides better SFC than would constant

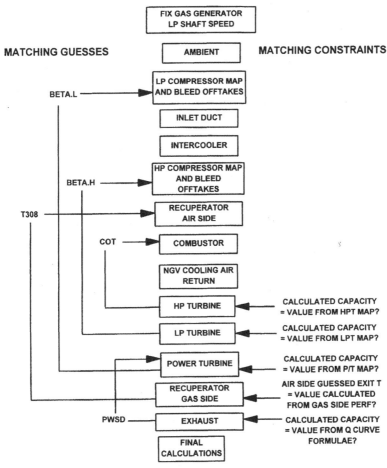

Notes:

Matching guesses shown require initial values, updated by each pass through the relevant loop.

Power turbine speed is calculated from the load characteristic, e.g. ship's propeller cube law.

For simplicity turbine cooling flow returns downstream of the first NGV are not shown.

Recirculating bleeds require additional guesses and matches at return station for flow quantity and temperature.

A variable area power turbine nozzle would allow additional iteration, e.g. to fixed power turbine inlet temperature.

Fig. 7.13 Matching diagram, intercooled, recuperated turboshaft, twin spool gas generator with free power turbine.

recuperator inlet temperature (T6) while avoiding the excessive T6 levels that a constant SOT algorithm would give.

7.3.4 *Turboshaft, two spool gas generator and free power turbine*

Figure 7.12 shows the matching diagram, with the engine run to fixed speeds for the LP compressor and free power turbine. There are four iteration loops, for the capacities of the three turbines and exhaust, requiring adjustment of the guessed values of COT, compressor betas, and power turbine output power. HP shaft speed is determined by reading the HP compressor map with the current values of flow and beta.

7.3.5 *Turboshaft, intercooled and recuperated*

Figure 7.13 shows the matching diagram, with the engine again run to fixed speeds for the LP compressor and power turbine. As for the preceding configuration there are four major iteration loops, for the capacities of the three turbines and the exhaust, requiring adjustment of the guessed values of COT, compressor betas, and power turbine output power. Again, HP shaft speed is determined by reading the HP compressor map with the current values of flow and beta. As per section 7.3.3 an additional iteration loop is required for the recuperator air side exit temperature.

The intercooler is a 'straight through' calculation, not normally requiring iteration but complex if intermediate liquid transport loops are employed. If the power turbine incorporates a variable area nozzle, its setting may either be specified directly or used in an additional outer iteration to achieve some other specified parameter. This is usually power turbine entry temperature (T46), as per section 7.3.3.

7.3.6 *Subsonic turbojet, single spool*

Figure 7.14 shows the matching diagram, with the engine run to fixed compressor speed. The main iteration is to satisfy the capacities of the turbine and nozzle, via adjustment of the guessed values of SOT and compressor beta respectively. This matching scheme is basically as per a single spool gas generator and free power turbine, per section 7.3.2.

7.3.7 *Subsonic turbofan, two spool*

Figure 7.15 shows the matching diagram with the engine run to fixed fan speed. Unlike other configurations, the first iteration is to satisfy the bypass nozzle capacity via adjustment of the matching guess for bypass ratio. There are three further iterations, for the flow capacities of the two turbines and the core nozzle, requiring adjustment of the guessed values of COT, HP compressor beta, and fan beta respectively.

If a mixer and common nozzle are employed rather than separate jets, then the requirement for equal static pressures at the mixer chute exit plane is a matching constraint, replacing that of the second propelling nozzle capacity. The matching scheme then becomes as follows.

Matching guess	Matching constraint
Bypass ratio	Equal mixer exit static pressures
HP compressor beta	LP turbine capacity
SOT	HP turbine capacity
Fan beta	Common nozzle capacity

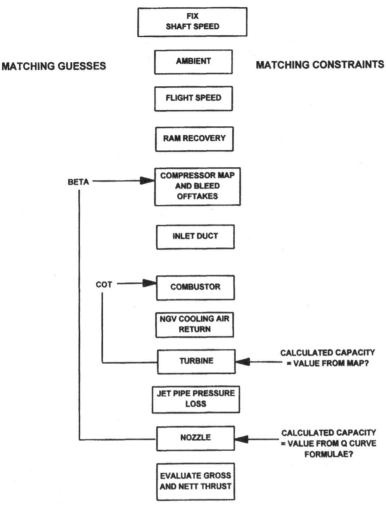

MATCHING GUESSES

MATCHING CONSTRAINTS

FIX SHAFT SPEED

AMBIENT

FLIGHT SPEED

RAM RECOVERY

BETA → COMPRESSOR MAP AND BLEED OFFTAKES

INLET DUCT

COT → COMBUSTOR

NGV COOLING AIR RETURN

TURBINE ← CALCULATED CAPACITY = VALUE FROM MAP?

JET PIPE PRESSURE LOSS

NOZZLE ← CALCULATED CAPACITY = VALUE FROM Q CURVE FORMULAE?

EVALUATE GROSS AND NETT THRUST

Notes:
Matching guesses shown require initial values, updated by each pass through the relevant loop.
For simplicity turbine cooling flow returns downstream of the first NGV are not shown.
Recirculating bleeds require additional guesses and matches at return station for flow quantity and temperature.

Fig. 7.14   Matching diagram, subsonic turbojet, single spool.

7.3.8   *Supersonic turbojet or turbofan*

Supersonic operation produces a choked intake. It also requires that an afterburner be fitted, if only for transonic acceleration. Variable intake geometry is essential to allow efficient operation at a range of Mach numbers (which must be down to zero for a manned aircraft), and variable propelling nozzle area is also essential to provide acceptable compressor working lines and nozzle performance.

For intakes, variable areas are provided for blow in, spillage and throat. Modelling utilises graphs of pressure recovery or efficiency (see Chapter 5) versus area setting, flight Mach number and delivered flow capacity. The last effect is because pressure adjusts between the throat and the engine as engine inlet capacity changes. Engine physical mass flow is then fixed

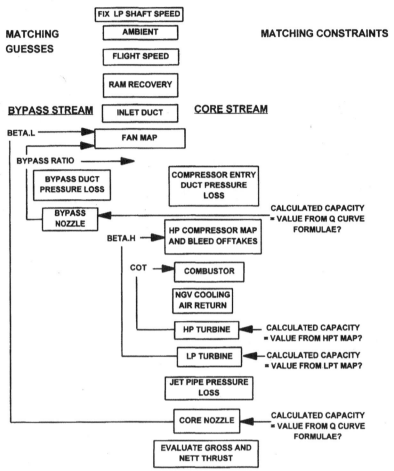

Notes:

Matching guesses shown require initial values, updated by each pass through the relevant loop.

Recirculating bleeds require additional guesses and matches at return station for flow quantity and temperature.

For simplicity turbine cooling flow returns downstream of the first NGV are not shown.

Separate maps may be employed for fan tip and root.

Diagram is for separate jets – mixed exhaust requires matching on equal stream static pressures.

Fig. 7.15 Matching diagram, subsonic turbofan, twin spool.

by the intake throat, and pressure recovery requires an additional iteration with the first compressor beta. A combustion calculation for a given afterburner fuel flow determines propelling nozzle inlet temperature, and an iteration on compressor speed is used to satisfy propelling nozzle capacity. The propelling nozzle thrust- and discharge coefficients are loaded as graphs versus the variable area setting.

For a single spool afterburning turbojet the matching scheme is as follows.

Matching guess	Matching constraint
HP compressor beta	Intake recovery
HP compressor speed	Nozzle capacity
SOT	HP turbine capacity

7.3.9   *Ramjet*

Though this is the simplest mechanical configuration of all, the area is quite specialised and few companies are active; accordingly only a brief summary is presented here. To achieve efficient operation over a wide Mach number range variable intake and/or exhaust geometry would be desirable, however in the expendable applications to date (i.e. missiles) this has not been cost effective. Owing to their choked operation both the intake and exhaust are complex, and must be rigorously modelled.

Intake

The intake design depends heavily on the level or range of flight Mach number required. Any intake with a supersonic inlet Mach number requires a convergent–divergent or 'con–di' configuration, as above Mach 1 diffusion requires a reduction in flow area. Mach 1 is achieved at the throat, via various possible shock systems. The throat is therefore choked, and its area fixes the air mass flow for all levels of fuel flow. Downstream of the intake throat, subsonic diffusion occurs to a low enough Mach number to allow stable combustion.

For modelling, appropriate graphs of throat pressure recovery factor versus flight Mach number are used. Throat conditions set the mass flow, and the subsonic diffuser pressure loss varies to match propelling nozzle capacity.

Combustion

The Mach number here is still significantly higher than in conventional combustors, as to avoid prohibitive supersonic wave drag the engine diameter must not exceed that of the intake. As described in section 5.21, various flameholders are required to sustain combustion, and the chemical combustion efficiency is relatively low. Cooled liners are also required for mechanical integrity. The temperature profiles produced by both the cooling air and the uneven combustion are accounted as a further inefficiency on the ability to produce thrust.

For modelling graphs of chemical- and thrust combustor efficiency versus loading are used. An ideal mean nozzle entry temperature is used for the continuity and thrust calculations (i.e. based on 100% combustion efficiency), and the thrust then factored down by the efficiency value.

Propelling nozzle

The hot gas is discharged through a con–di propelling nozzle, which at the high pressure ratios encountered is mandatory. Again, the diameter constraint restricts the nozzle design, and the exit area is inevitably far smaller than that to 'run full' and achieve full flow acceleration (see section 5.13.2).

For modelling the throat, area and a discharge coefficient are used together with intake mass flow, fuel flow and temperature to calculate a pressure ratio. As stated the combustion efficiency is then applied to the gross thrust obtained from Formulae F5.13.12 and F5.13.16.

7.3.10   *Components*

Chapter 5 describes all engine components in detail, including how off design performance may be represented via the parameter groups presented in Chapter 4. The calculations for each individual component within an off design engine performance model are described below. The combination of these component calculations for each engine configuration was presented in the preceding parts of section 7.3, via flow diagrams describing the matching calculations. These explanations are supported by sample calculations.

Ambient conditions

Inputs: Altitude and deviation from ISA conditions
Calculations: Look up ISA tables in Chapter 2, or use Formulae F2.1–F2.3.
Outputs: Ambient pressure and temperature

Flight conditions
Inputs: Flight speed or Mach number, and ambient conditions
Calculations: Derive Mach number if necessary, using charts or formulae from Chapter 2.
Evaluate total temperature and total pressure using Formulae F3.31 and F3.32.
Outputs: Free stream total temperature and pressure, also theta and delta

Intake ram recovery
Inputs: Ambient and free stream total pressures, ram recovery factor
Calculations: Evaluate dynamic head by subtracting ambient pressure from free stream total.
Apply Formula F5.13.9 (or Formula F5.13.11 if expressed as intake efficiency).
Outputs: Engine inlet total pressure and temperature

Subsonic installation/engine entry ducts and other ducts
These calculations are performed after those for the downstream compressor, as only then is the air mass flow known.
Inputs: Loss coefficient; inlet W, T and P
Calculations: Evaluate pressure loss from Formulae F3.33 and F5.13.4 or F5.13.6.
Outputs: Exit W, T and P

Supersonic installation/engine intakes
Inputs: Loss coefficient; leading edge mass flow, total pressure and temperature, flight Mach number
Calculations: Apply Formula F5.14.4 for shock system and F3.33, F5.13.4 or F5.13.6.
Outputs: Exit W, T and P

LP compressor
Inputs: Inlet T and P; speed, guessed beta
Calculations: Evaluate referred speed.
Look up map for $W\sqrt{T}/P$, pressure ratio and efficiency using beta and referred speed.
Apply scale factors and deltas if required, as per Formula F5.2.5.
Apply Re corrections to map values as per Formulae F5.2.1 and F5.2.2.
Evaluate absolute inlet mass flow and exit pressure.
Evaluate exit temperature from Formula F5.1.4.
Evaluate power via Formula F5.1.2.
Outputs: Exit W, T and P; power, efficiency

HP compressor
Inputs: Inlet W, T and P; guessed beta
Calculations: Apply scale factors and deltas if required, per Formula F5.2.5.
Apply Re corrections to map values per Formulae F5.2.1 and F5.2.2.
Look up map for pressure ratio, efficiency and referred speed using beta and $W\sqrt{T}/P$.
Evaluate absolute mass flow and exit pressure.
Evaluate exit temperature from Formula F5.1.4.
Evaluate power via Formula F5.1.2.
Outputs: Exit W, T and P; power, efficiency

Fans
Inputs: Inlet W, T and P; speed, guessed beta and bypass ratio
Calculations: Evaluate referred speed.
Look up map for $W\sqrt{T}/P$, pressure ratio, efficiency.

Apply scale factors and deltas if required, per Formula F5.2.5.
Apply Re corrections to map values per Formulae F5.2.1 and F5.2.2.
Evaluate fan root pressure ratio and efficiency, if separate fan root maps are not available.
Evaluate absolute mass flow, exit pressure.
Evaluate exit temperature from Formula F5.1.4.
Evaluate power via Formula F5.1.2.

Outputs: Exit W, T and P; power, efficiency

Cooling air offtakes

Inputs: Fraction of cooling air extraction, fraction of compressor work if bleed from intermediate stage

Calculations: Evaluate cooling flows as the fractions times the source gas path station mass flows.
Evaluate downstream gas path air flows by subtracting the cooling flows removed.
Source gas path station total temperature and pressure are unchanged.
For interstage bleed adjust compressor drive power if necessary, via Formula F5.15.1.

Outputs: Downstream W, T and P; revised compressor drive power

Combustors

Inputs: Inlet W, T and P; guessed combuster exit T, COT (which is SOT if NGV is uncooled); cold and hot pressure loss coefficients

Calculations: Calculate combustor loading from Formula F5.7.2.
Obtain efficiency from correlation versus loading (e.g. Formula F5.7.8).
Evaluate fuel air ratio using appropriate choice from Formulae F3.37–F3.40.
Evaluate fuel flow using above and inlet air mass flow.
Evaluate exit gas flow by adding fuel flow to inlet air flow.
Evaluate pressure loss from Formulae F5.7.9 and F5.7.10.

Outputs: Exit W, T and P; fuel flow

Combustors – alternative if fuel flow supplied (e.g. for matrix iteration)

Inputs: Inlet W, T and P; fuel flow, cold and hot pressure loss coefficients

Calculations: Evaluate fuel air ratio.
Evaluate exit gas flow by adding fuel flow to inlet air flow.
Evaluate exit temperature by iteration using Formula F3.41 (this is SOT if NGV is uncooled).
Evaluate pressure loss from Formulae F5.7.9 and F5.7.10.

Outputs: Exit W, T and P

Cooling air returns

Inputs: Inlet W, T and P; cooling flow W and T

Calculations: Evaluate station air flows by simply adding the cooling flows to the gas path values.
Evaluate mixed temperature via an enthalpy balance, as per Formula F5.15.2.
Total pressure is unchanged.

Ouputs: Exit W, T and P

Parasitic losses and power offtakes

Inputs: Customer or accessory power extraction, shaft speeds, pressures and temperatures near disc faces

Calculations: Evaluate bearing losses using Formulae F5.17.1–F5.17.4, as appropriate for bearing types.

Evaluate disc windage using Formula F5.17.6 for all appropriate disc faces.

Add up all power losses on each shaft.

Outputs: Total power loss on each shaft

Turbines

Inputs: Inlet W, T and P; speed, power

Calculations: Evaluate referred speed.

Power output is that of the driven compressor plus losses and offtakes from that shaft.

Evaluate calculated $W\sqrt{T}/P$ from inlet conditions.

Evaluate work parameter DH/T from power, inlet temperature and mass flow.

Look up map for matching constraint value of $W\sqrt{T}/P$, and for also efficiency and exit swirl angle.

Apply scale factors and deltas if required, per Formula F5.10.4.

Apply Re corrections to map values per Formulae F5.10.1–F5.10.3.

Evaluate exit temperature from Formula F3.15 or F3.16.

Evaluate exit pressure from rearranging Formula F5.9.4.

Outputs: Exit W, T and P; efficiency

Note: If a power turbine incorporates a variable area nozzle then it becomes an additional independent variable, and a suite of maps must be used.

Heat exchangers

The component supplier will normally have specific computer code, and may provide this; otherwise 'best fit' correlations should be obtained versus physical or corrected flow. Formulae F5.24.1, F5.24.3 and F5.24.4 give simple correlations of off design effectiveness and pressure losses for a recuperator or regenerator, while F5.24.2 and F5.24.5 give them for an intercooler.

Air side

Inputs: Inlet W, T and P; for a regenerator the 'carry over' and leakage flow fractions

Calculations: Apply guessed temperature rise.

Apply pressure loss, which can normally be calculated knowing temperature rise.

For a regenerator, deduct 'carry over' and leakage flows.

Outputs: Exit W, T and P

Recuperator or regenerator gas side

Inputs: Air and gas side inlet W, T and P

Calculations: Evaluate air side temperature rise from effectiveness, defined by Formula F5.23.1.

Evaluate gas side pressure loss.

Evaluate gas side temperature drop from enthalpy balance using Formula F3.15 or F3.16.

Outputs: Gas side exit W, T and P; air side exit W, T and P

Propelling nozzles (thrust engines)

Inputs: Nozzle exit plane W, T and P; geometric area

Calculations: Evaluate nozzle pressure ratio, as the ratio of nozzle exit plane total pressure to ambient. (If nozzle is choked then exit velocity will be the choking value and there will be additional pressure thrust.)

Evaluate nozzle effective area as the discharge coefficient times the geometric area.

Look up map for thrust and discharge coefficients, CV and CD.

Calculate nozzle $W\sqrt{T}/P$ from inlet conditions.

Evaluate matching constraint value of nozzle $W\sqrt{T}/P$ using Q curve Formula F3.33.

Evaluate gross thrust as the sum of momentum thrust and pressure thrust using Formula F5.13.12 or F5.13.13 with thrust coefficient and geometric area.

Outputs: Nozzle capacity, gross thrust

Engine exhaust area (industrial engines)

Inputs: Exhaust exit plane W, T and P; geometric area

Calculations: Evaluate pressure ratio, as the ratio of exhaust exit plane total pressure to ambient.

Look up map for discharge coefficient, CD.

Calculate exhaust exit $W\sqrt{T}/P$ from inlet conditions.

Evaluate effective area as the discharge coefficient times the geometric area.

Evaluate matching constraint value of exhaust exit $W\sqrt{T}/P$ using Q curve Formula F3.33.

Outputs: Exhaust capacity

Momentum drag

Inputs: Flight conditions, engine inlet air flow

Calculations: Evaluate true air speed VTAS using Formula F2.15.

Evaluate momentum drag as simply air flow times VTAS, Formula F5.13.12.

Outputs: Momentum drag

Final calculations

Inputs: Gross thrust, momentum drag, shaft power, fuel flow

Calculation: Evaluate net thrust as gross thrust less momentum drag, Formula F5.13.12.

Evaluate thrust SFC as the ratio of fuel flow and net thrust, Formula F6.8.

Evaluate shaft SFC as the ratio of fuel flow and shaft power, Formula F6.7.

Evaluate shaft thermal efficiency using Formula F6.9.

Outputs: Net thrust, thrust SFC, shaft SFC, shaft thermal efficiency

7.4   Geometric variation: modelling and effects

7.4.1   Variable intakes

In supersonic applications throat area is varied by hydraulic rams, to achieve flow diffusion via several oblique shock waves as described in Chapter 5. Modelling utilises graphs of intake recovery factor or efficiency versus flight Mach number and area setting. These graphs include both the shock loss and that for the downstream diffusing section.

7.4.2   Variable stator vanes (VSVs)

As described in Chapter 5, these are employed on one or more of the front stages of many axial compressors. They are closed at low speed to reduce the mass flow passed by the front stages for that given speed, which raises the surge line. The working line does not change, being set by downstream components, however the compressor speed does increase for a given flow and pressure ratio. (This may provide a second-order benefit to turbine efficiency.)

The VSV stages are almost always ganged in some optimised fashion, and scheduled versus compressor referred speed. In this case a *composite* compressor map may be used, as each referred speed line reflects a unique geometry. If, alternatively, the VSV scheduling varies, for example aircraft engines have different schedules for low altitude, high altitude and for transient operation, then VSV angle is added as an extra independent variable. Effectively several compressor maps are then used, each for a defined VSV angle, with flow, pressure ratio and efficiency read off via interpolation.

7.4.3   Blow off valves (BOVs)

The impact of BOVs on compressor performance is discussed in Chapter 5. Steady state they are employed mainly at low power to drop compressor working lines, and if extracted from an intermediate compressor stage, to raise surge lines. Transiently they may be opened at higher speed to maintain satisfactory IP or LP compressor surge margin on a decel, or in power generation to reduce power turbine overspeed on a drop load.

BOVs downstream of a compressor do not affect the map, only the matched operating point on it. Where BOVs extract flow from between compressor stages, the compressor map changes with the number of BOVs open; in particular at low speed the surge line rises. If the BOVs are controlled on a unique schedule versus referred speed, a *composite* map may be used with each referred speed line reflecting its own bleed setting. If, alternatively, the BOV scheduling varies, BOV configuration is added as an extra independent variable. Several compressor maps are then used, each for a defined BOV configuration, with interpolation between them.

BOV flows are usually modelled as a fixed percentage of the gas path flow at the station from which they were extracted. Where BOVs are choked, modelling as a fixed WRTP is an alternative. Where low pressure ratio means BOVs are unchoked, detailed modelling of the downstream manifolds etc. may be undertaken, using appropriate graphs, to accurately predict BOV pressure ratios and hence flows.

BOV switch points are normally scheduled versus compressor referred speed, with appropriate transient biasing. Hysteresis is employed to prevent BOVs from repeatedly opening and closing; this is the difference between a higher closure point on an accel and a lower opening point on a decel. In between these points the BOV configuration depends on whether the engine has been previously accelerated or decelerated. This is input to a steady state model via a switch, and if this indicates it has been accelerated the higher switch point values should be used. For modelling, the fact that BOVs are either off or on means they have a discontinuous effect on cycle matching calculations. To allow convergence the BOV configuration must be 'frozen' based on initial guesses of switching criteria, and switching criteria rechecked once converged values have been obtained from the engine calculation.

7.4.4   Turbine variable area nozzles (VANs)

As described in section 7.1, VANs are employed in recuperated engines, and usually controlled to maintain a scheduled value of power turbine entry temperature. A suite of turbine maps is therefore required, with VAN angle as an extra independent variable. The scheduling does not neatly follow turbine referred speed and allow any more compact form of map.

7.4.5   Variable propelling nozzles

These are essential for afterburning turbojets and turbofans, which give sufficiently high jet velocities for supersonic flight. With the afterburner lit the nozzle area must be increased to avoid compressor surge problems. To a first order, area is increased in proportion to the square root of jet pipe temperature. This maintains compressor working lines by maintaining the core turbine expansion ratio.

Modelling is fundamentally as described in section 7.3.10 with thrust- and discharge coefficient curves and a capacity match. The only addition is the variation in geometric nozzle area; this affects the capacity and pressure thrust, and may be an additional independent variable for the thrust- and discharge coefficient curves.

7.4.6 *Changes in geometry due to inlet temperature (T1)*

Very few performance models are elaborate enough to cover T1 effects explicitly. Chapter 4 describes how they can change engine performance levels, via geometry changes such as blade untwist and tip clearance variation. Compressor blade untwist results in higher levels of flow and pressure ratio, and may occur due to higher physical speed levels than for the rig map. Any tip clearance increase reduces efficiency, and for compressors may significantly lower the surge line.

Compressor blade untwist is accounted by scaling the characteristic to match engine test data, via factoring the map's referred speed values down. Tip clearance changes mainly occur transiently, due to changes in both temperature and speed and their main effect is on compressor surge lines. Accordingly they are rarely directly incorporated in performance modelling and accounting is usually via the stability stack up described in Chapter 8, where surge margin threats are added up for critical manoeuvres and the design working line level set accordingly. For turbines the effect is simply on efficiency; the most sophisticated engines incorporate active tip clearance control, where the cooling air to the turbine casings is metered to control thermal growths and hence tip clearances. In this instance the modelling requirement is that of the cooling flow.

7.5   Engine scaling and different working fluids

Chapter 4 comprehensively describes the impact on referred parameter groups of engine and component scaling, and changes in working fluid. The generic off design characteristic charts provided in section 7.1 address to a first order both engine scaling and working fluid changes, as the referred parameters shown can equally be interpreted as scaling parameters or full non-dimensional groups.

7.6   Off design matching: physical mechanisms

Understanding, and being able to predict, the effect on a cycle of design changes is assisted by considering the physical processes involved. This section describes and explains several key phenomena associated with engine off design matching.

7.6.1 *Part power or thrust*

For open shaft power cycles reducing power requires lower fuel flow, which reduces turbine temperatures and causes the spools to slow down (except for a single shaft power generation engine). The engine cycle therefore has reduced pressure ratio and SOT. From the design point diagrams in Chapter 6, the cycle effect almost invariably means worse thermal efficiency, hence SFC also becomes immediately worse. Furthermore, at very low referred speeds component efficiencies degrade radically and handling bleeds are often extracted.

For thrust engines part power cycle thermal efficiency degrades quickly for the same reasons as above. However, the reduced exhaust velocities and temperatures improve propulsive efficiency, hence SFC initially improves before worsening at the lowest thrust levels, when falling thermal efficiency outweighs this. The shape of the curve depends on the Mach number, as described in section 7.1.

At low power most engines employ blow off valves (BOVs); BOV opening produces very large effects on the engine cycle, as anywhere from 5 to 25% of flow is bled off. In consequence turbine temperatures rise significantly at a power or thrust level, and SFC gets worse. The latter effect is greatly reduced in a recuperated engine, where the higher recuperator inlet temperature increases heat recovery.

As described earlier, for turbofan engines, bypass ratio rises as the engine is throttled back because the core compressor capacity falls more quickly than that of the bypass nozzle. Low thrust and/or Mach number levels result in unchoked nozzles, and hence a high fan working line. These effects are described in section 7.1.

7.6.2   Single spool turbojet

For a single spool turbojet the parameters fixing the compressor working line are the capacities of the turbine and the propelling nozzle. If turbine capacity is reduced, the compressor exit capacity at any speed must also be reduced. This corresponds to a higher working line as lower flow and higher exit pressure both reduce compressor exit capacity.

If the final nozzle area is reduced, this also results in a higher compressor working line. The reduced turbine expansion ratio must be offset by increased turbine inlet temperature, directly proportional to turbine work, at any compressor speed. For a fixed level of turbine capacity the increased temperature has the same effect on compressor flow and exit pressure as reducing turbine capacity.

7.6.3   Multiple spool engines

When two or even three spools are employed, the components interact to produce distinctive trends when any component's performance changes.

Though high pressure ratios are possible for a single spool compressor, by using many stages and several variable stator rows, engines often use multiple compressors driven by their own turbines. Driving an HP compressor with its own HP turbine allows a higher speed for these stages. This overcomes the effect of the elevated inlet temperature in raising the work and speed required to achieve any pressure ratio. In addition the HP shaft speed remains higher at part load than if it were part of a single shaft. The LP compressor working line therefore remains more acceptable down to lower power or thrust, because the downstream fast running HP compressor can pass more flow. The HP speed remains high because the HP turbine expansion ratio is fixed over a wide power or thrust range by the 'turbine between choked nozzles' effect, described below.

Effect of turbine and nozzle capacities

If a turbine nozzle is choked its capacity is fixed, independent of referred speed. If the downstream turbine NGV or propelling nozzle is also choked then the first turbine's expansion ratio are also fixed. Hence *if a turbine rotor operates between choked nozzles, then that turbine's capacity, expansion ratio and hence work parameter remain sensibly constant.* (If a turbine is designed to choke first in the rotor then referred speed does slightly influence capacity, as described in Chapter 5, but this effect is far smaller than if the turbine were unchoked.) The benefit of multiple spools on compressor matching occurs because the HP turbine and the LP turbine downstream remain choked for much of the operating range, hence HP turbine expansion ratio cannot change and HP shaft speed remains high.

Changing turbine capacities changes the *work split* between them. For example reducing LP turbine capacity increases LP turbine expansion ratio and hence work, while reducing that of the HP turbine. Hence for a fixed fuel flow, the LP speed will increase and the HP reduce. The HP compressor working line is raised, as per that of a single spool turbojet in section 7.6.2 above. The LP compressor working line is also raised, due to the reduced downstream capacity of the slower running HP compressor; this illustrates an important result that *the strongest influence on a compressor working line is the capacity of the driving turbine.*

Reducing HP turbine capacity or that downstream of the LP turbine has a broadly converse effect relative to reducing LP turbine capacity. For a turboshaft with a two spool gas generator and free power turbine, reducing the capacity of each of the turbines has the following effects at constant power when all turbines are choked:

	LP shaft speed	LP working line	HP shaft speed	HP working line	SOT
HP turbine	Lower	Lower	Raise	Raise	Raise
LP turbine	Raise	Raise	Lower	Raise	Lower
Power turbine	Lower	Lower	Raise	Raise	Raise

Such effects are encountered even for fixed engine geometry. For thrust engines, a core propelling nozzle unchokes at low thrust and low Mach number, reducing the core compressor speed and causing the fan to operate at increased bypass ratio. For shaft power engines, increasing the power turbine speed at a power reduces its capacity, as explained in Chapter 5.

Effect of component efficiencies

If a compressor or turbine has reduced efficiency then that spool speed will decrease. For example, as described above, the HP turbine has a fixed expansion ratio from which to develop drive power. If the required power at a speed increases due to low compressor efficiency, or the power produced reduces due to low turbine efficiency, a lower speed will be attained.

The magnitude of the effect of a component efficiency change is high where the amount of power on that spool is high relative to the other spools. For example, a high bypass ratio large civil turbofan has most power on the LP spool, and a two spool gas generator has an increasing fraction of power on the HP spool at low power levels.

Operation at referred speeds higher than the design rating level normally results in reduced compressor efficiencies, which outweigh any benefit of increased cycle parameters and cause SFC to get worse. This effect applies both to thrust and shaft power engines, and for the former is additional to the effect of worsening propulsive efficiency. The effect of component efficiencies on part power operation is discussed in section 7.6.1.

7.6.4   Effect of core flow changes

If at the same fuel flow one component is changed such that physical air flow through the turbines is decreased, then fuel air ratio is increased and the engine will run hotter. For instance, for a two spool gas generator with free power turbine this might be via a reduced LP shaft speed due to reduced LP turbine efficiency, or increased power turbine speed which reduces its capacity. For a high bypass ratio civil turbofan any deficiency in the core, as opposed to the LP system, is shown by the engine running hot due to reduced core air flow.

7.7   Exchange rates

Exchange rates relate changes in component performance levels to changes in engine parameters. Chapter 6 describes *design point exchange rates* and provides typical levels. Chapter 11 describes *analysis exchange rates* for engine test data analysis. Various forms of off design performance exchange rates may be defined, depending on which *base* parameters are chosen to be held constant:

● *Synthesis*, or *constant geometry* are by far the most common, and show the real effect of making only the subject change with the rest of the engine hardware unchanged. They may be evaluated by simply running the engine performance model with the single change made to a component efficency or capacity etc.

- *Fundamental* are where any other consequences of rematching on other component maps are excluded. All other component efficiencies, and turbine capacities, are held constant as the subject change is made. This shows the effect of only the subject change on the cycle, rather than that of rematching, as pressure ratio and SOT change.
- *Constant cycle* are where the design point calculation is repeated and all the *other* components are scaled in size to maintain as far as possible constant temperatures and compressor working lines (i.e. efficiencies and pressure ratio). This new geometry is then run at the desired off design operating condition. This shows the effect of the component efficiency change isolated from the other component rematching caused by changes in efficiencies of other components and of the cycle.

The most appropriate type of exchange rate depends on the stage of engine development in terms of the feasibility or otherwise of changing other components. For each of these exchange rates the actual values are specific to each engine type and even to each engine design, due to differing component worksplits etc. To determine accurately the effects of changes, either exchange rates should be obtained for a similar engine type and design, or preferably the engine's own off design performance model should be run. For a given engine design the exchange rates are usually non-linear, hence tables are often produced showing say the effects of both a 1% and a 5% change.

One important consequence of these non-linearities is that the effects of multiple component changes even assessed *individually* in a model *do not* always add up to the total effect they give if *all* implemented in the model. To ensure consistency the effects should be input in a consistent order as cumulative changes to the model, but this does not absolutely define the 'real' effect of each individual change. Such difficulties illustrate the reason why fundamental and constant cycle exchange rates are also used.

7.8   Ratings and control

Ratings define allowable power or thrust levels versus ambient temperature and altitude. These are set by the control system governing to specified values of certain measured *rating parameters*. Rating levels are designed to ensure satisfactory engine life, while achieving the power or thrust required by the application at key operating conditions. For example, an aircraft requires defined thrust levels at takeoff, climb, cruise, flight idle and ground idle.

Water and steam injection have been variously employed in aero and industrial applications to change the shape of ratings curves, as described in Chapter 12.

7.8.1   *Standard definitions*

Figure 7.16 summarises standard rating definitions for aero and landbased engines. These are drawn from References 1 and 2. Another standard term is *flat rating*, where constant power or thrust is offered over a range of ambient temperature. This is clearly valuable to an operator but a profile of ambient temperature versus time must be defined to allow prediction of an *integrated* engine service life.

7.8.2   *Maximum power or thrust*

Maximum ratings correspond to a maximum allowable value of some limiting parameter for engine integrity. Figures 7.17 and 7.18 illustrate maximum typical ratings curves for an industrial turboshaft and an aero turbofan respectively. In general, temperatures or mechanical speeds limit operation on hot days, and referred speed or pressure levels on cold days. The control algorithm achieves the transition between the different rating parameters schedules by selecting fuel flow on a 'lowest wins' basis; this means the limit 'hit first' is active.

Power lever angle (PLA)	Operation	Civil	Military
100	Augmented (Water injection or reheat)	Emergency	Maximum
90		Maximum	Maximum continuous
60		Minimum	Minimum augmented
55		Wet takeoff	—
55	Non augmented	—	Maximum
50		Dry takeoff	Intermediate
45		Maximum continuous	Maximum continuous
40		Maximum climb	—
35		Maximum cruise	—
21		Flight idle	Idle (flight)
20		Ground idle	Idle (ground)
15	Reverse thrust	Idle	Idle
5		Maximum	Maximum reverse

(a) Aero-engines, as per AS681 REV E

Class	Power rating	Hours per year	Starts per year
A	Reserve peak	Up to 500	
B	Peak	Up to 2000	
B	ISO Standard peak load	2000	500 Average
C	Semi-base	Up to 6000	
D	Base	Up to 8760	
D	ISO Standard base load	8760	25 Average

(b) Landbased engines, as per ISO 3977

Fig. 7.16 Ratings definitions.

7.8.3 *Part power or thrust*

A part load rating is generally set to be some fraction of maximum rating, broadly to achieve a set percentage of defined engine capability.

At part throttle conditions power or thrust could be set by an 'open loop' fuel schedule, however assessing the achieved power or thrust level via some measured parameter is more useful and accurate. For industrial and marine engines 'closed loop' control on measured power may be employed, using a torquemeter. Some older engines govern to specified values of LP shaft speed, the rating being defined as a graph versus ambient temperature. For aero-engines it is usual to govern to *exhaust pressure ratio (EPR)*, which is nozzle entry pressure divided by inlet total pressure; this relates well to engine thrust at that flight condition.

7.8.4 *Idle*

This is the minimum feasible level of power or thrust, produced when the ideal requirement of the application is actually zero. For all engines, determination of what is 'feasible' reflects

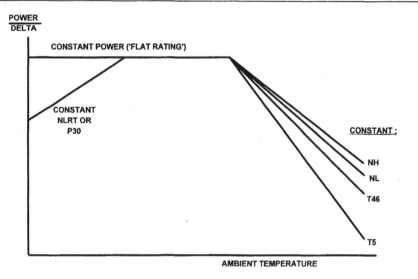

Notes:
Lines indicate possible maximum ratings.
Part power ratings would be parallel to these.

Fig. 7.17   Typical ratings curves, landbased engines.

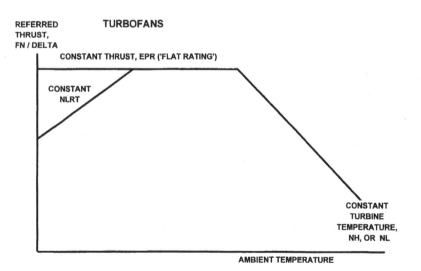

Notes:
Lines indicate possible maximum ratings.
Part power ratings would be parallel to these.
DELTA = P1/101.325 PSIA.
EPR = Exhaust nozzle pressure ratio, which is used as a cockpit thrust indicator.
For choked nozzles the choice of turbine temperature to hold constant makes little difference.

Fig. 7.18   Typical ratings curves, flight engines.

several engine issues, including combustor stability, engine temperatures and surge margin levels. In addition certain applications impose their own particular requirements:

- Aero-engine requirements are dominated by achievable accel times to higher power or thrust. This results in an increased thrust setting for flight- versus ground idle, to address the go around case. The idle rating may be implemented via a combination of speeds, referred speeds, or pressures depending on the position in the flight envelope.
- For marine engines idle power dissipation is sometimes a major issue, and engines may be required to operate with the power turbine rotor locked.
- Power generation engines remain at synchronous output speed when at synchronous idle, to facilitate reloading the alternator. A sub-synchronous idle is also used to ease the thermal transients on start-up.

For landbased engines idle settings are usually governed to fixed physical or referred speed, while for aero-engines P30 may also be used. At idle the controller transitions between the various rating parameters by selecting fuel flow on a 'highest wins' basis, the higher value being safer for stable engine operation.

Formulae

All component formulae required are presented in Chapter 5.

F7.1 **Update of matching guesses in matrix matching model = fn(matrix of errors in matching constraints, inverted matrix of partial derivatives of matching constraints with respect to matching guesses)**

$$DX = Y * ((dY/dX)^\wedge - 1)$$

F7.2 **Updated matching guess for SOT in nested loops iteration (K) = fn(old guessed SOT (K), map value of HP turbine capacity (kg $\sqrt{K}$/s kPa), calculated value of HP turbine capacity (kg $\sqrt{K}$/s kPa)**

$$T41new = T41old * ((W\sqrt{T}/P.map)/(W\sqrt{T}/P.calc))^\wedge 2$$

F7.3 **Updated matching guess for compressor beta in nested loops iteration = (old guessed beta, Qcurves value of final nozzle capacity (kg$\sqrt{K}$/s kPa), calculated value of final nozzle capacity (kg$\sqrt{K}$/s kPa))**

$$Beta.new = Beta.old * (W\sqrt{T}/P.Qcurves)/(W\sqrt{T}/P.calc)$$

(i) For multiple spool engines other component capacities may be relevant, e.g. LP turbine.
(ii) More complex numerical algorithms may be used to speed up convergence.

Sample calculations

C7.1 **Find the referred and actual SFC, P3 and air mass flow rate at 2000 m on a −40 8C day and 20% referred power, as a percentage of the corresponding values at ISO conditions and 100% power for a single spool turboshaft operating at synchronous speed.**

From Chart 2.1, PAMB at 250 m = 98.362 kPa.

(i) Referred parameters

From Chart 7.1:

> Referred PW = 20%
> Referred SFC = 200%
> Referred P3 = 92%
> Referred W1 = 110.5%

(ii) Actual parameters

Substituting into Formulae F2.11 and F2.12:

> DELTA = 98.363/101.325
> DELTA = 0.971
>
> THETA = 233.15/288.15
> THETA = 0.809

Evaluate actual parameter values using the parameter groups presented in Chapter 4:

> PW = 20 * SQRT(0.809) * 0.971
> PW = 17.5%
>
> SFC = 200%
>
> P3 = 92 * 0.971
> P3 = 89%
>
> W1 = 110.5 * 0.971/SQRT(0.809)
> W1 = 113.7%

C7.2 Find the referred and actual power, SFC, P3 and SOT at ISO conditions and 86% gas generator speed, as a percentage of the corresponding values at ISO conditions and 100% power for a single spool gas generator with free power turbine for (i) cube law and (ii) synchronous operation.

Since the throttled back operating condition is at ISO, actual parameter values = referred parameter values.

(i) Cube law operation

From Chart 7.2a:

> Referred P3 = 51%
> Referred T41 = 75%

From Chart 7.2b at 86% gas generator referred speed on a curve where shaft power is proportional to power turbine speed cubed:

> Referred Shaft Power = 35% corresponding to 70% power turbine speed
> Referred Shaft SFC = 126%

(ii) Synchronous operation

The gas generator operating condition is unchanged, hence parameter values are as per section (i).

From Chart 7.2a at 86% gas generator referred speed with the power turbine operating at synchronous speed of 100%:

> Referred Shaft Power = 21%
> Referred Shaft SFC = 164%

Hence it is immediately apparent how SFC at part power is significantly worse when operating at synchronous speed as opposed to a cube law.

C7.3 **Find the referred and actual net thrust, SFC, P3 and air mass flow rate at 10 000 m on a MIL 210 hot day, 0.8 Mach number and 70% referred speed, as a percentage of the corresponding values at (i) ISA SLS conditions and (ii) 100% referred speed for a single spool turbojet.**

From Chart 2.11 THETA = 0.952 and from Chart 2.10 DELTA = 0.4.

(i)   Referred parameters
From Chart 7.6:

  Referred FN = 10%
  Referred SFC = 160%
  Referred P3 = 42%
  Referred W1 = 60%

(ii)   Actual parameters
Evaluate actual parameter values using the parameter groups presented in Chapter 4:

  FN = 10 * 0.4
  FN = 4%

  SFC = 130 * SQRT(0.952)
  SFC = 127%

  P3 = 42 * 0.4
  P3 = 16.8%

  W1 = 60 * 0.4/SQRT(0.952)
  W1 = 24.6%

  N = 70 * SQRT(0.952)
  N = 68.3%

C7.4 **Conduct an off design performance calculation for a turbojet engine at 11 000 m, 0.8 Mach number ISA and 100% referred speed. The engine is as per design point sample calculation C6.1. Excerpts from the compressor and turbine maps which have been designed to provide the capacities, pressure ratio, CPdT/T and efficiencies of this design point are presented below.**

Compressor map

N/$\sqrt{}$T (%)	W $\sqrt{}$T/P(kg $\sqrt{}$K/s kPa)			PR			Efficiency (%)		
	95	100	105	95	100	105	95	100	105
BETA									
1	38.3	43.0	46.8	22.6	27.0	32.0	82.5	81.0	80.5
2	39.3	43.8	47.9	22.0	26.6	30.8	84.0	82.9	82.0
3	40.6	45.2	48.4	20.8	25.5	29.0	83.2	82.2	81.5
4	41.6	46.1	48.9	19.0	24.3	27.1	82.5	81.2	79.0
5	42.3	46.6	49.3	17.0	21.5	24.4	79.5	78.0	76.5

Turbine map

N/$\sqrt{\mathrm{T}}$	W$\sqrt{\mathrm{T}}$/P(kg$\sqrt{\mathrm{K}}$/s kPa)			Efficiency (%)		
	90	100	110	90	100	110
CPdT/T						
0.25	4.68	4.68	4.68	89.0	89.5	89.3
0.30	4.68	4.68	4.68	90.0	90.6	90.5
0.35	4.68	4.68	4.68	90.5	90.6	90.5
0.40	4.68	4.68	4.68	90.2	90.3	90.0

Notes:
Design point speed is 5000 rpm.
Hence 100% compressor N/$\sqrt{\mathrm{T}}$ = 5000/SQRT(244.4) = 319.8 rpm/$\sqrt{\mathrm{K}}$.
Hence 100% turbine N/$\sqrt{\mathrm{T}}$ = 5000/SQRT(1400) = 133.63 rpm/$\sqrt{\mathrm{K}}$.
The flow through this calculation is as per Fig. 7.14.

(i)   Calculate pressure loss coefficients
Using data from C6.1 and Formula F5.14.2:

> K12 = 0.005/(100 * 244.4^0.5/34.47)^2
> K12 = 2.4E-06

> K331 = 0.02/(100 * 695.1^0.5/857.5)^2
> K331 = 2.12E-03

> K314 = 0.03/(100 * 695.1^0.5/840.35)^2
> K314 = 3.05E-03   (For simplicity only the cold loss considered, F5.7.9)

> K57 = 0.01/(101.64 * 1010.5^0.5/187.8)^2
> K56 = 3.37E-05

(ii)   Conduct engine off design calculations
Free stream conditions
Use Q curve Formulae F3.31 and F3.32 to derive free stream total temperature and pressure:

> T0/216.7 = (1 + (1.4 − 1)/2 * 0.8^2)
> T0 = 244.4 K

> P0/22.628 = (244.4/216.7)^3.5
> P0 = 34.47 kPa

Look up compressor map
Guess BETA = 4

Now look up map with 100% N/$\sqrt{\mathrm{T}}$, and BETA = 4:

> W$\sqrt{\mathrm{T}}$/P = 46.1 kg$\sqrt{\mathrm{K}}$/s kPa
> PR = 24.2
> Efficiency = 81.2%

Intake
Total temperature is unchanged along duct, and pressure loss is evaluated from F5.14.2 and K12 derived in (i):

> T2 = T0 = 244.4 K

> P0D2Q0 = 2.4E-06 * 46.1^2
> P0D2Q0 = 0.51%

$P2 = 34.47 * (1 - 0.0051) * P1$
$P2 = 34.29\,kPa$

$W2 = 46.1 * 34.29/244.4^{\wedge}0.5$
$W2 = 101.1\,kg/s$

$W1 = W2 = 101.1\,kg/s$

Compressor

$T3 - 244.4 = 244.4/0.812 * (24.2^{\wedge}(1/3.5) - 1)$
$T3 = 691.4\,K$

$P3 = 24.2 * 34.29$
$P3 = 829.8\,kPa$

$W3 = 101.1\,kg/s$

$PW2 = 101.1 * 1.005 * (691.4 - 244.4)$
$PW2 = 45\,418\,kW$

Compressor exit diffuser

Total temperature is unchanged, derive exit pressure by applying loss coefficient derived in (i) and F5.14.2:

$T31 = T3 = 691.4\,K$

$P3D31Q3 = 2.12E\text{-}03 * (101.1 * 691.4^{\wedge}0.5/829.8)^{\wedge}2$
$P3D31Q3 = 0.022$

$P31 = 829.8 * (1 - 0.022)$
$P31 = 811.5\,kPa$

$W31 = 101.1\,kg/s$

Combustor

Apply percentage loss coefficient via F5.7.9 to derive P41:

$P31D4Q31 = 3.05E\text{-}03 * (101.1 * 691.4^{\wedge}0.5/811.5)^{\wedge}2$
$P31D4Q31 = 0.033$

$P4 = 811.5 * (1 - 0.033)$
$P4 = 784.7\,kPa$
$P41 = P4 = 784.7\,kPa$

Guess $T4 = T41 = 1450\,K$.

Calculate FAR to station 41, i.e. after the cooling air that will do work has mixed in before the turbine, using F3.40:

$FAR = 1.15 * (1450 - 691.4)/0.999/43100$
$FAR = 0.020$

$WF = 0.020 * 101.1$
$WF = 2.02\,kg/s$

$W4 = 101.1 + 2.02$
$W4 = 103.1\,kg/s$

Look up turbine map and iterate by reguessing T41 until the map capacities and calculated values are equal.

No cooling air is considered hence parameters at the SOT station 41 are equal to those at station 4. Equate power to that of compressor and use F5.9.2 to derive exit temperature:

PW41 = 45 418/0.995
PW41 = 45 646 kW

45 646 = 103.1 * 1.15 * (1450 − T5)
T5 = 1064.6 K

Evaluate CPdT/T and N/$\sqrt{}$T and then look up turbine capacity map:

CPdT/T = 1.15 * (1450 − 1064.6)/1450
CPdT/T = 0.306 kJ/kg K

N/$\sqrt{}$T = 5000/SQRT(1450)
N/$\sqrt{}$T = 131.3 rpm/$\sqrt{}$K
N/$\sqrt{}$T = 131.3/133.63
= 98.2%

W$\sqrt{}$T/Pmap = 4.68 kg $\sqrt{}$K/s kPa

Now calculate capacity from gas path parameters resulting from earlier guesses of BETA and SOT:

W$\sqrt{}$T/Pcalc = 103.1 * 1450^0.5/784.7
W$\sqrt{}$T/Pcalc = 5.00 kg $\sqrt{}$K/s kPa

Compare map and calculated values and reguess SOT using F7.2:

Error = (5.00 − 4.68)/4.68 * 100
Error = 6.8%

T41new guess = 1450 * (4.68/5.00)^2
T41new guess = 1270 K

After going back to where T41 was guessed in the combustor and repeating this loop with T41 guess = 1270 K:

FAR = 0.015

WF = 1.56 kg/s

W4 = W41 = 102.6 kg/s

T5 = 883.1 K

W$\sqrt{}$T/Pmap = 4.68 kg $\sqrt{}$K/s kPa
W$\sqrt{}$T/Pcalc = 4.66 kg $\sqrt{}$K/s kPa

Error = 0.4%

This iteration loop should be repeated until Error <0.05%, however for brevity this example will proceed with the Error at 0.4%.

Complete turbine calculations
First look up efficiency from the map:

CPdT/T = 0.350 kJ/kg K
N/$\sqrt{}$T = 105%

E41 = 0.9055

1270 − 883.1 = 0.9055 * 1270 * (1 − P41Q5^(−1/4))
P41Q4 = 5.16

P5 = 784.7/5.16
P5 = 152.1 kPa

W5 = W4 = 102.6 kg/s

Jet pipe
Total temperature is unchanged, derive exit pressure by applying loss coefficient from (i) together with F5.14.2:

T7 = T5 = 883.1 K

P5D7Q7 = 3.37E-05 * (102.6 * 883.1^0.5/152.1)^2
P5D7Q7 = 0.014

P7 = 152.1 * (1 − 0.014)
P7 = 150.0 kPa

W7 = W5 = 102.6 kg/s

Propelling nozzle
P7Q9 = 150.0/22.628
P7Q9 = 6.629

Via the Q curves presented in Chart 3.8 or Formulae F3.32 and F3.33 the nozzle is choked and hence Q = 39.6971 kg $\sqrt{K}$/s m^2 kPa. Now calculate Q from gas path conditions and the propelling nozzle effective area as fixed in C6.1 at 0.439 m^2 and hence the error in nozzle capacity:

Qcalc = 102.6 * 883.1^0.5/0.439/150.0
Qcalc = 46.30 kg $\sqrt{K}$/s m^2 kPa

Error = (46.3 − 39.6971)/39.6971 * 100
Error = 16.7%

Now reguess BETA:

BETAnew guess = 4 * 39.6971/46.3
BETAnew guess = 3.43

The calculation process now goes back to where the compressor map is looked up and the process is repeated until the error in propelling nozzle capacity is <0.05%. The compressor map may be linearly interpolated. For each new guess of BETA the inner loop must be repeated with T41 being reguessed until the calculated turbine capacity is equal to that from the map.

For brevity the iterations will not be repeated here but they result in:

BETA = 3.42
PR = 25
W$\sqrt{T}$/P compressor = 45.58 kg $\sqrt{K}$/s kPa
W1 = 100 kg/s
T4 = T41 = 1400 K

Remaining propelling nozzle and final calculations
These are as for the design point sample calculation C6.1.

Note: The above example was an off design calculation at the same operating condition as the design point evaluated in C6.1. The same off design calculation process may now be performed for any rotational speed, altitude, flight Mach number, etc.

A full model would include the air system, gas properties by one of the more rigorous methods in Chapter 3 and also Reynolds number corrections etc. as described in Chapter 7.

Charts

Chart 7.1   Referred performance charts, single spool turboshaft or turboprop.

Note:
Physical shaft speed is fixed at the synchronous value.

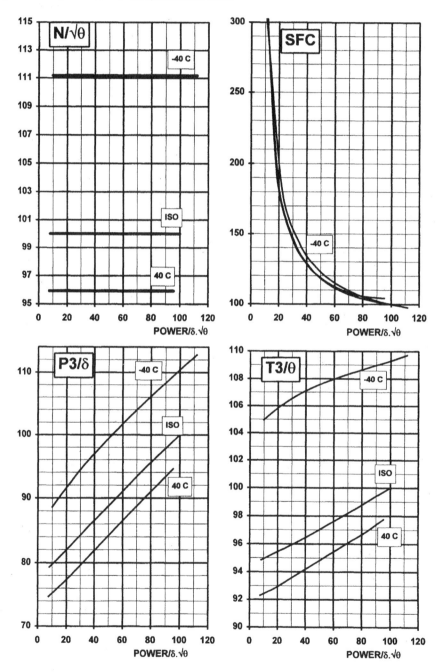

Chart 7.1 *contd.*

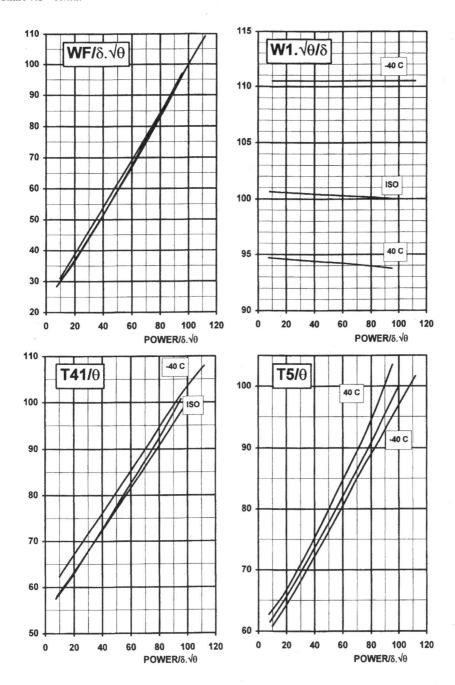

Chart 7.2   Referred performance charts, turboshaft or turboprop, single spool gas generator with free power turbine.

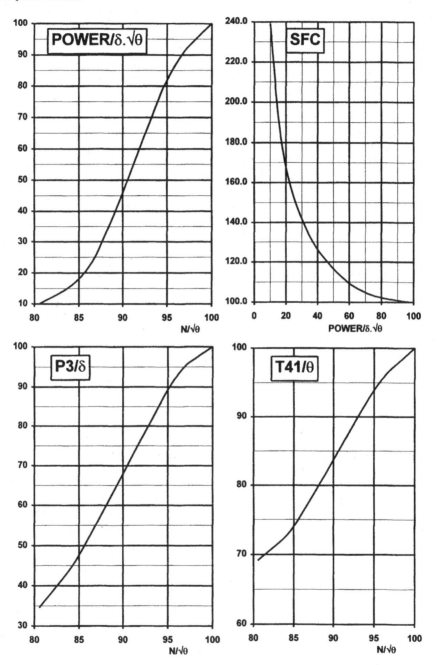

(a)   Referred performance charts

Notes:
Data shown reflect synchronous power turbine.
Cube law operation has lower temperature and SFC at low power.

Chart 7.2a   *contd.*

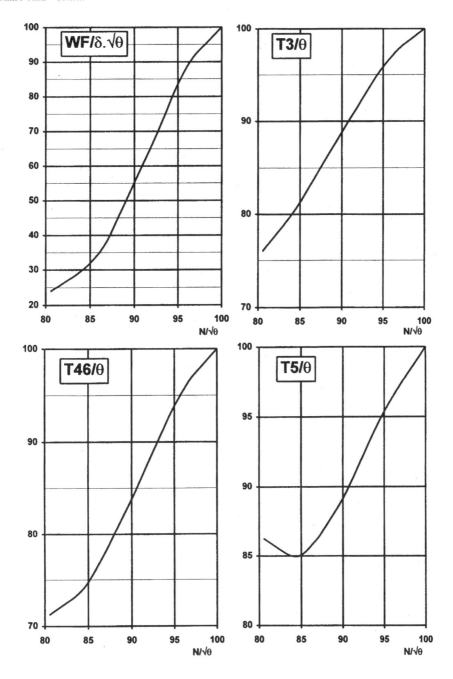

Chart 7.2 *contd.*

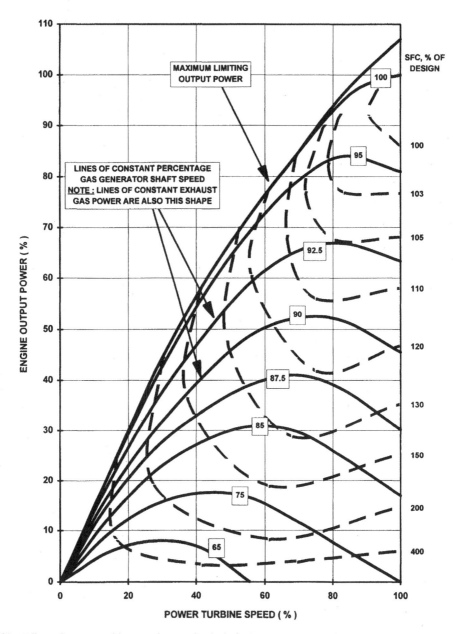

(b)   Effect of power turbine speed on turboshaft performance.

Notes:
Shaft speed shown is for single shaft gas generator.
Shaft power = exhaust gas power × power turbine isentropic efficiency.

Chart 7.2   *contd.*

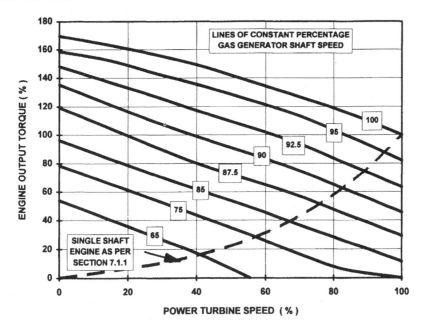

(c)   Torque versus output speed and power.

Note:
Shaft speed shown is for single shaft gas generator.

Chart 7.3   Referred performance charts, recuperated turboshaft, single spool gas generator with free power turbine.

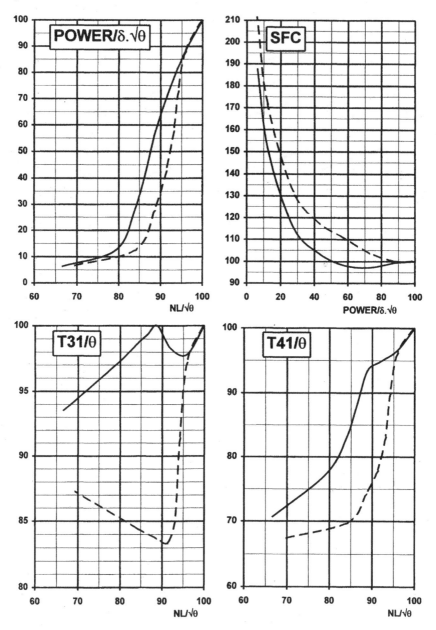

Notes:
Solid line is for power turbine with variable area nozzle.
Broken line is for fixed geometry.
Data shown reflect synchronous power turbine.
Cube law operation has lower temperature and SFC at low power.

Chart 7.3 *contd.*

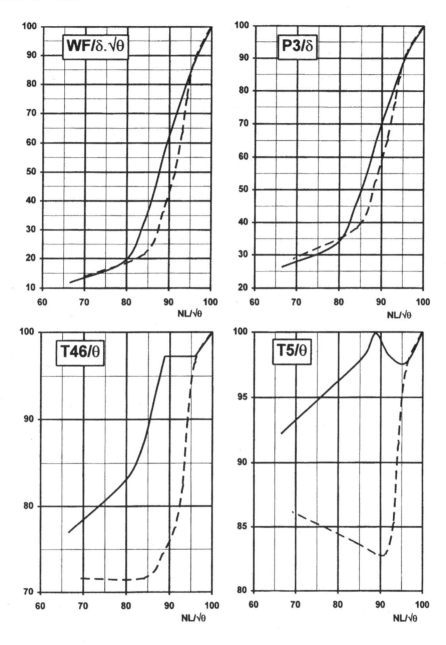

Chart 7.4   Referred performance charts, turboshaft or turboprop, twin spool gas generator with free power turbine.

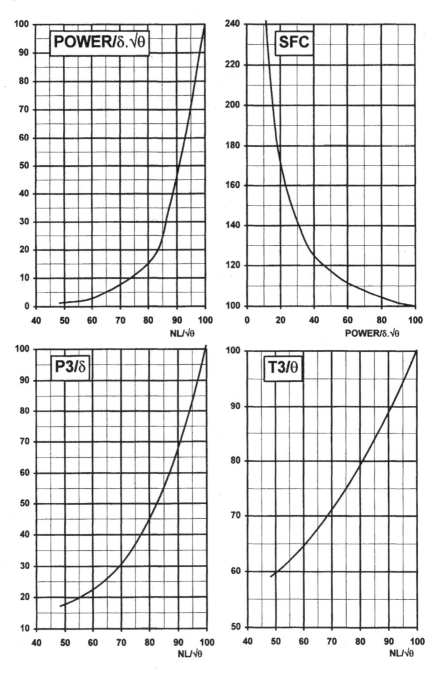

Chart 7.4 *contd.*

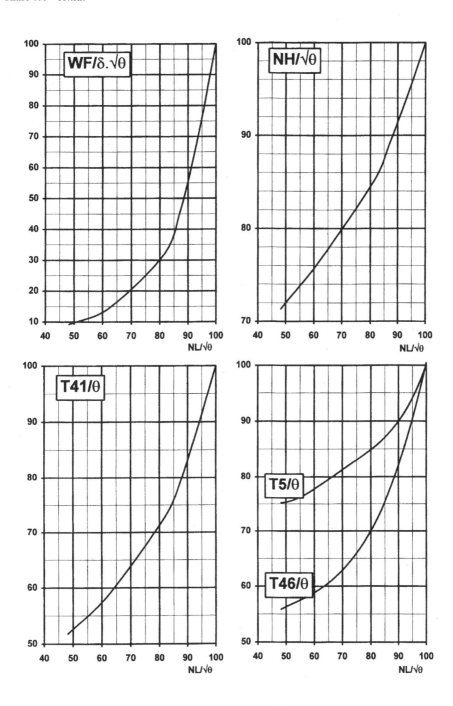

Chart 7.5   Referred performance charts, intercooled, recuperated turboshaft, twin spool gas generator with free power turbine.

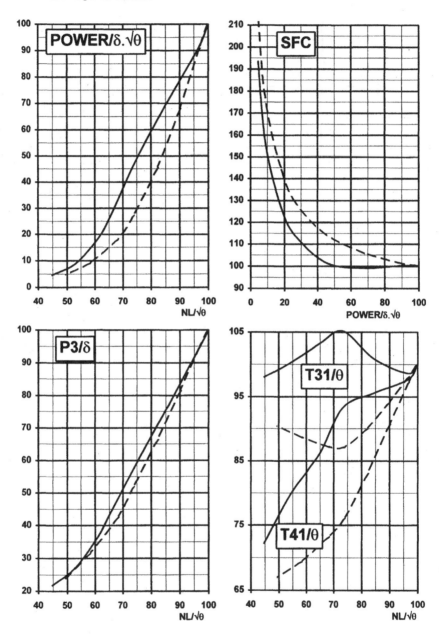

Notes:

Solid line is for power turbine with variable area nozzle.

Broken line is for fixed geometry.

Charts use constant ratio of intercooler sink temperature to ambient. Other ratios produce further families of charts.

Chart 7.5   *contd.*

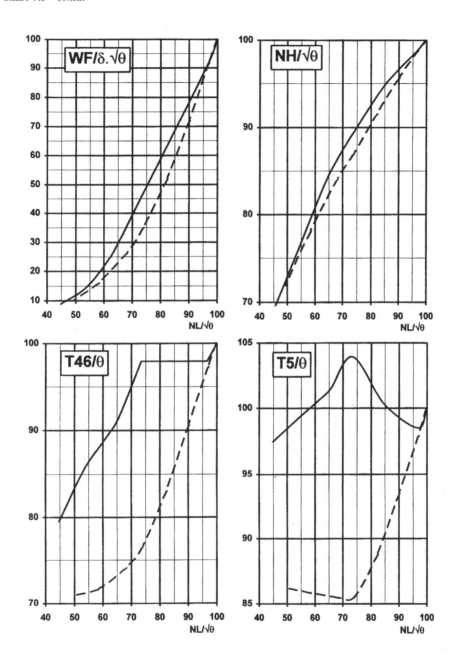

Chart 7.6 Referred performance charts, subsonic turbojet, single spool.

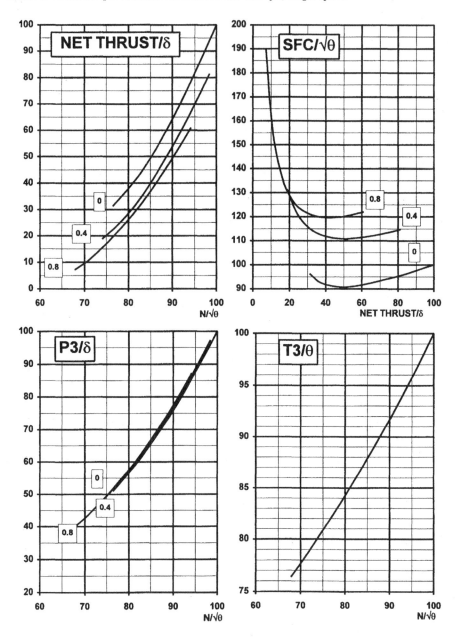

Notes:
Net thrust and net thrust SFC follow separate lines at different Mach numbers due to momentum drag and pressure thrust.
If instead the gross thrust parameter is plotted, a coincident line will result where the propelling nozzle is choked.

Chart 7.6 *contd.*

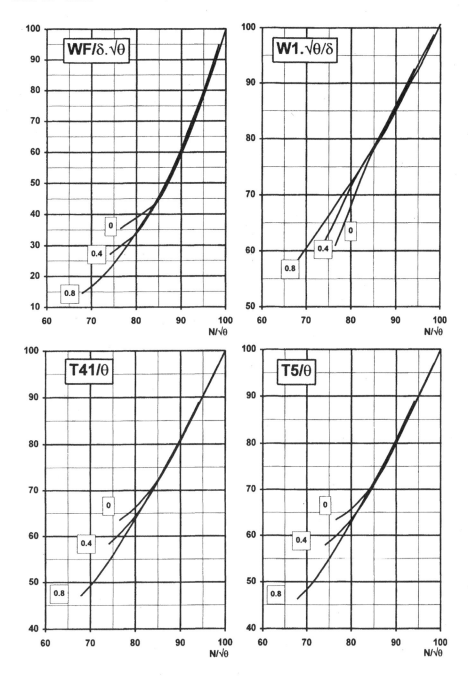

Chart 7.7 Referred performance charts, subsonic turbofan, twin spool.

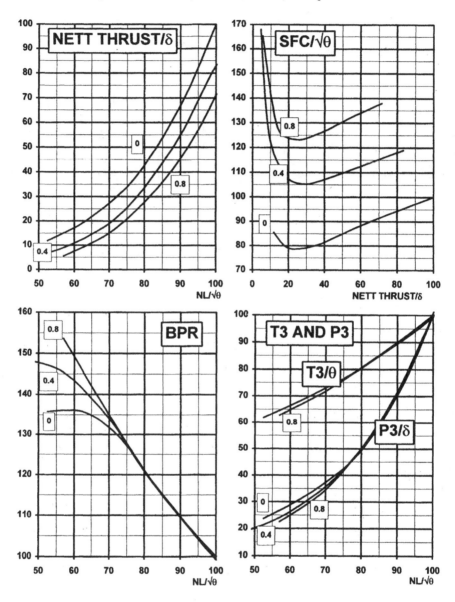

Notes:

All plots show lines of constant Mach number, values 0, 0.4 and 0.8.

Data are for separate exhaust jets. Chapter 5 discusses the effects of mixing.

Nett thrust and nett thrust SFC follow separate lines at different Mach numbers due to momentum drag and pressure thrust.

If instead the gross thrust parameter is plotted, a coincident line will result where the propelling nozzle is choked.

Chart 7.7   *contd.*

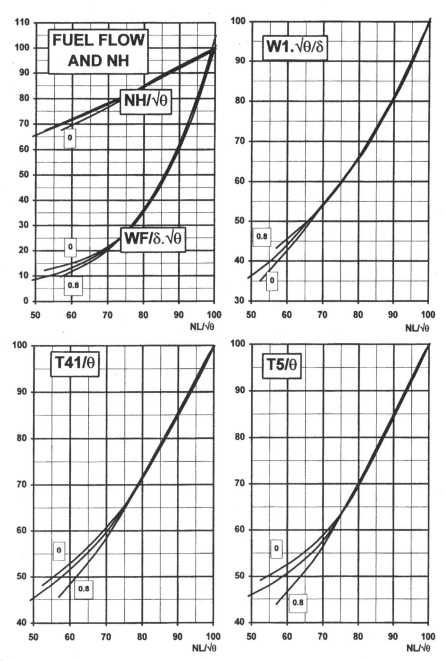

Notes:
All plots show lines of constant Mach number, values 0, 0.4 and 0.8.
Data are for separate exhaust jets. Chapter 5 discusses the effects of mixing.

Chart 7.8   Matrix iteration matching model, turboshaft or turboprop, single spool.

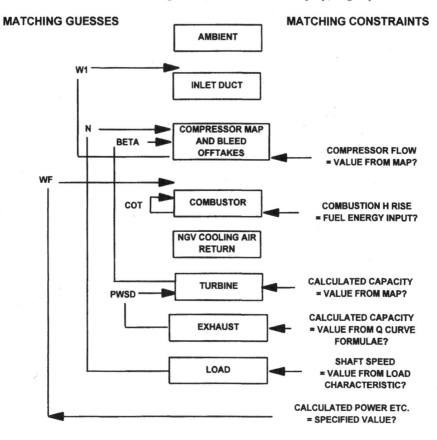

Notes:

Lines to matching guesses indicate main numerical dependencies.

Overall iteration may specify any parameter, not just output power.

Recirculating bleeds require additional guesses and matches at return station for flow quantity and temperature.

For simplicity turbine cooling flow returns downstream of the first NGV are not shown.

Chart 7.9 Matrix iteration matching model, turboshaft or turboprop, single spool gas generator with free power turbine.

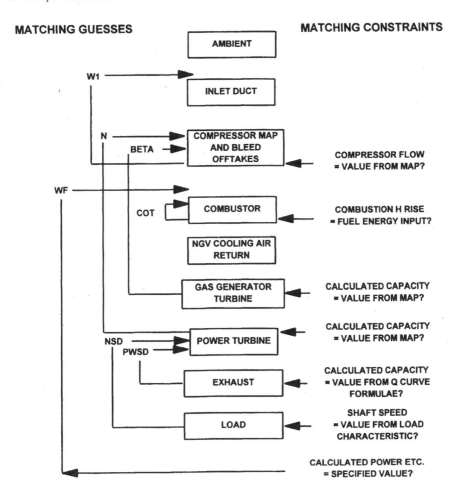

Notes:
Lines to matching guesses indicate main numerical dependencies.
Overall iteration may specify any parameter, not just output power.
Recirculating bleeds require additional guesses and matches at return station for flow quantity and temperature.
For simplicity turbine cooling flow returns downstream of the first NGV are not shown.

Chart 7.10 Matrix iteration matching model, recuperated turboshaft, single spool gas generator with free power turbine.

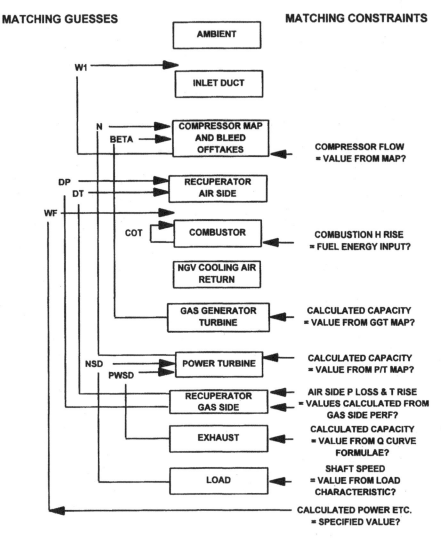

Notes:

Lines to matching guesses indicate main numerical dependencies.

Overall iteration may specify any parameter, not just output power.

A variable area power turbine nozzle would allow additional iteration, e.g. to fixed power turbine inlet temperature.

Recirculating bleeds require additional guesses and matches at return station for flow quantity and temperature.

For simplicity turbine cooling flow returns downstream of the first NGV are not shown.

Chart 7.11 Matrix iteration matching model, turboshaft or turboprop, twin spool gas generator with free power turbine.

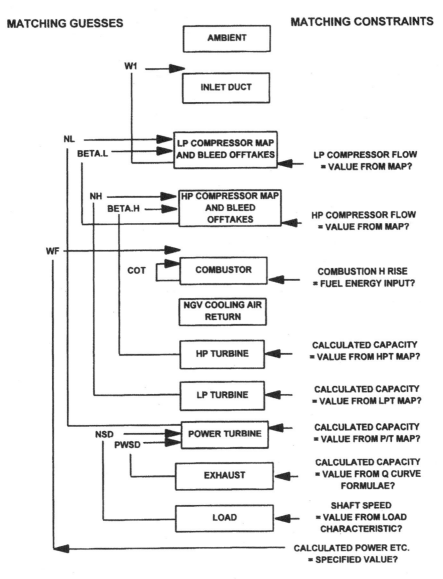

Notes:

Lines to matching guesses indicate main numerical dependencies.

Overall iteration may specify any parameter, not just output power.

Recirculating bleeds require additional guesses and matches at return station for flow quantity and temperature.

For simplicity turbine cooling flow returns downstream of the first NGV are not shown.

Chart 7.12 Matrix iteration matching model, intercooled, recuperated turboshaft, twin spool gas generator with free power turbine.

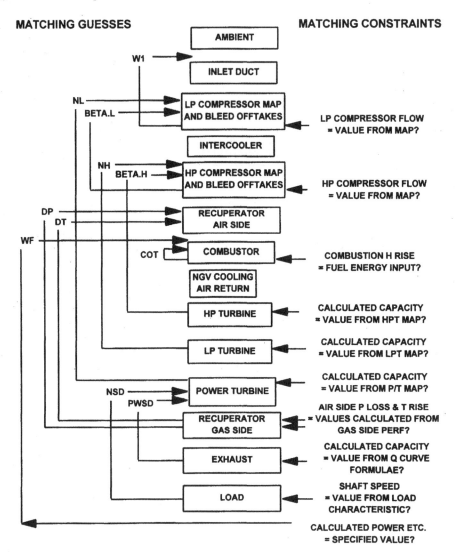

Notes:

Lines to matching guesses indicate main numerical dependencies.

Overall iteration may specify any parameter, not just output power.

A variable area power turbine nozzle would allow additional iteration, e.g. to fixed power turbine inlet temperature.

Recirculating bleeds require additional guesses and matches at return station for flow quantity and temperature.

For simplicity turbine cooling flow returns downstream of the first NGV are not shown.

Chart 7.13   Matrix iteration matching model, subsonic turbojet, single spool.

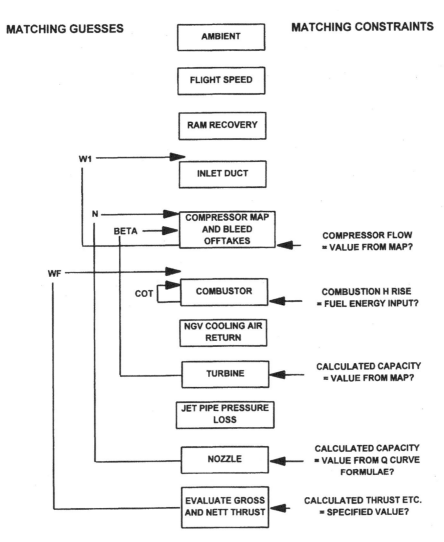

Notes:

Lines to matching guesses indicate main numerical dependencies.

Overall iteration may specify any parameter, not just thrust.

Recirculating bleeds require additional guesses and matches at return station for flow quantity and temperature.

For simplicity turbine cooling flow returns downstream of the first NGV are not shown.

Chart 7.14   Matrix iteration matching model, subsonic turbofan, twin spool.

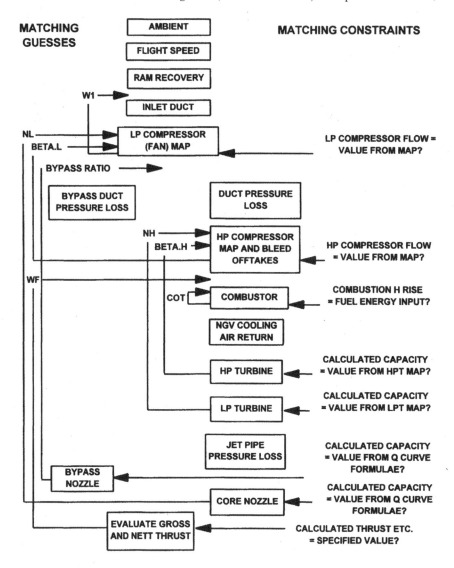

Notes:

Lines to matching guesses indicate main numerical dependencies.

Overall iteration may specify any parameter, not just thrust.

Recirculating bleeds require additional guesses and matches at return station for flow quantity and temperature.

For simplicity turbine cooling flow returns downstream of the first NGV are not shown.

Separate maps may be employed for fan tip and root.

Mixed exhaust requires matching on equal stream static pressures.

References

1. SAE (1974) *Gas Turbine Engine Steady State and Transient Performance Presentation for Digital Computers*, SAE Publication – AS681, Society of Automotive Engineers, New York.
2. BSI, ISO (1992) *Guide for Gas Turbines Procurement*, BS3863/ISO 3977, British Standards Institution, London, International Organization for Standardization, Geneva.

Chapter 8
Transient Performance

8.0   Introduction

Transient performance deals with the operating regime where engine performance parameters are changing with time. This chapter covers transient performance above idle, while Chapter 9 covers sub idle transient performance, or *starting*. Chapter 11 describes transient performance testing.

Engine operation during transient manoeuvres is often referred to as *handling*. It will be apparent from this chapter that control system design and transient performance are inseparable, since the engine responds to the schedules of fuel flow, variable geometry, etc. implemented via the control system.

8.1   The fundamental transient mechanism

Engine performance while operating transiently differs from that described for steady state operation in Chapter 7. This section illustrates the key differences via descriptions of an accel manoeuvre for each of the main engine configurations.

8.1.1   *Turbojets and turbofans*

If a single spool turbojet is operating steady state and the control system suddenly increases fuel flow, then due to the increased temperature the turbine power output increases. This now exceeds that required to drive the compressor, auxiliaries and mechanical losses. Hence for the transient case there is *unbalanced power* available as shown by Formula F8.1. The unbalanced power produces spool acceleration, the fundamental mechanics being as per Formulae F8.2–F8.4. Air flow, pressures, temperatures, etc., and hence thrust or shaft power, all increase as the spool accelerates. This acceleration continues until the steady state condition corresponding to the new fuel flow is achieved. Conversely for a deceleration the unbalanced power is negative and spool speed reduces accordingly.

For multi-spool turbojets and turbofans the process is similar with unbalanced power available on all spools. For a given engine all spool speeds, thrust, pressures, etc. achieve their new steady state values in similar times.

8.1.2   *Free power turbine shaft power engines*

Load applied on a cube law, e.g. ship's propeller

For this engine configuration the gas generator behaves as per the turbojet described in section 8.1.1. As the gas generator accelerates it delivers additional *exhaust gas power* (defined in Chapter 6) to the free power turbine, in excess of that required to drive the load. Exhaust gas power increases at a similar rate to the gas generator spool speeds. The power turbine acceleration time then depends upon its inertia, as well as the inertia and loading characteristics of the driven equipment; its accel time may be significantly longer than that of the gas generator. As shown by Chart 7.2b the gas generator may be at 100% speed and exhaust gas power, while the power turbine is at low speed but providing a high torque. Here the power turbine torque and power output are greater than the propeller load, and the unbalanced power accelerates the combined inertias. The propeller power absorption increases until the new steady state power turbine speed is achieved, where the power turbine power and the propeller load are equal.

Load applied at synchronous speed, e.g. electrical generation

Where the free power turbine is driving an electrical generator it will be running at fixed *synchronous* output speed. If a step increase in load is applied, power turbine speed initially falls due to the load being greater than power turbine power. The control system then increases fuel flow and the gas generator accelerates, providing increased exhaust gas power and hence power turbine power. The power turbine spool then accelerates back to synchronous speed, with power turbine power again equal to the load.

8.1.3   Single spool shaft power engine

In this configuration a single spool gas turbine drives an electrical generator directly, at synchronous speed. If there is a step increase in load the engine speed falls and the control system increases fuel flow to recover synchronous speed.

The transient response is significantly faster than for a free power turbine engine. This is because with a single spool the higher inertia minimises the speed reduction, and there is no lag while a gas generator accelerates to increase exhaust gas power to accelerate a free power turbine. Furthermore the single spool engine has superior capability to cope with a sudden loss of load. Its higher inertia reduces the tendency to overspeed, hence fuel can be reduced more slowly with less risk of weak extinction. Also there is a single compressor that stays close to synchronous speed, and no LP compressor whose transient working line moves towards surge during this manoeuvre.

Whilst this configuration is good for maintaining constant speed at varying power levels it cannot provide the converse, constant power at varying speed. It therefore does not suit mechanical drive applications, as discussed in Chapter 7.

8.2   Transient performance manoeuvres

Manoeuvres comprise changes in engine power or thrust level, which the control system basically achieves by altering the levels of fuel flow.

8.2.1   Slam accelerations and decelerations

Figure 8.1 shows the typical response versus time of gas generator performance parameters to a *slam* (step) increase, or decrease in *power lever angle* (PLA). In the simplest control system each level of PLA corresponds to a given *speed demand*. Following the step increase in PLA to initiate an *accel* (acceleration) the speed demand is far higher than the actual engine speed. The control system responds by increasing fuel flow at a defined limiting rate until the demanded speed is achieved. The overfuelling is typically between 20 and 100% of the steady state value for the current speed. Owing to the additional fuel flow the turbine produces more power than the compressor requires. For a *decel* (deceleration) the opposite occurs. The compressor working lines differ from steady state operation, as shown in Figs 8.2 and 8.3. Sections 8.5 and 8.6 discuss operability concerns related to compressor behaviour.

The high temperatures associated with a slam accel are of such short duration that they do not affect the creep or oxidation life. However as transient times are reduced so is the cyclic

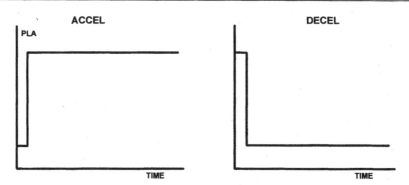

(a)   Power lever angle versus time

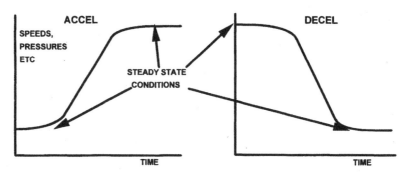

(b)   Engine parameters versus time

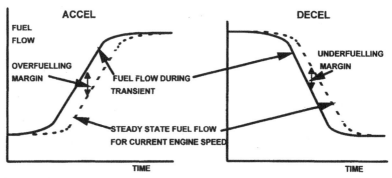

(c)   Fuel flow versus time

Fig. 8.1   Engine performance parameters versus time during slam accels and decels.

life, due to the severe thermal stresses induced. One cycle is usually defined as a start, a holding period at idle, an accel to full thrust or power and eventually the corresponding engine shut down. Though the main damage is that from starting, fast accel times also contribute.

8.2.2   Slow accels and decels

Whenever longer engine response times than those for slam manoeuvres are acceptable to the application, PLA and hence fuel flow are changed at a slow rate. This greatly eases the operability concerns described later, as well as increasing engine cyclic life.

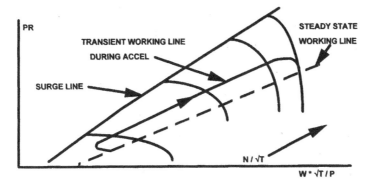

(a) HP compressor for turbojet, turbofan or free power turbine – one, two or three spool

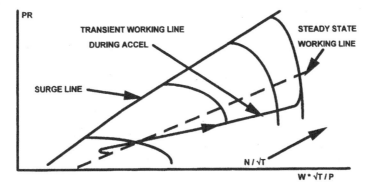

(b) LP or IP compressor, or fan, for turbojet, turbofan or free power turbine turboshaft/ turboprop – two or three spool

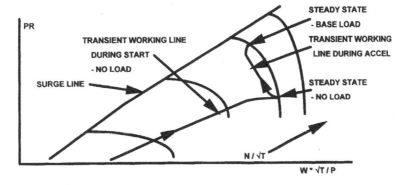

(c) Single spool turboshaft or turboprop with the load driven directly from the gas generator

Fig. 8.2 Transient working lines during accel manoeuvre.

8.2.3 *The hot reslam or Bodie*

The *hot reslam* is a particularly severe manoeuvre described in Fig. 8.4 and is only used in service during an emergency. It is also referred to as a *Bodie*, being named after a US air force pilot who first used the manoeuvre during engine flight trials. First the engine is held at a high power condition for at least 5 minutes to ensure the carcass has *soaked* to its hot condition.

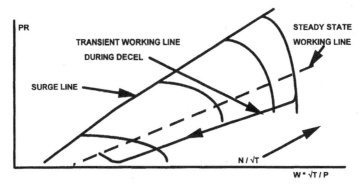

(a)   HP compressor for turbojet, turbofan or free power turbine – one, two or three spool

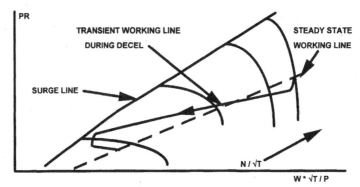

(b)   LP or IP compressor, or fan, for turbojet, turbofan or free power turbine turboshaft/ turboprop – two or three spool

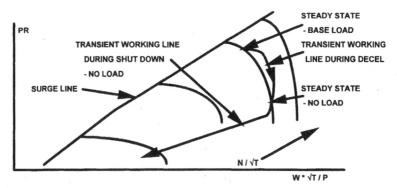

(c)   Single spool turboshaft or turboprop wth the load driven directly from the gas generator

Fig. 8.3   Transient working lines during deceleration.

A slam decel to around idle is followed immediately by a reslam back to high power, allowing no time for the carcass to thermally soak at the low speed.

   In the combustor and turbines heat soakage is akin to additional fuel flow, and in the HP compressor it lowers the surge line as described in section 8.4.4. The adverse impact on the transient HP compressor working line is also shown in Fig. 8.4 and described further in section 8.5. This manoeuvre is used during engine development programmes to give the engine harder operation than it will normally see in service to search for any potential surge margin deficiencies.

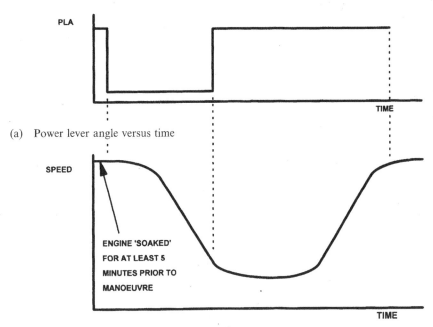

(a) Power lever angle versus time

(b) Engine speed versus time

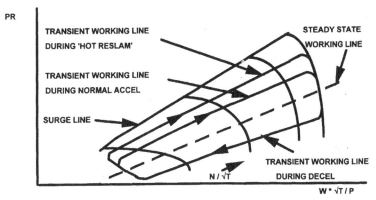

(c) HP compressor transient working line

Fig. 8.4   The hot reslam or 'Bodie' manoeuvre.

8.2.4 *Cold start-accel*

Except in military applications the *cold start-accel* would only be used in service in an emergency, but it is used during development testing to search for potential surge margin shortfalls. The engine is left overnight such that it soaks down to the ambient temperature. It is then started to idle and immediately slam acceled to the maximum rating. Under normal circumstances the engine would be soaked for at least 5 minutes, and often up to 30 minutes, at idle to allow the carcass temperatures to rise gradually.

The cold start-accel is particularly severe in that it maximises the difference in thermal growth of the discs and blades relative to the casings as described in section 8.4.3, and consequently gives the highest tip clearances. As described in Chapter 5, higher tip clearances significantly degrade compressor surge lines. Typically the largest tip clearances, and hence the point of minimum surge margin, occur between 30 and 80 seconds after the engine has achieved maximum rating.

8.2.5   Shaft breakage

In the rare event of a gas generator shaft failure, rapid overspeed will follow as the power of the turbine is decoupled from its driven compressor, leaving a low inertia and a very large power to accelerate it. Similarly for shaft power engines a power turbine shaft failure would produce a power turbine acceleration rate faster than for a drop-load since the generator inertia, as well as the load, would be instantly lost. These transient phenomena must be considered during the engine design process, as a disc burst is unacceptable for safety and could occur in well under a second.

To prevent overspeed the control system must instantaneously close the main fuel valve and open bleed valves. This is the one instance where surge is desirable, as it will inhibit the flow to the turbines. However if wrongly triggered the consequences of such control system actions are severe, especially for aero applications. In addition the sensors and actuators to achieve them quickly enough may be heavy and expensive. In many instances reliability data is used to prove that shaft failure is sufficiently rare that no control capability is needed, or alternatively test data are used to prove that no disc burst occurs if a shaft does fail.

8.2.6   Emergency shutdown

A standard engine shut down comprises a slow decel to idle, a short stabilisation period (of up to around 5 minutes) and then closure of the main fuel valve causing engine run down. However the control system (or external systems) may signal an *emergency shut down* from any operating condition if a fault endangering human safety or the engine is detected, such as an overtemperature or overspeed. Here the main fuel valve is closed instantaneously, and possibly bleed valves opened. This is a severe manoeuvre with respect to LP compressor surge margin as described in section 8.5. Even during an emergency shut down surge is unacceptable.

For engines with large volumes, such as recuperators or dry low emissions combustors, a fast decel reduces HP compressor surge margin. This is because the air mass stored in the pressurised volume flows out of it, reducing the mass flow through the HP compressor. This is described further in section

8.2.7   Drop load or propeller out of water

For power generation engines a *drop load*, i.e. a sudden loss of electrical load, may occur due to electrical failure downstream. This is distinctly different from an emergency shut down as, due to loss of load, fuel must be pulled off more rapidly to avoid overspeed, and the engine must be decelerated to idle to be available immediately for re-loading. Hence the drop load is a challenge for combustor stability, and also for multiple spool engines it is more adverse with respect to LP compressor operability.

In a marine application it may be necessary to deal with the propeller coming out of the water in heavy seas, depending upon the vessel design. This is a similar manoeuvre to a drop load in power generation. Alternatively the ship's controller may limit engine power, and hence power turbine speed, when in sea states where this may occur. The power turbine acceleration then does not risk overspeed. The period before the propeller re-enters the water is short, only around 1–2 seconds.

8.2.8   *Bird or water ingestion*

As stipulated by the air worthiness authorities aircraft engines are required to ingest a number of birds, or a given amount of water/ice, and function safely both during this event and afterwards. Owing to the high importance of engine operation, transient performance during the ingestion event is often the subject of much transient modelling, or analysis of transient test data from such a test during development. This is discussed further in Chapter 11.

8.3   Engine accel and decel requirements

For each engine application there are time requirements for key transient manoeuvres; indicative levels for each application are summarised below. Time zero for accel and decel times is at the instant that the PLA is changed. Typically the gas generator accel is timed to 98% speed corresponding to around 95% thrust or exhaust gas power. All times must be achieved free from any of the operability concerns discussed in section 8.5. In addition the engine must meet the required cyclic life.

The modulation of accel and decel times for a given engine as altitude, ambient temperature and flight Mach number vary is discussed in section 8.10.

8.3.1   *Power generation*

Free power turbine engine

For simple cycle applications connected to a grid the power turbine and generator are at constant synchronous speed, and are typically loaded over a 2 minute period. Hence the gas generator must accel from idle to 95% power in 2 minutes. Similar times are required for gas generator decel when unloading. While the gas generator would be capable of significantly faster times these are not necessary for the application and hence the 2 minute transients are employed to increase engine cyclic life. In combined cycle or CHP applications the generator will be loaded or unloaded over times in excess of 15 minutes due to steam plant limitations.

For independent power producers not connected to a grid more rapid power changes are required as there are fewer load devices, which may be switched on and off instantaneously. Generally a 50% load change must be accommodated with only a 2% change in frequency, and hence power turbine speed, requiring a 50% change in exhaust gas power in around 5 seconds. The precise gas generator accel or decel time depends on the inertia of the power turbine and generator.

Furthermore, in both applications, the gas turbine also has to deal with more severe changes in generator torque produced by the electrical load. For a drop load as discussed in section 8.2.7, electrical generator design rules usually permit a 10% overspeed, while 15% is usual for the power turbine. Typically a gas generator would have to decelerate from rated power to within 5% of idle power in less than 3 seconds, though the time depends upon the inertia of the power turbine and generator. For electrical short circuits, or malsynchronisation with a grid, momentary torques are produced which are several times larger than the torque at full load.

Single spool engine

Loading and unloading rates are as for a free power turbine engine. These involve smaller changes in spool speed, and are easier to achieve. The engine is less prone to operability problems during drop loads, as discussed in section 8.1.3.

8.3.2   *Gas and oil pumping*

Here typical loading rates are 2 minutes from idle to 95% of rated power. The gas generator accel requirement is similar as at these slow rates there is negligible lag of the power turbine speed. Unloading rates are also around 2 minutes from rated power to within 5% of idle power output. Owing to the volume packing (see section 8.4.2) of natural gas in the pipeline, which is hundreds of kilometres long, the engine accel time has little effect on the overall rate of change of gas transmission.

The drop load case discussed for power generation engines does not apply here as it is normally impossible to remove the load instantly from the compressor in a pipeline. The power turbine output shaft failure case.

8.3.3   *Automotive*

For a family saloon car the vehicle acceleration time requirement from zero to 100 km/h is typically 15 seconds, hence the engine gas generator must accelerate from idle to 95% of maximum rating in less than 3 seconds at ISO conditions. The power turbine accel time is as that of the vehicle since it is geared to the wheels.

For a high performance *super car* the gas generator accel requirement reduces to less than 1 second to achieve vehicle acceleration from zero to 100 km/h in typically less than 6 seconds. For automotive gas turbine projects to date this has proved difficult to achieve. For hybrid vehicles (see section 1.4.7) however this is readily attainable since the engine has a single spool operating at fixed speed with the advantages described in section 8.1.3. In addition the battery power can be used to supplement the gas turbine.

For a truck, the vehicle acceleration times are far slower and hence a gas generator accel time of less than 5 seconds is usually acceptable. Tanks have surprisingly arduous acceleration requirements, having to accelerate from zero to 30 km/h in less than around 7 seconds to move one vehicle length very quickly when under attack. Individual tank designs differ radically, hence there is no unique requirement for gas generator accel time, however an indicative value would be around 3 seconds at ISO conditions.

In general, decel time requirements are similar to avoid overly demanding brake capability and wear. For engines with variable power turbine NGVs vehicle deceleration may be aided by moving them beyond axial to provide negative torque.

8.3.4   *Marine*

Under normal operation a typical accel time requirement for a marine engine gas generator is 30 seconds from idle to 95% of rated power. Usually the same time would be required on a decel from rated power to within 5% of idle. The accel or decel time for the power turbine then depends upon its inertia, as well as that of the drive train and propeller. It would be between 40 and 60 seconds for the 30 second gas generator accel time; often the propeller cannot be accelerated any faster, to avoid *cavitation* and consequent erosion of the blades. Typically the vessel would attain its new speed within 10 seconds of the power turbine reaching its rated speed.

Further transient requirements, as described in section 8.2.7, may be imposed depending on the application. If the vessel designer specifies that the propeller out of the water case be considered the gas generator decel time is dependent upon the inertias of the power turbine, drive train and propeller. Generally less than 2 seconds from rated power to within 5% of idle would be required. If the engine is driving an electrical generator then a similar gas generator decel time is required in the event of a drop load.

8.3.5   *Civil aircraft*

To enable the aircraft to *go around* in the event of an aborted landing, airworthiness requirements, such as those of the American FAA, state that the aircraft must be able to achieve a climb gradient of at least 3.2%, 8 seconds after demanding takeoff thrust. This typically requires that the engine must accelerate from *flight idle* to 95% thrust in less than around 8 seconds at 4500 m, which is the highest likely runway altitude. To achieve this time at 4500 m usually 5 seconds is required at sea level, indeed this is a further airworthiness requirement. These accel times must be achieved with the maximum allowable customer bleed and power extractions. In addition the airframe manufacturer will usually stipulate accel times from minimum idle to 98% speed of less than 15 seconds throughout the takeoff envelope, and less than 30 seconds at extreme altitude.

Airworthiness requirements stipulate decel times in the case of an aborted takeoff to enable the aircraft to stop on the runway within a safe distance. This usually requires decel times giving 75% of the thrust change between takeoff and minimum idle, in less than 7 seconds up to an altitude of 4500 m. This means that 4.5 seconds must be achieved at sea level. The airframe manufacturer may also stipulate that 90% of the thrust change between takeoff and ground idle be achieved in less than 11 seconds at 4500 m and less than 7.25 seconds at sea level. Times of 1 second less may also be stipulated relative to flight idle.

The above aircraft requirements apply equally to turboprops and thrust engines. The engine transient requirements depend heavily on the capabilities of any variable pitch propeller mechanism, and also on engine layout such as single spool versus free power turbine. For the former the rotational speed change is small and hence power changes are relatively easy to accomplish.

8.3.6 *Military aircraft*

Military aircraft demands vary significantly dependent upon the application. However an accel time from minimum idle to 98% speed of less than 4 seconds at sea level with maximum customer bleed and power extraction is usually the most demanding. The fastest decel time requirement is also around 4 seconds from takeoff thrust to 75% of the thrust change between takeoff and flight idle.

8.3.7 *Helicopter*

Again time requirements between minimum idle and full power are around 4 seconds at sea level. The energy stored in the fast moving rotor blades, together with blade pitch control, allows a faster change in downwards thrust. It also allows both free power turbine and single spool engine configurations to be used; in the latter case power changes can be effected without the lag of engine spool acceleration.

The storage of energy in the rotor is also utilised during *autorotation*, where after engine failure the helicopter effectively glides with the rotor windmilling, and slows the descent before landing by adjusting blade pitch.

8.3.8 *Ramjet propelled aircraft*

Operation responds strongly to vehicle acceleration and deceleration, due to the dominant effect of flight Mach number on cycle parameters. Engine operation cannot be entirely split from that of the vehicle. Since a ramjet has no rotating turbomachinery and hence no inertia or compressor surge issues to consider, thrust increase due to a change in fuel flow occurs in less than 1 second. This allows for fuel valve response and combustor delay. Thrust reduction is slower, to guard against combustor weak extinction.

8.4   Transient performance phenomena

There are a number of phenomena particular to transient performance which must be considered in addition to the steady state considerations described in Chapter 7. These phenomena are summarised below, and References 1–4 provide further background. Their impact on transient performance, and guidelines for modelling them, are presented later in this chapter.

8.4.1   *Heat soakage*

During transient operation there are significant net heat fluxes between the working fluid and the engine metal, unlike for steady state operation where there is negligible net heat transfer. For example, due to an accel from idle to full power or thrust the engine carcass must soak to a new higher steady state operating temperature, which absorbs typically 30% of the excess fuel energy. This net heat transfer from the working fluid to the metal is termed *heat soakage* and has a significant effect on engine performance.

Where heat exchangers are employed the impact of heat soakage during transient operation can be dramatic, due to the large thermal inertias. Of conventional engine components the combustor has the largest effect, due to its large surface area, thermal mass and temperature changes.

8.4.2   *Volume packing*

During steady state operation the mass flow entering a given volume, such as a duct, is equal to that leaving. This is no longer true under transient operation as the pressure, temperature and hence density of the fluid change with time. This is known as *volume packing* and can have a notable impact upon an engine's transient performance, especially for the largest volumes such as ducts and heat exchangers. For most other engines the combustor has the largest volume and is the primary concern, though other components must also be considered for fast transients.

8.4.3   *Tip clearance changes*

During an accel the thermal growth of the compressor or turbine discs is slower than the pressure- and thermal growth of casings, causing blade tip clearances to be temporarily increased. The converse is true during a decel which can lead to *rubs*. This change in compressor geometry affects its map, the main issue being lower surge lines, as discussed in section 5.2.12. There is also a second-order reduction in flow and efficiency at a speed.

8.4.4   *Heat transfer within multi-stage components*

Where a single map is used to model a multi-stage component such as an axial flow compressor, net heat transfer will have a second-order effect upon the map during a transient. This is due to its effect upon gas temperature through the component and hence stage matching, as it changes the referred speed and hence flow capabilities of the rear stages.

8.4.5   *Combustion delay*

There is a time delay between the fuel leaving the injector and actually burning to release heat within the combustor (section 8.11.6). For steady state performance this is irrelevant, however for transient performance it should be considered.

8.4.6   *Control system delays and lags*

Hydro-mechanical components of the engine control system such as fuel valves, variable inlet guide vane (VIGV) actuation rings, etc. take a finite time to move to new positions *demanded* by the controller during a transient. This finite time may comprise a *delay*, where there is no movement for a given time, and/or a *lag* where the device is moving but lagging behind the demanded signal. In addition, control system sensors measuring parameters such as pressures and temperatures will show delays and lags relative to real engine conditions.

8.5 Operability concerns

During transient manoeuvres engine operation is inherently more prone to undesirable *events* than when running steady state. These must be avoided by engine and control system design as they may endanger the engine application or even human life, and are unacceptable to an engine operator. Such events are caused by *operability* issues, which are discussed in this section.

8.5.1 Transient working line excursions

During transient operation compressor working lines deviate significantly from their steady state position. The consequence of the transient working line crossing the surge/stall line is either surge, rotating stall or locked stall. These are described in detail in Chapter 5, and must be avoided in service where at all possible. (Rotating stall during starting, before light off, is an exception as discussed in Chapter 9.)

The following are apparent from Fig. 8.2, which presents working line deviations during accels.

- The HP compressor transient working line is above that for steady state during the accel for turbojets, turbofans and free power turbine engines. This is true whether the gas generator is one, two or three spool.

   The only exception is for a recuperated free power turbine engine with variable power turbine NGVs (not shown on Fig. 8.2). The steady state schedule is normally overridden during an accel, and the vanes held fully open giving a lower transient working line.
- The LP compressor (or fan) and IP compressor working lines of two or three spool engines initially move a little towards surge, and then rapidly fall significantly below the steady state working line.
- For a single spool engine governed to a fixed speed for power generation surge margin reduces as the engine is loaded. There is also some reduction in speed during the transient until the governor responds by increasing fuel flow.

The fundamental reason for the HP compressor moving towards surge on an accel is that the HP turbine capacity (W405$\sqrt{}$T405/P405) must be satisfied. It will normally be choked, and hence not increase with rising expansion ratio. As fuel flow is increased to initiate the accel, T405 increases almost instantaneously. The HP compressor inertia prevents an instantaneous change in speed and the compressor responds by moving up its constant speed line. This reduces air flow and increases pressure to balance the higher T405 to satisfy the HP turbine capacity. As the overfuelling is maintained the spool then accelerates and is normally controlled such that the accel working line runs approximately parallel to the surge line. Indeed whether or not the HP turbine is choked this fundamental effect occurs.

Initially the LP or IP compressor working line shows a small rise up its constant speed line during an accel to match the reduction in HP compressor flow. As the HP compressor then accelerates and can swallow more flow the LP and IP working lines are 'dragged' below their steady state levels. The amount that the LP or IP working line falls below the steady state level is a function of how much greater its inertia is than that of the HP spool. As will be seen from Fig. 8.3, the opposite of the above occurs during a decel. This is for precisely the converse reasons.

These transient working line excursions increase as the accel or decel times are reduced. Variable guide vanes, and compressor bleed valves are employed to 'manage' these working line excursions, and/or affect the surge line, as described in Chapter 5. Surge margin requirements are provided in section 8.7.

8.5.2   *Impact of transient phenomena on working line excursions*

Two examples of how the transient phenomena described in section 8.4 may produce additional effects are illustrated below.

During the Bodie manoeuvre described in section 8.2.3 the propensity to surge is increased relative to a normal accel. Figure 8.4 shows the HP compressor working line, which for the decel is entirely standard. During the reaccel it is higher than for a normal accel because the engine carcass has not had time to soak down to the low power, low temperature condition. There is therefore a high heat flux from the metal during the reaccel, which acts like additional fuel flow. As stated, for a normal accel the metal does the opposite and *absorbs* heat from the gas stream.

During a fast decel volume dynamics can dramatically impact compressor working line excursions. For engines with large volumes, such as recuperators or certain DLE combustors, the HP compressor working line actually *rises* because as pressure reduces more air must leave the large volume than enters it. This increased flow into the HP turbine is greater than the fall in T405, hence the ratio of compressor exit flow and pressure must fall, which is a higher working line.

8.5.3   *Combustor stability*

Combustor stability is discussed in Chapter 5. Typical steady state operation is shown superimposed upon the combustor stability diagram in Chart 5.7. To facilitate engine deceleration the fuel flow must be lowered relative to that for the corresponding steady state speed, by around typically 20–50%. Also, as described in section 8.5.1, the HP compressor exit flow at a speed is then increased. This means that the combustor operates at a lower fuel air ratio than normal, which risks weak extinction. The worst case will normally be just prior to achieving idle during a slam decel. The likelihood of weak extinction is accentuated by a fast decel requirement, a low idle setting or high altitude operation. The last effect is due to high combustor loading rather than a low fuel air ratio.

Fuel air ratio during decels can be predicted by a transient model, or measured during engine testing. A first-order guide is that it should be maintained at a value greater than 0.008 under transient operation. Hence steady state idle fuel air ratio should be around 0.01 to allow adequate underfuelling margin. These guidelines are based upon the total combustor air flow as opposed to just that in the primary zone. Primary zone equivalence ratio should be greater than 0.35 at idle and 0.28 during a decel.

Except during starting, rich extinction is never encountered during an accel as the amount of fuel required is higher than that which would normally cause HP compressor surge or overtemperature. Should weak extinction occur then it must be detected immediately by the control system and the fuel shut off. This avoids injecting additional fuel which may ignite outside the combustion system. Flame out is normally accompanied by a low 'pop' sound, and may be detected by a drop in spool speeds and measured temperatures. For cannular systems partial flame out produces a step increase in temperature 'spread'.

8.5.4   *Run down*

As discussed in section 8.10, the control system will implement an accel fuel flow schedule which sets the engine maximum allowable transient fuel flow. Under normal operation this is significantly higher than the steady state demand, allowing a margin for overfuelling for acceleration. However as the engine deteriorates in service the steady state fuel flow required increases. If the required steady state idle fuel flow rises above the limit set by the accel schedule, then the control system will prevent the required steady state flow being metered and the engine will *run down*. This is a particularly serious case for an aircraft engine on approach.

8.6 Surge, rotating stall and locked stall – the events and their detection

8.6.1 *Surge*

The compressor map and surge line, and the surge phenomenon are described in Chapter 5. Figure 8.5 shows leading performance parameters versus time during a surge event occurring shortly after a throttle re-slam (Bodie), for a two spool gas generator. For this example the combustor has remained alight and three surge cycles occur until the engine fully *recovers*. If the combustor should flame out during a surge then the engine will run down. For aero applications the engine must then be re-started as soon as possible.

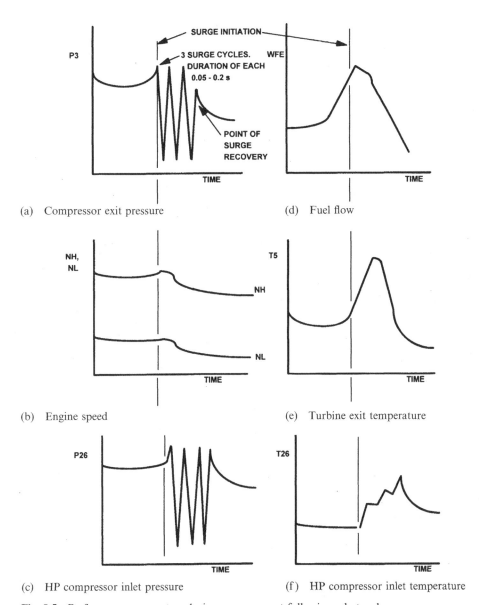

(a) Compressor exit pressure

(d) Fuel flow

(b) Engine speed

(e) Turbine exit temperature

(c) HP compressor inlet pressure

(f) HP compressor inlet temperature

Fig. 8.5 Performance parameters during a surge event following a hot reslam.

At the instant of the first surge compressor delivery pressure reduces dramatically, due to the loss of pressure rise through the compressor. Compressor delivery pressure then recovers to close to its original value, and another surge occurs immediately afterwards. The surge is repeated again but this time on recovery compressor delivery pressure is significantly lower due to a combination of having opened handling bleed valves (not shown), and reduced fuel flow. This enables the engine to recover fully from the surge sequence. Depending upon engine size and configuration each surge cycle will last for between 0.05 and 0.2 seconds. Mass flow is not shown on Fig. 8.5, however it shows the same reactions to surge and recovery as compressor delivery pressure. Partial flow reversal occurs, indeed flames from the combustor often reach the engine intake.

Most engine control systems would detect the rapid reduction in compressor delivery pressure and act to reduce fuel flow, as shown in Fig. 8.5. However fuel flow continues to rise for a finite time after the first surge event as there is a delay before the control system detects the change, and then a lag for the hydro-mechanical fuel system components to react.

The rate of acceleration of engine speeds reduces shortly after the surge event, and soon they are both decreasing due to the surging and reducing fuel flow. Turbine exit temperature continues to rise through the surge sequence until fuel flow has been reduced significantly. HP compressor inlet temperature increases during a surge due to hot gas moving forward through the engine, on surge recovery it reduces. HP compressor inlet pressure rises sharply at the instant of the first surge, due also to high pressure hot gas moving forward, and then cycles as per the compressor delivery pressure. On each cycle the LP or IP compressor will also surge a short time (milliseconds) after the HP compressor, as its pressure ratio is increased. Similarly a collapse in LP compressor pressure ratio, due say to entering rotating stall, will usually cause HP compressor surge due to raising *its* pressure ratio.

On an engine test bed surge is readily apparent due to the loud 'bang' associated with each shock wave and the flames visible in the intake. If necessary the operator will then quickly pull the throttle back to reduce fuel flow, and open handling bleed valves.

The engine design and development programme will ensure that for engines in service surges will be rare. However they may still occur due to isolated incidents such as 'bird strike' for an aero-engine. Relying upon manual intervention is not sufficient as the bang may not be audible to a pilot flying a large aircraft, or for an industrial engine being operated from a remote control room. However if testing shows that the standard controller does not recover the compressor immediately from all surges, then a surge detection system must also be implemented which will then automatically open handling bleed valves. The most common detection mechanism is by measuring the rate of change of compressor delivery pressure divided by its absolute value. Typically a rate of reduction of more than 500%/s (or 150%/s for large combustor volumes) will be a surge, whereas a lower rate will be a normal handling manoeuvre such as a slam decel, or engine shut down.

8.6.2 *Rotating stall*

As described in Chapter 5, rotating stall results in a pocket of stalled blades at the front of a multi-stage axial flow compressor rotating at approximately half engine speed. Its presence only results in modest changes in engine performance, with turbine temperatures increased due to the reduced compressor flow, pressure ratio and efficiency at a speed. A very strong reason to avoid it is the consequent blade excitation which may lead to high cycle fatigue damage. During engine starting rotating stall above light off is a major issue, as discussed in Chapter 9.

Steady state operation in rotating stall is accompanied by a low hum which may be heard from an engine, which is far less obvious than the dramatic bang associated with a surge. The best way to detect rotating stall during engine development is via compressor pressure measurements using close coupled fast response transducers. The frequency of perturbations is between 40 and 70% of engine rotation. To avoid pneumatic lag the line between the transducer and tapping should be no more than 1 metre long. The dynamic signal from this transducer may be displayed on an oscilloscope and spectral analysis employed to distinguish frequencies from broad band noise. Again the engine design and development programme should ensure that rotating stall does not occur above light off when in service. However if it does then its detection is difficult and the experience of the pilot or operator is relied upon to detect starting problems, or the engine running hot to maintain a speed, power or thrust.

8.6.3 *Locked stall*

Like rotating stall, locked stall is accompanied by a light hum. However, as described in Chapter 5 it is far more severe in that compressor flow, pressure ratio and efficiency reduce to approximately half their normal values at a speed. Locked stall is far more obvious than rotating stall in that speeds will reduce and the engine will overheat rapidly. It must be shut down immediately by the pilot, operator or the control system to avoid mechanical damage.

8.7 Surge margin requirements and the surge margin stack up

Surge margin is defined by Formula F8.5; this is the internationally accepted SAE definition, though others have been used in previous years. The minimum steady state surge margin required will depend upon the engine configuration and application requirements. The power or thrust level at which the minimum surge margin occurs will also vary. During the engine concept design phase the steady state model can be used to predict surge margin 'pinch points'.

For each engine application the worst operating conditions and transient requirements vary and it is not possible to cover all combinations here. Generally surge margin stack ups are conducted at these key operating conditions. Once the required margins there have been achieved, the values resulting at some other single operating condition may be compared for different engine types. This is usually at ISA sea level static, maximum rating.

8.7.1 *The surge margin stack up*

The required surge margin is evaluated from a *surge margin stack up* where a range of issues, including transient working line excursions, must be addressed at the worst operating condition. These issues are listed below, together with typical values for a civil aero-engine HP compressor at ISA SLS and rated power or thrust. The required surge margin is calculated by adding the arithmetic sum of the systematic deviances, to the root sum square of the random variances, as per Formula F8.6. For example:

New production engine to engine working line variation	$0 \pm 1.5\%$
New production engine to engine surge line variation	$0 \pm 4.0\%$
In service working line deterioration	-2.0%
In service surge line deterioration	-4.0%
Control system fuel metering, VIGV positioning, etc.	$0 \pm 1.0\%$
Reynolds number effects	-1.0%
Intake distortion	-1.0%
Transient allowance	-12.0%
Total	<u>$20\% \pm 4.4\%$</u>
Surge margin required	24.4%

8.7.2 *Typical surge margin requirements at ISA SLS, maximum rating*

The required surge margin at ISA, SLS and maximum rating varies greatly, being dependent upon accel and decel times required, engine configuration, whether centrifugal or axial compressors are applied, whether bleed valves or VSVs are employed at part load, etc. The levels listed below are a first-order guide.

	Fan	LP/IP compressor	HP compressor
Power generation		15–20	15–20
Gas and oil		10–15	15–20
Automotive		15–20	20–25
Marine		10–15	15–20
Civil aero	10–15	15–20	20–25
Helicopter		15–20	20–25
Military fighter	15–20	20–25	25–30

For a fan, the biggest single contributor to the requirement is inlet distortion where up to 5% surge margin must be allowed.

8.8   Parameter groups and transient performance

Steady state referred and quasi-dimensionless parameter group relationships are discussed in Chapters 4 and 7. During an engine transient any one additional parameter group must be stipulated, beyond those required for steady state performance, to define all others. Parameter groups are discussed in relation to transient performance in section 4.3.6, with the turbojet as an example. For that transient example two parameter groups must be stipulated, in addition to flight Mach number, to define all others. Where the propelling nozzle is choked the relationships are independent of flight Mach number.

These fundamental referred or quasi-dimensionless parameter group relationships are employed in defining engine transient control strategies, as described in section 8.10 below. In both the control strategy examples one referred parameter group is scheduled versus referred speed; with two referred parameters thereby set so are all others. Hence compressor pressure ratios and referred inlet flows are defined at a referred speed during the transient no matter what the ambient temperature, ambient pressure, etc. Thus transient working lines are unique at all ambient conditions. Similar scheduling is employed on a decel so that referred fuel air ratio and combustor Mach number also have unique transient trajectories, to guard against weak extinction.

Engine deterioration, and customer extraction of power and air bleed, change the 'engine hardware' and hence the referred parameter relationships. The control strategies must account for this.

8.9   Scaling parameter groups and transient performance

The impact on transient performance of scaling an engine is described in section 4.3.4. Accel and decel times would theoretically increase by the linear scale factor as an engine is scaled up, but in practice they do not change significantly. The same scaling parameter groups are used in the control system to transiently schedule fuel flow, VIGVs, bleed valves, etc. after an engine has been scaled. This means that surge margins are maintained at approximately the same values after scaling. Combustor weak extinction capability improves marginally after scaling up due to reduced combustor loading, as described in section 5.7.2.

8.10   Control strategies during transient manoeuvres

8.10.1   *The digital controller*

Most modern gas turbine engines employ *digital control systems* which allow far more complex control strategies than their all *hydro-mechanical* predecessors. The digital controller is effectively a computer packaged to withstand the engine environment. For aircraft the package is extremely sophisticated, being mounted on the engine casing and capable of withstanding lightning strikes etc. For industrial engines it is often a rack mounted system remote from the engine. In both cases some hydro-mechanical equipment must be retained with the digital controller for actuation of fuel valves, VSVs, etc.

Measurements are taken of key engine performance parameters such as speed, temperatures and pressures. These analogue signals then go through *analogue to digital converters* before being transmitted to the digital controller. The digital controller reads these inputs, executes one pass through its algorithms, and sends output signals. Typically this sequence is repeated every 10 to 30 ms. The digital controller output signals are in mV, and are amplified to cause stepper motor movement to facilitate fuel control valve actuation, hydraulic switching to move VSV actuators, etc.

The control algorithms are extremely complex, the biggest design challenge being posed by engine transient manoeuvres. For these the digital controller stores data from previous passes, from which rates of change may be derived. Engine transient performance, and control system design and development, are inseparable.

8.10.2   *Requirements for fuel control schedules*

The *control philosophy* is implemented via a set of algorithms in the digital controller. These and the schedule levels must be set such that the response times described in section 8.3 are achieved, while avoiding the engine instabilities described in section 8.5. The accel schedule must be set high enough to avoid run down and also achieve the required accel times for the highest fuel flow engine situation, while avoiding HP compressor surge. The decel schedule must be set low enough to achieve the required decel times, but high enough to avoid weak extinction and LP or IP compressor surge.

The two most common control philosophies are described in the following sections. The *referred fuel flow* method has traditionally been used in hydro-mechanical control systems. However with the advent of digital controllers the *NDOT* technique has become practical, and has been adopted almost exclusively for aircraft due to the advantages described here. Land-based and marine engines, which do not have such arduous accel or decel time demands, have generally retained the referred fuel flow approach even when digital controllers are employed. (As shown in Appendix A, the ARP 755 nomenclature for spool acceleration rate in rpm/s is NU. However NDOT is the term predominantly used in industry when discussing control system design.)

In both cases a controller normally has a *look up schedule* of engine speed demand versus input PLA. Following a PLA slam, as per Fig. 8.1, the difference between demanded and measured actual speed causes the controller to increase fuel flow and meter it to the limiting levels set by the accel schedule, as described below.

8.10.3   *Referred fuel flow versus referred speed*

Figure 8.6 shows the referred fuel flow versus referred speed control philosophy. An accel fuel schedule is coded into the digital controller as a look up table of $WF/P3\sqrt{T3}$ versus $NH/\sqrt{T26}$. The controller measures NH, T26, P3 and T3 and uses these to evaluate the fuel flow to be metered, from the schedule. At the start of the transient a *jump and rate* algorithm limits the rate at which the fuel valve is allowed to step and to move until it reaches the accel schedule. A decel is implemented similarly.

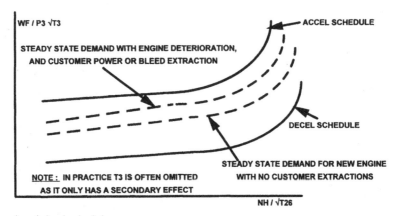

(a)   Accel and decel schedules

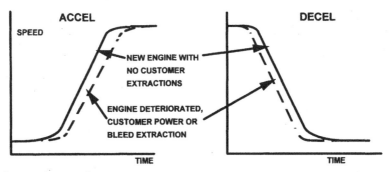

(b)   Engine speed versus time

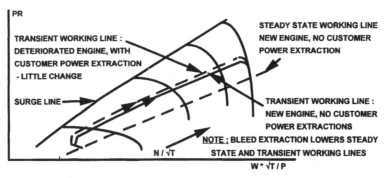

(c)   HP compressor transient working line

Fig. 8.6   Referred fuel flow versus referred speed transient fuel scheduling.

Figure 8.6 also shows some of the salient features of this control philosophy. A new engine, without customer bleed or power extraction, will accelerate faster and decelerate slower than a deteriorated engine with extractions. This is due to the latter case requiring a higher steady state fuel flow; this decreases the overfuelling and increases the underfuelling provided by the accel and decel schedules respectively. Hence the accel schedule must be designed such that the deteriorated engine just makes accel time requirements with bleed and power extractions. Conversely the decel schedule level must be designed such that decel times are achieved for a new engine without bleed or power extraction.

Figure 8.6 also shows that with this control philosophy power extraction or engine deterioration have little or no adverse impact upon the HP compressor transient working line, since transient referred fuel flow and hence working lines are unchanged. Bleed extraction anyway lowers both the steady state and transient working lines.

Accel and decel times increase with altitude by approximately the ratio of ambient pressure to that at ISA sea level, or DELTA. This is because as described in section 8.8, by fixing both referred speed and referred fuel flow at each point during the transient then all other referred parameter groups are fixed. Hence referred unbalanced torque and referred spool accel rate (the group NU in Chart 4.1) are also fixed. It is apparent from Chart 4.1 that *actual* values decrease with P1. One other important result is that they are independent of T1.

The impact of flight Mach number is to reduce accel times, but decel times increase. This is because as shown in Chapter 7, steady state referred fuel flow reduces with Mach number at a referred speed due to the effect of ram pressure on unchoked nozzles. However the accel and decel schedules are unchanged and overfueling increases on an accel, whereas underfuelling decreases on a decel.

The following table summarises the effects of various engine conditions on response time and working line excursions for referred fuel flow versus referred speed control.

	Accel time	Transient HP surge margin	Decel time	Transient LP surge margin
Deterioration	Slower	Little change	Faster	Worse
Power extraction	Slower	Little change	Faster	Worse
Bleed extraction	Slower	Better	Faster	Little change
Flight Mach number	Faster	Better	Slower	Better

8.10.4   *Referred rate of change of speed (NDOT) versus referred speed*

The referred rate of change of speed (NDOT) versus referred speed transient fuel schedule is illustrated in Fig. 8.7. Normally an accel schedule, which is a limiting referred fuel flow schedule, is employed as a 'catcher' in parallel.

Employing NDOT control ensures accel and decel times are always the same for a given operating condition, independent of engine deterioration, customer extractions, etc. This is favoured for aircraft, particularly for twin engines which will always respond at the same rate so that the pilot does not have to address yaw along the runway. However the transient compressor working line excursions are greater with power extraction and engine deterioration.

Again transient times increase with falling P1 but are sensibly independent of T1. The impact of flight Mach number is to reduce both accel times and decel times. This is because referred NDOT is maintained, P1 has increased, and hence actual NDOT has increased. Any effect of unchoked nozzles is compensated by fuel flow changes.

The following table summarises the effects of various engine conditions in response time and working line excursions for NDOT control.

	Accel time	Transient HP surge margin	Decel time	Transient LP surge margin
Deterioration	No change	Worse	No change	Little change
Power extraction	No change	Worse	No change	Little change
Bleed extraction	No change	Little change	No change	Better
Flight Mach number	Faster	Better	Faster	Better

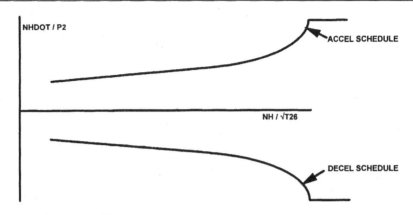

(a)   Accel and decel schedules

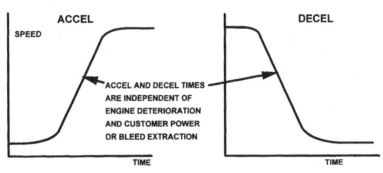

(b)   Engine speed versus time

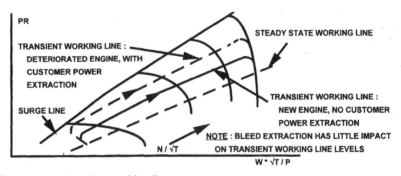

(c)   HP compressor transient working line

Fig. 8.7   Referred NDOT versus referred speed transient fuel scheduling.

8.10.5   *Variable geometry scheduling*

VSVs, VIGVs and handling bleed valves are usually scheduled versus the referred speed of the relevant compressor. This means that the resulting 'composite' compressor map is still unique, as at each referred speed it has a unique geometry. More elaborate maps are also used, as discussed in Chapter 5.

8.11 Transient performance and control models

Transient performance and control models are now an essential design tool for engine detailed design and development. Usually the thermodynamic matching model and one variant of real time model are employed. The different types of transient model and their uses are described here. Except for during the earliest design phase all models include the control system as well as the engine, to ensure fuel flow etc. are scheduled in a meaningful manner.

8.11.1 *Thermodynamic matching transient performance and control model*

Figure 8.8 illustrates the configuration of this type of transient performance and control model for a turbojet. It comprises an engine thermodynamic model coupled to the digital control algorithms, and subroutines to model the hydro-mechanical equipment. For this model there is no hardware, unlike others described later.

The core of the thermodynamic matching element is in fact the steady state model run to specified speeds as described in Chapter 7. At time zero a steady state match point is run which is exactly as per the steady state model. The digital control algorithms will almost certainly require 'historic values' from previous time steps. All of these historic values are now initialised to the steady state values of the respective parameters.

Time is then incremented and the model is switched into transient mode, with subroutines activated to model transient phenomena such as heat soakage, volume dynamics, etc. An introduction to modelling techniques for these phenomena is presented in section 8.11.6. Also unbalanced power is introduced onto the shaft to allow calculation of shaft acceleration. Unbalanced power is utilised in an additional iteration loop around the outside of the steady state model. In this loop the unbalanced power on the shaft is guessed (matching guess) and the calculated fuel flow is compared with that set externally by the control algorithms (matching constraint). Iterations continue with the guessed unbalanced power being adjusted until the fuel flow calculated by the engine model matches that set by the control algorithms.

Once this *matched* point has been achieved, time is incremented and a new speed calculated from the original speed, unbalanced power and known spool inertia via Formula F8.4. Values corresponding to measurements that the control system takes from the engine are then fed into the control system subroutine after suitable lags and delays have been applied (see section 8.11.6). The control algorithms are normally implemented in the same computer language as the engine model, but following the same logic flowpath as the machine code routines implemented in the real controller. This control routine then provides fuel flow, VIGV position, etc. as would be scheduled by the real controller. Lags and delays are applied to model the hydro-mechanical control hardware and hence fuel flow etc. are provided to the thermodynamic matching model. The whole process is then repeated and so a transient manoeuvre is marched through. Sample calculation C8.1 illustrates this process for a turbojet.

As for the steady state models described in Chapter 7, the 'matrix solution' may be employed to speed up the convergence process, as opposed to nested iteration loops shown in Fig. 8.8. The matching guesses (guessed value) and matching constraints (known value) for the turbojet example are listed below.

Steady state		Transient	
Guesses	Constraints	Guesses	Constraints
SOT	HP turbine capacity	SOT	HP turbine capacity
BETA	Propelling nozzle capacity	BETA	Propelling nozzle capacity
		Unbalanced power	Fuel flow from control system

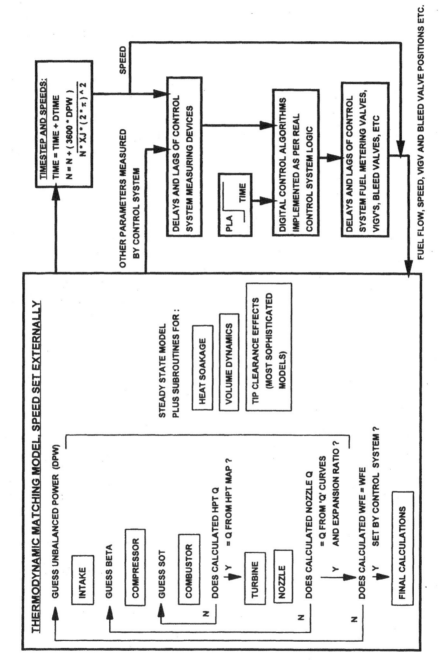

Fig. 8.8   Thermodynamic matching transient performance model.

The overall calculation process is the same for all other engine configurations. For each engine type the matching guess of unbalanced power, and the matching constraint of fuel flow as set by the controller, are added to the steady state matching schemes described in Chapter 7. For a two spool engine the HP spool speed is also calculated from its unbalanced power and shaft inertia. In the steady state model HP speed is a matching guess, whereas in the transient model the HP spool unbalanced power replaces it. For a two spool turbojet matching guesses and constraints are listed below.

Steady state		Transient	
Guesses	Constraints	Guesses	Constraints
SOT	HP turbine capacity	SOT	HP turbine capacity
LP BETA	Propelling nozzle capacity	LP BETA	Propelling nozzle capacity
HP speed	LP turbine capacity	HP unbalanced power	LP turbine capacity
		LP unbalance power	Fuel flow from control system

Ideally the time increment for the model should be set to that of the real digital controller which will be between 10 and 30 ms, to avoid first-order errors in control system historical parameters, lags, delays etc. For initial models comprising the engine with fuel flow etc. input from rudimentary schedules, time increments of up to 0.25 s may be used with small loss in accuracy in the calculated engine performance parameters.

Reference 5 provides aerospace recommended practices for this type of transient model. It contains the most accurate representation of *engine* transient performance parameters and consequently its main uses are:

- Design of engine control philosophy
- Design of control schedules for fuel flow, VIGVs, variable area propelling nozzles, etc. to achieve engine transient requirements, while maintaining adequate surge and weak extinction margins
- Examination of engine transient performance during the design phase, prior to an engine being available for transient testing
- Prediction of engine transient performance in extreme corners of the operational envelope where engine testing may not be practical. For aero-engines Airworthiness Authorities will accept such model data as proof of safety, once the model has been aligned to test data at more convenient operational conditions.
- Examination of manoeuvres which may be too expensive or impractical to test, such as shaft breakage

Even when using the fastest computers available (at the time of writing) the thermodynamic matching model described above will not usually run in *real time*. Convergence of the matched point will take longer than the available time increment of 10–30 ms. To achieve real time execution a thermo-dynamic matching model must be grossly simplified, with a consequent loss in accuracy. Hence a standard thermodynamic engine model cannot be combined with a *hardware* control system. Furthermore, the iterative method produces noise on the output parameters, due to solutions at each time step falling at random places within the permitted tolerance band. This noise may prevent assessment of system stability, due to the perturbations produced.

8.11.2   *Real time transient performance models*

Figure 8.9 shows the configuration of a real time transient performance model, combined with a hardware control system. The hydro-mechanical systems are shown in hardware as well as the digital controller. This layout, often referred to as an *iron bird* for aero applications, requires hydraulic pressure, electrical power, etc. to be available so that all of the hydro-mechanical systems can be actuated. At earlier stages in a project these systems may be modelled in *software* with only the digital controller in hardware.

Reference 6 is the aerospace recommended practice for real time models. All real time engine models utilise as many look up tables as possible, as opposed to polynomial expressions which are slower for a computer to execute. Real time models have a multitude of uses:

- Development of the control hydro-mechanical hardware independent of an engine. This is a big cost saving, and also means that the control hardware can be tested early in the development programme before an engine is available.
- Proving the machine code digital controller algorithms before utilising them on a real engine. Hence the machine code implementation of the control philosophy, designed using the matching model, can be validated to ensure that there are no translation errors which may have endangered an engine.
- 'Passing off' production control systems
- Providing the engine model for *flight simulators* for training pilots
- Assessment of control system and overall stability, providing there is no iteration and consequent noise on output parameters

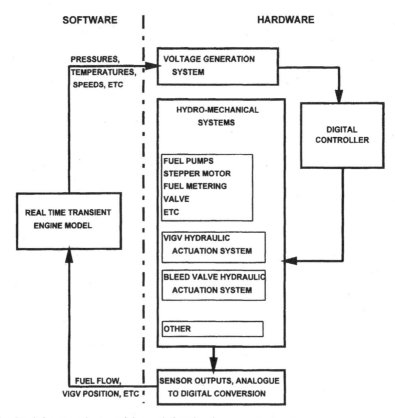

Fig. 8.9   Real time transient model coupled to hardware control system.

8.11.3 *Real time aerothermal transient performance model*

The *aerothermal model* is the most accurate of the many real time engine models for modelling engine performance transient parameters. Reference 7 presents a detailed description of this model type. Figure 8.10 shows the component breakdown and station numbers employed for a two spool turbojet, while Fig. 8.11 shows the computation procedure and formulae required. As per the matching model a steady state point is run and then the model switches to transient mode and runs points at intervals of time corresponding to the digital controller update time frame.

The volumes between the turbomachinery are important in the calculations. Figure 8.11 shows that after time has been incremented from time t − 1 to time t, the first calculations are the computation of P26, P45 and P6 using values of dP/dt calculated from volume dynamics at the end of the previous point at time t − 1. Speeds at time t are now also calculated using unbalanced powers, speeds from the previous point and the spool inertias.

The computation then moves to the combustor, calculating T4, P4 and P3 using the fuel flow scheduled from the control system, and mass flows and temperatures from the previous time step. Next in the HP turbine the component map as per the matching model is employed, but capacity and efficiency are looked up using the known expansion ratio and corrected speed. Hence T45 and W4 are calculated. The computation then loops back through the combustor three times to update the W4 used to time t as opposed to time t − 1.

Next, intake calculations are performed as per steady state routines, and then the LP compressor map read using the known pressure ratio and referred speed yielding W2, W26 and T26. This process is repeated for the HP compressor map to give T3 and W3. The computation then loops back through the combustor three times to update W3 and T3 to values at time t as opposed to values at t − 1. If these loops slow the computation down unacceptably then they may be omitted with little loss of accuracy.

The LP turbine map is then read using the known expansion ratio and corrected speed, and hence T6, W45 and W5 are evaluated. The propelling nozzle capacity and hence W6 can then be calculated from its expansion ratio, effective area and Q curve formulae. The expansion ratio is P6 divided by ambient pressure, or the critical value if the nozzle is choked.

For each of the three intercomponent volumes dP/dt is then calculated, using the mass flows entering and leaving the volumes as calculated via the above. Finally the spool unbalanced powers are evaluated, after which time is incremented again for the next transient point.

Relative to the matching model this method has superior execution time, as it does not require iteration, though there is some loss in accuracy. It is the most commonly used model

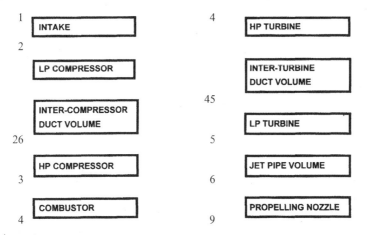

Fig. 8.10 Components and station numbers for aero-thermal real time model of two spool turbojet.

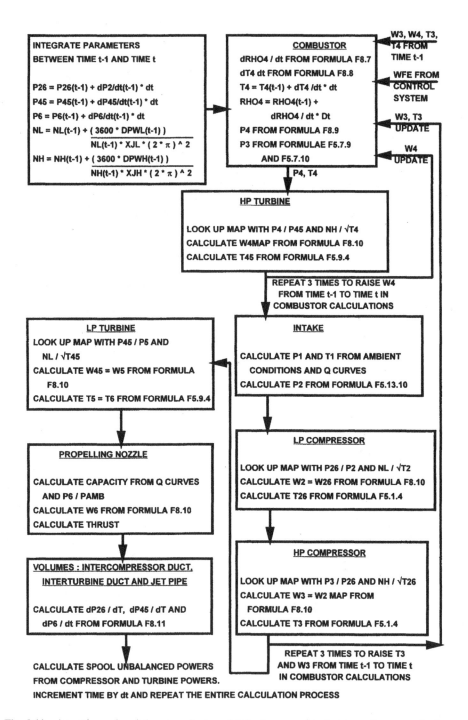

Fig. 8.11   Aero-thermal real time transient model for two spool turbojet – volumes method.

for engine control system hardware development, due to it being the most accurate real time engine model. The other model types described below are more suited to flight simulators, where lower accuracy is permissible.

8.11.4 *Real time transfer function transient performance model*

In this type of real time model key performance parameters are related to fuel flow using the transfer function equation, Formula F8.12. The time constants must be derived from outputs from the transient matching model, or from engine test data. The output from the model is limited to those parameters for which transfer functions have been derived. This model is generally less accurate than the aerothermal real time model, however it is simpler and faster. Though sometimes used for control system development it is more common in flight simulators. Here auxiliary system parameters such as for the oil or starter system are required in addition to engine performance parameters, and can easily be modelled in this way.

8.11.5 *Real time lumped parameter transient performance model*

In this type of real time engine model a large matrix of steady state and transient performance parameters is generated from aero-thermal models or test data. These parameters are divided into *states* reflecting energy storage (such as spool speeds) and *inputs* such as fuel flow. Partial derivatives of state variables are obtained with respect to all other parameters. Formula F8.13 is then used together with the matrix to evaluate performance parameters versus time during a transient. This model type is again more suited to flight simulators but is not as commonly used as the transfer function method above.

8.11.6 *Modelling transient phenomena*

A comprehensive guide to modelling the transient phenomena discussed in section 8.4 is beyond the scope of this book and hence only a simple introduction is provided here. The methodology is described below, with sample calculations C8.2 and C8.3 providing a basic illustration. References 1–4, 8 and 9 provide detailed explanations.

Heat soakage
To model heat soakage accurately the heat nett heat flux to or from the metal component must be calculated using Formulae F8.14–F8.16. Component geometric data, thermal masses and heat transfer coefficients are required. The latter must be calculated from correlations based on parameters such as Reynolds number and Prandtl number, and are difficult to evaluate accurately. Hence empirical correlation factors are often applied to the heat transfer coefficients to align the model to transient engine test data once it is available.

Formula F8.17 provides a method for simple models for engines which do not involve heat exchangers as published in Reference 1. Here all heat soakage is considered to take place in the combustor and a time constant is used to calculate the heat soakage versus time. For crude calculations effective overfuelling, i.e. fuel burnt rather than fuel metered, may be reduced by 30% throughout an accel transient.

Volume dynamics
Volume dynamics should be accounted for ducts, the combustor and heat exchangers. Formula F8.18 allows the change in mass flow leaving the volume, relative to that entering, to be calculated.

Compressor performance
Changing tip clearances and interstage heating may significantly lower the surge line, depending on the design of compressor. An empirical approach is often most appropriate, via

engine testing as described in Chapter 11. Otherwise the changes to the compressor map are of second order and may be ignored for simple transient performance models. References 3 and 4 describe these phenomena in detail and provide guidance for modelling.

Combustion delays
The time delay from fuel being injected to releasing heat in a combustor is of the order of 1–2 ms. In either case Formula F8.19 can easily be employed.

Control system delays and lags
Formula F8.19 is used for delays in movement of control system hardware components, and Formula F8.20 is used to model lags. The delays and time constants need to be provided by control system component manufacturers.

Formulae

All formulae presented here employ specific heat and gamma. For rigorous calculations enthalpy should be used. Chapter 3 provides the methodology required.

F8.1   Unbalanced power (kW) = fn(compressor/turbine mass flow (kg/s), temperature change (K), CP (kJ/kg K), mechanical efficiency (fraction))

$$DPW = W4 * CP * (T5 - T4) - W2 * CP * (T3 - T2)/ETAM$$

(i)   Note that for steady state operation DPW is zero and hence:

$$W4 * CP * (T4 - T5) = W2 * CP * (T3 - T2)/ETAM$$

(ii)   These formulae are for the case of no turbine cooling air to produce additional temperature drops.

F8.2 and F8.3   Fundamental mechanics equations (N m, kg m^2, rad/s^2, W, rad/s)

F8.2   TRQ = XJ * NDOT
F8.3   PW = TRQ * N

F8.4   Spool acceleration rate (rpm/s) = fn(unbalanced power (W), polar moment of inertia (kg m^2), speed (rpm))

$$NDOT = DPW/(XJ * N * (PI * 2/60)^2)$$

F8.5   Surge margin (%) = fn(PRsurge, PRworking line)

$$SM = 100 * (PRsurge - PRworking line)/PRworking line$$
(i)   Pressure ratios are at constant referred flow.

F8.6   Surge margin required (%) = fn(systematic surge margins (%), surge margin variances (%))

$$SMrequired = SMsystematic + SQRT\left(\sum(SMvariances^2)\right)$$

F8.7   Rate of change of density at combustor exit (kg/m^3 s) = fn(combustor inlet and exit mass flow (kg/s), fuel flow (kg/s), volume (m^3))

$$dRHO4/dt = (W3 - W4 + WF)/V$$

F8.8   Rate of change of T4 (K/s) = fn(CP3, CP4, CV (W/kg K), combustor inlet and exit mass flow (kg/s) and temperature (K), fuel flow (kg/s), fuel lower heating value (J/kg), rate of change of density at exit (kg/m³ s), volume (m³))

$$\text{dT4/dt} = (\text{CP3} * \text{T3} * \text{W3} - \text{CP4} * \text{T4} * \text{W4} + \text{WF} * \text{LHV})$$
$$/(\text{CV} * \text{T4} * \text{V} * \text{dRHO/dt})$$

F8.9   Combustor exit pressure (kPa) = fn(Mach number, gamma, gas constant (kJ/kg K), density (kg/m³), temperature (K))

$$\text{P4} = (1 + (\gamma - 1)/2 * \text{M4}^\wedge 2)^\wedge (1/(\gamma - 1)) * \text{R} * \text{RHO4} * \text{T4}$$

F8.10   Mass flow (kg/s) = fn(capacity (kg √K/s kPa), temperature (K), pressure (kPa))

$$\text{W} = \text{Q} * \text{P/T}^\wedge 0.5$$

F8.11   Rate of change of pressure in a volume (kPa/s) = fn(Mach number, gamma, gas constant (kJ/kg K), mean total temperature (K), inlet and exit mass flow (kg/s), volume (m³))

$$\text{dP/dt} = (1 + (\gamma - 1)/2 * \text{M}^\wedge 2)^\wedge (1/(\gamma - 1)) * \text{R} * \text{T} * (\text{Win} - \text{Wout})/\text{V}$$

F8.12   Transfer function equation: speed at time t = fn(fuel flow at time t and t − 1, speed at t − 1, lead time constant, lag time constant)

$$\text{NL(t)} = \text{WFE(t)} * (1 - \text{TClead/TClag}) + (1 - \text{TClead/TClag})$$
$$* (\text{NL(t} - 1) + (\text{NL(t)} - \text{WFE(t} - 1)) * (1 - \text{e}^\wedge(\text{dt/TClag})))$$

F8.13   Lumped parameter equation – example

$$\partial\text{NL/dt} = (\partial\text{NLdot/dNL}) * (\text{NL} - \text{NLb}) + (\partial\text{NLdot/}\partial\text{NH})$$
$$* (\text{NH} - \text{NHb}) + (\partial\text{NLdot/}\partial\text{WFE}) * (\text{WFE} - \text{WFEb})$$

(i)   An engine's steady state and transient response is characterised by stored matrices of steady state parameter values and of their partial derivatives with respect to other parameters.

(ii)   The number of pertinent parameters to consider is chosen depending on the engine configuration; these are classified as either *inputs* (e.g. *WFE)* or *states* (energy storage parameters, e.g. NH).

(iii)   For all states partial derivatives are evaluated with respect to all other states and inputs.

(iv)   The values are obtained from a thermodynamic model or test data, at various base operating points 'b'.

(v)   In this example, variation of LP spool acceleration is obtained from partial derivatives of NLdot with respect to NL, NH and WFE.

F8.14 to F8.16   Heat soakage = fn(heat transfer (kW), heat transfer coefficient (kW/m² K), gas and metal temperatures (K), gas mass flow (kg/s), area of metal (m²), mass of metal (kg), CP of metal (kJ/kg K))

F8.14 $\text{QU} = \text{h} * \text{A} * (\text{Tgas} - \text{Tmetal})$
F8.15 $\text{dTgas} = -\text{QU/(W} * \text{CPgas})$
F8.16 $\text{dTmetal/dt} = \text{QU/(Mass.metal} * \text{CPmetal})$

(i)   In Fomula F8.14 Tgas and Tmetal are at time t. Hence for a given heat transfer coefficient QU is derived and then dTgas and dTmetal are evaluated for the given timestep dt from Formulae F8.15 and F8.16 respectively.

F8.17 Approximate heat soakage = fn(max heat soakage (kW), time (s), time constant (s))

$$QU = QUmax * e^\wedge(-t/TC)$$

(i) Time constant ranges from 5 s for a 200 kg mass engine to 40 s for a 2 tonne mass engine.

F8.18 Rate of mass storage in a volume (kg/s) = fn(gamma, Mach number, gas constant (kJ/kg K), mean temperature (K), mean pressure (kPa), volume (m³))

$$Win - Wout = V * dP/dt/(((1 + (\gamma - 1)/2 * M^\wedge2)^\wedge(1/(\gamma - 1)) * R * T)$$

dP/dt is calculated from the known values of P at time t and time t − 1.

F8.19 Delayed value of parameter = fn(delay (s), value.old, time (s))

$$Value(time = T) = Value(time = (T - delay))$$

For example, combustion delay of 80 ms. WFburnt = WFmetered at time = T − 0.08.

F8.20 Lagged value of parameter = fn(lagged value at previous time step, time step (s), time constant (s))

$$Lagged.value.(time = T) = (Lagged.value(Time = T - DT)$$
$$+ Actual.value(Time = T))/(TC + DT)$$

(i) Example: actual value of a gas temperature, and lagged value a thermocouple reading.
(ii) Example: demanded value of engine spool speed, and lagged value the real rpm.

Sample calculations

C8.1 A single spool turbojet with a polar moment of inertia of 0.1 kg m² is at its idle speed of 15 000 rpm when fuel flow is increased as per the accel schedule. The thermodynamic matching model as described in section 8.11 is run to simulate the transient manoeuvre and the resulting unbalanced power from the first transient match point is calculated to be 16.1 kW. The time increment is 50 ms, calculate rotational speed for the second transient match point.

$$\textbf{F8.4} \quad NDOT = DPW/(XJ * N * (PI * 2/60)^\wedge2)$$

Apply F8.4:

$$NDOT = 16\,100/(0.1 * 15\,000 * (PI * 2/60)^\wedge2)$$
$$NDOT = 979 \, rpm/s$$

Calculate speed for second transient match point:

$$N = 15\,000 + 979 * 0.05$$
$$N = 15\,049 \, rpm$$

C8.2 For the above case find the difference between combustor inlet and outlet mass flow for the first iteration of the second transient match point. The combustor volume = 0.006 m³.

First iteration for second transient match point:
Mean total pressure = 203.4 kPa
Inlet Mach number = 0.35
Mean total temperature = 675 K

Converged value from first transient match point:
Mean total pressure $= 200\,\text{kPa}$

F8.18 $W1 - W2 = V * dP2/dt/(R * T)/(1 + (\gamma - 1)/2 * M^2)^\wedge(1/(\gamma - 1))$

Apply F8.18:

$\text{Wout} - \text{Win} = 0.006 * (203.4 - 200)/0.05/(0.28705 * 675)/(1 + (1.333 - 1)/2$
$\qquad * 0.35^\wedge 2)^\wedge(1/(1.333 - 1))$
$\text{Wout} - \text{Win} = 0.002\,\text{kg/s}$

C8.3   **The fuel flow demand during the first 120 ms of an accel is listed below. The fuel system has a time constant of 10 ms and there is a 2 ms delay between fuel being metered to the combustor and it burning. The transient performance and control model has an update time of 20 ms and fuel flow demanded by the control system for the first six time steps is as below. Evaluate the fuel burnt at 120 ms.**

Time (ms)	0	20	40	60	80	100	120
Fuel demand (kg/s)	1.0	1.05	1.10	1.15	1.20	1.25	1.30

First calculate fuel flow metered versus time
Substituting values into F8.20 at 20 ms, recognising that time zero is a steady state condition where fuel flow metered is equal to fuel flow demanded:

$\text{WFmetered}_{20\text{ms}} = ((\text{WFmetered}_{0\text{ms}} * 0.01)$
$\qquad\qquad + (\text{WFdemand}_{20\text{ms}} * 0.02))/(0.01 + 0.02)$
$\text{WFmetered}_{20\text{ms}} = ((1.0 * 0.01) + (1.05 * 0.02))/(0.01 + 0.02)$
$\text{WFmetered}_{20\text{ms}} = 1.033\,\text{kg/s}$

Now repeat the above for the ensuing time steps:

Time (ms)	0	20	40	60	80	100	120
Fuel metered (kg/s)	1.0	1.033	1.077	1.126	1.175	1.225	1.275

Apply the delay
Owing to the combustion delay, fuel is burnt 2 ms after it is metered. Hence at 120 ms 1.270 kg/s of fuel is burnt (linear interpolation utilised).

References

1. B. Thomson (1975) Basic transient effects of aero gas turbines, *NATO AGARD Conference Proceedings No. 151*, AGARD, Brussels.
2. N. R. L. MacCallum (1981) *Further Studies of the Influence of Thermal Effects on the Predicted Acceleration of Gas Turbines*, ASME 81-GT-21, ASME, New York.
3. P. F. Neal (1982) Mechanical and thermal effects on the transient and steady state component performance of gas turbine engines. *NATO AGARD Conference Proceedings No 324 – Engine Handling*, AGARD, Brussels.
4. N. R. L. MacCallum (1982) Axial compressor characteristics during transients. *NATO AGARD Conference Proceedings No 324 – Engine Handling*, AGARD, Brussels.
5. SAE (1989) *Gas Turbine Engine Transient Performance Presentation for Digital Computers*, SAE Publication – ARP1257, Society of Automotive Engineers, Warrandale, Pennsylvania.
6. SAE (1996) *Gas Turbine Engine Real Time Performance Model Presentation for Digital Computers*, SAE Publication – ARP 4148, Society of Automotive Engineers, Warrandale, Pennsylvania.

7. B. D. MacIsaac and H. I. H. Saravanamuttoo (1974) *A Comparison of Analog, Digital and Hybrid Computing Techniques for Simulation of Gas Turbine Performance*, ASME 74-GT-127, ASME, New York.

8. R. A. Onions and A. M. Foss (1982) Developments in the dynamic simulation of gas turbines. *NATO AGARD Conference Proceedings No 324 – Engine Handling*, AGARD, Brussels.

9. B. L. Koff (1982) Designing for fighter engine transients. *NATO AGARD Conference Proceedings No 324 – Engine Handling*, AGARD, Brussels.

Chapter 9
Starting

9.0 Introduction

Starting is one of the most technically challenging aspects of gas turbine performance. It covers engine operation from when the operator or pilot selects a start, through to stabilisation at idle. The fundamentals of transient performance presented in Chapter 8 are all applicable to the start regime, and as for above idle transient performance, starting is inseparable from control system definition. For starting, processes additional to those for above idle transient performance must be addressed such as cranking, ignition and combustor light round. The combustor light off challenge must never be underestimated. Light off at the full range of engine conditions must be addressed by a combustor rig test programme.

Both *starts* and *restarts* occur on the ground for all engine types, and for aircraft engines restarts also occur in flight. The background to the starting capability required by each gas turbine engine application is discussed in Chapter 1, and Chapter 11 describes start testing. The engine configurations referred to herein are described before Chapter 1.

9.1 The fundamental starting process

This section describes the fundamental starting process as generically applicable to all engine types and circumstances. It also introduces the manner in which shaft speeds, torque and power, and compressor working lines vary during a start. Section 9.2 describes starting features specific to each engine type.

9.1.1 *Start phases*

The key phases of a start are briefly defined below. Each is then comprehensively described in the ensuing sections.

Dry cranking: The engine HP shaft is rotated by the *starter* with no fuel being metered to the combustor.

Purging: This ensures that there is no fuel from previous operation or failed start attempts in the engine gas path or exhaust that may ignite and cause damage. The engine is *dead cranked* at the maximum speed the starter can sustain, which *purges* any fuel into the atmosphere.

Light off: Fuel is metered to the combustor, and *ignitors* are energised. This causes *ignition* locally within the combustor, and then *light around* of all the burners.

Acceleration to idle: This is achieved via a steady increase in fuel flow, and continuing starter assistance.

Thermal soakage: Engines are often held at idle to allow the carcass to thermally soak to the new temperature to preserve cyclic life.

Figure 9.1 shows rotational speeds versus time for both a free power turbine turboshaft engine and an aero turbofan during a start, and Fig. 9.2 presents the torque and power balance on the HP spool.

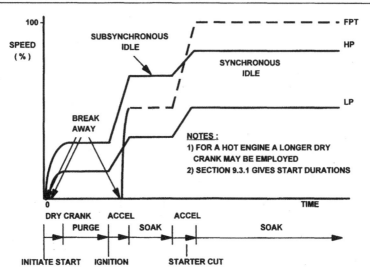

(a)   Two spool gas generator plus free power turbine powergen engine

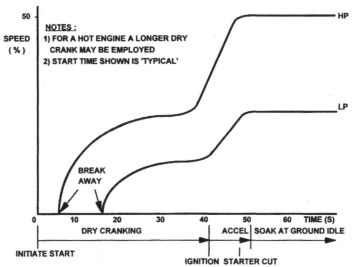

(b)   Two spool turbojet or turbofan

Fig. 9.1   Key engine start phases and speeds versus time.

The two curves shown in Fig. 9.2 are commonly referred to as the *engine resistance/ assistance*, and the *starter assistance*. The former is the net engine unbalanced torque or power output during the start, negative values being *resistance* and positive values being *assistance*. By standard convention the starter motor torque or power is shown as a negative value even though it is an assistance. This allows a reader to quickly see the *net unbalanced assistance torque or power*, which is the vertical distance between the two curves. Formulae F9.1–F9.4 show the relationships between spool speeds, torque, unbalanced power and acceleration, based on fundamental mechanics. These formulae match those given for above idle transient performance (Formulae F8.1–F8.4) except for the inclusion of starter output power.

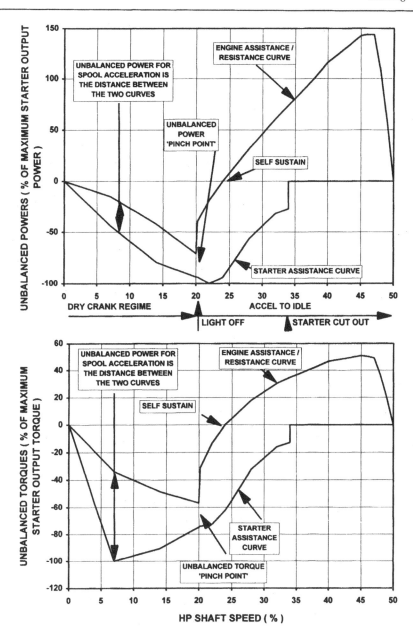

Fig. 9.2 Torque and power on the HP spool during starting.

Figure 9.3 shows LP and HP compressor maps in the start regime, with typical start working lines. Also shown are the *rotating stall* and *hang* boundaries, which are described comprehensively in section 9.5.

9.1.2 *Dry cranking*

The purpose of dry cranking is to develop sufficient pressure and mass flow in the combustor to permit light off. At start initiation the starter is energised and applies torque to the HP

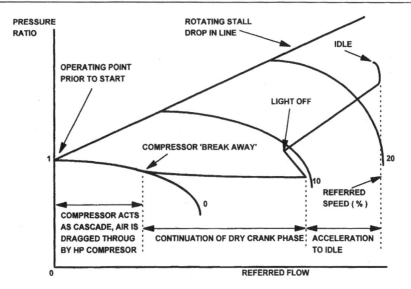

(a)   Fan, LP or IP compressor

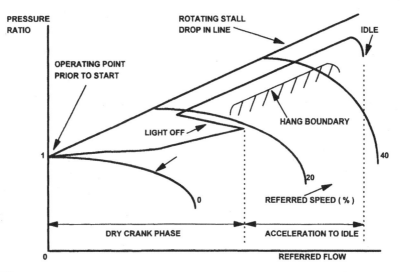

(b)   HP compressor

Fig. 9.3   Compressor transient working lines during starting.

spool. To minimise shock torque loads torque may be applied gradually, for example via slow opening of air valves for a turbine starter. The HP spool then rotates and accelerates due to the excess starter assistance power. The airflow induced by the HP compressor causes the LP spool and if applicable eventually the free power turbine to *break away* from the stiction of the oil at the bearings.

As is apparent from Fig. 9.2, starter torque peaks shortly after start initiation, while starter power typically peaks at about 50% of the idle speed. In the dry crank phase the engine provides a resistance on the HP spool which increases with cranking speed, the turbine output power being less than that taken by the compressor, auxiliaries, bearings and disc windage.

The relative magnitudes and methods for evaluating the constituents of engine resistance are provided in section 9.9.1.

Figure 9.3 shows that the HP compressor working line always has a pressure ratio greater than 1, though at the very lowest cranking speeds it may be in rotating stall. The LP compressor behaviour changes as follows:

- Before breaking away it acts as a *cascade*, with a positive flow and a pressure loss, but no work and hence no change in total temperature.
- After break away it acts as a *paddle* or *stirrer* for most of dry cranking, with a pressure loss but absorbing shaft work and hence producing a total temperature rise.
- Before the top of crank it achieves compressor operation with the work input producing a pressure rise as well as the total temperature rise.

At all times the LP compressor power input is provided entirely by the LP turbine. It is usual to have handling bleed valves open during starting to lower working lines and, for interstage bleeds, to raise the surge line as described in Chapter 5. The achievement of adequate driving pressure ratio for the bleed valves to pass flow is crucial, levels should be assessed early in the design phase. At low LP spool speeds LP compressor delivery bleed valves actually suck air in.

Invariably the starter cranks the HP spool rather than the LP, this being the most efficient way to provide the combustor mass flow and pressure to enable ignition and light around. Starting is eased for engines which have a high fraction of their pressure ratio developed by the HP compressor, as accelerating these stages via direct input shaft power avoids energy loss to the low efficiency of the LP turbine at these conditions. A related effect is that a relatively high LP spool inertia lengthens the time for that spool to accelerate, and hence the whole start sequence. The alternative of cranking an LP spool would incur worse pressure losses upstream of the combustor in the HP compressor.

For aircraft engines free windmilling prior to an in-flight start, as discussed in section 9.2.6, can be considered analogous to dry cranking.

9.1.3 *Purging*

Purging is required for all starts and restarts with gas fuel, and may be used for liquid fuels following a failed start or emergency shut down. Where required purging typically lasts for 1–10 minutes depending on engine size and the type of unburnt fuel to remove. The dead crank phase where purging is performed does not appear on Fig. 9.2. There the combustor has been lit before the HP spool has accelerated to a point where the engine nett resistance equals the starter assistance.

9.1.4 *Light off – ignition and light around*

Here the ignitors are activated and a constant *light off flat* of fuel flow is metered to the combustor by the control system. The light off flat may be as low as 300 kW for a small RPV turbojet and up to 5000 kW for a large turbofan. Once fuel has ignited local to the ignitor, the flame must propagate and stabilise circumferentially around the combustor. The requirements for successful light off are discussed in section 5.8.5, with Chart 5.7 showing combustor ignition envelopes. The HP speed at which the combustor conditions are suitable for light off must be found from a combination of modelling, combustor rig testing and finally engine testing. Light off is a key combustor condition with many practical issues to be overcome such as the atomisation of kerosene or highly viscous diesel on cold days. As discussed in section 5.8.5 combustor efficiency is typically 10–50% at light off, which usually occurs between 15% and 25% HP speed. The lower values are generally for high altitude relight. Chart 5.5 shows combustion efficiency versus loading and may be used to first-order accuracy in the start regime.

Figure 9.2 shows that on light off there is a step reduction in engine resistance, however starter assistance is still required to continue HP spool acceleration. Turbine power output is still usually less than the sum of compressor input power, bearing and windage losses, and auxiliary requirements.

Figure 9.3 shows that both compressor working lines also show a step upwards due to light off, with the HP compressor being the closest to the rotating stall drop in line. It is essential that the starter motor size, hence HP speed at the top of crank, and the light off flat are chosen such that the HP compressor does not go into rotating stall. The control system usually detects ignition and light around by means of thermocouples placed rearwards in the turbines. If light off does not occur within some specified time (e.g. 10 seconds) the control system aborts the start, shuts off fuel flow and commences a purging phase.

9.1.5 *Acceleration to idle*

Fuel flow is steadily increased, causing the engine to accelerate towards idle very much as per the above idle accels described in Chapter 8. The starter motor continues to provide *crank assistance* well after light around. As speed increases the engine assistance eventually dwarfs that of the starter, which *cuts out* before idle via *de-energisation* and *declutching*.

As shown on Fig. 9.2, the engine resistance crosses the 'X' axis and becomes *assistance* shortly after light off. This point is called *self sustain* and theoretically if the starter motor were cut the engine could operate there steady state. However combustor exit temperature profiles make this impractical with respect to turbine life, and hence this speed must be passed through transiently. During acceleration fuel flow is scheduled such that the compressor working lines run approximately parallel to the rotating stall drop in line. The control strategies for scheduling fuel flow during this phase are discussed in section 9.7.

Many combustion systems employ separate, lower flow injectors for starting. This is because at low fuel flows the main injectors may not produce adequate atomisation; here the combustor stability and efficiency depend strongly on which injectors are in use. In many cases below a certain threshold the main burner system may in fact pass no flow at all. The point of *changeover* between systems must be chosen with these issues in mind, to avoid extinction, stall or hang.

On reaching idle, fuel flow is cut back and the engine assistance/resistance becomes zero; no unbalanced power is required for steady state idle operation. The idle point on the HP compressor map is below the transient start working line, whereas for the LP compressor it is higher. As described in section 9.3, various levels of idle can be defined.

Heat soakage, as described in Chapter 8, can have a very significant impact on working lines for hot restarts or cold soaked starts. For an immediate restart following shut down heat transfer from the carcass is akin to additional fuel flow, and will push the compressor towards rotating stall and increase turbine temperatures. In addition, the compressor surge lines are lowered, as described in section 9.5.1. Conversely, after a prolonged cold soak, heat transfer to the carcass is akin to reduced fuel flow and may drive the engine towards hang.

9.2 Start processes for major engine types and applications

9.2.1 *Power generation*

Figure 9.1 shows that for powergen applications there are two levels of idle – *sub synchronous* and *synchronous*. For both these alternator power output is zero and power turbine output power is only that for auxiliaries, plus bearing and windage losses. Sub synchronous idle is utilised during starting, with output speed around 50% of synchronous speed. A dwell here allows the engine to thermally soak at very low speeds and temperatures, reducing thermal stresses and improving cyclic life. After this fuel flow is increased and the engine accelerates to synchronous idle, where 100% output speed is achieved and the alternator connected to the grid. Again a dwell is allowed to enhance cyclic life. The wide-spread use of natural gas fuel means a purge is always required at the top of crank to avoid the risk of explosion.

9.2.2   Gas and oil

For gas and oil applications a free power turbine configuration is almost universal. Idle is at only around 40–50% of free power turbine maximum speed, as it remains connected to the gas compressor which to a first order provides a cubic load line. Unlike synchronous idle in powergen, no intermediate idle condition is required.

9.2.3   Marine engines

During the start phase the free power turbine speed is low, and below 25% of maximum at idle. If the propeller is of variable pitch design it is set to no load to prevent ship motion. If the propeller is of fixed pitch it is de-clutched from the power turbine, or locked by a brake mechanism. Often even if it is not locked the power turbine remains stationary up to idle anyway, due to significant stiction in the drive train.

9.2.4   Automotive engines

The process is analogous to that for a marine engine with the free power turbine de-clutched from the drive train.

9.2.5   Heat exchanged engines

Intercoolers

Intercoolers normally remain active during the start phase, avoiding the need for one more sequencing operation. Furthermore, during starting the intercooler anyway usually makes little difference to HP compressor inlet temperature as there is almost no temperature rise through the upstream LP compressor. One exception is for marine vessels in arctic waters, where the sea water sink temperature may be up to around 35 8C warmer than the ambient air. Here the intercooler could become an 'inter warmer' and provide beneficial heating of the air to aid fuel vaporisation. Any risk of water freezing in the heat exchangers must be considered, however.

Recuperators

Owing to the relative flow volumes and temperatures involved, recuperators are normally able to be bypassed on the air side but not on the gas side. There are arguments in favour of both active and bypassed starting:

- An active recuperator suffers less low cycle fatigue as admitting cold air to a hot recuperator when it becomes active causes thermal shock.
- A bypassed recuperator benefits compressor surge margin as this gives a lower pressure drop even for a cold start. For a restart it also avoids the effect of the variable heat returned upstream of the combustor from the hot recuperator metal; this heat is akin to additional fuel.

One compromise is to crank with the recuperator bypassed, avoiding the greatest risk of not clearing rotating stall, and then to transition the recuperator to active during the accel to idle. Here the variable heat return can be compensated via adjusting fuel flow.

9.2.6   *Manned aircraft thrust engines*

Two idle levels are defined, listed in order of increasing thrust level:

(1)   Ground idle: used when braked on the runway
(2)   Flight idle: used during descent, it offers decreased accel times in case high thrust is
      needed quickly for a go around

Figure 9.2 also shows engine speeds versus time for a start to ground idle for a two spool
turbojet or turbofan. A dwell here, which eases thermal stress, is usual but not mandatory.

Though not part of normal operation in-flight restarting capability is vital. The requirements are covered later with Fig. 9.4 showing a typical restart envelope for a civil turbofan.
The restart process is similar to ground starting for the *starter assist* portion of the envelope.
In the *windmilling* portion the starter is not employed and cranking is achieved from the ram
effect of high flight speed, as described in Chapter 10. The ranges of altitude and flight Mach
number over which each method can achieve successful starting is determined by combustor
light off capability, compressor stall, and hang boundaries, as described below. For both
regimes as flight Mach number is increased start time reduces, and the HP compressor
working line is lower.

Starter assist

The left-hand side of the starter assist envelope is largely defined by low combustor mass flow
and loading, which lie outside the ignition envelope. To be at an acceptable level of fuel air
ratio, fuel flow must be so low that even if ignition is achieved the engine will hang. After light
off rotating stall is also a concern as at the low Mach number the windmill assistance
and speeds are low, meaning that high fuel air ratio is required to accelerate the engine to idle.
This and the unchoked nozzles resulting from low Mach number, give a high HP compressor
working line. The right-hand side of the starter assist envelope is dictated by the combustor
Mach number exceeding that at which allows light off.

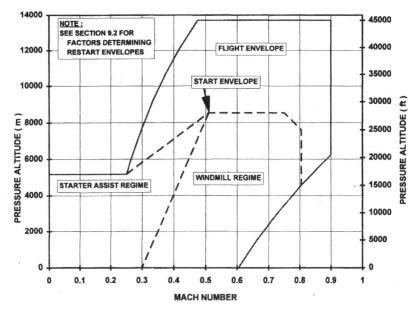

Fig. 9.4   Manned aero thrust engine restart envelopes.

Windmill

In the windmill envelope the starter is not used. Beyond the starter assist envelope reversion to free windmilling allows combustor light off by reducing air flow and preventing excessive combustor Mach numbers. The left-hand boundary is in fact limited by low rotational speed, which causes high temperatures or even low fuel pressure from the driven pump. The right-hand boundary of the windmill start envelope is dictated by excessive combustor mass flow and loading, as for the starter assist envelope. The variation of combustor inlet mass flow, temperature and pressure under free windmilling conditions is described in Chapter 10.

9.2.7   *Aircraft turboprops and turboshafts*

The start process is similar to power generation engines, but with no sub synchronous idle. The propeller is usually set to a fine pitch (i.e. blades point circumferentially) to reduce thrust and the aircraft brakes used to prevent forward motion. The alternative of the propeller being feathered (blades point axially) is usually avoided as excessive torque results on the power turbine output shaft.

9.2.8   *Subsonic missile, drone or RPV*

Here applications use only either ground or in-flight starts, and no restarting. Ground starting is generally as per manned aircraft, though as described in section 9.8 more novel starter systems may be used.

In-flight starting is necessarily fast and often utilises an aerodynamically profiled blank over the engine intake to ease combustor ignition and light around. As the vehicle is launched the ignitors are activated, the light off fuel flow flat is metered, and the blank is blown off simultaneously. Engine start testing must use an altitude test facility to simulate the launch ram effect.

9.2.9   *Ramjets*

Ramjets must be started in flight. As described in Chapter 7, the strong dependence of ramjet performance on flight Mach number makes vehicle acceleration a crucial part of the engine start process. A supersonic intake may require its own *starting,* for example via acceleration beyond the design flight Mach number, to establish the internal shock pattern. Combustor Mach numbers are relatively high in a ramjet, dictated by size constraints, hence a sheltered pilot zone is required to ensure stability. During starting pilot fuel must be admitted and then lit, and once ignition has been detected main fuelling commenced. Ignition is quick, around 0.25 seconds from fuel on to maximum thrust.

9.3   Engine start requirements

Each application has start time requirements at ISO, or ISA sea level static, and throughout the operational envelope. Start times are stated from the instant the operator or pilot selects a start, to a speed of either 2% lower than that for idle or to full power. Achievement of fast start times is restricted by the operability issues discussed in section 9.5, and by the detrimental impact of high thermal stresses on cyclic life. Indeed to allow thermal stabilisation engines are often held at idle for a significant period following a start.

9.3.1   *Power generation*

Typical start time requirements at ISO for the major categories of power generation engines are as below:

- Standby/emergency generators: 10–30 seconds
- Peak loppers: 2 minutes
- Mid merit and base load aero-derivatives: 10–15 minutes
- Heavyweight simple cycle base load: 30 minutes–1 hour
- Heavyweight combined cycle: 1–4 hours

These time requirements generally increase with engine size. The relatively long start times permissible for heavyweight engines in simple cycle allow a low cost heavy construction, such as solid rotors or large thick discs and thick casings. Rapid starts would induce prohibitively high transient thermal stresses in such components. For combined cycle the achievable times are even longer due to the thermal response of the steam plant. The above times are for starts to full power, and include a dwell at idle for thermal soakage as discussed in Chapter 8. A longer dwell is utilised on colder days, and is a major reason for the wide ranges quoted. Purging times to remove any unburnt gas fuel are additional.

Usually the start envelope is the same as the environmental envelope as per section 2.1.6. A *black start* capability is often required, where the gas turbine must be capable of being started at the press of a button with no external electrical power supply to the building in which it is housed. This requires solutions such as battery power.

At sea level start times up to 25% longer than that at ISO are usually acceptable for hot or cold days. In addition the start time is commonly inversely proportional to ambient pressure, hence at 3000 m and 15 8C the allowable start time will be 1.45 times that at sea level.

9.3.2   *Gas and oil pumping*

Depending on ambient temperature, start times of 15–30 minutes are usual; such relatively long times being to ease thermal stresses in the power turbine, which is normally of 'heavyweight' construction. In addition, for emergencies a rapid start capability of 4 minutes plus purge time is required. Again the engine must be able to start anywhere in the environmental envelope as per section 2.1.6. The normal start time may vary with ambient conditions as per a power generation engine.

9.3.3   *Automotive gas turbines*

At ISO start times of 15 seconds are desirable, however up to 30 seconds may be acceptable for trucks. While this is longer than for piston engines it is acceptable to most customers. Again the engine must be able to start anywhere in the environmental envelope as described in section 2.1.7. Start time may vary with ambient conditions as per a power generation engine.

9.3.4   *Marine engines*

For most marine engine installations a start time of less than 90 seconds throughout the operational envelope is mandatory. Engines are often started in harbour, hence the more extreme environmental envelope described in section 2.1.8 should be considered.

9.3.5   *Manned aircraft thrust engines*

Civil airworthiness requirements, such as the American Federal Airworthiness Regulations, do not state required start times. The requirement is that the 'engine must be capable of ground starting at all ambient temperatures and corresponding oil temperatures, and anywhere in the restart envelope declared by the manufacturer without damage to the engine'. Civil airframe manufacturers usually demand a start time at ISA sea level static of less than 1 minute, and less than 2 minutes anywhere else in the ground start envelope. Ground starting is usually required up to 5000 m. The corresponding environmental envelope is as per section 2.1.9.

Restarting in flight is a very important engine capability for all aircraft, as occasionally engines do *flame out*. In one famous incident all four engines on a Boeing 747 flamed out in a volcanic cloud over Indonesia. They were successfully restarted and the aircraft landed safely.

Figure 9.4 presents a typical restart envelope required for a turbofan powered civil aircraft. Again, the corresponding environmental envelope is as per section 2.1.9. The restart envelope is smaller than the flight envelope, as discussed in section 9.2.6 combustor relight is not possible above the Mach number levels shown.

For a military fighter start times of around 10–20 seconds are required to idle, with at least a minute normally spent warming the engine carcass before selecting full thrust. The restart envelope is similar to civil aircraft, any thought of restart at high supersonic Mach number being irrelevant as after an engine flame out the aircraft would rapidly decelerate.

9.3.6 *Manned aircraft turboprops and turboshafts*

As for civil thrust engines a start time of less than 1 minute is usually required for ground starting turboprops. The restart envelope is usually declared as per the flight envelope, but due to reduced windmilling the starter assist portion extends to higher flight Mach numbers than for a turbofan. For restarting, times of up to 90 seconds are usually acceptable.

Typically for helicopter engines, both start and restart times of less than 1 minute are required throughout the operational envelope. The restart envelope also corresponds to the flight envelope, but the starter is always employed as only negligible windmilling assistance is available.

9.3.7 *Subsonic missile, drone or RPV*

Many RPVs have ground start capability, as distinct from *ground launch* discussed below. Ground start times of typically less than 1 minute are required at ISA sea level static. Ground start capability is required throughout the full environmental envelope as per section 2.1.9, with start times of up to twice that at ISA sea level static usually being acceptable for hot and cold days. A ground started RPV would almost never have in-flight restart capability.

In flight starting may be required, after either a rocket assisted ground launch or release from a parent manned aircraft. Start time requirements of less than 15 seconds are usual, to provide thrust quickly to maintain flight. The specific start design issues related to in flight starting are described in section 9.2.8. Again, in flight restarting would not normally be required.

9.3.8 *Supersonic missile using a ramjet*

A missile propelled by a ramjet must be launched using a rocket to develop sufficient ram pressure for the combustor to be lit. The start time is mainly that of missile acceleration. If launched from an aircraft such a missile must commence operation at up to 20 000 m. The full environmental envelope must be considered as per section 2.1.9. In flight restarting is not practical, as high Mach number must be maintained to allow ramjet operation.

9.4   The impact of ambient temperature and pressure

9.4.1 *Cold days*

The impact of a cold day on the start process is complex and it is a key design point for the combustor, starter sizing and compressor operability. In practice cold day start time ranges from being comparable to that at ISO to up to twice as long. The individual effects are discussed below.

For the various landbased shaft power engines the off engine fuel, oil and starter system components are often housed within spaces maintained at or above around 0 8C. This is either machinery space outside the engine enclosure, or the enclosure itself.

Oil viscosity

As described in Chapter 13, oil viscosity rises approximately exponentially as temperature is decreased. As per Formulae F5.17.1–F5.17.4, bearing resistance increases with oil viscosity and hence engine resistance rises dramatically on a cold day. It is rare that oil is heated prior to a start, though one exception is for air launched RPV engines where the vehicle is carried underneath a parent aircraft. Here the oil can soak down to very low temperature, and due to the very short start time requirements the oil tank and pipes to the bearings may be heated electrically. For manned applications dry cranking is frequently employed before attempting a start to ensure the oil in the bearings is warm enough to give adequate lubrication.

Fuel

Chapter 13 shows that kerosene and diesel fuel viscosity rise exponentially as temperature reduces. Indeed, diesel fuel atomisation is problematic below 5 8C, and 'waxing' occurs around −20 8C, depending on the additives used. This increase in viscosity dramatically reduces the atomisation quality of the fuel, hence reducing light off capability and combustion efficiency. Cold day light off is therefore another key combustor design point and rig testing must be conducted accordingly. For natural gas fuel viscosity is not a concern, however checks should be made with respect to the dew point if the local blend contains significant amounts of water.

Starter

Starter assistance may be affected significantly by cold ambients, depending on the system and where it is mounted. The impact of ambient temperature on each of the major start assistance systems is discussed in section 9.8.

Fuel scheduling

The final effect on engine assistance-resistance is via the fuel scheduling. Owing to low cycle fatigue and thermal shock a limit will be placed on combustor exit temperature for turbine mechanical integrity. This may allow additional fuel to be metered by the control system on a cold day, so long as the compressor rotating stall limit is not reached.

9.4.2   Hot days

The hot day is another key design point for starting. As for cold days the situation is complex and start time in relation to other ambient conditions varies depending upon the complete start system. Oil and fuel viscosity are marginally lower than for an ISO day, and hence the effects discussed for a cold day are not present. However engine assistance is often reduced relative to an ISO day as fuel flow must be lowered to prevent excedance of the combustor exit temperature limit for mechanical integrity. Again, as for a cold day, the starter assistance varies dependent upon the system configuration.

9.4.3   Ambient pressure

As per the parameter groups described in Chapter 4, engine assistance-resistance is proportional to ambient pressure, and hence start time is inversely proportional to it. The impact on the starter system depends heavily on its configuration. Also as described in section 5.8, ignition is more difficult at low ambient pressure and hence the lowest level at which starting is required must be considered in the combustor design and rig test programme.

9.5 Operability issues

Setting the fuel and handling bleed valve schedules to achieve ignition and light around, and overcome hang, while avoiding rotating stall or over temperature is complex. As shown on Fig. 9.3, there may only be a small 'window' between stall and hang in which to set the fuel schedule. These phenomena are discussed in the sections below.

During the design phase all operability issues must be considered. However, due to its nature, starting prediction is necessarily imprecise and hence considerable engine testing is required to ensure starting is free of operability problems throughout the start and restart envelopes.

9.5.1 *Rotating stall*

The phenomena of compressor rotating stall and locked stall are described in section 5.2.6. Locked stall does not normally occur in the starting regime, however rotating stall is a concern. As shown by Fig. 9.3, it is the HP compressor start working line which is most crucial for multi-spool engines; the LP and IP compressor working lines are relatively benign. The LP compressor influences the HP compressor inlet conditions in terms of whirl and pressure level however, and the latter may reduce interstage handling bleed extraction from the HP compressor.

It is common that during the early parts of dry cranking the HP compressor is in rotating stall. This is acceptable because the starter is sized to overcome the increase in overall power absorption due to reduced compressor efficiency. Also the pressures are so low that there is insufficient excitation to cause high cycle fatigue damage. Stall should drop out at very low speed, though there can be appreciable hysteresis between the stall drop in and drop out lines. At and above combustor light off rotating stall is unacceptable. To maintain spool acceleration with the lower compressor efficiency would require excessive turbine entry temperatures.

To address rotating stall during development testing either the fuel schedule must be lowered, or compressor bleed or whirl schedules adjusted. As described in Chapter 5, increasing delivery handling bleeds will lower the working line, while closing VIGVs or VSVs will raise the surge line. Taking increased compressor interstage bleed will do both.

One major effect during a restart is that of heat soakage on compressor surge margins, whether for a single spool engine or an HP compressor. Heat soakage in the combustor or turbine is akin to additional fuel flow, causing faster acceleration and a higher working line. Furthermore, the heat soakage within the front compressor stages reduces the referred speed of the rear stages. This in turn reduces the rear stage flow capacities, driving the front stages towards stall and lowering the overall compressor stall line.

During a cold soaked start compressor tip clearances are increased, and surge lines lowered, by slower thermal growth of the discs than the casings. The tip clearances may take several minutes to stabilise, and this is a further reason for a dwell at idle.

9.5.2 *Combustor ignition and light around*

Combustor ignition envelopes are described in section 5.8. As stressed earlier, ignition is a key design point for the combustor and must be considered during the combustor design and rig testing phases. If on an engine test ignition is not initially satisfactory then ignitor position and combustor geometry can be changed. Failing this combustor inlet conditions can be altered via changing handling bleed valve scheduling, the starter gear ratio or in extreme circumstances the starter itself.

If ignition is satisfactory but light around cannot be achieved it is unlikely that the problem is related to inlet conditions. Usually either a fuel injector or, for cannular systems, 'interconnector' geometry is at fault.

9.5.3   *Hang*

Hang is the condition where spool acceleration is negligible, as crank assistance plus turbine output power only just exceed that required to drive the compressor and overcome mechanical losses. It often occurs because speed is too low, due to insufficient crank or windmill assistance, or to too low a fuel schedule. Usually after a short period of hang the start must be aborted or mechanical damage may occur due to the high levels of turbine temperatures and patternation.

To overcome hang the first step is to increase crank assistance. If this is not possible then the fuel schedule may be increased, however rotating stall, and the turbine entry temperature limit, may restrict this. Any handling bleeds open during a start also increase compressor input power, and hence fuel flow must be further increased to overcome hang. This situation is further complicated by the consideration of cold day, hot day and altitude starting as discussed in section 9.4.

Confusingly the effects of rotating stall – high temperature and low acceleration – are also sometimes referred to as 'hang'. For clarity the distinction between stall and hang is preserved herein.

9.6   Starting and parameter groups

Section 8.8 states that for above idle transient performance the values of two parameter groups, as well as flight Mach number if nozzles are unchoked, must be stipulated to fix all others. During starting the referred starter power must be stipulated in addition. The more pronounced impact of 'real' effects such as heat soakage, the lack of choked nozzles, and a low and varying combustion efficiency further complicate the referred parameter relationships. They nevertheless form the basis of control strategies for starting, as described in section 9.7.

Similar scaling rules apply to starting as discussed in section 8.9 for above idle transient performance. Where an engine is linearly scaled upwards virtually all the start system must be resized. The engine's starter power requirement becomes either that given by the square of the linear scale factor, to achieve the same dead crank blade tip speed, or else that given by the cube of it to achieve the same acceleration times. The latter requirement will be reduced if scaling produces less than the theoretical increase in engine inertia, as discussed in section 4.3.4. The starter input gear ratio should also be addressed, as scaling the starter may produce differing effects on its speed and power characteristics.

9.7   Control strategies during start manoeuvres

The start fuel schedule usually comprises three parts, the *light off flat*, *start accel schedule* and *idle governor*. Figure 9.5 shows the typical resultant fuel flow versus referred HP compressor speed. Fuel scheduling for engine acceleration during starting must be low enough to avoid rotating stall and over temperature, but high enough to avoid hang, while achieving the desired start time. Scheduling for deceleration is not necessary below idle as on shut down from idle the fuel cock is simply closed.

The light off flat is a constant fuel energy flow with time, initiated at a speed up to 10% above that of rotating stall drop out. For starter assisted in flight restarts fuel metering may have to begin at a lower HP speed, as rapid spool acceleration may mean the ignition window is missed. As described in section 9.1.4, fuel flow is metered during the ignition phase to achieve ignition and then combustor light around. Its magnitude will be determined during the combustor rig test and is typically such that the primary zone equivalence ratio is around 0.35–0.75. As shown by Fig. 9.5, the value is usually modulated with ambient temperature and pressure such that its absolute magnitude is increased on a cold day and reduced at altitude. On a −40 8C day fuel flow may be 5–10% higher than for an ISO day.

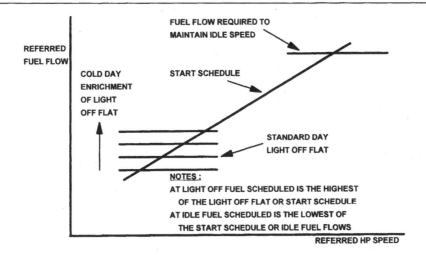

Fig. 9.5 Referred fuel flow versus referred speed start fuel schedule.

Light around is detected via exhaust thermocouples. After this, referred fuel flow is usually scheduled versus referred speed. An alternative is 'NHdot' control, where fuel flow is modulated to maintain a rate of change of HP spool speed versus referred speed, as per above idle transient performance discussed in Chapter 8. The fundamental comparisons between these techniques are all as per section 8.10. The only difference in the start regime is that the transient performance phenomena discussed in sections 8.4 and 9.6 have a more pronounced effect. Significant engine testing is required to adjust empirically predicted theoretical schedules.

As the engine approaches the idle condition the idle governor overrides the accel schedule and the steady state idle condition is achieved. The various idle ratings are discussed in Chapter 7.

As mentioned, VIGVs are usually fully closed in the start regime to give the highest rotating stall line, while bleed valves are usually open to give the lowest working line and for interstage bleeds also to raise the rotating stall line. Sequencing of ignitors and the starter motor by the control system are discussed in section 9.1.

The most common control strategy for a hot restart is to dry crank the engine until the exhaust thermocouples indicate that the level of heat soakage is acceptable. Another less robust strategy is to reduce the fuel schedule using the exhaust thermocouple reading. For gas fuelled industrial engines the purge cycle provides sufficient cooling.

9.8   Starter system variants and selection

Many different starter systems are utilised for the range of gas turbine applications. It is vital to consider the whole start system during design and engine testing since assistance is a function of both the primary start device and its power source. For instance, if the primary start device is a DC electric motor, then the performance of the battery stack which drives it must also be considered throughout the start envelope. For an air turbine the off design performance of the APU powering it, and the pressure loss of the supply ducting, must be considered. The performance of start systems is complex, and hence only the fundamental issues of those most commonly used are discussed here. References are provided for readers who wish to pursue further detail.

9.8.1 *Air turbine*

Air turbines are the most common starters in large aircraft and marine applications because high pressure air is usually available, and they have the lowest system weight and size. In addition, engines pumping natural gas have traditionally used a turbine for start assistance, but driven using high pressure natural gas tapped off the pipeline and exhausted up a tall stack.

As shown in Fig. 9.6, the air turbine module comprises an inlet valve, intake, usually a single stage axial flow turbine and an exhaust. The turbine cranks the engine HP spool via the starter's own epicyclic reduction gear, a centrifugal clutch, the engine accessory gearbox, a radial quill shaft and finally a bevel gear.

For aircraft engines the compressed air is supplied to the turbine from either an on-board APU, a *ground start cart* or by *cross bleed* from another engine. A three way valve and ducting enable all to be available for a given installation. The ground start cart is usually a diesel engine driving a load compressor, or a gas turbine, mounted on a mobile platform. Cross bleed from another engine or the APU must be employed for in flight re-starting. Reference 1 covers types of APUs and their performance characteristics throughout their operational envelope. For marine engines air may be provided by either a high pressure 'accumulator' tank or bottles, a diesel or gas turbine auxiliary engine, or cross bleed from another main engine.

At ISA SLS typical air turbine inlet conditions using an APU are around 480 K, and 170–270 kPa. As shown by Reference 1, available pressure and mass flow fall as ambient temperature is increased. At 40 8C starter assistance is approximately 60% of that at −40 8C. For an accumulator the air turbine inlet temperature is usually as per ambient conditions, and the pressure is constant at all ambient temperatures, hence less starter assistance is achieved on a cold day. The system must be evaluated at all corner points of the start envelope, however sea level cold day is often the primary system design point.

9.8.2 *Batteries and electric motor*

This type of system is preferred for APUs, ground start RPVs, small turboprops, engines for helicopters, automotive engines, and power generation applications. For the last where a black start capability is required the engine must be started at regular intervals, even if output power is not required, to charge the batteries.

Figure 9.7 shows the major components of an electric motor start system. This comprises batteries and an electric motor driving the HP spool via a clutch, gearbox and bevel gear. Reference 2 describes this system in detail, including its performance at all ambient conditions.

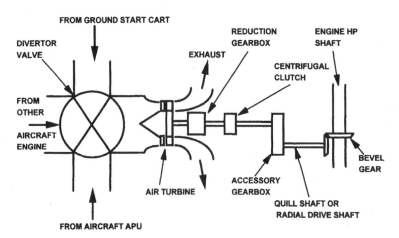

Fig. 9.6 Major components of air turbine start system.

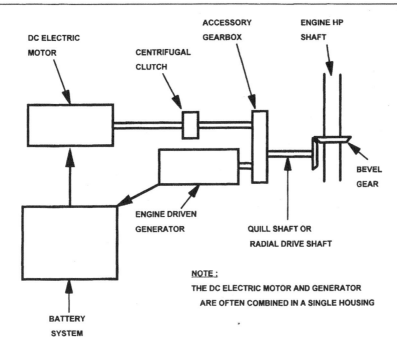

DC ELECTRIC MOTOR

CENTRIFUGAL CLUTCH

ACCESSORY GEARBOX

ENGINE HP SHAFT

BEVEL GEAR

ENGINE DRIVEN GENERATOR

QUILL SHAFT OR RADIAL DRIVE SHAFT

BATTERY SYSTEM

NOTE :
THE DC ELECTRIC MOTOR AND GENERATOR
ARE OFTEN COMBINED IN A SINGLE HOUSING

Note:
The DC electric motor and generator are often combined in a single housing.

Fig. 9.7   Major components of electric start system.

Lead acid batteries of 12 or 24 V combined with a series wound DC motor are almost exclusively used, the motor often being also a DC generator for above idle operation. The DC motor has the most suitable torque characteristics as well as low weight and volume. Its power and torque output are approximately proportional to volume.

Figure 9.8 illustrates battery performance, and shows how initial output power decreases significantly at low temperature. Reference 2 also shows how it deteriorates with time, output may reduce by 50% in 5 minutes if maximum power is extracted, whereas it may only decay by 10% in 30 minutes if lightly loaded. Furthermore, battery output degrades with storage time and cyclic use. These considerations emphasise how all elements must be considered to design a complete 'start system' which will meet the application's starting requirements.

A further use of electric motors is that of the AC variety for some industrial engines. Here, a fluid coupling or *torque converter* is often employed to match the motor's torque/speed characteristic to the cranking load.

9.8.3   *Hydraulic motor*

Hydraulic start systems are common for power generation applications. An electric motor powered by batteries, or another engine, drives a hydraulic pump which provides high pressure hydraulic fluid to the starter to crank the engine. This system is particularly suited to where there is more than one engine in an installation, as one electric motor and one hydraulic pump can supply hydraulic fluid to all the engines.

The hydraulic system is being considered for more gas and oil applications as exhausting substantial amounts of natural gas from a turbine starter (see section 9.8.1) is now unacceptable for environmental reasons in many parts of the world. However such systems do require an adequate power supply to be available for the electric motor.

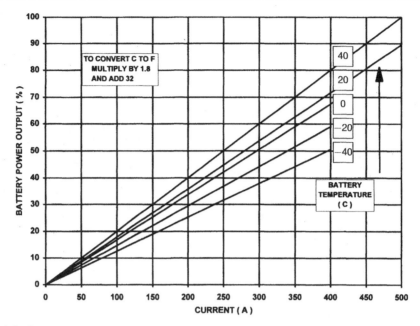

Fig. 9.8   Battery power output versus current and temperature.

9.8.4   *Impingement*

Pyrotechnic cartridges are used almost exclusively for 'expendable' RPVs such as missiles or decoys which are started in flight and need no re-start capability. This is the lowest weight, and most economical system. High pressure accumulator air impingement is often employed for emergency generating sets that may have a start time requirement as low as 10 seconds. Here a more conventional start system would be extremely large and expensive to achieve the required start time.

As detailed in Reference 3, this type of start system employs 2–4 stationary nozzles in the turbine casing, set at almost 908 to either the HP turbine blading, or occasionally to additional slots on the disc face. Once the start is initiated either the pyrotechnic cartridge is detonated or the accumulator valve is opened. Impingement of high pressure gas or air discharging from the nozzles provides crank assistance. As shown by Reference 3, impingement turbine efficiency is only of the order of 20–40%, improving as the number of impingement nozzles is increased.

9.8.5   *Starter sizing*

Sizing a starter system is difficult due to the imprecision in quantifying the many contributory phenomena, and is normally only done during the detailed design phase of an engine programme. One of the start modelling techniques described in section 9.9 will be employed, the level of complexity being dependent mainly upon available resource. Hot day, cold day and altitude start requirements should be considered, as well as those at ISO and any restart requirements. Owing to the high level of imprecision a reasonable degree of pessimism is used in the analysis, to ensure that the system selected is capable of starting the engine. During the engine development programme the starter size, gear ratio, or its power supply may be refined to ensure satisfactory starting throughout the start envelope.

To illustrate the order of magnitude for starter power output some indicative levels are listed below:

200 daN RPV turbojet	1.5 kW	25 MW marine engine	100 kW
200 kW APU	4 kW	45 000 daN turbofan engine	250 kW
1.2 MW tank engine	8 kW		

9.9   Start and control models

It is extremely difficult to produce sophisticated models like those for operation above idle described in Chapter 8. Operation is on uncertain, steep regions of component maps, and achieving numerical convergence is difficult. Accordingly the simpler *extrapolation model* type is more commonly used.

9.9.1   *Extrapolation model*

This is the simplest model type to build and provides a good tool to size the starter motor such that start time requirements can be achieved. Sample calculation C9.1 describes the technique, which is ideally suited to 'spreadsheet' analysis packages. This method is particularly effective for single spool engines. It can also be used for multi-spool systems with the calculation only performed on the cranked HP spool, and hence accuracy degraded. Data for each of the power elements described below must be derived and tabulated versus HP speed. Then, by applying Formulae F9.1 and F9.4 at discrete time intervals, speed can be integrated versus time from zero to idle. The analysis routine may be repeatedly run with different starter motor power inputs and gear ratios until the desired start time is achieved. The complete process must be repeated for each ambient condition for which starting is to be assessed.

Compressor input and turbine output powers
The steady state matching model (per Chapter 7) is run at a series of points above idle, and to as low a power or thrust as it will converge in the sub idle regime; bleed valves, VIGVs, etc. are set as they would be in the start regime. To simulate the unbalanced power causing acceleration a power offtake is applied at each point such that the compressor surge margin is at the minimum allowable, or the engine is at its maximum allowable temperature. From these runs compressor input power is tabulated and then extrapolated following a cube law versus speed (Formula F9.5) to zero input at zero speed. Turbine power is also tabulated and then extrapolated as a cube law but only to the point of light off. This point must be estimated and will be between 15 and 25% HP speed. Below light off turbine output power is greatly reduced as it is proportional to inlet temperature; typically output power at the top of crank is 25% of its lit value at the same engine speed. From this unlit, top of crank point turbine output power may again be extrapolated following a cube law versus speed through zero output power at zero speed as per Formula F9.5.

Bearing and windage losses
These are calculated using Formulae F5.17.1–F5.17.4 and F5.17.6. If the engine is of single shaft configuration directly driving a generator, then this too will have losses which should be obtained from the manufacturer.

Oil and fuel pump power requirements
At ISO these power requirements are relatively small, however on cold days if there is no heating of the oil or fuel prior to start then they become significant. Again, a cube law (Formula F9.5) may be used to calculate the pumping power at and below idle; pump bearing frictional losses are ignored as they are relatively small.

Gearbox
If the engine has a directly driven gearbox then its efficiency, as defined by Formula F5.17.10, may be evaluated from Formula F5.18.1 throughout the start regime.

Start system

As described above, the start system power output is also tabulated versus speed; the full calculation process is repeated with a scale factor on it adjusted until the required start time is achieved. The output power versus speed characteristic will be available from start system manufacturers. It is important not to consider just the starter motor characteristic but also that of the device powering it such as a battery or APU.

System polar moment of inertia

The polar moment of inertia of all items geared to the engine HP spool should be *referred* to it before summation, using Formula F9.6 which accounts for the effect of gear ratio on torque and speed. In most cases that of the turbomachinery overshadows all other devices. One exception is for a single spool engine *directly* driving a generator, the inertia of which is simply added to that of the turbomachinery.

9.9.2   Thermodynamic matching model

The start thermodynamic matching and control model is an extension of the above idle model described in section 8.11.1. However it is challenging to produce an accurate or reliable model as the impact of real effects such as heat soakage is more pronounced and difficult to predict, component maps are generally inaccurate, and combustion efficiency may be significantly less than 100%. However, despite these difficulties, start thermodynamic transient matching models can be built and provide a valuable design tool for understanding the start and control mechanism for a given engine. This is particularly true for aero-engines where restarting at a variety of flight conditions is required.

The main additional features relative to the above idle model are the incorporation of starter power input, component maps loaded based on different variables as described in Chapter 5, and a relaxation of the matching tolerances. By this means investigation of engine behaviour well below light off is feasible.

9.9.3   Real time models

As per above, idle models described in section 8.11 real time engine start models are also required for testing control system hardware and for flight simulators. Owing to the difficulties of accurate modelling in the start regime these models usually comprise tabulations of key parameters versus time during a start at a range of ambient or flight conditions.

Formulae

F9.1   Unbalanced power (kW) = fn(compressor/turbine mass flow (kg/s), temperature change (K), CP (kJ/kg K), starter assistance power (kW), power offtakes (kW))

$$DPW = W4 * CP * (T4 - T5) - W2 * CP * (T3 - T2) + PWstarter - PWpar$$

(i)   Engine unbalanced power is termed *engine assistance-resistance*.
(ii)   Power offtakes are bearings, windage, accessories, etc.

F9.2 and F9.3   Fundamental mechanics equations relating: torque (Nm), polar moment of inertia (kg m^2), shaft acceleration (rad/s^2), power (W), shaft speed (rad/s)

F8.2 $TRQ = XJ * NDOT$
F8.3 $PW = TRQ * N$

F9.4 Spool acceleration rate (rpm/s) = fn(unbalanced power (W), polar moment of inertia (kg m^2), speed (rpm))

$$NDOT = DPW/(XJ * N * (PI * 2/60)^\wedge 2)$$

F9.5 Cube law: component power absorption (W) = fn(datum power (W), datum speed (rpm), speed (rpm))

$$PW = PWdatum * (N/Ndatum)^\wedge 3$$

(i) The datum condition for compressors and turbines is discussed in section 9.9.1.
(ii) That for accessories can be any known operating point, e.g. full power or engine idle.

F9.6 Referred inertia for gear ratio (kg m^2) = fn(inertia (kg m^2), gear ratio)

$$XJR = XJ * GQ^\wedge 2$$

(i) Gear ratio, GQ, is the ratio of the driven speed (e.g. accessories) to the driving speed (e.g. HP spool).

Sample calculations

C9.1 Calculate the start time for a 200 daN single spool RPV turbojet. The compressor input power and turbine output power have been extrapolated through the sub idle regime for the prevailing ISO ambient conditions as per section 9.9.1. The values are tabulated below along with the starter motor assistance curve from the manufacturer. The oil is at ambient temperature of 15 8C and its dynamic viscosity is 0.3 kg/m s. The full power fuel and oil pump power requirements are 50 and 500 W respectively, and the ball and roller bearing pitch diameters are both 45 mm. Windage acts on one disc face of diameter 0.3 m – consider air to be at ambient conditions with density = 1.225 kg/m^3. Assume an air temperature of 288.15 K for viscosity.

Oil viscosity = 0.3 kg/s m
Polar moment of inertia = 0.005 kg m^2
Light off is at 10% speed.
Oil flow = 0.001 gallons/minute
Starter cut out = 40% speed
100% rotational speed = 40 000 rpm
Idle = 40% speed

Compressor		Turbine		Starter assistance	
N (%)	Power (kW)	N (%)	Power (kW)	N (%)	Power (kW)
0	0			0	0
5	0.065			5	0.5
10	0.52	10	0.68	10	0.9
15	1.755	15	2.295	15	1.25
20	4.16	20	5.44	20	1.5
25	8.125	25	10.625	25	1.25
30	14.04	30	18.36	30	0.8
35	22.295	35	29.155	35	0.4
40	33.28	40	43.52	40	0.1
45	47.385	45	61.965	45	
50	65	50	85	50	

F9.1   Unbalanced power (kW)

$$DPW = W4 * CP * (T4 - T5) - W2 * CP * (T3 - T2)/ETAM + PWstarter$$

F9.5   Cube law: component power absorption (kW)

$$PW = PWdatum * (N/Ndatum)^3$$

F5.17.1   Ball bearing power loss (kW)

$$PW = (8.45E\text{-}03 * D^{3.95} * N^{1.75} * VISoil^{0.4} \\ + 1.358E\text{-}03 * N * D * Qoil)$$

F5.17.2   Roller bearing power loss (kW)

$$PW = (3.6E\text{-}03 * D^{3.95} * N^{1.75} * VISoil^{0.4} \\ + 6.613E\text{-}04 * N * D * Qoil)$$

F5.17.6   Disc windage power (kW)

$$PW = 4.28E\text{-}12 * RHO * D^2 * Urim^3$$

(i) *Evaluate engine assistance-resistance*

Rotational speed (rpm)	Compressor (kW)	Turbine (kW)	Roller bearing (kW)	Ball bearing (kW)	Windage (kW)	Oil pump (kW)	Fuel pump (kW)	Assistance-resistance (kW)
	F9.5	F9.5	F5.17.2	F5.17.1	F5.17.6	F9.5	F9.5	F9.1
0	0		0	0	0	0	0	0
2 000	0.065		0.0028413	0.0067812	0.0003149	6.25E-06	0.0000625	−0.0750061
4 000	0.52		0.0094827	0.0224822	0.0024892	0.00005	0.0005	−0.5550041
4 001	0.52	0.68	0.0094868	0.0224919	0.0024911	0.0001688	0.0016875	0.1236739
6 000	1.755	2.295	0.0191949	0.045391	0.0083445	0.0001688	0.0016875	0.4652133
8 000	4.16	5.44	0.0316592	0.0747596	0.019687	0.0004	0.004	1.1494941
10 000	8.125	10.625	0.0466738	0.1101142	0.0383137	0.0007813	0.0078125	2.2963046
12 000	14.04	18.36	0.0640933	0.1511138	0.0660149	0.00135	0.0135	4.023928
14 000	22.295	29.155	0.0838056	0.1974952	0.104576	0.0021438	0.0214375	6.450419
16 000	33.28	43.52	0.1057203	0.2490459	0.1557779	0.0032	0.032	9.6942559
18 000	47.385	61.965	0.1297619	0.3055889	0.221398	0.0045563	0.0455625	13.873132
20 000	65	85	0.155866	0.3669732	0.3032106	0.00625	0.0625	19.1052

(ii)   Evaluate unbalanced torque versus rotational speed

Rotational speed (rpm)	assistance-resistance (kW)	Starter assistance (kW)	Unbalanced power (kW)	Unbalanced torque (N m)
0	0	0	0	2.5
2 000	−0.0750061	0.5	0.4249939	2.0302255
4 000	−0.5550041	0.9	0.3449959	0.8240348
4 001	0.1236739	0.9	1.0236739	2.4444699
6 000	0.4652133	1.25	1.7152133	2.7312314
8 000	1.1494941	1.5	2.6494941	3.1642048
10 000	2.2963046	1.25	3.5463046	3.3881891
12 000	4.023928	0.8	4.823928	3.840707
14 000	6.4505419	0.4	6.8505419	4.6750741
16 000	9.6942559	0.1	9.7942559	5.8484808
18 000	13.873132		13.873132	7.3636584
20 000	19.1052		19.1052	9.126688

Note: Torque at time zero has been taken to be the starter torque extrapolated back to time zero.

(iii)   Integrate speed versus time

Time (s)	Unbalanced torque (N m)	Accel rate (rpm/s)	Rotational speed (rpm)
0	2.5	500	0
2	2.5	500	1 000
4	2.3571668	471.43337	2 000
6	2.0302337	406.04674	2 942.867
8	1.5533562	310.67123	3 754.96
10	1.0114315	202.28629	4 376.303
12	2.6523386	530.46771	4 780.875
14	2.635177	527.0354	5 841.811
16	2.6358846	527.17693	6 895.881
18	2.702131	540.42621	7 950.235
20	2.8337522	566.75043	9 031.088
22	3.0365319	607.30637	10 164.59
24	3.3229821	443.06429	11 379.2
26	3.7137807	495.17076	12 265.33
28	4.0536191	540.48255	13 255.67
30	4.4880596	598.40795	14 336.64
32	5.0280815	670.41087	15 533.45
34	5.7061263	760.81685	·16 874.27

Note: Curve fits of torque versus speed for pre and post light off were used.

References

1.   C. Rodgers (1983) *The Performance of Single Shaft Gas Turbine Load Compressor Auxiliary Power Units*, AIAA, New York.
2.   L. A. Fizer (1984) *Electric Direct Current Starter Motor for Gas Turbine Engines*, SAE Technical Paper 841569, SAE, Warrandale, Pennsylvania.
3.   C. Rodgers (1984) *Fast Start System for a 200 kW Gas Turbine Generator Set*, SAE Technical Paper 841568, SAE, Warrandale, Pennsylvania.

Chapter 10
Windmilling

10.0 Introduction

Windmilling occurs when air flowing through an unlit engine causes spool rotation. This phenomenon applies mostly to aircraft engines, where it is caused by ram pressure. Examples include when an engine has flamed out during flight, or an unmanned air launched vehicle is being carried by a parent aircraft prior to launch. The direction of rotation is the same as for normal operation. Under certain conditions windmilling also occurs for landbased and marine engines. *Free windmilling* is where all the engine spools are free to rotate. *Locked rotor windmilling* is where the HP spool is mechanically prevented from rotating.

A knowledge of key performance parameters during windmilling is essential:

- To ensure successful light or relight, appropriate combustor design requires knowledge of combustor entry pressure, temperature and mass flow.
- The aircraft systems designers must know how much power may be extracted from the engine, if any.
- To understand bearing lubrication requirements, and how engine auxiliaries will perform, rotational speeds are required.
- The aircraft designers must know the engine drag (i.e. negative thrust) during windmilling. Drag is caused by air slowing down as it passes through the engine.

This chapter describes the fundamentals of the windmilling process. Data are provided allowing first cut estimates of the magnitudes of the above parameters. However it must be stressed that the parameter values will be heavily reliant upon the particular engine design in question, and so computer model predictions or preferably test data should be obtained at the earliest opportunity. This is particularly true for more complex configurations such as multi-spool turbofans.

10.1 Turbojet windmilling

10.1.1 *The turbojet windmilling process*

Figure 10.1 shows schematically how the usual non-dimensional relationships can be extended to the windmill regime. Referred fuel flow and mass flow are shown versus referred speed for a high and low flight Mach number; a locus of windmill points is apparent at the windmill condition of zero fuel flow. Mass flow increases with flight Mach number, as the higher the Mach number the higher the referred speed. Operation on the lower parts of the curves may not be practical since the combustor is likely to weak extinct in this regime, causing the engine to decelerate to the windmill point. Other referred parameter groups may be plotted in this manner. Finally, Fig. 10.1 shows the locus of operating points for free and locked rotor windmilling on the low speed compressor map.

Charts 10.1 and 10.2 present pressure and temperature ratios at key stations through a turbojet while windmilling. The compressor behaviour depends on flight Mach number:

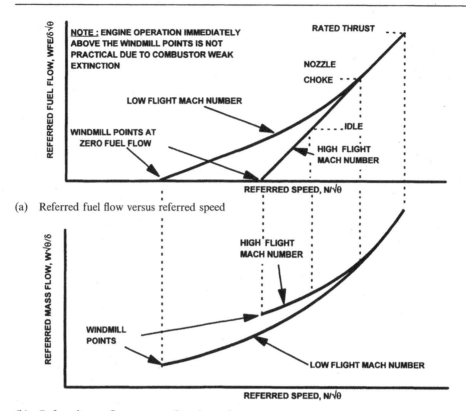

(a)   Referred fuel flow versus referred speed

(b)   Referred mass flow versus referred speed

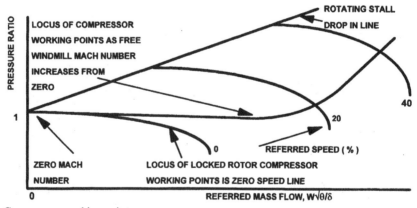

(c)   Compressor working points

Fig. 10.1   Turbojet windmilling: referred parameter group relationships.

- As flight Mach number is increased from zero, compressor pressure ratio initially falls from the value of one. The temperature ratio is greater than one however, and hence the compressor is operating as a *stirrer* or *paddle*. This mode of operation is described in more detail in Chapter 5. For a multi-stage axial flow compressor some front stages may actually perform as a turbine but overall the machine has a nett work input.
- As Mach number increases further, pressure ratio increases and eventually exceeds one. This is particularly true for high design pressure ratios and multiple axial stages. Here the

compressor overall, and certainly the back stages, function in the true compressor fashion. Typically the higher the design pressure ratio, the higher the windmill pressure ratio at a given flight Mach number. For a given design pressure ratio centrifugal compressors tend to have a lower pressure ratio at a flight Mach number, due to the absence of back stages able to operate normally.

During steady state windmilling the compressor always absorbs power overall. Otherwise the shaft would accelerate as there is no other significant power absorption mechanism, bearing and windage losses being small. The turbine is able to supply the compressor input power because there is an expansion ratio remaining after all pressure drops due to ducts, combustor and possibly compressor(s) have been deducted from the ram pressure ratio. The combustor, intake, exhaust and other ducts impose pressure losses with no change to total temperature.

The expansion ratio across the propelling nozzle is small and is typically independent of turbojet design pressure ratio. Total temperature at the nozzle is slightly lower than the ram temperature. This small drop is due to any work extracted by the bearings and engine auxiliaries, the effects of compressor and turbine work otherwise cancelling each other.

Reference 1 discusses the turbojet windmilling process further. Sample calculation C10.1 shows how the information presented herein may be employed to estimate key parameters for a windmilling turbojet.

10.1.2   *Free windmilling combustor inlet conditions*

Estimates of combustor inlet pressure and temperature may be taken from Charts 10.1 and 10.2 described above. Combustor entry mass flow increases with the windmill flight Mach number, and may be estimated either via Chart 10.3 or using the ESDU curves reproduced in Chart 10.4. Chart 10.3 includes typical curves of referred mass flow versus flight Mach number for high and low design point values of flow per unit area. Values from Chart 10.3 should be factored by the ratio of combustor entry to engine inlet mass flows at the design point. The method for using the ESDU curves, which are from Reference 2, is described in section 10.2.2.

10.1.3   *Free windmilling drag*

The drag of air passing through the engine is termed *internal drag*, and reflects the momentum reduction between free stream and nozzle exit. Since the propelling nozzle is not choked the standard relationship for thrust may be simplified for windmilling as shown by Formula F6.4. Chart 10.3 also shows typical curves of referred internal drag versus flight Mach number for high and low design point values of flow per unit area. Again the area used is that at the first compressor face including any nose bullet. Internal drag is simply approximately proportional to engine mass flow, because as shown earlier nozzle expansion ratio and total temperature at a given flight Mach number are approximately equal for all turbojet designs, and hence also jet velocity.

Internal drag may also be estimated using ESDU curves reproduced in Chart 10.5. The method for using these curves is described in section 10.2.3.

In addition to internal drag a *spillage drag* occurs as air which does not enter the engine is accelerated by the static pressure field around the nacelle. Spillage drag increases gradually with Mach number. For a turbojet spillage drag is typically 10–20% of the internal drag. References 3 and 4 describe this further.

10.1.4   *Free windmilling rotational speeds*

Chart 10.6 shows typical levels of referred rotational speeds for single and two spool turbojets versus flight Mach number. In general engines with centrifugal compressors rotate at lower speeds than those with axial turbomachinery, as described in section 10.1.1 above. For two spool engines the HP spool rotates at higher percentage referred speeds than the LP system, due partly to the lack of the usual temperature rise across the LP compressor.

10.1.5 *Customer power extraction and the locked rotor condition*

Chart 10.7 shows referred power offtake versus referred rotational speed for lines of constant flight Mach number. The parabolic form of these curves is as per that presented in Chapter 7 for an industrial engine where output speed is being controlled by the speed of the load. When no power is extracted the engine is at its *free windmilling speed*. As power extraction increases, speed falls and a point of maximum power extraction is reached. If the load constrains the engine spool to run at lower speeds, due to it demanding high torque, then the power available decreases. This family of curves may be used to estimate the maximum power extraction available at a given rotational speed. The zero rotational speed case is the locked rotor condition where the level of torque is called the *stall torque*.

Chart 10.7 also presents the effect of power extraction on key windmill parameters. Typical levels of pressure ratio, referred mass flow and drag for the locked rotor condition are presented along with the free windmilling levels. For single spool engines with a locked rotor compressor delivery temperature is equal to the ram temperature since the compressor is acting as a cascade with no work extraction. Engine operation with useful power extraction will lie between the free windmilling and locked rotor levels for all parameters.

Chart 10.7 applies equally to single and two spool turbojets. For the latter customer power extraction is usually from the HP spool due to mechanical arrangement considerations. When the HP spool is locked the LP system typically rotates at 25% of its free windmilling speed.

10.2 Turbofan windmilling

10.2.1 *The turbofan windmilling process*

The bypass duct presents the path of least resistance to the ram pressure at the fan face, hence most flow takes this path. This means that the fan demands high work input and therefore the core must match such that a high expansion ratio is available for the LP turbine. The HP turbine therefore has a low power output, and the HP compressor has a significantly lower pressure ratio than in an equivalent turbojet.

10.2.2 *Combustor inlet conditions (Formula F10.1)*

For turbofans, compressor delivery pressure during windmilling falls progressively as design point bypass ratio is increased. For a turbofan of 5:1 bypass ratio a factor of around 0.6 must be applied to compressor delivery pressure relative to a turbojet of the same design point pressure ratio. To a first order this factor varies linearly with bypass ratio. Correspondingly compressor delivery temperature is lower than for a turbojet.

Engine inlet mass flow may be estimated using the ESDU curves (Reference 2) reproduced in Chart 10.4. These present the ESDU mass flow function (Formula F10.1) versus ISA SLS takeoff specific thrust, for free and locked rotor windmilling. To estimate core flow, bypass ratio in the windmilling operating regime must be known. A turbofan with a design bypass ratio of 5:1 has a windmilling bypass ratio as high as 80:1. For lower design point bypass ratios that in windmill will be proportionally lower. Finally, the same ratio of core inlet flow to combustor inlet flow as the design point should be assumed. As described in Reference 2, these curves are derived from a combination of first principles and empirical data.

10.2.3 *Windmilling drag (Formula F10.2)*

Internal windmilling drag may be estimated using the ESDU curves in Chart 10.5. First a theoretical internal drag coefficient is shown versus ISA SLS takeoff specific thrust, for free and locked rotor windmilling. Chart 10.5 also shows an empirically derived delta, which must be debited from the theoretical drag coefficient, as a function of ISA SLS takeoff specific thrust and flight Mach number. The resultant drag coefficient may then be converted into drag using Formula F10.2. Reference 2 describes a derivation from first principles.

At low bypass ratios percentage spillage drag is similar to that of a turbojet. However as bypass ratio is increased it becomes more significant, internal drag being proportionately lower than for a turbojet, and front face (nacelle) area larger. At a design bypass ratio of 5:1 spillage drag is approximately equal to internal drag. The evaluation of spillage drag is comprehensively described in References 3 and 4.

10.2.4 *Rotational speeds*

Chart 10.8 shows typical levels of LP and HP spool referred speeds versus flight Mach number at a bypass ratio of 5:1. The LP spool is at a significantly higher speed than the HP, due to the large power absorption of the fan itself. This is in marked contrast to Chart 10.6 for a two spool turbojet. Intermediate bypass ratios lie between these two charts.

10.2.5 *Customer power extraction*

For the reasons described in section 10.2.1 the maximum referred power extraction available from the HP spool decreases rapidly as bypass ratio is increased. Indeed at a bypass ratio of 5:1 negligible power extraction is available. As for two spool turbojets it is unusual to have customer power extraction mechanically available from the LP spool.

10.3   Turboprop windmilling

The dominant parameter determining turboprop windmilling behaviour is the propeller pitch. This, and the potential for a range of different propeller efficiencies matched to single spool and free power turbine variants, means that it is not practical to condense turboprop windmilling onto a small number of generic charts. Hence only a general description of the turboprop windmilling process is provided.

10.3.1 *The windmilling process for a single spool turboprop*

Reference 5 provides examples of windmill tests in an altitude facility on two single spool turboprops, with the propeller pitch set for maximum windmill rotational speed. In this configuration the propeller acts as a turbine, dropping pressure and temperature, and hence producing shaft power. The result is that 100% referred rotational speed is achieved at a flight Mach number of less than 0.4, and substantial customer power extraction is available. The compressor pressure ratio is greater than one for all flight Mach numbers as it is driven by both the propeller and turbine. As shown by Fig. 10.2, the compressor working line is lower than the no load line during normal operation, due to zero fuel flow and hence zero combustor temperature rise. For the 100% referred speed case pressure ratio is approximately 25% of its ISO takeoff design point value. This pressure ratio, less combustor and duct pressure losses, is available for expansion across the turbine. Referred air mass flow is approximately twice that of a turbojet at the same flight Mach number.

The drag is predominantly created by the propeller. At high flight speeds the magnitude of the drag would approach that of cruise thrust in normal operation. In actual flight situations such a large drag makes it impractical to operate with the propeller pitch as above. Hence engines are fitted with a 'reverse torque switch' in the gearbox, which senses windmill operation by the change in direction of torque due to the propeller driving the engine (as opposed to vice versa). The control system then *coarsens* the pitch to the *feathered* position where the propeller blades are parallel to the direction of flight, preventing any engine rotation and ensuring drag is minimal.

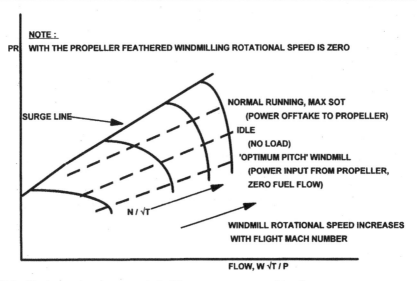

Fig. 10.2   Single spool turboprop windmilling: compressor working lines.

10.3.2   *The windmilling process for a free power turbine turboprop*

In this instance the propeller must be feathered immediately since it will otherwise over-speed the power turbine. There is no connection to the compressor to absorb the output power from the propeller and the power turbine itself. Only a small fraction of the available power is absorbed by auxiliaries, bearings and windage.

10.4   Industrial engine windmilling

Owing to wind conditions around an industrial installation it is possible to have a modest ram pressure on the engine inlet leading to windmilling. This is particularly true for exposed installations such as oil rigs at sea, where the highest ram pressure effect corresponds to around 0.1 Mach number. For free power turbine engines the gas generator may rotate at up to 5% referred speed, hence periodic bearing lubrication is required. The free power turbine will usually not rotate if it is permanently connected to the load. For single spool engines permanently connected to the load the bearing stiction is usually too high to allow wind-milling rotation.

For engines with very long installation exhausts convection occurs after a hot engine shut down. The 'chimney effect' can draw air through the engine via a depressed static pressure at the engine exhaust. This leads to rotation of the gas generator, and of the free power turbine if it is declutched from the load. The effect decays as the exhaust soaks back to ambient temperature. Peak referred rotational speeds are 1–2%. For single spool engines if such windmilling does not occur *barring* of the rotor may be required after shut down, to prevent the rotor hogging due to natural convection making the top of it hotter.

10.5   Marine engine windmilling

Marine engine intakes are usually well sheltered within the superstructure to avoid sea spray ingestion. This means that it is rare to get any ram effect on the front of the engine induced by local weather conditions. As for industrial engines, a chimney effect may occur after a hot engine shut down, as there may be very long installation exhausts passing through a number of deck levels prior to the funnel. Marine engines almost without exception have a free power turbine, and this is often declutched from the ship's gearbox.

The most prevalent cause of marine engine windmilling occurs for twin engined installations with a common gearbox. When one engine is shut down it is declutched, however viscous oil drag in the clutch system can cause the power turbine to rotate at up to 5% referred speed. The induced air flow may then cause the gas generator to rotate at up to 2% referred rotational speed. The speeds of the windmilling engine are a function of the output speed of the driving engine. Lubrication systems must be designed to cater for this mode of operation.

10.6   The effect of ambient conditions

For aero-engines the pressure reduction due to increasing altitude gradually reduces the levels of the referred parameters at a given flight Mach number. For a turbojet a 5000 m increase in altitude may reduce referred speed by approximately 1% at 0.8 Mach number. This is because the frictional load of bearings and engine auxiliary power offtake remain constant with altitude, whereas the available shaft power is proportional to the reducing ambient pressure.

Provided that the oil is maintained at a reasonable temperature level, say above 0 8C, ambient temperature will not have a significant impact on the referred parameter relationships. However if the oil is allowed to soak down to lower ambient temperatures then its increased viscosity will have a dramatic impact upon bearing drag and hence engine windmill performance. Its effect is like that of customer power extraction.

The charts presented in this chapter are nominally for 5000 m, warm oil with no customer power offtake.

10.7   Scaling an engine

The parameter groups for mass flow, power and drag are presented on the charts herein as scaling parameters, including the physical dimension term. That for speed is simply the normalised referred form, allowing use of an appropriate datum value. This permits the impact of linearly scaling an engine on actual drag, mass flow, etc. to be evaluated.

10.8   Windmill testing

Data provided in this chapter enables first cut estimates of engine performance during windmill operation. However to obtain accurate data for a given engine, a windmill test must be conducted. Ideally this would be in an altitude test facility (see Chapter 11) with the capability of imposing an inlet ram pressure. The instrumentation employed should have sufficient accuracy in the windmill operating regime, for example low pressure levels require low range transducers. Typically performance data is taken at Mach number increments of 0.2 for two altitudes, and derived referred parameters plotted versus Mach number.

Alternatively, windmill data may be recorded during flight testing, though the quality of data may not be as good as that from an altitude test facility, particularly with air mass flow and drag not directly available.

10.9   Windmill computer modelling

The windmilling regime may be modelled by off design engine matching programs, as described in Chapter 7. However data generated should be treated with caution, particularly if the model has not been calibrated with windmill test data, as component maps can be suspect at very low referred speeds. Nevertheless during the design phase such models can give an invaluable insight into how the engine will perform, and provide estimates for the key parameters discussed above which are essential for the engine design.

The extrapolation of component characteristics into the low rotational speed regime, the form in which the component maps should be used, and other important facets of sub idle modelling are described in detail in Chapter 5 and Chapter 9. During windmill operation fuel flow is zero and hence not a variable. For a single spool turbojet the matching parameters are as follows.

Matching guesses	Matching constraints
Rotational speed	Turbine capacity
Compressor operating point (*Beta*)	Propelling nozzle capacity

Reference 1 provides a non-matching approach to turbojet windmill modelling.

Formulae

F10.1   ESDU mass flow function = fn(mass flow (kg/s), gas constant (J/kg K), total temperature (K), area (m²), total pressure (Pa), Mach number)

Wflow fn = W1 * SQRT(R * T1)/(A2 * P1 * fn(M))

where

fn(M) = SQRT(γ) * M/(1 + 0.5 * (γ − 1) * M^2)^(0.5 * (γ + 1)/(γ − 1))

(i)   The area A2 is that at the compressor face and the total temperature and pressure are the ram values at the flight Mach number.

F10.2   Drag coefficient = fn(internal drag (N), air density (kg/m³), true air speed (m/s), compressor face area (m²))

CDwm = Fdrag/(0.5 * RHO * VTAS^2 * A2)

(i)   The area A2 is that at the compressor face and the total temperature and pressure are the ram values at the flight Mach number.

Sample calculations

C10.1   A single spool turbojet with an axial flow compressor of design point pressure ratio of 20:1 and inlet area of 0.1 m² is free windmilling at 5000 m, MIL 210 cold day and 0.6 flight Mach number. Takeoff specific thrust at ISA SLS is 450 N s/kg and 5% of inlet air bypasses the combustor. Find approximate values of the parameters listed below:

Combustor inlet pressure temperature and mass flow
Rotational speed
Engine drag
Maximum auxiliary power extraction available

F10.1   Wflow fn = W1 * SQRT(R * T1)/(A2 * P1 * fn(M))

where

$$fn(M) = SQRT(\gamma) * M/(1 + 0.5 * (\gamma - 1) * M^\wedge 2)^\wedge (0.5 * (\gamma + 1)/(\gamma - 1))$$

F10.2 $CDwm = Fdrag/(0.5 * RHO * VTAS^\wedge 2 * A2)$

Inlet conditions, true air speed and ambient density
From Chart 2.11, for MIL STD 210 cold day, THETA = 0.848.
From Chart 2.10 DELTA = 0.69.
From Chart 2.3 RHOREL = 0.65.
Note these could also be calculated via Chart 2.1, F2.12 and F2.11:

T1 = 288.15 * 0.88
T1 = 253.5 K

P1 = 101.325 * 0.69
P1 = 69.9 kPa

RHO = 0.65 * 1.225
RHO = 0.796 kg/m^3

From Chart 2.15 VTAS = 360 kt = 185 m/s.

Scaling parameter groups
From Charts 10.1, 10.2 and 10.3:

Pressure ratio = 1.0
Temperature ratio = 1.021
Mass flow function = 7.4 kg$\sqrt{K}$/s m^2 kPa   (Median flow per unit area)

From Chart 10.10.6:

Referred rotational speed = 29%

From Chart 10.3:

Internal drag function = 80 N/kPa m^2 (Median drag per unit area)

From Chart 10.7:

Shaft power = 0.05 kW/m^2 kPa $\sqrt{K}$

Note: Speed would drop to 15%.

Actual parameters
P3 = 1.0 * 69.9
P3 = 69.9 kPa

T3 = 1.021 * 253.5
T3 = 258.8 K

W1 = 7.4 * 69.9 * 0.1/253.5$^\wedge$0.5
W1 = 3.25 kg/s

W3 = 3.25 * 0.95
W3 = 3.1 kg/s

Drag = 80 * 69.9 * 0.1
Drag = 559 N

PWcustomer max = 0.05 * 0.1 * 69.9 * 253.5$^\wedge$0.5
PWcustomer max = 5.6 kW

Evaluate mass flow and drag from ESDU curves for comparison
From Chart 10.4:

>   ESDU mass flow function $= 0.213$

Substituting into F10.1 and F10.2:

> $f(M) = SQRT(1.4) * 0.6/(1 + 0.5 * (1.4 - 1) * 0.6^2)^{\wedge}(0.5 * (1.4 + 1)/(1.4 - 1))$
> $f(M) = 0.7099/1.072^{\wedge}3$
> $f(M) = 0.576$

> $W1 = 0.213 * 0.1 * 69900 * 0.576/SQRT(287.05 * 253.5)$
> $W1 = 3.18 \, kg/s$

> $W3 = 3.18 * 0.95$
> $W3 = 3.0 \, kg/s$

From Chart 10.5:

>   CDwm theoretical $= 0.246$
>   CDwm delta $= 0.08$
>   CDwm $= 0.246 - (-0.08)$
>   CDwm $= 0.166$

Substituting into F10.2:

> $Drag = 0.166 * (0.5 * 0.796 * 185^{\wedge}2 * 0.1)$
> $Drag = 226 \, N$

Charts

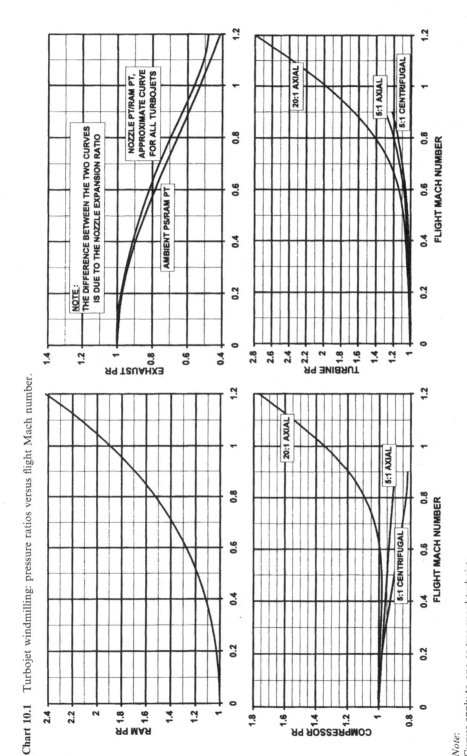

Chart 10.1 Turbojet windmilling: pressure ratios versus flight Mach number.

Note:
Curves apply to one or two spool turbojets.
The shapes of curves are rigorous; the levels are indicative only, being dependent on engine design.

Chart 10.2   Turbojet windmilling: compressor and overall temperature ratios versus flight Mach number.

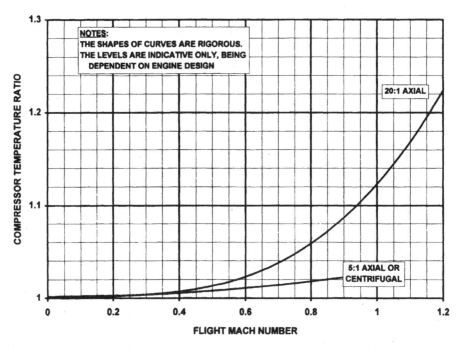

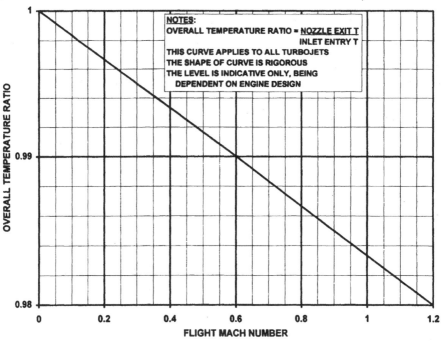

Chart 10.3   Turbojet windmilling: mass flow and internal drag versus flight Mach number.

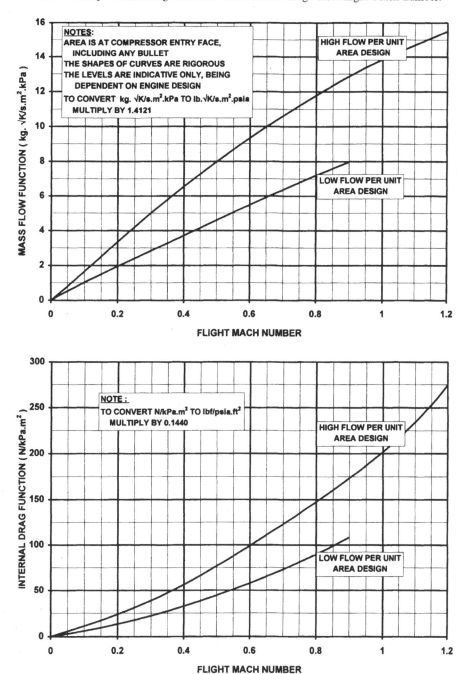

Chart 10.4 Turbojet and turbofan windmilling: ESDU mass flow function versus specific thrust.

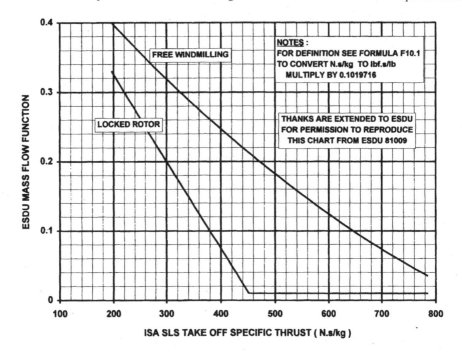

ISA SLS TAKE OFF SPECIFIC THRUST (N.s/kg)

Chart 10.5 Turbojet and turbofan windmilling: internal drag coefficient versus specific thrust and Mach number.

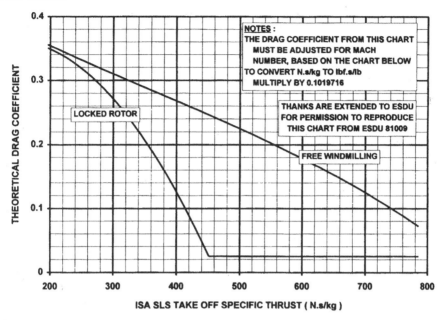

(a) Theoretical drag coefficient

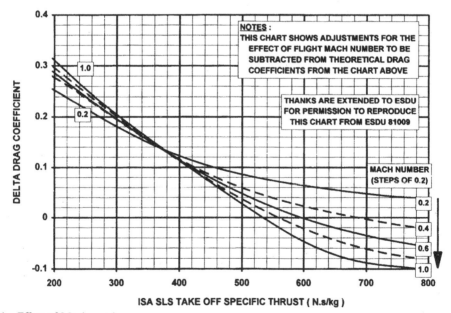

(b) Effect of Mach number

Chart 10.6 Turbojet windmilling: rotational speeds versus flight Mach number, single and two spool engines.

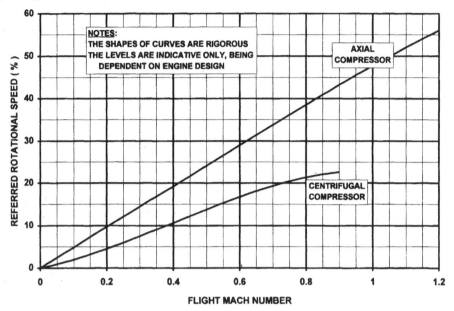

(a)   Single spool turbojet

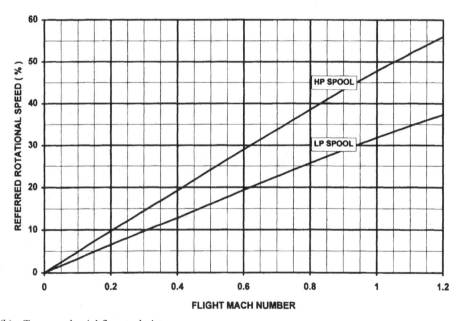

(b)   Two spool axial flow turbojet

Chart 10.7   Turbojet windmilling: effect of power extraction.

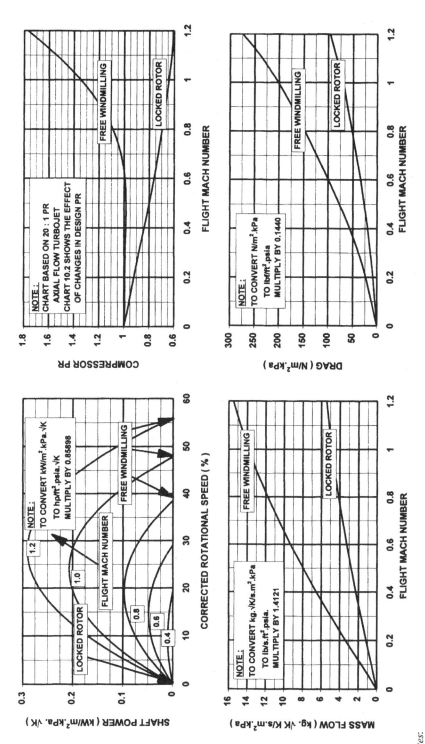

Notes:

Graphs show scaling parameter groups: area is at compressor face including any bullet.
The shapes of curves are rigorous; the levels are indicative only, being dependent on engine design.

Chart 10.8 Turbofan windmilling: rotational speeds versus flight Mach number.

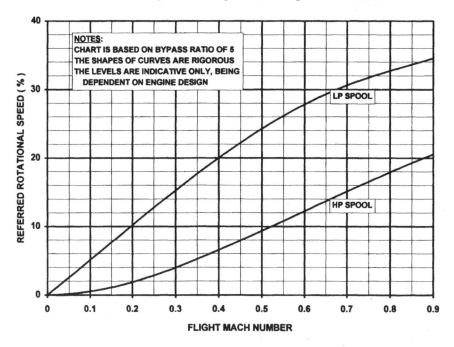

References

1. Z. Q. Shou (1980) *Calculation of Windmilling Characteristics of Turbojet Engines*, ASME 80-GT-50, ASME, New York.
2. ESDU (1981) *Estimation of Windmill Drag and Airflow of Turbojet and Turbofan Engines*, ESDU 81009, Performance Volume 4, ESDU, London.
3. ESDU (1984) *Estimation of Spillage Drag for a Wide Range of Axisymmetric Intakes at M Less Than 1*, ESDU 84004, Performance Volume 4, ESDU, London.
4. ESDU (1984) *Estimation of Drag due to Inoperative Turbojet and Turbofan Engines using Data Items Nos 81009 and 84004*, ESDU 84005, Performance Volume 4, ESDU, London.
5. L. F. Wallner and H. J. Welna (1951) *Generalisation of Turbojet and Turbine Propeller Engine Performance in Windmilling Condition*, NACA RM E51J23, Cleveland, Ohio.

Chapter 11
Engine Performance Testing

11.0 Introduction

Typically the detailed design and development programme for an engine takes 3–7 years from inception to service entry. Designing 'right first time' is not practical for such a high technology product. Development comprises individual component tests followed by hundreds of hours of engine testing, based on which many design modifications are introduced. The resulting production engine standard will then comply as closely as possible with the original specification.

After service entry, *production acceptance* or *production pass off* testing of each individual production engine is common practice, ensuring that it meets key *acceptance criteria*. This test is the final check on component manufacture and engine build quality prior to delivery to the customer.

This chapter outlines various types of engine test, and provides details of test beds, instrumentation and analysis methods. Engineers from all disciplines involved in engine development or production acceptance must understand the fundamentals of performance testing technology, as it is central to both processes.

11.1 Types of engine test bed

This section describes the configuration of engine test beds, and provides design to guidelines to ensure that:

- Engine inlet flow is uniform. Any distortion will affect compressor performance (see section 5.2.10) and will give an erroneous mass flow measurement. At worst, vortices may be shed from the floor or walls causing high cycle fatigue failure of compressor blades.
- There is no reingestion of hot exhaust gas. Should this occur the resulting distorted inlet temperature profile will again affect compressor performance. It will also prevent accurate measurement of inlet temperature, which is essential for referral of measured engine performance.
- For thrust engines, that the static pressure field around the engine is as close as practical to that of free stream conditions. As described below the thrust reading must be corrected for this effect: the smaller the effect the less scope for error in the correction.
- The static pressure distribution at the propelling nozzle exit plane allows accurate determination of a mean value. Otherwise the thrust measurement will be in error, and the engine performance may be affected.

As described in section 11.2, many other parameters are measured besides those mentioned above. For these accuracy depends on good measurement practice rather than overall test bed design.

11.1.1 *Outdoor sea level thrust test bed*

This is illustrated in Fig. 11.1, and consists basically of an open air stand supporting an engine and providing thrust measurements. The effects of cross wind on entry conditions are negated

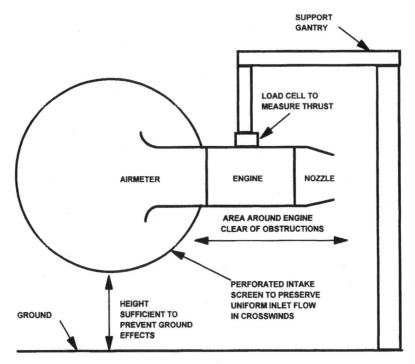

Notes:
To minimise noise disturbance outdoor test beds are normally sited in remote areas.
Climatic conditions also influence choice of location.

Fig. 11.1 Outdoor sea level thrust test bed.

by a large mesh screen fitted around the engine inlet. The immediate test bed area is free of obstructions to the air flow, to ensure the validity of the thrust and air flow readings. This is the most definitive thrust test bed, as for indoor test beds the thrust and air flow measurements are corrupted by the flow field generated by the sidewalls. Outdoor test beds are sited in remote areas, to minimise the environmental disturbance of the noise produced. Because of the resultant logistic difficulties and the impact of adverse weather conditions, indoor testing is preferred in most countries, with measurements *calibrated* versus outdoor facilities.

11.1.2 *Indoor sea level thrust test bed*

Here a similar engine arrangement to that of Fig. 11.1 is mounted indoors, as shown in Fig. 11.2. The air flowpath to the engine is crucial, as flow disturbance must be minimised. The engine nozzle efflux enters a *detuner*, which exhausts hot gases and provides sound attenuation.

For a given engine the measured thrust may be up to 10% less than the value that would be recorded on an outdoor test bed. This is due to unrepresentative static pressure forces acting on the engine and cradle, caused by the velocity of air within the cell passing around the engine. This air is *entrained* into the detuner by the *ejector* effect of the engine jet; it prevents hot gas re-ingestion and also cools the detuner. Furthermore, if the engine final nozzle is unchoked the test bed configuration can cause a rematch due to the local static pressure distribution at the nozzle exit. An indoor test bed gives useful all-weather availability, but unless the test bed is purely functional it should be calibrated against an outdoor test bed, as discussed in section 11.3, to determine the effects of the static pressure field and rematch.

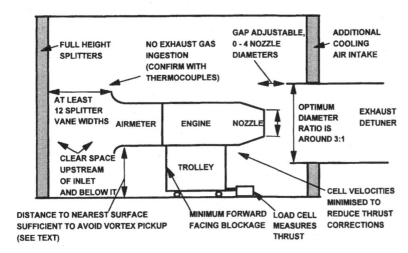

Note:
Sea level gas generator test bed is similar but without thrust measurement.

Fig. 11.2   Indoor sea level thrust test bed.

Key guidelines for designing an indoor test bed are presented below. These minimise the measured thrust deficiency described above, and also prevent extreme flow fields such as vortices from entering the airmeter. Owing to the complex flows within a test bed, some adjustability and subsequent experimentation may be required.

- Air should enter the test bed via a full height intake at the front, with *splitters* for flow straightening and noise attenuation. Additional side intakes should only be used if size constraints make them absolutely necessary.
- The distance from the inlet splitters to engine airmeter inlet should exceed a dozen splitter widths.
- Rear intakes, passing up to 10% of the main intake flow, may be employed to provide additional detuner cooling air whilst minimising flow velocities past the engine.
- To avoid vortex pickup the height from the airmeter centreline to the floor or walls should exceed 5 airmeter throat diameters for an approach velocity of 0.01 times the throat velocity, and 2 throat diameters for 0.1 times the throat velocity. This may influence the choice of cell height and width.
- The space upstream of and around the airmeter should be unobstructed, to avoid flow disturbances which would impair the flow measurement.
- The air velocity within the cell flowing past the engine casing should not exceed around 10 m/s. This may also influence the cell dimensions, though it is heavily dependent on the entrained flow, which is discussed below.
- The engine cradle and slave equipment should present minimum forward facing area, to reduce thrust corrections.
- Recirculation of hot exhaust gas to the engine intake must be avoided; during test bed commissioning this should be confirmed using suitably located thermocouples.
- If the detuner turns flow vertically upwards it should incorporate a cascaded bend to minimise pressure loss.
- The detuner inlet diameter should be around three engine nozzle diameters. Increasing this increases the entrained flow.

- The axial gap between the detuner and the engine nozzle should be adjustable; values of at least 2 and ideally 3 or more engine nozzle diameters are recommended. Increasing this does not affect the entrained flow, but reduces thrust loss and rematching by avoiding pressure disturbances around the nozzle exit plane.
- The *entrainment ratio*, which is the ratio of entrained air flow to engine propelling nozzle flow, may be calculated from measured temperatures and a simple enthalpy balance using Formula F11.1. A good design target is 3:1 for turbojets, and lower for turbofans as engine flow is higher and discharge temperature lower.
- During test bed commissioning flow visualisation using smoke should be employed as a further check on the quality of the test bed aerodynamics. This should confirm the absence of hot gas recirculation or vortex pickup, and determine suitable positions for cell static pressure measurement.

11.1.3 *Indoor sea level jet bed for turboshaft gas generator tests*

This test bed is employed when a turboshaft engine *gas generator* is tested with a final nozzle and exhaust detuner instead of its power turbine. The theoretical *exhaust gas power* is calculated based on gas generator exit flow, pressure and temperature, and is the power that would be available from a power turbine of 100% efficiency with no duct pressure losses (Formulae F3.15, F3.19–F3.21). Such testing is necessary when the power turbine either remains with the installation, or is supplied by another company. As this applies mainly to landbased engines, an altitude chamber would not normally be required.

Key points are as follows:

- The accurate measurement of air mass flow is crucial to the power calculation. As for indoor thrust beds, flow disturbances near the airmeter must be avoided as well as vortex pickup.
- Hot gas recirculation to the intake must be avoided, as for thrust engines.
- The effect of the exhaust detuner is less critical, as the slave nozzle exit area may be 'trimmed' to ensure capacity adequately matches that of the power turbine.

11.1.4 *Indoor sea level shaft power bed*

This is used for turboprop or turboshaft engines, with output power measured directly. Figure 11.3 shows the key features of such test beds. The main differences from an indoor thrust bed are as follows.

- Air may be ducted directly to the engine from ambient rather than it flowing though the test cell, with flow measured outside the test bed at entry to the ducting. The test bed configuration does not affect measured air flow and hence measured performance; the building is only there to provide protection from adverse weather conditions.
- Alternatively air may enter the test bed through splitters and then into the engine as per a thrust bed; here test bed configuration does affect the measured air flow and the rules provided in section 11.1.2 should be adhered to.
- The exhaust flow has a low velocity and is ducted directly to atmosphere with no detuner required. Entrainment effects therefore need not be considered.

On shaft power test beds some device must absorb the engine output power, providing suitable characteristics of load versus speed. There are several possibilities:

- For a turboprop, an aircraft propeller may be fitted on the test stand.
- An alternator may be used to generate electrical power, to be either dissipated in electrical resistance banks or passed to a grid system. The latter is appealing environmentally, but

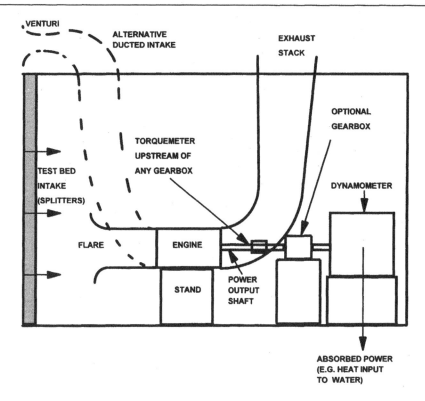

Notes:

If flare used for mass flow measurement, positioning guidelines are per Fig. 11.2.

Shaft power test bed has almost no entrained air flow.

If a gearbox is used to match power turbine speed to dynamometer then torque/power measurement must exclude gearbox losses.

Fig. 11.3 Indoor sea level shaft power test bed.

usually impractical during an engine development programme. Set up costs are high, rotational speed is tied to grid frequency, and intermittent operation may be unacceptable to a grid operator.

• A *dynamometer* absorbs power over a range of power and speed combinations, and often also measures torque. In the hydraulic type a vaned rotor and stator arrangement pumps water through the vanes. The power absorbed heats the water, which must either be cooled or a fresh supply provided. Valves control the water level within the dynamometer, which changes the power absorbed at any given speed and allows for various power/speed laws. Torque measurement utilises a load arm and weighing system on the external casing, which is freely mounted on bearings. The input torque is transmitted via the water and any bearing friction.

11.1.5 *Altitude test facility (ATF)*

Thrust or shaft power test beds may be housed within an *altitude test facility* (ATF), which reproduces the inlet conditions resulting from altitude and flight Mach number. Figure 11.4 shows the key features of an ATF. Unlike a sea level test bed the plant must provide a continuous airflow even without the engine operating, to maintain reduced pressure and temperature.

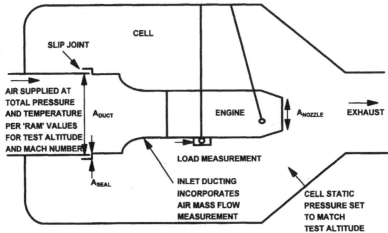

Notes:
$$\text{GROSS THRUST} = \text{LOAD} + A_{\text{SEAL}} * (PS_{\text{SEAL}} - PS_{\text{CELL}}) + A_{\text{DUCT}} * (PS_{\text{DUCT}} - PS_{\text{CELL}})$$
$$+ W_{\text{DUCT}} * V_{\text{DUCT}}$$
$$\text{NET THRUST} = \text{GROSS THRUST} - W_{\text{DUCT}} * V_O$$

W_{DUCT}, LOAD, PT_{DUCT}, T_{DUCT}, PS_{CELL} and PS_{SEAL} are measured directly.
V_{DUCT} and PS_{DUCT} are calculated from the measurements using Q curves (see Chapter 3).
V_O is flight velocity, calculated from static temperature and flight Mach number.

Fig. 11.4   Altitude test facility (ATF).

Figure 11.5 illustrates two possible layouts of ATF plant. To simulate both ambient conditions and flight Mach number engine inlet total pressure and temperature must be controlled to the *ram* (free stream total) values for the altitude and Mach number. Also, the static pressure at the nozzle exit plane must be set to that of the test altitude. These parameters are mostly sub-ambient at altitude, hence common features of the various types of ATF are substantial pressure reduction, chilling and drying capabilities, and recompression of discharge air back to ambient. Accurate measurement of thrust is complex, as discussed in section 11.2.7.

The other possibility for testing at flight conditions is a *flying test bed* as described in section 11.1.6, where the engine is mounted on an aircraft. The main advantages of the ATF are:

- A full range of ambient and flight conditions may be tested in one geographical location
- Better instrumentation, including direct measurement of air mass flow and thrust
- High availability, independent of weather conditions

11.1.6   *Flying test bed*

A flying test bed is also often used for major aero-engine programmes. Typically a four engined aircraft is modified to mount a single, new development engine at one berth. Compared with an ATF the advantages are:

- Better simulation of functional effects such as carcase loads and inlet distortion
- Lower capital cost

However as mentioned there are no direct measurements of thrust and mass flow. These must be calculated as follows.

- Propelling nozzle thrust coefficient and capacity are obtained from rig and engine tests, ideally in an ATF.
- Nozzle entry total pressure and temperature are measured directly, with sufficient coverage to obtain valid average data.

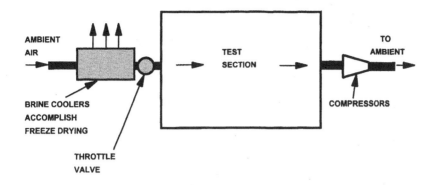

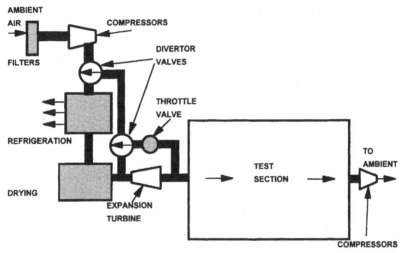

Note:
Different valve settings are used depending on the pressure and temperature levels required.

Fig. 11.5   Altitude test facility (ATF): possible plant layouts.

- Nozzle mass flow may now be calculated, along with exit velocity, using Formulae F3.33 and F3.35.
- Any air offtake and nacelle ejector flows are estimated from design data.
- Fuel flow is measured directly.
- Inlet air flow may now be calculated, from Formula F11.2.
- Net thrust is the difference between the total exit and inlet momentum, and if the nozzle is choked any pressure thrust must be added, as shown in Formula F11.3.

11.2   Measurements and instrumentation

Engine tests use differing amounts and sophistication of instrumentation, depending on their purpose. Many development tests require detailed performance investigation, hence pressures and temperatures are measured at virtually every station, as well as power or thrust, shaft speeds, fuel and air flow, etc. At the other extreme, for production pass off or endurance testing only a minimum of measurements are taken beyond those of the production control system, such as ambient conditions, power or thrust level, and fuel flow.

This section provides background to the instrumentation used for each possible measured parameter. Indicative accuracies and coverage requirements are provided, along with a summary of how the instruments work.

Engine testing is very expensive, hence to ensure good quality data is obtained the importance of the following cannot be overemphasised.

- The test bed and all instrumentation must be properly calibrated, as per section 11.3.1.
- Test planning should include careful specification of instrumentation requirements.
- Key measurements must be repeatedly checked during the testing, to ensure data is valid. Engine removal from the test bed must be delayed, and testing repeated, if necessary.

For all the above an understanding of likely accuracy levels is required.

11.2.1   *Pressures*

Pressures are measured for a number of reasons:

- Determination of overall engine performance requires ambient or cell pressure so that parameters can be referred back to standard conditions as described in Chapter 4.
- Engine station pressures help define component performance, e.g. pressure ratios, surge margins and flow capacities.
- Mass flow measurement is based on the local difference between total and static pressure levels, as described in section 11.2.5.

Figure 11.6 illustrates the main elements of a typical pressure measuring system, and Reference 1 thoroughly covers pressure measurement methods. Local small holes called *tappings* allow the engine gas stream pressure to reach a measuring device outside the engine, via fine *capillary tube* of 1–2 mm diameter. Reference 2 gives a recommended tapping design. Generally at least three circumferential locations are used at a station, to increase coverage and to allow error detection by comparison of readings. A leaking line usually reads low, though thought should be given as to whether lines pass through higher pressure regions. An alternative is to *gang* the lines from several tappings together into a manifold and then read the manifold pressure as a *pneumatic average*. Though this method is relatively inexpensive a leak in any one pressure line is difficult to detect. The measuring devices used are described below.

Manometers

For pressures below around 2 bar older test beds have used water or mercury *manometers*, where the height of a column of liquid in a glass tube is read visually. Small corrections are applied for the temperature of the liquid column, as per Formulae F11.4 and F11.5. For a well designed system accuracy is around 0.25%. As automatic data recording has become more prevalent manometers have virtually disappeared.

Transducers

Modern test beds use a *transducer*, where a pressure difference causes movement of a diaphragm, which is converted to an electrical signal. The other side of the diaphragm may be at ambient pressure or a vacuum, giving *gauge* and *absolute* readings respectively. The diaphragm movement is converted to a voltage, which is read by the data logging system. For many transducers the conversion uses an *energising voltage* and a resistive straingauge on the diaphragm. Alternatives are piezo-electric, which generate their own voltage, or inductive. *Calibration curves* relate the electrical signal to pressure levels, and are obtained by either a *dead weight tester* which applies a known force and hence air pressure, or comparison with other calibrated transducers. Transducer designs are optimised for various pressure ranges;

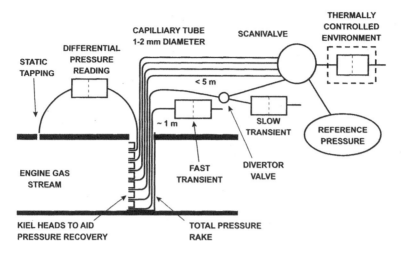

(a)   System schematic

Notes:
Total pressure heads may also be mounted on vane leading edges.
Static tappings are often placed in same plane as rakes.
Fast transient transducer may have water jacket to aid thermal stability.
Transducers illustrated are connected to ambient unless otherwise shown. Versions are also available with internal vacuum.

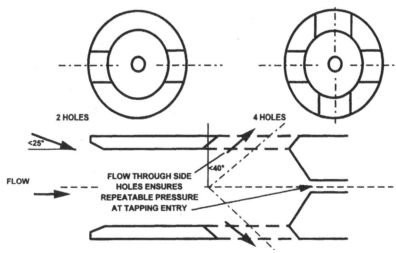

(b)   Kiel head details

Fig. 11.6   Pressure measuring system.

selection should ensure operation is in the most linear part of the range, typically 10–90% of full scale. Transducer temperature should be controlled as this affects the straingauge resistance; even *compensating circuitry* does not fully eliminate the effect. Accuracies quoted herein are for a controlled transducer temperature.

Steady state, typical accuracies are around 0.1% of full scale for the basic transducer, however a good overall accuracy is 0.5% for engine pressures. This figure allows for calibration drift, hysteresis, engine stability and pressure profiles, transducer non-linearity, and drift in the voltage supply. Cell static pressure is less prone to engine effects hence an accuracy of 0.25% is obtainable.

For steady state testing, many pressure tappings are read in turn via *scanivalves*, where rotating valvery connects a transducer to each tapping in turn. For best accuracy the transducer is kept in a temperature controlled environment, and known *reference pressures* are also read one or more times per scan giving continual, automatic update of the transducer calibration. With best practice overall accuracy can be around 0.25% as most transducer error effects are eliminated, leaving those due to engine pressure profiles and stability during the time taken to complete a scan. Such a system is unsuitable for transient use due to intermittent reading and volume packing of the pressure lines. The latter effect means around 4 minutes stabilisation is required before an accurate pressure reading can be obtained.

Comments on pressure measurements at the various engine stations are presented below.

Ambient pressure – barometers

Barometers are used to measure ambient pressure, with an accuracy of around 0.1%. They fall into two main categories:

(1) An *aneroid* barometer consists of a dial and pointer controlled by an evacuated metal cylinder with corrugated sides and a spring action. Changes in ambient pressure cause movement which is read on the dial.
(2) A mercury in glass or *Fortin* barometer uses a mercury column in a closed glass tube, evacuated at the closed, upper end. The height of the column is read visually, Formulae F11.5–F11.7 provide corrections to readings for temperature, latitude, and differential elevation.

Formula F2.1 can be used to determine the effect of changes in altitude on ambient pressure, as for example a barometer may be located an appreciable number of metres above or even below an engine intake.

Test cell static pressure

This is required for all engines where the intake is inside a test bed, and is measured in at least two places of low cell velocity. Usual locations are on the side walls in the plane of the nozzle exit, at the same height as engine centreline. The instrumentation comprises the open end of a 1–2 mm diameter capillary tube surrounded by a perforated 'pepper pot', which removes the effects of incident velocity. The tube is then connected to either a transducer or water manometer.

Engine static pressures

If the axis of a tapping is perpendicular to the flow direction then it will read static pressure, as no dynamic head will be recovered. Reference 2 describes key geometric features of a wall tapping. These give a more accurate reading than side or rearward facing tappings on immersed probes, as the presence of the probe disturbs the flow. Static pressure tappings may be used in place of total pressure readings (see below) if a calibration has already been obtained versus the total pressure reading. Such calibrations become tenuous however if there is swirl angle variation, as this changes the local Mach number.

Engine total pressures

To achieve full recovery of the stream dynamic head, and hence measure total pressure, the pressure tapping is mounted in a probe which points its axis towards the direction of gas flow. If the flow angle varies by more than ±58, a *Kiel head* should be employed, as illustrated in Fig. 11.6. This uses a chamfered entry to recover effectively the stream dynamic head for incidences of up to ±258. Above gas temperatures of around 1300 K total pressure probes are not normally viable, as they will require cooling and hence become so large that associated pressure drops are prohibitive.

Coverage requirements depend on how well understood the pressure uniformity is at a station, and should be agreed with the relevant component designer. Many radial and circumferential locations may be addressed via either multiple heads on vane leading edges, or several multihead *rakes* inserted into the gas stream. The heads are often placed on centres of equal flow area to assist in data averaging as per section 11.8.2. Calibration of rig versus engine instrumentation standards may also be undertaken. Coverage in the cold end (compressors) should be at least three off rakes with one to five heads, depending on engine size. If large radial or circumferential non-uniformities are likely then more coverage may be employed, such as downstream of an aero-engine fan where up to ten heads are used. For the hot end there is often significant swirl hence greater coverage may be needed. Around six rakes, or preferably instrumented vane leading edges, are suggested with two to five heads depending on engine size.

Total pressure rakes and wall static tappings may be used in combination, set in the same plane. Static tappings complete definition of the total pressure profile, as static and total pressures are equal at the walls. In addition, if the static pressure is reasonably uniform, which requires low swirl angle, the difference between total and static pressure indicates flow velocity via Formulae F3.10 and F3.11.

Placement of total pressure rakes should consider obvious sources of error such as wakes downstream of struts. Good practice is to derive calibration factors between fitted instrumentation and that giving fuller coverage as listed above.

Differential pressures

Normally pressure is measured as the difference between the gas stream and ambient, and an absolute pressure level obtained by addition of ambient pressure to the gauge reading. To read a pressure *difference* between two points, both sides of a transducer may be connected to tappings at the engine stations in question. This allows use of a more precise, lower range transducer, and avoids large inaccuracies due to the subtraction of similar numbers. One disadvantage is that recalibrating such a transducer is not possible without disconnecting the instrumentation, unlike for scanivalve systems.

Transient pressures

For transient testing dedicated pressure transducers, designed to optimise transient response, are required for each tapping. This provides a continuous reading, unlike a scanivalve system. The transducers are mounted local to the engine to minimise line volumes, and hence allow fast response to pressure changes; they may be water jacketed to enhance thermal stability. Line length limits are around 5 m for ordinary handling and 1 m for faster transients such as fuel spiking (see section 11.5.2). In the former case a divertor valve may be employed to allow the same tapping to be read by the steady state scanivalve; for the shorter line length space does not permit this. Typical scan rates range from 10 to 500 scans per second. Absolute accuracies are lower for dedicated transient transducers than for a scanivalve system, around 1.5% of full range, and the transducers are more subject to drift.

A calibration curve should be run at the start of each working day to provide a comparison with the steady state instrumentation. In addition a transient manoeuvre should be performed to check for lag due to divertor valve faults.

Dynamic measurements of pressures

Dynamic measurements of pressure address high frequency pressure perturbations, rather than 'dynamic pressure' as in 'velocity head'. These measurements are employed to detect flow instabilities such as rotating stall or rumble which can occur in the compression and combustion systems. The accuracy in determining amplitude is low, around 10% of range, as the design is optimised for response and the thermal environment is normally uncontrolled. Probes such as the *kistler* (resistive) and *kulite* (piezo) varieties are utilised, which *incorporate* pressure transducers – the low volume allows response to the very high frequencies involved. The kistler probe is more vulnerable to vibration but can tolerate temperatures up to 350 8C, which is 80 8C higher than the kulite. The signal may be available in the control room on an oscilloscope and is normally recorded in analogue form on tape. Later examination of amplitudes and frequencies helps pinpoint the cause or at least onset of instability. High frequency phenomena may also be shown qualitatively by noise on ordinary transient pressure signals.

11.2.2   *Temperatures*

Measurement of temperatures provides the following information on engine and component performance.

- Determination of overall engine performance requires inlet temperature so that parameters can be referred back to standard conditions.
- The temperatures local to a component are required to define its performance, i.e. efficiency and flow capacity.
- Temperatures are required to ensure the engine is not operated beyond limits stipulated for mechanical integrity.
- Mass flow measurement utilises temperature levels.

Temperature measurement is complex. It is vital to follow the good design and working practices outlined herein, otherwise significant inaccuracies may result. Reference 3 provides a comprehensive description.

Temperature readings more closely reflect total rather than static conditions (see Chapter 3), as a rake in the gas stream brings the gas to rest on its surface. In fact it is impossible to measure purely static temperature. The fraction of the dynamic temperature recovered is termed the *recovery factor* (Formula F11.8). As for pressure measurement, temperature rakes employ Kiel heads where swirl angle variation is greater than 58, to ensure the dynamic temperature is recovered over a wide range of incident flow angles. Little error is normally incurred if rake recovery factors are taken as a constant value of 0.94, assuming well designed probes.

Figure 11.7 illustrates the main elements of a typical temperature measuring system. The following sections describe the instruments used.

Resistance bulbs thermometers (RBT)

Here temperature is measured via changes in the resistance of a heated material. Platinum is frequently used, hence the common alternative expression *PRT* (platinum resistance thermo-meter). In theory resistance thermometers are suitable for temperatures up to around 1000 K, and may give a high accuracy of potentially around 0.1 K if carefully calibrated. They are however comparatively delicate, and rarely used actually within an engine; the most common uses are for air inlet temperature and to measure *reference temperature* in thermocouple systems described below.

Snakes

Snakes are resistance thermometers many metres long sometimes used to measure average inlet temperature, and may for example be strung out over the inlet debris guard or splitter. One disadvantage is that the large physical size makes accurate calibration impossible, hence a preferred alternative is multiple resistance bulbs. Indicative overall accuracy for a snake is 1–2 K.

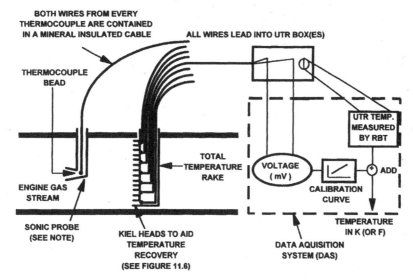

BOTH WIRES FROM EVERY
THERMOCOUPLE ARE CONTAINED
IN A MINERAL INSULATED CABLE

ALL WIRES LEAD INTO UTR BOX(ES)

THERMOCOUPLE
BEAD

UTR TEMP.
MEASURED
BY RBT

TOTAL
TEMPERATURE
RAKE

VOLTAGE
(mV)

CALIBRATION
CURVE

ADD

ENGINE GAS
STREAM

SONIC PROBE
(SEE NOTE)

KIEL HEADS TO AID
TEMPERATURE
RECOVERY
(SEE FIGURE 11.6)

DATA AQUISITION
SYSTEM (DAS)

TEMPERATURE
IN K (OR F)

Notes:
Test bed inlet temperature is measured by multiple RBTs, or on older beds a 'snake'.
It is not possible to measure static temperature, as most of the dynamic temperature is always recovered.
UTR = universal temperature reference
RBT = resistance bulb thermometer
Sonic probes increase heat transfer to the thermocouple by accelerating flow past the dead and to overboard.
Aspirated probes use an auxiliary air ejector to accelerate flow past probe.
Thermocouples may also be mounted on vane leading edges.

Fig. 11.7   Temperature measuring system.

Thermocouples

If two dissimilar metal wires are connected at a *junction*, and the loose ends maintained at some *reference temperature*, a voltage is generated dependent on the temperature difference between the junction and the reference. Typically the junction is a welded bead of up to 1.1 times the wire diameter. Thermocouples are less accurate than RBTs, but more robust. The loose ends' temperature is maintained by either a *UTR* (uniform temperature reference) box, whose own temperature is measured by resistance bulb thermometers, or an *Icell* (ice cell). Single pieces of wire should be used between the hot and cold ends, otherwise measurement uncertainties increase by around 2 K per extra junction. With single pieces *batch wire calibration* is applicable, where a calibration is obtained of a number of thermocouples made from a particular batch of cable. Providing the results agree, this calibration is applicable to other thermocouples made from that same batch.

Thermocouples which employ different wire materials produce different voltage curves. Standard curves are defined for different thermocouple types as per Reference 4, and different *classes* of thermocouple lie within different tolerances of these curves. *Class 1* thermocouples will lie within the greater of 1.5 K or 0.4% of the standard curve, and *class 2* within 2.5 K or 0.75%. At higher temperatures drift and hysteresis together contribute a further 3 K inaccuracy, despite heat treatment to improve thermoelectric stability. In addition typically 1 K error will be contributed by both circuitry, and conduction effects around the hot junction. Combining all these effects by root sum squaring gives overall system errors at 1000 K of 5.6 K and 8.2 K for class 1 and 2 respectively. *Type K* thermocouples use chromel–alumel (Ni–Cr/Ni–Al) wire. *Type N* thermocouples (Nicrosil/Nisil, Ni–Cr–Si/Ni–Si) were introduced around 1990 to provide longer life and improved stability over type K. Overall system errors at 1000 K are reduced to 4.2 K and 7.6 K.

In siting thermocouples radiation from adjacent surfaces must be avoided, otherwise the temperature measured is not that of the gas stream. For locations where radiation may be severe, *shielded* thermocouples are employed, which use up to four concentric thin tubes surrounding the thermocouple bead.

The application of the above devices to measuring temperatures at key engine stations is described below.

Air inlet temperature

Recommended coverage is at least 3 RBTs mounted on the intake debris guard, and more if non-uniform inlet temperature profiles are suspected. The test bed layout should be adjusted to ensure that the difference between the readings is less than 1 K, otherwise it is difficult to be sure that the true average temperature is being measured, and the temperature profile may fundamentally affect engine performance. Snakes or even thermocouples may also be used, however this results in lower accuracy as described above.

Cold end (compressor) temperatures

Usually thermocouples are employed, and accuracy is as described above. Coverage should be at least three points circumferentially, with rakes having one to five heads radially depending on engine size and the expected radial temperature profile. The heads are usually placed on centres of equal area to provide uniform coverage and assist in data averaging. An aero-engine fan is a special case; due to the temperature profiles and relatively low temperature levels rakes with up to ten heads are employed.

Hot end (turbine) temperatures

Temperature measurement is significantly more difficult for turbines, for two main reasons:

(1) Above temperatures of around 1300 K, the mechanical integrity of a probe becomes an issue, requiring bulky, cooled designs which are highly intrusive. Such measurements are rarely attempted.
(2) At combustor exit, and to a decreasing extent rearwards through a turbine system, there is severe temperature 'patternation' causing both circumferential and radial profiles. This is due to having discrete fuel injection points within the combustion system and cooling air influx downstream. To obtain a thermodynamically valid average temperature from a finite number of readings may be impractical.

For both reasons the temperature at combustor exit cannot be measured, and measurements are rarely possible at exit from any first HP turbine stage. For measurement stations further downstream, the minimum coverage required is *at least* eight locations circumferentially, and three to five thermocouple heads radially, depending on engine size. Patternation introduces a further error beyond the thermocouple inaccuracies described above. Rather than using centres of equal area, head placement is often biased towards the walls, where the temperature gradient is steepest.

Transient temperatures

One further important thermocouple property is response time. Physically large thermocouples take time to respond to temperature changes, due to thermal inertia, and are unsuitable for transient development testing. Response is governed mainly by the time constant of the junction itself. For development testing physically small junctions are employed, mounted to minimise conduction and radiation.

For control system instruments, large robust 'production' devices are required. Here the time constant of the thermocouple can be allowed for in the control algorithms, or heat transfer increased by increasing the flow past the thermocouple junction. Where the pressure ratio to ambient exceeds around 1.2 a *sonic probe* is used as shown in Fig. 11.7. A small flow is extracted from the gas path through a venturi surrounding the thermocouple bead. The flow accelerates to choked conditions at the bead and then diffuses before being dumped overboard. Otherwise *aspiration* is employed, where higher pressure air is injected to draw flow past the bead by an ejector effect.

11.2.3   *Liquid fuel energy flow*

Measuring fuel flow fulfils two essential purposes:

(1)   It is vital for calculating thermal efficiency and SFC.
(2)   Calculating temperature levels, hence lives, of the combustor and HP turbine requires fuel flow, as these temperatures cannot be measured (see section 11.2.2).

In all cases the method involves the measurement of *volumetric* fuel flow, conversion to fuel *mass* flow based on the actual fuel density, and finally deriving energy flow using the fuel heating value (FHV). To obtain the density and FHV periodic laboratory analysis of fuel samples is required. This must be done at least once per fuel batch delivery, and even each day during key performance testing. Formula F11.9 shows how fuel energy flow is calculated from the volumetric flow, fuel heating value and specific gravity.

Volumetric flow

For liquid fuels three main instruments may be used:

(1)   A *bulkmeter* measures volumetric flow over a time period, using pistons connected to a rotating crankshaft. Volumetric flow rate is directly proportional to rotational speed as this determines the rate at which the piston volumes are filled and emptied. Bulkmeters also require calibration for fuel viscosity which has a second-order effect of up to 1%. Hence fuel temperature and fuel type such as diesel or kerosene must be allowed for. Bulkmeters are almost mandatory for steady state testing and are very accurate if engine operation is stable, around 0.25% or less. They are unaffected by inlet flow profile or swirl and hence no upstream flow conditioning is required.
(2)   A *turbine flow meter* indicates instantaneous volumetric fuel flow rate, via a calibration versus rotational speed. Again fuel viscosity must be allowed for. Initial accuracy is good, around 0.5%, but drifts by up to 1% which is more than other instrument types, due to wear on vane tips and in the bearings. Turbine flow meters are affected by inlet flow distortion hence a minimum settling length of ten pipe diameters must be allowed upstream, with a flow straightener such as a colander plate at inlet to this settling length.
(3)   A *glass bottle and stopwatch* may be used for steady state testing of small engines. This involves timing the visible consumption of a known fuel volume, and gives an accuracy of around 1%. The method is not suitable for automatic data recording.

Because fuel flow is so vitally important at least two of these devices should be used in series. Figure 11.8 shows the main elements of a fuel flow measuring system, including typical installation requirements.

Density

To measure density of a fuel sample a *hydrometer* is used, where a graduated scale on a float is read visually. The 'units' are usually *relative density* or *specific gravity* which, as shown by Formula F11.10, is simply the ratio of the actual density (mass per unit volume) to a standard value of 1000.0 kg/m^3. The accuracy of measurement is around 0.1%.

Actual density varies with temperature, as shown by Formula F11.11, hence temperature readings are required for both the laboratory sample test and the fuel supply to the engine. For the latter, resistance bulb thermometers are placed in the fuel pipes, ideally ten pipe diameters away from the flow meter to prevent possible flow disturbances corrupting the flow measurement.

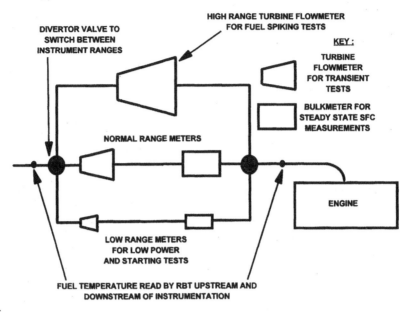

Notes:

Distances between meters and pipe bends etc. should be at least ten pipe diameters to prevent flow disturbances from corrupting the measurements.

Steady state at least two meters of some kind should be used in series to confirm the measurement accuracy.

Fuel samples should be taken for analysis at least once per fuel batch and daily during key steady state testing.

RBT = resistance bulb thermometer

Turbine flow meters require upstream flow straightening, as described in the text.

Fig. 11.8   Fuel flow measuring system.

Fuel heating value

Fuel heating value is also obtained via laboratory analysis of samples. The main methods are:

- Bomb calorimeter, in which a fuel sample is burnt and the heat release measured directly from the temperature rise
- Analysis of composition, via spectroscopy. Here the percentages of carbon and hydrogen are used to calculate the energy that would be released in combustion, via Formula F11.12.

Accuracy of the above methods is around 0.1%. If no laboratory analysis is available, a fuel heating value for kerosene only may be calculated from the density using Formula F11.13.

Transient fuel flow measurement

For measuring volumetric fuel flow transiently only a turbine flow meter is suitable as its speed gives an instantaneous flow rate. The usable range is around 10–100% of full scale, so different instruments are required for starting and fuel spiking, with ranges of around 20% and 300% of that for normal running. Fuel density and lower heating value are simply the same numbers as for steady state. Both a bulkmeter and flow bottle are unsuitable as they rely on fuel flow being steady.

One backup measurement that may be particularly useful transiently is via the combustion fuel injector *flow number*. Given the upstream fuel pressure measurement, flow is simply the flow number times the square root of the pressure drop, Formula F11.14. This is useful during fuel spiking, described in section 11.5.2, where fuel flow changes so quickly that even a turbine flow meter reading may lag due to inertia. The method may be checked against the main instruments based on the previous steady state condition.

11.2.4 Gas fuel energy flow

The measurement process is similar to that for liquid fuels, with fuel energy flow evaluated from volumetric flow, density, and lower heating value.

Volumetric flow

A bulkmeter cannot be used to measure gas fuel flow as it is *compressible*, i.e. its density changes with pressure, and the positive displacement principle upon which a bulkmeter relies is only applicable to incompressible liquid fuels. One of two techniques is employed as outlined below.

As for liquid fuel a turbine flow meter may be utilised, with similar installation rules and measurement accuracies. It is not necessary to calibrate for viscosity.

An orifice plate may also be utilised, and provided that its geometry adheres exactly to Reference 2 an accuracy of ±1% can be achieved. The installation criteria are also described in Reference 2, with a flow straightener at the measuring section inlet, and typically upstream and downstream straight lengths of pipe of ten and five diameters respectively. However orifice plates have a number of significant disadvantages relative to a turbine flow meter:

- The accuracy is only maintained for a fuel flow *turn down ratio* of around 4:1, whereas for a turbine flow meter over 10:1 is common.
- The pressure drop of an orifice plate is higher, which must be allowed for in the fuel supply system.
- It is not possible to achieve the required manufacturing tolerances for engines of less than around 5 MW.

Calibration may utilise a series of *sonic nozzles* in parallel. This is an extremely sophisticated system and would normally be used in a laboratory rather than an engine test bed.

Density

The fact that gas fuel is compressible makes evaluation of its density complicated. The most accurate method of determining gas fuel density is from its pressure, temperature, gas constant and *compressibility* via Formula F11.15. Static pressure should be the average of values measured around five diameters upstream and downstream of the volumetric flow measurement device; this upstream location is purposely after the flow straightener pressure losses. Temperature should be the average of measurements taken upstream of the flow straightener and approximately five diameters downstream of the volumetric measuring device using resistance bulbs thermometers. For the gas constant and compressibility term (z) it is essential to analyse a sample using a gas chromatograph and then utilise formulae presented in References 5 and 6. The last measure is necessary as in fact very few gas supplies do not have some variability in the gas composition, and relatively small changes can significantly affect the gas constant and compressibility. Failure to do this can lead to inaccuracies of up to ±5%.

Taking all of the above measures will lead to accuracies for gas fuel mass flow of around ±1%. This is the best accuracy achievable, and is not vastly changed whether combining these effects is via root sum square (for 95% confidence) or arithmetic addition (for 100% confidence). The main inaccuracy is in volumetric flow, at best around ±0.75%, while pressure is only accurate to around ±0.25–±0.5%. Temperature has little effect, being only ±0.1 K if an RBT is used, while the compressibility and gas constant should be accurate if the formulae are used correctly.

Fuel heating value

The fuel heating value can be evaluated to within 0.1% from the gas chromatograph output and the formulae presented in Reference 5. Formula F11.9 enables fuel energy flow to be calculated from volumetric flow, density and fuel heating value, exactly as per liquid fuel.

11.2.5 Air mass flow

Measurement of engine inlet mass flow is vital for various reasons:

- For thrust engines inlet mass flow determines momentum drag, and hence net thrust and SFC.
- For any engine temperature levels in the combustor and HP turbine can only be deter-
- mined by calculation from air mass flow, combustor inlet temperature and fuel energy flow.
- Determining compressor surge margins and component flow capacities requires mass flow to be known.
- For shaft power engines tested without a free power turbine, air mass flow is required in the calculation of exhaust gas power.

Total and static pressures are read in a duct of known area, usually where contraction occurs and increases the dynamic head, along with total temperature. Mass flow is calculated using either Formula F11.16 or Q curves (see Chapter 3). Measurements may also be taken at other stations, and where direct measurement is not possible calculations are performed based on flow continuity using design air system assumptions.

Airmeters

As shown in Fig. 11.9, two similar instruments exist for measuring engine inlet flow:

(1) A *flare* is a short duct with an entry bellmouth, fitted immediately in front of the engine. This imposes only a low pressure drop on the engine.
(2) A *venturi* is a contraction in a longer upstream duct, followed by a diffusing section. It is placed well upstream of any bow wave effects caused by the engine. Geometry dictates that the venturi throat generally gives a larger *depression* – the difference between total and static pressure – than a flare. This depression may then be more accurately measured down to low flow rates.

The term *airmeter* is applied to either device. Both require the use of a discharge coefficient (CD), which is the ratio of effective flow area to geometric area (Formula F11.17). This is less than unity because first, velocity is less than the ideal value, due to skin friction, i.e. total pressure is lost between the upstream measured value and the throat; and second, the flow does not fill the whole of the throat due to boundary layer growth. If the engine face is very close, such as with a flare, bow wave effects may add to this. The discharge coefficient may be determined by several possible methods:

- Cross calibration with an existing airmeter, that has been calibrated on the same engine type and test bed (see section 11.3)
- Evaluation of throat conditions via a high coverage rake, with many total pressure heads
- Use of a standard airmeter geometry and CD calculation such as given in Reference 2
- Theoretical predictions of flow behaviour using *CFD* (computational fluid dynamics) methods

At least six to twelve throat static tappings should be used and read individually to allow error detection; the number depends on size. For a flare total pressure should be that in the cell. For a venturi there may be some pressure loss upstream of the throat, hence total pressure

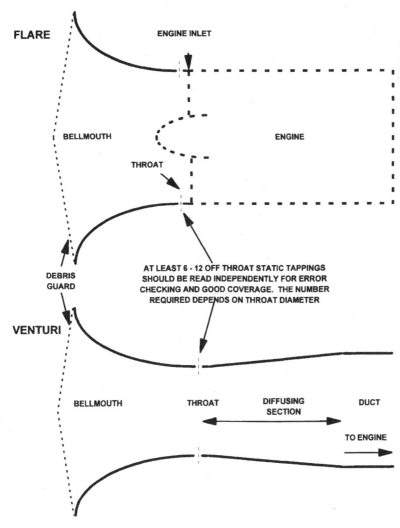

FLARE

ENGINE INLET

BELLMOUTH

ENGINE

THROAT

DEBRIS GUARD

AT LEAST 6 - 12 OFF THROAT STATIC TAPPINGS SHOULD BE READ INDEPENDENTLY FOR ERROR CHECKING AND GOOD COVERAGE. THE NUMBER REQUIRED DEPENDS ON THROAT DIAMETER

VENTURI

BELLMOUTH   **THROAT**   **DIFFUSING SECTION**   **DUCT**

TO ENGINE

Notes:
Upstream temperature measurement normally uses multiple RBTs, mounted on test cell inlet splitters or debris guard.
Upstream pressure measurement is either that for cell static pressure, ambient (barometric), or total probes mounted on the debris guard.
RBT = resistance bulb thermometer
Airmeter throat should be sized such that Mach number does not exceed 0.7.
Reference 2 gives guidelines on airmeter design.

Fig. 11.9   Engine inlet air mass flow measurement.

must also be measured or the loss allowed for by calibration. Total temperature measurement ideally uses RBTs, as discussed in section 11.2.2. Overall, a typical value for accuracy is around 0.5% of airflow, providing the tappings remain clean.

Turbofan core flow
For a turbofan the mass flow actually entering the core is of greatest interest, and may be determined in various ways as discussed in section 11.8.3.

Air system

Once a value has been established for first compressor inlet mass flow, the air system design data for cooling and leakage flows are initially assumed to obtain values for mass flows at other stations. The methods discussed in section 11.8.3, or direct measurements of individual air system flows, are used to check this assumption.

Other measurements

For all engine stations, including entry, measurement of total and static pressures and total temperature may provide an indicative mass flow measurement. This may be calibrated against some true measurement, either directly or using air system data depending on the location.

Transient air mass flow

For transient testing a *flow probe* is often fitted, especially at HP compressor entry. This reads a difference between total and static pressures via a dedicated differential transducer, with short length *balanced* pipework to remove aerodynamic noise. The mass flow is calculated using an effective area and calibration versus steady state data based on inlet flow readings from the airmeter. If the airmeter is a close coupled flare then it may be read transiently, using dedicated transducers with 1 m or 5 m line lengths.

11.2.6   Injected steam flow

Steam may be injected into an engine to reduce emissions and increase power, as discussed in Chapter 12. The supplied steam flow must be measured directly, as least for engine development. This is normally accomplished using a *vortex shedding flow meter*. Here a bluff body within the flow vibrates due to shedding of wakes (just as does a telephone wire in the wind). The frequency of vibration is roughly proportional to the volumetric flow. Such a device is calibrated to determine the exact dependency, versus a choked (*critical*) nozzle.

11.2.7   Thrust

For aero-engines, measuring thrust is essential for three main reasons:

(1)   This is a vitally important, fundamental design goal and will determine whether or not the engine/airframe combination can meet the desired mission.
(2)   It is required to calculate SFC.
(3)   Any inferred component lifing information requires a thrust level to be meaningful.

Figures 11.1, 11.2 and 11.4 show the basic layout of sea level and altitude thrust test beds. An engine is mounted in a cradle and restrained axially via *load cells*, which measure the axial force required. These contain springs and are often preloaded and mounted in opposition to maintain stiffness. Standard practice is to calibrate before and after a test run, often by hanging weights, and use the mean as valid during the test. An indicative accuracy level for the load cell is within 0.25% reading; to convert this to thrust requires test bed calibration, which is discussed in section 11.3. Basically, for indoor test beds the impact of the static pressure field on the engine relative to that in an infinite atmosphere must also be evaluated. This leads to an overall thrust accuracy of around 1%.

In an altitude chamber additional effects contribute to measured thrust as shown in Fig. 11.4. The inlet ducting incorporates a slip joint to prevent axial force transmission to the engine; this joint has a seal area larger than the upstream duct. The actual gross thrust produced by the engine is the measured load cell force plus inlet momentum drag and pressure loads acting on the duct and seal areas. Typical accuracy is around 1.5%. For both test beds, measurement of mass flow is also important, to corroborate predictions of momentum drag at flight conditions.

11.2.8 *Shaft speeds*

Shaft speed measurements provide valuable component information:

- Anticipated turbomachinery performance is strongly influenced by speed, especially for compressors.
- Turbine lives and shaft critical speeds are crucially dependent on speed, and for aero-engines shaft speed levels are a certification issue.
- For shaft power engines output speed is of vital importance, and is also used in the calculation of output power.

A *phonic wheel* and pick up are normally used. This method is highly accurate and reliable, and hence available even as a production measurement. The passing of teeth cut in a wheel, integral with the shaft, is sensed by an electromagnetic coil. The pulses generated are converted to a shaft speed, knowing the number of pulses per revolution. Two main methods may be used for this:

(1) A *clock and pulse counter* suits steady state operation, and as implied counts the number of pulses over a timed period. Accuracy is around 0.1%, assuming engine operation is stable.
(2) A *frequency to DC converter* suits transient operation, and gives an readout of instantaneous shaft speed. Again accuracy is around 0.1%.

11.2.9 *Engine output shaft torque and power*

These are highly interrelated as power is simply the product of torque and rotational speed. The reasons to make such measurements are exactly as discussed in section 11.2.7 for thrust.

A *torquemeter* forms part of a load carrying shaft, and measures torque by sensing the relative rotation of two ends of a length of shaft. Shaft torque must be measured at an appropriate point along the output shafting, as shown in Fig. 11.3, to exclude any losses which are not part of the engine supply. Figure 11.10 illustrates one method where phonic wheels, which also measure speed, are attached to both the shaft and an outer unloaded tube. By this means the teeth are in the same plane, and changes in the waveform picked up indicate angular displacement, and hence torque. Overall torquemeter accuracy is around 0.5–1%, depending on engine stability. An alternative method utilises strain gauges to measure shaft twist, however such devices are delicate and have not been widely successful.

As mentioned in section 11.1.4, a hydraulic dynamometer may also measure torque. Input torque is transmitted to the casing by the water and any bearing friction; the casing is freely mounted externally on bearings to allow torque measurement via a load arm and a weighing system.

11.2.10 *Humidity*

Humidity must be measured so that test data can be referred to either dry or ISO conditions, for evaluation on a defined basis. There are three main methods of humidity measurement:

(1) *Capacitance sensors* utilise changes in the dielectric constant of a water absorbing material.
(2) *Chilled mirrors* utilise changes in reflected light levels as different relative humidity results in different amounts of condensation.
(3) *Wet and dry bulb thermometers* use an air draught to evaporate water from a wet lint on one of the bulbs, and reduce that bulb's temperature reading. Humidity is found from the temperature difference between the wet and dry bulbs, normally via tables such as those in Reference 7. The traditional method was simple glass thermometers mounted on a 'football rattle' frame, though ducted fans and other instruments are now used.

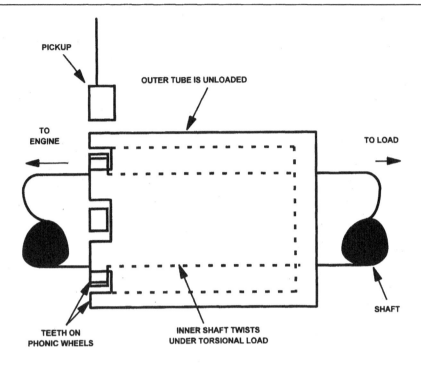

Notes:
Twist of inner shaft changes angular relationship of phonic wheels.
Changes in pickup waveform allow computation of shaft torque.
The type illustrated is known as a 'phase displacement meter'.

Fig. 11.10 Shaft torque measurement.

For all these methods an indicative accuracy level is the greater of 5% relative humidity or 0.2% specific humidity. Chapter 2 provides conversions between relative and specific humidity for various ambient conditions and Chapter 12 discusses the effect of humidity on engine performance.

11.2.11 *Geometric parameters*

Variable stator vanes are common on axial compressors, and recuperated engines often have variable turbine nozzles. To measure vane rotation directly, around three off *RVDTs* should be used. An RVDT is a 'rotary variable displacement transformer', in which the relative movement of electric coils changes the mutual inductance. Accuracy is good, around 0.58, assuming all vanes are held at the same angle. If alternatively the *linear* displacement of the actuator moving the ring is measured, via an LVDT, then other effects are relevant, such as ring tolerances and any slack in the total mechanism. The accuracy of measured vane position is then entirely dependent on the system design.

Turbine throat areas are subject to manufacturing variation and are normally measured geometrically on build, with a repeatability of around 0.1%; method (and hence definition) adjustments may give apparent 'step' changes however. The conversion factor between area and capacity is derived from the turbine aerodynamic design data.

For thrust engines propelling nozzle areas are required. Defining and measuring certain areas may not be trivial, for example, on a turbofan the bypass nozzle plane may contain struts, or if a mixer is employed *its* area is required.

11.3   Test bed calibration

Engine performance measured on indoor test beds will differ from that in an 'infinite atmosphere', for the reasons described in section 11.1. *Test bed calibration* enables data from an indoor test bed to be adjusted to reflect infinite atmosphere conditions. In addition, engines of a given mark may be tested on more than one test bed, each with a different configuration and hence different *calibration factors*.

For shaft power engines where air is ducted from ambient as per typical installations, and the measurement made at duct entry, the test bed configuration does not normally need to be calibrated for. All that remains is the usual requirement for all instrumentation to be functioning satisfactorily.

Test bed definitions

Test bed *approval* is a formal process, involving regulation by the appropriate airworthiness authority for aircraft engines, and high customer involvement for other engine types. The following basic definitions apply, though other similar terms are also used:

- 'Gold' standard test bed
  This is the datum for *cross calibration* of all other beds for that engine type. Such calibration may be performed directly or via calibration of an intermediate 'silver standard test bed.
- 'Silver' test bed
  This has been directly cross calibrated against the gold standard test bed. It may be used for cross calibration of other beds for that engine type.
- 'Bronze' test bed
  This has been calibrated against a silver test bed. It may be used for performance testing, but not for calibrating other test beds.
- Functional test bed
  This test bed has not been calibrated for performance purposes, and provides results that are indicative only. Such a test bed may be useful for endurance testing or demonstrating the functionality of a production engine, but not its performance.

A cross calibration exercise normally involves comparing results obtained running the same engine on two test beds, without any transportation of test bed slave equipment. The benefits of cross calibration are dubious for engine configurations where transportation between test beds would require significant rebuild.

11.3.1   *Calibration of gold standard test bed*

For thrust engines, the gold standard bed for the first engine of a type is declared by calibration versus an outdoor test bed, i.e. infinite atmosphere. For shaft power engines direct calibration of the mass flow measurement is required, by methods discussed in section 11.2.5.

11.3.2   *Steps in test bed cross calibration*

The object of cross calibrating a test bed is to reproduce the performance measurements obtained on the gold standard test bed, after referral to standard conditions, such as ISA sea level static. The steps involved are:

- All test bed instrumentation is calibrated.
- An *uncertainty analysis* is performed to determine the likely errors in all measurements and derived parameters accumulating from all sources. Reference 8 provides guidance on the methods, though clearly much is specific to a particular test bed.

- Calibration curves (i.e. sequences of throttle settings) are run on both test beds, consisting of around a dozen stabilised points from idle to maximum rating.
- Ideally a '1–2–1' sequence is used with the gold or silver bed as '2'; here any faults may be detected based on history. For example, the curve of inlet depression versus mass flow should be well known, and any deviation during the calibration run should be detectable. For the bed to be calibrated there may be no such history, hence having two runs proves repeatability. Other test sequences may be employed where high confidence levels make the cross calibration more of a confirmatory exercise.
- The physical configuration of the bed being calibrated is adjusted during the last test if necessary, with respect to how it affects air flow and hence thrust measurement. This includes inlet and exhaust details, engine position, and auxiliary equipment in the cell. Section 11.1.2 gives guidelines on desirable configuration features.
- Finally calibration factors are derived and are tabulated, versus speed say, for use in all future testing. These are the differences in thrust, air flow and station temperatures and pressures.

11.3.3 *Test bed audit*

Test bed calibration exercises are expensive due to the engine running involved.. Once a test bed has been calibrated the process is rarely repeated unless bed configuration is changed, or there is an inexplicable change in production engine performance. However to maintain test bed approved status it should be audited, typically biannually. Important features to review in an audit are as follows:

- Trend plots of engine and overall parameters, as described in section 11.7.1
- Instrumentation calibration status
- Original test bed calibration against a gold or silver test bed and back to infinite atmosphere
- Test bed configuration status against that formally recorded during the original calibration. Such items as physical layout, and instrumentation types and serial numbers should be examined
- Quality controls for data reduction software
- Actual working practices versus those in the engine test schedules
- Any significant events or changes noted since the previous audit
- Completion status of any 'actions' from previous audits, or recorded since then

11.4   Steady state development testing

11.4.1 *Specific performance tests*

By the end of development the achieved performance standard must be demonstrated. Contractual implications mean both the customer and the engine manufacturer will be keenly interested. During development the standard is achieved via specific performance tests, which have the following aims:

- To validate the steady state performance model, or else establish which components do not achieve the predicted or rig-based, bid level and hence require development effort.
- To determine the effect of component modifications implemented during a development programme. Ideally these are tested singly, in a 'back to back' sequence, however this will not be the case if time is short or modifications are *known* to be of benefit.
- To derive control schedules for ratings and to optimise part load performance. For example having both blow off valves and variable vanes may allow many possible operating settings at part power for a given surge margin level.

A typical test is a performance curve consisting of around five to ten stabilised points. These cover the range from the minimum power or thrust level important for the application (usually idle) to maximum rating. Stabilisation times before taking readings range from around 4 to 20 minutes, the latter if large heat exchangers are utilised. Shorter times do not allow scanivalve pressure readings to stabilise.

For aircraft engines similar performance tests are conducted in an altitude test facility or flying test bed, to simulate the conditions encountered at cruise accurately. Owing to the time spent, altitude cruise is the most important condition for SFC.

11.4.2 *Windmilling*

Windmilling is described comprehensively in Chapter 10. For aircraft engines drag, auxiliary power capability and combustor inlet conditions are measured over a range of Mach numbers and altitudes in either an altitude test facility or flying test bed. Windmill testing is not normally conducted for other engine types.

11.4.3 *Support of engine component development*

During engine development various tests are necessary where evaluating performance is not the goal but defining its level is vital to meet the primary test objective. For interest, examples of such tests are described below.

Ingestion tests for aircraft engines
Ingestion events cannot be avoided for aircraft thrust engines since adequate filtration is impractical. Testing is a regulatory requirement to ensure satisfactory engine behaviour, as outlined below.

- For *bird ingestion* tests, a number of dead birds are fired simultaneously into the engine when operating at maximum rating. The airworthiness requirements are either to maintain 75% thrust for a certain time after the impact, or shut down safely, depending on the size of bird.
- For *water ingestion*, the test requires static ingestion of up to 4% water, to simulate rain. However to address the reduced fan centrifuge effect at flight speeds additional tests are normally carried out, which spray the water either at high pressure or even downstream of the fan.
- For *hail ingestion*, tests involve firing 25 mm and 50 mm hailstones at the engine in a specified pattern.
- At low airflows ice may build up in front of and even on the fan, and be shed and ingested; for *ice ingestion* the test is to ingest the ice which would build up if the anti-icing system were not switched on for 2 minutes after entering icing conditions. Ice has similar thermodynamic effects to water ingestion, but raises additional concerns about mechanical integrity.

Thermal paint
This records the metal temperatures of key 'hot end' components such as combustors and turbine blades. Special paint is applied, which changes to various colours depending on the temperature reached; several paint grades are available to suit different temperature ranges. The engine is first run at low power to heat the carcass to ensure representative seal clearances, hence cooling flows, and then held for between 3 and 10 minutes at the desired high power condition. The paint normally takes 3 minutes to cure, longer times are utilised when engine stabilisation times are long, for example with a recuperated engine, though this must balance the risk of paint degradation. The paint is examined on engine strip after the test, or occasionally via borescope inspection. The indicated component surface temperatures are then related to the temperatures of the gas path and cooling air, assessed from the performance measurements and analysis.

Component temperatures

In addition to thermal paint tests, key component areas, such as disc faces, may be fitted with thermocouples to measure metal temperatures. This allows the assessment of many operating conditions during a build, and provides safety monitoring during high power running or, in case of cooling flow inadequacy, in early development.

Air system

Pressure and temperature readings may be taken in key air system chambers, to validate design data for cooling flows, cooling temperatures and pressure margins.

Straingauging

To assess the susceptibility of gas path components to resonant vibration, *straingauges* are cemented in key locations. These consist of fine wires embedded in a mounting plate; vibration causes the wires to stretch, which changes their electrical resistance. The test normally consists of slow acceleration and deceleration (around 4 minutes each) with various strain gauge selections being recorded. The frequency and amplitude of the measured signal characterises blade response, identifying potentially harmful resonances. Testing explores the full range of mechanical and referred rotational speeds that the engine type will encounter. Mechanical speed determines the frequency of excitation forces, referred speed produces flow regimes that may cause excitation, and pressure level determines amplitude.

Achieving the highest speeds is often a challenge with limited test hardware, though in-service ambient variability may be addressed via an altitude chamber. Very high cycle temperatures may be reached, often requiring builds with special sizes of turbine and propelling nozzles to provide alleviation.

Bearing loads

For this test special straingauged equipment is built in to measure the axial loads on shaft bearings. The engine is run over a representative operating range to allow assessment of bearing lives; stabilisation periods may be long (20 minutes and more) as air system pressure levels depend on seal clearances and hence disc growths.

Emissions

Exhaust gases are sampled, usually via a cruciform rake in the exhaust, and analysed using a *spectrometer* for levels of pollutants such as unburnt hydrocarbons, carbon monoxide and oxides of nitrogen. Such tests are increasingly important for contractual purposes, as legislation becomes more stringent.

11.4.4   Endurance and type testing

All engines complete some form of arduous endurance test before service entry. This may take the following forms.

- Aviation authorities such as the FAA stipulate a *type test* comprising 150 hours' cyclic operation, including SOT levels that are the highest that then may be permitted for any engine in service. The engine manufacturer decides this temperature level, based on hottest day/highest altitude takeoff, with allowances for engine variation, deterioration, etc. The last item in particular is flexible, but if the type test temperature gives low margins relative to new or overhauled engines, the costs of frequent maintenance will be incurred.
- For marine engines one such test has been the US Navy's *qualification test*. This typically consists of 3000 hours cyclic operation, 40% of which has the HP turbine temperature at the level required at 37.8 8C (100 8F) ambient temperature to achieve the power to be cleared. This level must be continually adjusted to address any deterioration, and is normally achieved during the test via 'customer' bleed extraction.

- For industrial engines, testing is normally specified by the engine manufacturer, and typically comprises 100–300 hours of cyclic testing with maximum SOT 10–20 K above that at base load.

11.5 Transient development testing

Transient performance testing aims to confirm that the engine can perform accels and decels (known as *handling*) within specified times. Potential operability problems include surge and weak extinction. Before a new engine variant is tested, transient performance work will normally be based on a transient model of the engine. This is validated using transient test data and can then be used to support testing and prove transient capability at extreme conditions that are impractical to test. Chapter 8 describes transient performance in detail.

11.5.1 *Handling*

The manoeuvres tested exceed the most severe to be met in normal service, in order to give high confidence that engines will behave satisfactorily. Examples include:

- Fast accel, which due to overfuelling gives a high HP working line.
- Slow accels and decels, defined to just avoid tripping the transient bias on handling bleeds. This gives high working lines.
- *Bodie* or *reslam*, as described in Chapter 8, where the engine is decelerated and then almost immediately reaccelerated. Heat soakage raises the working line beyond that of a normal accel, and also within the HP compressor drops the surge line.
- Accel following start-up from cold soak conditions. This gives the highest compressor tip clearances and hence low surge lines.

During these manoeuvres key parameters are measured which allow definition of speed versus time, transient working line excursions, weak extinction margins, etc. A suitable scan rate is 10–50 scans per second. For aircraft engines most work is done on sea level test beds, however some altitude chamber or flying test bed testing is essential to verify effects at the precise conditions to be encountered.

11.5.2 *Surge line measurement*

Surge lines are effected by phenomena such as engine structural loads creating asymmetric tip clearances, inlet distortion, or heat soakage within the HP compressor which lowers the surge line after a decel. It may be necessary to measure actual surge lines in an engine, or at least define a surge free region, rather than relying on rig or predicted data. Key parameters are measured so that the actual trajectory to surge can be defined. Methods of initiating surge are as follows.

For an HP compressor the most usual method is *fuel spiking*. Here the working line is raised by momentarily injecting excess fuel, between 100% and 400% extra over around 200 ms. A suitable scan rate is 100–500 scans per second. Either a slave rig is used, or the standard control system reprogrammed.

For LP compressors or fans the most usual method is to build a development engine with a reduced capacity LP turbine or bypass nozzle respectively, to raise the working line. Bleed extraction is used to enable stable operation at high power; the engine is then slowly decelerated until it surges. This process is repeated with various levels of bleed to map out the surge line.

Other less common techniques include *in bleeding* air downstream of the subject compressor, for example on small engines, or if applicable utilising heat exchanger bypass capabilities. The first surge on a build may increase compressor tip clearance, which raises operating temperatures and lowers the surge line. These effects should be quantified by further testing, and compressors examined by borescope to check for damage. The deterioration from subsequent surges is much less.

11.5.3 *Controller and engine operability tests*

As described in Chapter 8, transient performance and control are inseparable. The initial control strategy and algorithms are further developed during extensive sea level and altitude testing. One particular test of interest is surge recovery, where the controller must ensure that should an engine surge for any reason in service it will recover safely. This is demonstrated during development testing via a surge induced by one of the methods described.

11.5.4 *Starting tests*

Chapter 9 describes engine starting in detail. Specific tests addressing the light up region are:

- For an air starter, per aero-engine practice, measuring crank speed versus starter supply pressure, including that available in service such as from an aircraft APU.
- Starter performance verification. With zero fuel flow, the starter is cut from steady state cranking conditions; knowing the HP shaft inertia the initial deceleration rate indicates the cranking torque.
- Exploration of a matrix of crank speeds and light up flows, identifying the light off 'window' and rotating stall boundaries.
- Sensitivity to variable stator vane positions and bleed flows.

Between light up and idle, tests are required to determine the upper and lower limits of acceptable fuelling. Too much fuel results in compressor stall, which is affected by the bleed and whirl levels. Too little fuel results in *hang*, where at low speed some limitation is reached, such as over temperature, surge or rundown. This may be explored by reducing fuel flow at various starter power levels.

For industrial, marine and automotive engines start testing is usually conducted at the prevailing ambient conditions for the sea level test bed. Problems that may arise in service on cold or hot days are then dealt with as they arise. For aircraft engines cold and hot day start tests must be included in the development programme. Methods include taking the engine to a cold or hot climatic region, chilling the engine in a cold room, or using large electrical heaters at engine inlet. Furthermore altitude re-start capability compliant with the windmill and starter assist envelopes must be proven by testing in an altitude test facility or flying test bed.

A further test for all engine types is a hot restart, shortly after shutdown with the carcass still at high temperature. As well as the effect on seal clearances, heat in the gas path lowers surge lines and raises working lines.

11.5.5 *Engine failure investigation*

Engine mechanical failures during development invariably have a transient dimension. In addition to any transient performance data logging, key mechanical parameters such as vibration, speeds and pressures must be recorded either analogue or at least at 1000 scans per second to cater for such an event. To avoid storing immense amounts of data it may be overwritten, say every 30 minutes.

In the event of a failure this log may be interrogated to distinguish cause and effects. One example is to determine whether a surge occurred before (and hence caused) or after (and hence resulted from) mechanical damage to a compressor. If the surge came first, the amplitude of noise or vibration with a frequency corresponding to the spool speed will increase *after* the step decrease in compressor delivery pressure and step increase in broad band vibration caused by the surge.

11.6 Application testing

Further development testing is often conducted with a new engine installed in its application, prior to production release. This enables the following key issues to be addressed.

- Overall performance levels are measured in the application, as opposed to a test bed where calibration factors are required as described in section 11.3. This is particularly important for aircraft thrust engines, where test bed effects are most significant. Such testing usually has high commercial importance, as it is the final determination of whether or not the engine/airframe combination meets its performance guarantees.
- Use of 'real' hardware for accessories and installation ducting, rather than any slave test bed items.
- Engine mechanical integrity faced with representative structural loads, vibration levels, ambient conditions and operating profiles.

11.7 Production pass off

11.7.1 *Production pass off test*

It is common practice that each engine about to be delivered to a customer, whether new production, overhauled or repair, undergoes a *pass off* test. The inevitable variation in manufacturing and build dimensions, even within permitted tolerances, results in variation of component and overall engine performance as described in Chapter 6. A typical test sequence is as follows:

- Starting: Whilst an unavoidable test this is nonetheless vital.
- *Running in*: Here rotor seals are made to cut gradually into the static linings, by say a series of increasingly rapid accelerations, extended dry cranking, or dwells at increasing power or thrust levels.
- Performance curve: Here assessments are made of SFC, thrust or power versus temperature, working lines, shaft speeds, etc.
- Handling tests: Accels and decels are carried out to prove the engine meets its specification requirements.
- Control system set up: Any required control systems stops, trims and limiters are set. Trims are adjustments to measurement signals to address engine to engine variations.

For practicality an exception is made for heavyweight gas turbines however.

11.7.2 *Acceptance criteria*

There are various limiting levels which engine performance parameters must meet for pass off. These usually include demonstration of guaranteed power or thrust without exceeding set temperature and speed limits, and achieving guaranteed SFC at this power or thrust. The acceptance criteria are usually stated at some standard conditions, to which measured data is referred as outlined in section 11.8.3. Engines may initially fail their pass off test due to faults with instrumentation or less usually with the build.

In addition to performance parameters certain mechanical integrity issues are also addressed, such as fuel leaks, vibration, oil consumption, measured temperature spread, etc.

11.7.3 *Engine performance trends*

Changes in component manufacture or build practice may cause increasing numbers of engines to fail the pass off test. Identifying these changes early amongst the general scatter is challenging. Though some special instrumentation may be fitted for the pass off test, more modern

Engine no. (−)	SOT (K)	SOT-1200 (K)	CUSUM (K)	Engine no. (−)	SOT (K)	SOT-1200 (K)	CUSUM (K)
1	1206	6	6	17	1203	3	21
2	1201	1	7	18	1199	−1	20
3	1197	−3	4	19	1205	5	25
4	1214	14	18	20	1212	12	37
5	1204	4	22	21	1195	−5	32
6	1208	8	30	22	1211	11	43
7	1190	−10	20	23	1202	2	45
8	1196	−4	16	24	1207	7	52
9	1205	5	21	25	1214	14	66
10	1192	−8	13	26	1190	−10	56
11	1207	7	20	27	1201	1	57
12	1202	2	22	28	1210	10	67
13	1201	1	23	29	1207	7	74
14	1196	−4	19	30	1213	13	87
15	1193	−7	12	31	1196	−4	83
16	1206	6	18	32	1211	11	94

(a) Tabular test data and CUSUM

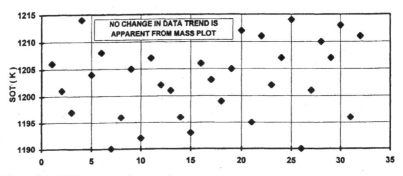

(b) Mass plot: SOT versus engine number

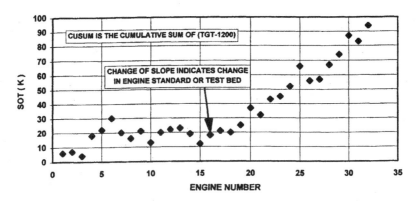

(c) CUSUM plot: SOT versus engine number

Fig. 11.11   Test data trending: 'mass' and CUSUM plots.

practice is to rely on that of the production control system. Instrumentation is therefore often too sparse or inaccurate to suit the conventional analysis methods described in section 11.8. Though these may be employed additional methods are necessary, which emphasise comparison with other engines. Figure 11.11 shows an example where pass off data from a production run is plotted for LP turbine inlet temperature:

- Mass plots: These show pass off performance parameters of interest plotted versus engine number over a time period. They truly define mean engine performance and scatter, and are useful for detecting measurement errors and 'rogue' engines.
- CUSUM plots: These plot the cumulative sum of the difference between each engine's temperature (say) and an arbitrary value within its scatter band, versus engine number. Changes of slope of the mean line indicate changes of trend that may be almost impossible to detect from the mass plot. Formula F11.18 defines this.

Reference 9 provides guidance on gas turbine acceptance tests, including such details as instrumentation accuracies.

11.8   Test data analysis

11.8.1   *Steps in test data analysis*

Test data requires considerable processing before engineering conclusions may be drawn:

(1)   The instrument signals must be converted from say millivolts (mV) to engineering units of K, kPa, etc.
(2)   Erroneous readings must be detected.
(3)   Where there are a number of readings at a station they must be averaged.
(4)   The implied performance of both the engine and individual components must be computed. An error band analysis is essential for key parameters.
(5)   The evaluated component and engine performance levels must be compared with expectation.
(6)   The reasons for any differences must be ascertained.

These steps do not always occur in this ideal order. Measurement errors are not necessarily easily detectable, for example they may be systematic and apply to all readings at one station. The analysis process may take a long time to 'finish' as new facts alter the understanding. Furthermore, certain methods involving cycle matching combine steps 4 and 5.

11.8.2   *Initial error detection and data averaging*

It is essential before using measured data to carry out error detection, for which some techniques are listed below.

- Compare rake head readings at the same radial immersion. Low pressure values usually indicate leaks, and any outlying temperature readings are probably in error.
- Plot readings on each rake versus immersion. This gives a finer check on validity as the 'rake profiles' should show a repeatable pattern, relative both to each other and to historical data, especially for compressor rakes.
- Check that any total to static pressure ratios are sensible.
- Compare cell and barometric pressures, especially for altitude chamber tests.
- Plot air meter static tappings versus circumferential position and compare with historical data.
- Compare fuel flow calculated from the different volumetric devices arranged in series.

• Instrumentation readings should be consistent with other readings near that station, such as from the controller or air system instrumentation.

Rake data requires averaging to form a single thermodynamic mean value at each instrumented station, as represented by the performance model. Such an average is usually either *area weighted* or *mass weighted*, as defined by Formulae F11.19 and F11.20 respectively; Formula F11.21 shows how heads may be placed on centres of equal area to simplify the averaging. Mass weighting is the ideal, but as it requires an accurate knowledge of the static pressure profile, which is difficult when swirl is present, area weighting is more usual. Either way, for the average to be meaningful any faulty heads must be detected and eliminated.

11.8.3   *Test bed analysis (TBA) calculations*

These compute engine and component performance levels, e.g. thrust, SFC, isentropic efficiencies, directly from the measurements. Calculations proceed sequentially through the engine, evaluating almost all parameters at each station. This requires assumptions for bleeds, pressure losses and power offtakes. Turbine exit conditions are assessed based on work, measured in the compressors and any power offtakes. Sample calculation C11.1 illustrates the approach.

TBA calculations may be performed either at the tested engine condition or at standard conditions such as ISA sea level or ISO, as per Chapter 2. In the latter case the measurements are referred using theta (θ), delta (δ) and the appropriate non-dimensional groups as described in Chapter 4. To account for variation in gas properties due to changing temperature and fuel air ratio, theta is raised to exponents other than the standard 1.0 or 0.5, known as *theta exponents*. The actual values are obtained by running the engine synthesis model. For illustration, Fig. 11.12 presents a typical likely range. Sample calculation C11.1 includes a referral to standard conditions.

At this stage, instrumentation error detection is again essential, using the derived parameters. Errors may be detected based on considerations such as the following:

Parameter	Theoretical exponent	Typical exponent
Temperatures	1	0.97–1.03
Pressures	0	0.0–0.06
Fuel flow	0.5	0.64–0.76
Shaft speeds	0.5	0.48–0.50
Air mass flow	0.5	0.47–0.52

Notes:
Theta is defined as $T1/288.15\,K$ (where T1 is engine inlet temperature).
To refer measured data to standard conditions raise theta to the powers shown. Chapter 4 describes theta, non-dimensional groups, and their usage.
Differences between theoretical and typical values reflect changes in gas properties due to fuel air ratio and temperature changes, described in Chapter 3.
Theta exponents for a particular engine type are obtained by running a thermodynamic model.

Uses of theta exponents:
To convert engine performance parameters to what would be obtained at other inlet temperatures. For example, test data may be referred to standard conditions for comparison with expectation.

Fig. 11.12   Theta exponents for performance data referral.

- For a fixed geometry engine with fixed bleed configuration the plot of any parameter against another should be smooth. If not, plotting both against any third parameter should indicate which reading is in error.
- Calculated component efficiencies should not exhibit high scatter, and should of course lie below 100%. Genuine hardware defects may give values below rig levels.
- Calculated component flow capacities should match any available measured values, such as based on measured throat areas.
- For a thrust engine, propelling nozzle coefficients should be uniform versus thrust level and again should resemble expectation levels.
- If an airmeter is fitted the non-dimensional flow should fall on a unique plot versus depression, be it that of the cell, engine inlet or throat.
- Apparent 'reciprocal' component changes are usually due to a single measurement error, for example the pressure or temperature between two compressors.

One special case regarding TBA calculations is a turbofan, where the core mass flow is not measured directly but must be calculated. Several methods exist, and for convenience the two

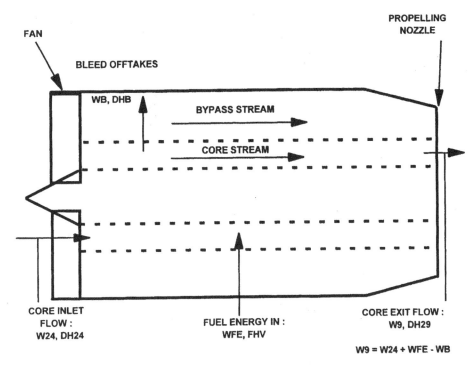

Heat (enthalpy) balance:
W24 * DH24 + WFE * FHV = WB * DHB + (W24 + WFE − WB) * DH9
Hence W24 = (WFE * (DH9 − FHV) + WB * (DHB − DH9))/(DH24 − DH9)

Notes:
Calculate specific enthalpy from temperature and fuel air ratio via formula F11.22
   Units are kJ/kg and are typically converted to be relative to a datum of 0 8C if a mean specific heat is utilised
The following parameters are measured directly:
   Bleed offtake flow, WB (kg/s)
   Fuel flow, WFE (kg/s)
   Core inlet and exit temperatures, T24 and T9 (K)

Fig. 11.13   Turbofan mass flow calculation: method 'A' core heat balance.

most commonly used are termed methods 'A' and 'B' herein. Method 'A' uses a core heat balance (see Fig. 11.13) and method 'B' employs a knowledge of HP turbine capacity. Both methods are illustrated in sample calculation C11.2. An alternative approach is a transient mass flow probe, which gives indicative rather than absolutely accurate flow measurements (see section 11.2.5).

11.8.4 *Uncertainty analysis*

For key parameters produced by the TBA calculations it is essential to evaluate the likely error band resulting from the measurement uncertainty. The first step is to evaluate the potential errors in a calculated parameter resulting from that in each contributory measurement individually; the second step is to combine the effects of these errors. The most common practice is a root-sum-square addition, for 95% confidence. For 99% confidence, simple arithmetic addition is used. Sample calculation C11.3 illustrates these two methods for measured power and reference 8 gives further guidance. The process is also called *error band analysis*.

11.8.5 *Comparison of test results with expectation*

It is usual to produce a *pre test* matching model of an engine going to test. For development tests the effects of measured throat areas and any other specific hardware are normally included, whereas for a production engine the expectation – and indeed the requirement – will remain constant. It may be simply expressed via charts relating the measured parameters to thrust or power at standard conditions.

Valid comparison of the test results with expectation requires that both are based on identical conditions of inlet pressure and temperature, installation losses, etc. In addition, for contractual purposes engine performance is normally required at some standard conditions, as stated. To achieve these aims there are two main approaches:

(1) Referral of measurements to standard conditions, as described in section 11.8.3. This provides engine performance data directly usable for contractual purposes.
(2) Running a synthesis model at the tested power or thrust and ambient conditions. This is more rigorous, and is mandatory for engines which do not behave non-dimensionally as discussed in Chapter 4. In this case the model is adjusted to match the test results, as described in section 11.8.6 below, and run again to predict performance levels at standard operating conditions.

For both approaches the model and test parameters are plotted versus some power setting parameter. For referred data it must be in a non-dimensional form.

11.8.6 *Evaluation of engine component performance*

Various methods exist for attempting to explain anomalies, which will be due to either unidentified measurement error or engine component performance differences from prediction. Test data analysis is notoriously difficult, as once dealing with 'real' data the possibilities for error are vast. The various methods are described below, and should preferably be used in combination.

Traditional methods

Engineering judgement and logical thought have long been used for test data analysis. Knowledge of what *could* be suspect is invaluable, and may be based on known potential causes of component deficiency, the state of engine hardware on strip, other knowledge of the test such as what checks were performed for leakage, etc. Comparison of predicted and TBA component efficiencies and capacities gives one view of component performance levels, though being vulnerable to measurement error and rematch effects it is not definitive. This traditional approach, together with some trial and error, may be used to determine changes to the component performance in the pre-test prediction model which allow it to reproduce all test measurements deemed valid across the power range.

Achieving alignment between a revised model and the measurements is assisted by tables of exchange rates. *Synthesis exchange rates* are described in Chapter 7 and show the change in engine performance parameters if a given component efficiency is in error by 1%. *Analysis exchange rates* show the error in a parameter from the TBA calculations due to a 1% error in each contributory measurement. The advantages of traditional methods are that the result of such an analysis carries high credibility, as the process is easily understood, and no additional setting up is required. The disadvantage is that unlike the more automated methods described below no immediate answer is available.

Cycle matching

Here an extended performance matching model is run at the tested conditions. It is used both to compare test data with expectation *and* to explain the results by changes in component performance. Chapter 7 describes performance matching models. For test analysis the matching is extended with measurement values as additional matching constraints, and component scale factors as additional matching guesses. The model thereby changes component performance to replicate the input measurement values.

The main advantage is that analysis is performed automatically, and at the tested condition; this is more accurate than referring test data. In addition engine station data may be produced with however few measurements being taken, parameters at other engine stations being calculated based on the predicted component performance. There are some disadvantages however. The method can only ascribe differences to component factors that are active in the matching, and the number of these is limited to the number of measurements; if some other component is at fault the answers are wrong. The matching iteration may not always work, or be ill-conditioned such that measurement errors even down to the repeatability level produce large, mutually cancelling changes in several components. Backup calculations are therefore required in on line applications.

Matching methods have been successfully used by manufacturers of large turbofans and by some, though not all, airliner manufacturers. Use for industrial engines is not yet widespread.

Methods based on probability

Here a computer program 'adjusts' *both* the measurement values *and* component performance levels so that calculated and measured station parameters match, using tables of exchange rates. By definition there is an infinite number of solutions. That given is *the most likely* solution, based on input values of the probability of error in components and measurements. Faced with no formal solution this gives a likely answer, and may produce useful suggestions. The solution is not unique however, and carries little or no weight without engineering substantiation.

11.8.7   Transient data

Here considerable processing is required to align instrument readings with the equivalent steady state parameters, given the reduced instrument coverage and transducer drift. Once this has been achieved, transient traces are compared with transient performance model predictions, and the model is then adjusted to achieve alignment. The model must already have a good alignment with engine *steady state* parameters, and adjustments are now made mainly in the modelling of heat soakage for transient operation, where real thermal masses and heat transfer coefficients are not normally known to great accuracy anyway.

All formulae presented here employ specific heat and gamma. For rigorous calculations enthalpy should be used. Chapter 3 provides the methodology required.

Formulae

F11.1   Test bed entrainment ratio = fn(entrained air flow (kg/s),
air temperature (K), engine exhaust gas temperature (K), mixed stack
exit temperature (K))

Wair/Wgas = (Hgas − Hmix)/(Hmix − Hair)

(i)   Temperatures must be measured values.
(ii) H is enthalpy, obtained from Formula F3.26 or 3.27 based on temperature and fuel
      air ratio.
(iii) Minor iteration may be required on fuel air ratio at stack exit to ensure accurate gas
      properties.

F11.2   Flying test bed, inlet air flow (kg/s) = fn(nozzle flow (kg/s),
offtake flows (kg/s), entrained flow (kg/s), fuel flow (kg/s))

W1 = W9 + WB − Wentr. − WF

F11.3   Flying test bed, net thrust (N) = fn(inlet air flow (kg/s), inlet velocity (m/s),
exit air flow (kg/s), exit velocity (m/s), nozzle static pressure (kPa), ambient
pressure (kPa), nozzle area (m^2))

FN = (W9 * VN) − (W1 * VTAS) + (PS9 − Pamb) * A9/1000

F11.4   Water manometer: effect of temperature on reading

DL = 0.21 * 10^−3 * DT

(i)   DL (%) is the change in the column height reading due to thermal expansion of water
      and glass.
(ii) DT (C) is a temperature change.
(iii) The manufacturer will state the water temperature at which the manometer was cali-
      brated. This correction accounts for the difference between that and the prevailing water
      temperature.

F11.5   Mercury barometer and manometer: effect of temperature on reading

DL = 0.18 * 10^−3 * DT

(i)   DL (%) is the change in the column height reading due to thermal expansion of mercury
      and brass.
(ii) DT (C) is a temperature change.
(iii) For example, 20 8C temperature increase changes the column height by 0.36%.
(iv) The manufacturer will state the temperature at which the manometer was calibrated.
      This correction accounts for the difference between that and the prevailing temperature.

F11.6   Mercury barometer: effect of latitude on reading

DL = 0.06604 − 0.134257142 * (cos(LAT))^2

(i)   DL (mm) is the change in the column height reading due to changes in the centrifugal
      force of the Earth's rotation.
(ii) LAT (deg) is latitude.
(iii) For example, the latitude difference between the central UK and Egypt changes the
      column height by 0.24%.
(iv) The manufacturer will state the latitude at which the barometer was calibrated. This
      correction accounts for the difference between that and the latitude at which it is used.

F11.7   Mercury barometer: effect of differential elevation on reading

$$DL = -8 * 10^\wedge - 6 * DELV$$

(i)   DL (mm) is the change in the column height reading due to changes in the centrifugal force of the Earth's rotation.

(ii)   DELV (m) is the difference in elevation above sea level (*not* pressure altitude).

(iii)   For example, 5000 m elevation difference changes the column height by 0.05%.

(iv)   The manufacturer will state the elevation at which the barometer was calibrated. This correction accounts for the difference between that and the elevation at which it is used.

F11.8   Recovery factor in temperature measurement = fn(total temperature (K), static temperature (K), measured temperature (K))

$$RF = (Tmeas - TS)/(T - TS)$$

F11.9   Fuel energy flow (kW) = fn(volumetric flow (m^3/s), fuel heating value (kJ/kg), density (kg/m^3))

$$QU = UVOL * FHV * RHO$$

F11.10   Fuel specific gravity = fn(actual fuel density (kg/m^3), standard density (kg/m^3))

$$FSG = RHOmeas/1000$$

(i)   This is also called 'relative density'.

(ii)   1000 kg/m^3 is the standard density of water at 4 8C.

F11.11   Fuel specific gravity = fn(sample specific gravity, actual fuel temperature (8C), sample temperature (8C))

$$FSG = FSGsample - (Tfuel - Tsample) * 0.00074$$

(i)   'Sample' refers to a fuel sample for which both the SG and temperature have been measured.

(ii)   To convert from FSG to actual density multiply by 1000 kg/m^3.

F11.12   Fuel heating value (kJ/kg) = fn(hydrogen content (% mass), sulphur content (% mass), FSG at 15 8C)

$$FHV = 37\,290 + 566 * Hyd - 330 * S - 2300 * FSG$$

(i)   The remainder of the composition is the percentage carbon content by mass.

(ii)   The formula is valid for FSG values between 0.79 and 0.83, hydrogen content 13–14.1%, sulphur content up to 0.3%. Over a wider range it provides indicative data only.

(iii)   To convert from FSG at 15 8C to actual density multiply by 1000 kg/m^3 and use Formula F11.11.

F11.13   Kerosene fuel heating value (kJ/kg) = fn(FSG at 15 8C)

$$FHV = 48142.3 - 548.05 * FSG - 6850 * FSG^\wedge 2$$

(i)   To convert from SG at 15 8C to actual density multiply by 1000 kg/m^3 and use Formula F11.11.

F11.14   Fuel flow (kg/h) = fn(injector fuel pressure (kPa), combustor pressure (kPa), flow number (kg/h $\sqrt{kPa}$))

$$WF = SQRT(PF - P4) * FLOW\_NO$$

F11.15   **Gas fuel density (kg/m^3) = fn(static pressure (kPa), gas constant (kJ/kg K), temperature (K), compressibility)**

RHO = P/(R * T * z)

F11.16   **Air mass flow parameter group Q (kg $\sqrt{K}$/s kPa m^2) = fn(gamma, total pressure (kPa), static pressure (kPa), gas constant (J/kg K), discharge coefficient, geometric area (m^2))**

Q = W * SQRT(T)/(Aeffective * CD * P)
  = 1000 * SQRT(2 * γ/((γ − 1) * R) * (PT/PS)^(−2/γ)
    *(1 − (PT/PS)^((1 − γ)/γ)))

F11.17   **Effective flow area (m^2) = fn(discharge coefficient, geometric area (m^2))**

AE = CD * Ageom

F11.18   **Next CUSUM = fn(previous CUSUM, next test data value, arbitary datum)**

CUSUMn = CUSUMm + (VALn − DATUM)

(i)   n is the current tested engine serial number.
(ii)   m is the previous tested engine serial number; usually m = n − 1.

F11.19   **Area weighted rake average = fn(number of heads, all head readings, weighting coefficients)**

For each radius where heads are placed, RingAve = Σ(HR)/Nrakes

(i)   RingAve is the *ring average*, the average of all valid heads at the radius considered.
(ii)   HR is the reading of each valid rake head at the radius considered.
(iii) Nrakes is the number of rakes at the station considered.

For each radius, AWC = (RADp^2 − RADm^2)/(RADo^2 − RADi^2)

(i)   AWC is the area weighting coefficient for the nth head.
(ii)   Heads are shown numbered in a radial sequence starting at the inner.
(iii) p is the number of the head at the next radius, p = n + 1.
(iv)   m is the number of the head at the previous radius, m = n − 1.
(v)   RADi is the annulus inner radius.
(vi)   RADo is the annulus outer radius.
(vii) For first and last heads, use annulus radii in place of RADm and RADp respectively.

Overall AWRA = Σ(RingAve * AWC)

(i)   AWRA is the overall area weighted rake average.
(ii)   RingAve is the ring average, the average of all valid heads at each radius.
(iii) AWC is the area weighting coefficient for each radius.

F11.20   **Mass weighted rake average**

Overall MWRA = Σ(RingAve * MWC)

(i)   MWRA is the overall mass weighted rake average.
(ii)   RingAve is the ring average, the average of all valid heads at each radius, per Formula F11.19.
(iii) MWC is the mass weighting coefficient for each radius.

For each radius, MWC = AWC * W/Wtotal

(i)   AWC is the area weighting coefficient per Formula F11.19.
(ii) W is the mass flow associated with the heads at that radius, using Formula F11.16.
(iii) Values for total pressure, temperature and static pressure are interpolated from other station measurements.
(iv) Wtotal is the sum of the mass flows for all radii.
(v)   Flow area is simply AWC times the annulus area.

$$A = AWC * PI * (RADo^2 - RADi^2)$$

(i)   RADi is the annulus inner radius.
(ii) RADo is the annulus outer radius.

F11.21   To place rake heads on centres of equal area, head radius (m) = fn(annulus radii (m), number of radial head positions)

For the first head:

$$RAD1 = SQRT(RADi^2 + (RADo^2 - RADi^2)/2 * N)$$

For the remaining heads:

$$RADn = SQRT(RADm^2 + (2 * n - 1) * (RADo^2 - RADi^2)/N)$$

(i)   RADi is the annulus inner radius.
(ii) RADo is the annulus outer radius.
(iii) Heads are shown numbered in a radial sequence starting at the inner.
(iv) n is the number of the head at the radius in question, n = 2 to N.
(v)   m is the number of the head at the previous radius, m = n − 1.

F11.22   Method 'A' core heat balance for turbofan, core mass flow (kg/s) = fn(fuel flow (kg/h), nozzle exit temperature (K), fuel LHV (kJ/kg), core bleed offtake flow (kg/s), bleed temperature (K), core inlet temperature (K))

CP at mean T form

$$W24 = (WF * (DT9 * CP.9 - FHV) \\ + WB * (DTB * CP.B - DT9 * CP.9))/(DT24 * CP24 - DT9 * CP.9)$$

(i)   This form is normally used with specific heat values based on the mean of the actual temperature and 0 8C, from Formulae F3.23 and F3.24, and DT as the temperature difference from 0 8C.
(ii) Minor iteration may be required to ensure valid fuel air ratio and hence gas properties.
(iii) Alternatively the following, standardised specific heat values may be used:
      Compressors: CP = 1.005 kJ/kg (as per g = 1.4)
      Turbines: CP = 1.150 kJ/kg (as per g = 1.333)
(iv) See sample calculation C11.2.

Rigorous form

$$W24 = (WF * (DH9 - FHV) + WB * (DHB - DH9))/(DH24 - DH9)$$

(i)   This uses specific enthalpy values from Formula F3.26 or F3.27; again the datum is often taken as 0 8C, though in this form any value will be correct.
(ii) Minor iteration may be required to ensure valid fuel air ratio and hence gas properties.

Note: In all cases a certain amount of heat will be lost from the engine, known as 'wild heat'. The amount depends on the engine design, but is usually between 0.5% and 1%.

Sample calculations

For all sample calculations presented herein fixed CP and γ have been used as below:

Compressors: CP = 1.005 kJ/kg, $\gamma = 1.4$, $\gamma/(\gamma - 1) = 3.5$
Turbines: CP = 1.150 kJ/kg, $\gamma = 1.333$, $\gamma/(\gamma - 1) = 4.0$

For engine test data analysis purposes this is an unacceptable error, either CP at mean T, or ideally the fully rigorous enthalpy/entropy polynomials presented in Chapter 3 must be used. Sample calculation C3.2 shows how these may be incoporated into the test data analysis calculations presented below.

Other simplifications are made for the examples presented here, such as the air system is not considered and the simple formula for combustor temperature rise (F3.40) is used. Furthermore, effects such as humidity are not considered. The reader may incoporate these based on sample calculations presented in other chapters.

C11.1   **Conduct (i) test bed analysis calculations for a turbojet, and (ii) the referral to standard conditions. The test data after converting to engineering units, error detection and averaging of multiple readings is as follows:**

Rotational speed	30 900 rpm
Ambient pressure	103.1 kPa
Cell depression	0.6 kPa
Engine inlet total temperature	294.2 K
Airmeter depression	5.1 kPa
(open limb of manometer is at ambient, negligible pressure loss).	0.95, 0.035 m²
Airmeter CD, A	
Compressor delivery pressure	458 kPa
Compressor delivery temperature	486 K
Fuel flow	0.133 litres/s
Fuel temperature	19 8C
Laboratory test fuel SG	0.821 at 14 8C
Laboratory test fuel LHV	43 150 kJ/kg
Turbine exit temperature	1170 K
(only measured at 4 points, 1 immersion)	
Turbine exit pressure	233 kPa
Measured thrust	2730 N

Other information:

Compressor exit diffuser pseudo loss coefficient	0.6
Combustor cold and hot pseudo loss coefficients	1.1, 0.06
Combustor efficiency from loading characteristic	0.998
Test bed calibration thrust loss relative to an outdoor test bed	2.3%

Consider there to be no air system extractions.

F11.17 $Q = W * SQRT(T)/(Aeffective * CD * P)$
        $= 1000 * SQRT(2 * \gamma/((\gamma - 1) * R)$
        $* (PT/PS)^{\wedge}(-2/\gamma) * (1 - (PT/PS)^{\wedge}((1 - \gamma)/\gamma)))$

F11.10 $FSG = RHOmeas/1000$

F11.11 $FSG = FSGsample - (Tfuel - Tsample) * 0.00074$

(i) *Calculate parameters at test ambient conditions*
Engine inlet conditions

$$P1 = 103.1 - 0.6$$
$$P1 = 102.5\,\text{kPa}$$

$$P1/PSairmeter = 102.5/(103.1 - 5.1)$$
$$P1/PSairmeter = 1.046$$

$$T1 = 294.2\,\text{K}$$

$$Q = 1000 * SQRT(2 * 1.4/(0.4 * 287.05)$$
$$* 1.046^{\wedge}(-1.486) * (1 - 1.046^{\wedge}(-0.2857)))$$
$$Q = 17.065\,\text{kg}\sqrt{\text{K}}/\text{s}\,\text{kPa}\,\text{m}^2$$

$$W1 = 17.065 * 0.035 * 102.5/294.4^{\wedge}0.5$$
$$W1 = 3.57\,\text{kg/s}$$

Compressor and intake combined
Apply F5.1.3 and F5.1.2:

$$P3/P1 = 458/102.5$$
$$P3/P1 = 4.47$$

$$E2 = 294.4/(486 - 294.4) * (4.47^{\wedge}(1/3.5) - 1)$$
$$E2 = 82.1\%$$

$$PW2 = 3.57 * 1.005 * (486 - 294.4)$$
$$PW2 = 687\,\text{kW}$$

Note: The intake could have been separated from the compressor by either having another total pressure measurement at the compressor face, or by using the design 'pseudo loss coefficient' to calculate P2.

Compressor exit diffuser
Total temperature is unchanged, derive exit pressure by applying percentage pressure loss:

$$T31 = T3 = 486\,\text{K}$$

$$P31 = 458 * (1 - 0.61 * (3.57 * 486^{\wedge}0.5/458)^{\wedge}2)$$
$$P31 = 449.9\,\text{kPa}$$

$$W31 = 3.57\,\text{kg/s}$$

Combustor
Apply F11.10 and F11.11 to find fuel flow:

$$FSG = 0.8221 - (19.3 - 14.1) * 0.00074$$
$$FSG = 0.8183$$

$$RHOfuel = 0.8183 * 1000$$
$$RHOfuel = 818.3\,\text{kg/m}^3$$

$$WF = 818.3 * 0.101E\text{-}03$$
$$WF = 0.0826\,\text{kg/s}$$

$$FAR = 0.0826/3.57$$
$$FAR = 0.0231$$

Apply F3.40:

$$0.0231 = 1.15 * (T4 - 486)/0.998/43\,150$$
$$T4 = 1351$$

Calculate P4 using F5.7.9 and F5.7.10:

$$P4 = 449.9 * (1 - 1.1 * (3.57 * 486^{0.5}/449.9)^{2}$$
$$-0.06 * (3.57 * 486^{0.5}/449.9)^{2} * (1351/486 - 1)$$
$$P4 = 433.3\,kPa$$

$$W4 = 3.57 + 0.0826$$
$$W4 = 3.653\,kg/s$$

Turbine
No cooling air is considered hence parameters at the SOT station 41 are equal to those at station 4. Equate power to that of compressor and use F5.9.2 to derive exit temperature:

$$PW415 = 687/0.995$$
$$PW415 = 690\,kW$$

$$690 = 3.653 * 1.15 * (1351 - T5)$$
$$T5 = 1186\,K$$

Note: The measured temperature = 1170 K, the calculated value is used here as the rake coverage is insufficient to measure T5 while properly accounting for the temperature profile resulting from combustor OTDF.

Derive E41 via F3.44:

$$P41Q5 = 433.3/233$$
$$P41Q5 = 1.86$$

$$1351 - 1186 = E41 * 1351 * (1 - 1.86^{(-1/4)})$$
$$E41 = 0.893$$

Final calculations
Correct measured net thrust by applying test bed calibration factor to measured value.
$$FN = 2730 * 1.023$$
$$FN = 2793$$

$$SFC = 0.0826 * 3600/2793$$
$$SFC = 0.106\,kg/N\,h$$

Notes: Many other parameters may be calculated using the methods provided for design point and off design performance calculations in Chapters 6 and 7. For instance, the pressures and temperatures through to the nozzle could have been calculated using a jet pipe pseudo loss coefficient. Where a parameter can be calculated in two ways it is good practice to do so, as this provides further checks on the accuracy of the measurements.

(ii)   Refer parameters to standard day conditions
$$THETA = 294.2/288.15$$
$$THETA = 1.021$$

$$DELTA = 102.5/101.325$$
$$DELTA = 1.0116$$

Refer to standard conditions using the parameter groups provided in Chapter 4:

$W1R = 3.57 * 1.021^{0.5}/1.0116$
$W1R = 3.566\,kg/s$

$T3R = 486/1.021$
$T3R = 476$

$P3R = 458/1.0116$
$P3R = 452.7\,kPa$

$WFR = 0.0826/(1.021^{0.5} * 1.0116)$
$WFR = 0.0808\,kg/s$

$FNR = 2793/1.0116$
$FNR = 2761\,N$

$SFC = 0.106/1.021^{0.5}$
$SFC = 0.105\,k/N\,h$

Note: For the above, the propelling nozzle has been considered to be unchoked and hence the full gross thrust parameter is not invoked. As described in section 11.8.3, 'non-standard' theta and delta exponents may be developed using the steady state performance model to cater for real effects.

C11.2 Calculate the core mass flow for a turbofan using (i) method 'A' and (ii) method 'B' based on the following measurements:

Core compressor inlet temperature	346.4 K
Core turbine outlet temperature	1001 K
Fuel flow	0.188 kg/s
Fuel LHV	43100 kJ/kg
Air bleeds leaving the control volume	0.171 kg/s
Air bleed temperature	671.3 K
HP turbine capacity	$0.318\,kg\,\sqrt{K}/s\,kPa$
Combustor inlet temperature	671.3 K
Combustor inlet pressure	1291 kPa
Combustor pressure loss coefficient	$0.82\,s^2.kPa^2/kg^2.K$

(i) Method 'A'
As per Fig. 11.13 and Formula F11.22:

$W24 * 1.005 * (346.4 - 273.15) + 0.188 * 43\,100$
$= 0.171 * 1.005 * (671.3 - 273.15)$
$\quad + (W24 + 0.188 - 0.171) * 1.15 * (1001 - 273.15)$
$73.62 * W24 + 8102.8 = 68.424 + 837.0 * W24 + 14.23$
$763.4 * W24 = 7020.1$
$W24 = 10.51\,kg/s$

(ii) Method 'B'
Guess combustor inlet mass flow $= 10.0\,kg/s$. Then apply F3.40:

$FAR = 0.188/10$
$FAR = 0.0188$

$0.0188 = 1.15 * (T4 - 671.3)/0.987/43\,100$
$T4 = 1366.7\,K$

$P4 = 1291 * (1 - 0.82 * (10.188 * 671.3^{0.5}/1291)^2)$
$P4 = 1247\,kPa$

Qcalculated = 10.188 * 1366.7^0.5/1247
Qcalculated = 0.302 kg√K/s kPa

Now calculate error between Qmeasured and Qcalculated:

Error = (0.302 − 0.318)/0.318 * 100
Error = 5.0%

Reguess combustor inlet mass flow:

W3guess = 10.0 * 0.318/0.302
W3guess = 10.53 kg/s

Now go back to the beginning of the calculation and repeat until the error in capacity is less than 0.05%. These iterations will result in:

W3 = 10.65 kg/s
T4 = 1324.3 K
P4 = 1240.9 kPa

Note: The values resulting from methods 'A' and 'B' should generally be within 1%. This corresponds to any measurement error and heat loss (*wild heat*), the latter being typically around 0.5–1%. These techniques can also be used for turbojets and turboshafts as a check on measurement accuracy of the airmeter.

**C11.3   Calculate the error in referred output power to (i) 95% and
(ii) 99% confidence resulting from measurement errors. The effects of the
relevant contributory measurements have been individually assessed by:**

**Defining the measurement error for each parameter
Changing each value individually by the above amount and calculating the impact upon power**

The results of the above are as follows:

Measurement	Resulting error in referred output power
Output torque	0.75%
Output speed	0.1%
T1	0.1%
P1	0.25%

(i)   Combine individual errors using 'root sum square' for 95% confidence
Error95 = SQRT(0.75^2 + 0.1^2 + 0.1^2 + 0.25^2)
Error95 = 0.80%

(ii)   Arithmetically add errors for 99% confidence
Error99 = 0.75 + 0.1 + 0.1 + 0.25
Error99 = 1.2%

References

1. ASME (1964) *Pressure Measurement*, ANSI/ASME PTC 19.2-1964, American Society of Mechanical Engineers, New York.
2. BSI, ISO (various) *Methods of Measurement of Fluid Flow in Closed Conduits*, BS1042, ISO 5167-1, British Standards Institution, London, International Organization for Standardization, Geneva.

3. ASME (1974) *Temperature Measurement*, ANSI/ASME PTC 19.3-1974, American Society of Mechanical Engineers, New York.
4. BSI (1993) *International Thermocouple Reference Tables*, BS 4937, British Standards Institution, London.
5. AGA (1981) *Gas Energy Metering*, Gas Measurement Manual Part 1, American Gas Association, Arlington, Virginia.
6. AGA (1985) *Compressibility and Supercompressibility for Natural Gas and Other Hydrocarbon Gases*, Transmission Measurement Committee Report No. 8, American Gas Association, Arlington, Virginia.
7. CIBS (1975) *CIBS Guide*, Part C, 1–2, Chartered Institution of Building Services, London.
8. ASME (1985) *Measurement Uncertainty. Instruments and Apparatus*, Part 1, ANSI/ASME PTC 19.1-1985, American Society of Mechanical Engineers, New York.
9. ISO (1985) *ISO 2314 Gas Turbines – Acceptance Tests*, International Organization for Standardization, Geneva.

Chapter 12
The Effects of Water – Liquid, Steam and Ice

12.0 Introduction

This chapter describes the effects of water in all forms – liquid, vapour and ice – on gas turbine engine performance. When only water vapour is present, as opposed to air as well, the gas is referred to as *steam*. There are several reasons why water may be present in the air used by a gas turbine engine:

- Ambient humidity
- Water or ice ingestion
- Water or steam injection

The above are additional to the water vapour produced by combustion. There the amount of water and carbon dioxide produced are based on fuel properties and fuel air ratio, accounted in performance calculations via modified gas properties as described in Chapter 3. The presence of any additional water beyond that produced by combustion impacts engine performance via several different effects:

- Changes in gas properties due to the presence of water vapour
- For liquid water or ice ingestion, solid or liquid water absorbs power and affects the compressor aerodynamics, which may also lower the surge line.
- Changes in temperature due to latent heat absorption or release in ice melting or local evaporation or condensation of water
- Increases in mass flow – for example water or steam injection into the combustor provides additional mass flow through the turbines relative to the compressors.

Performance modelling of phenomena associated with *all* forms of water ingestion or injection, including engine test data correction for humidity, are covered herein and relevant tables and charts are provided. For most gas turbine cycles the water vapour concentration is sufficiently low that the water/air mixture remains essentially a perfect gas. For the rare occasions where the water concentration is higher than 10% this is not so, and steam tables must be used.

12.1 Gas Properties

Chapter 3 describes three fundamental gas properties:

(i)   Specific heat at constant pressure, CP
(ii) Gas constant, R
(iii) Ratio of specific heats, gamma.

Formulae F3.2–F3.8 define the interrelationships between these parameters.

12.1.1 Changes in gas properties

The presence of water vapour changes the values of these gas properties, which can have a significant effect on the thermodynamic processes throughout the engine. Both CP and R increase significantly, whilst gamma diminishes more slowly; Chart 12.1 shows the variation of these parameters versus water vapour concentration. Chapter 3 explains these effects, which result mainly from the molecular weight of water being far lower than that of dry air. Formulae F12.1–F12.3 give the variation of CP, R and gamma with water content and Formula F3.23 gives polynomial coefficients for CP for both dry air and water vapour.

12.1.2 Effects on component performance

As explained in Chapter 4, engine components behave non-dimensionally, and their performance is mapped via various forms of parameter groups. Where the full dimensionless groups are used these maps are *unique even with varying amounts of water vapour present*. Quasi-dimensionless groups are also often used, where the gas properties are omitted and components mapped based on the gas properties of dry air. This form of the groups is useful as it relates the most easily observable engine parameters of flow, speed, pressure and temperature. Changes occur in these quasi-dimensionless maps due to the changes in gas properties described above. The use of such maps when water vapour is present is described in section 12.8.6.

For a compressor, the presence of water vapour lowers $W\sqrt{T}/P$ at constant $N/\sqrt{T}$ and PR. This is because the dimensionless group for speed, $N/\sqrt{(\gamma RT)}$, is reduced relative to dry air due to the increased R which outweighs the small reduction in γ. At this lower dimensionless speed the compressor passes a lower dimensionless flow, $W\sqrt{(RT)}/P\sqrt{\gamma}$. In addition, at this lower dimensionless flow the *quasi-dimensionless* flow $W\sqrt{T}/P$ is further reduced, again by the increased R.

For a turbine the presence of water vapour produces little or no change in $W\sqrt{T}/P$ at constant $N/\sqrt{T}$ and $\Delta H/T$. This is because, unlike a compressor, reduced speed normally either *raises* capacity or has no effect, as explained in Chapter 5. One further effect of humidity is to increase the specific heat in the turbines, reducing the temperature drop for a given expansion ratio. The impact of these effects on engine performance depends on where through the engine the water is introduced.

12.2 Humidity

12.2.1 Description

Humidity is water vapour naturally present in the atmosphere, as described in Chapter 2:

- *Specific humidity* is the ratio of water vapour to dry air by mass
- *Relative humidity* is specific humidity divided by the saturated value

Chart 2.7 shows the amount of water vapour present at 100% relative humidity for the extremes of ambient temperature versus pressure altitude. Specific humidity for a given level of relative humidity increases as air pressure reduces, since at a given temperature the water vapour pressure is constant. Charts 2.8 and 2.9 enable conversions between specific and relative humidity, while Formulae F2.8–F2.10 define these and the saturation vapour pressure for water in air.

12.2.2 Effects on engine performance

Figure 12.1 shows qualitatively the effect of humidity on leading engine parameters. Chart 12.2 presents generic exchange rates which may be utilised for first-order accuracy to predict the impact of humidity on key performance parameters. These have been derived by considering all

Parameter	Changes at		
	Fixed LP Speed	Fixed last turbine inlet temperature	Fixed power or thrust
LP speed	None	Increase	Increase
HP speed	Negligible	Increase	Increase
Mass flow	Decrease	Decrease	Decrease
Fuel flow	Decrease	Increase	Increase
Output power	Decrease	Increase	None
Net thrust	Decrease	Increase	None
SFC	Increase	Increase	Increase
Temperatures	Decrease	See Note 1	Decrease
Pressures	Decrease	Increase	Decrease

Notes:
(1)   Temperatures upstream of last turbine inlet have negligible change, those downstream decrease.
(2)   Changes shown are for physical, not referred parameters.

Fig. 12.1   Humidity: effect on leading performance parameters.

the full non-dimensional parameter groups to be constant at a given operating condition. The impact of humidity is greater than is immediately apparent, as *all* parameters must be factored. For example, at a specific humidity of 0.04 (100% relative humidity on a 368C day) shaft power increases by about 1% and LP turbine inlet temperature falls by about 1%. Hence if governing to fixed LP turbine inlet temperature the increase in power would be around three to four times this. The precise impact depends on the engine cycle.

Chart 12.2 may also be used for a first-order correction of engine test data from any humidity level to that for ISO conditions:

- Divide all parameters by the relevant factor from Chart 12.2 to obtain 'dry' performance
- Multiply all parameters by the factor for a specific humidity of 0.0064 (60% relative humidity at 288.15 K and 101.325 kPa)

12.3   Water injection

12.3.1   *Description*

Water injection has been employed for both aero and industrial engines to improve performance, utilising one of two injection points along the gas path:

- At first compressor entry – to boost power or thrust with minimal impact on SFC. Here a relatively small amount of water is used to lower the inlet temperature by evaporation.
- Into the combustor – primarily to reduce emissions for industrial engines. This also boosts power, but makes SFC worse.

Both practices are now largely superseded by other technologies. These are increased temperature capabilities and higher bypass ratios for aero-engines, and *dry low emissions* combustion technology for industrial engines, where low pollutant levels are achieved without the need for water injection.

Parameter	Changes at		
	Fixed LP Speed	Fixed last turbine inlet temperature	Fixed power or thrust
LP speed	None	Increase	Decrease
HP speed	Decrease	Negligible	Decrease
Mass flow	Increase	Increase	Increase
Fuel flow	Increase	Increase	Negligible
Output power	Increase	Increase	None
Net thrust	Increase	Increase	None
SFC	Decrease	Decrease	Negligible
Temperatures	Decrease	See Note 1	Decrease
Pressures	Increase	Increase	Increase

Notes:
(1) Temperatures upstream of last turbine inlet have negligible change, those downstream decrease.
(2) Changes shown are for physical, not referred parameters.
(3) The use of methanol as an antifreeze reduces SFC. If no liquid water persists to the combustor then this effect is slight.

Fig. 12.2 Compressor entry water injection: effects on leading performance parameters.

Compressor entry injection – aero-engines

As described in section 7.1.7, the takeoff condition, especially for hot days or high airfields, was particularly important for older engines with zero or low bypass ratio. Water was injected using spray nozzles, and prevented from freezing by adding methanol. The power or thrust gain primarily came from the lower inlet temperature and intercooling within the compressor rather than the additional mass flow of water. If the amount of water injected exceeded the amount that could evaporate within the compressor, further power boost was achieved by similar effects as produced by water injection into the combustor. The weight of tanks, pumps, etc. was a disadvantage, as was the need to supply purified water.

Compressor entry injection – industrial engines

Industrial engines operating in hot, arid conditions often employ an evaporative cooler upstream of the intake. This consists of a screen with water flowing over and the air flowing through it. The engine experiences a colder, more humid day, and hence can run to higher power and/or lower operating temperatures. Compared with aero-engines water consumption is lower, as only that which immediately evaporates is used. Such equipment would clearly have been impractical for aero-engines.

Combustor injection – industrial engines

Water injection into the combustor has been employed to reduce NO_x emissions, by lowering the temperature peaks in the combustor primary zone and hence the tendency for atmospheric nitrogen to dissociate. Side effects include increased power, but often increased emissions of carbon monoxide (CO) due to lower temperatures, especially for low pressure ratios and short residence times. In addition combustion becomes more uneven, which may induce mechanical problems due to temperature patternation and noise. Water to fuel ratios (WFR) between 1 and 2 have been employed; as combustor temperatures are very high saturation is not a problem. The large amount of water and associated plant is a substantial disadvantage, the water being purifed and pumped to the injection points via pipe work and manifolding. In addition, burners must incorporate water nozzles, and appropriate control capability is essential. As the water supply pipes etc. can be protected from low temperatures no methanol is employed. SFC is worse due to the significant amount of heat required to evaporate the water.

Steam injection, described in section 12.4 below, supplanted water injection for installations where steam plant could be used as it overcame the disadvantages associated with the chilling effect of liquid water. As stated, more modern industrial, and even marine, engines are moving instead to DLE combustion.

12.3.2 *Effects on engine performance*

For water injection pumping power is not an engine performance concern, as the liquid water is incompressible and hence the power requirement is small. Logistic and weight issues of the associated plant are significant, however.

Compressor entry

Figure 12.2 shows the main effects on engine performance of water injection at compressor entry. As outlined, compressor inlet temperature is reduced by water evaporation, producing effectively a colder day and 100% relative humidity. The maximum overall performance boost at fixed SOT on a hot day is:

- 20% power or thrust increase
- 5–10% shaft power SFC improvement

The SFC improvement is because in non-dimensional terms the engine suffering less from part load SFC deterioration. By coincidence the methanol prevents any increase in fuel requirement to maintain SOT (though unless liquid water persisted through to the combustor this effect would anyway be small). Saturation of the inlet air determines the feasible lowering of the inlet temperature, hence the initial humidity level has a strong limiting effect.

Figure 12.3 shows the effect on the compressor working lines, which are similar for both LP and HP compressors. At fixed mechanical speed reduced inlet temperature increases referred speed, causing operation to move further up the *same* compressor working line. At fixed SOT compressor referred speed increases further, again along the same working line.

Combustor

Figure 12.4 lists the effects on key engine parameters of water injection into the combustor. Typically at fixed SOT an engine with a combustor design not optimised for emissions would achieve 85% NO_x reduction but with up to a three-fold CO increase. For a 1:1 water fuel ratio (WFR) performance effects are:

- 10–20% power increase
- 6–8% worse SFC

The increase in power output is due both to the flow in the turbines exceeding that in the compressors, and the CP of the flow in the turbines being increased by the water. The latent heat of evaporation of water requires additional fuel flow, hence the worse SFC. At fixed SOT the compressor speeds are increased along with engine inlet mass flow.

Figure 12.3 shows that in both cases the HP compressor working line is *higher* as the HP turbine must pass the injected water vapour flow, raising compressor pressure ratio at a referred speed to maintain the required HP turbine capacity. Because the temperature drop in the turbines is reduced, different levels of power boost are achieved depending on which turbine inlet temperature is considered constant.

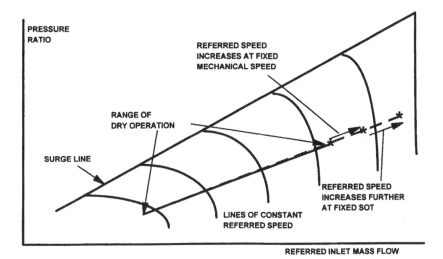

(a)   Water injection upstream of compressor

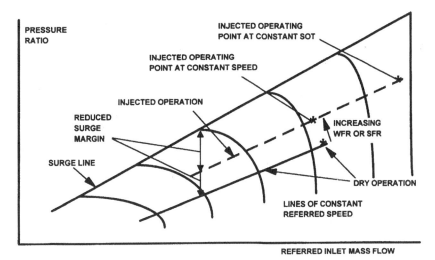

(b)   Water or steam injection into combustor

Notes:
For a single shaft engine with combustion injection, speed would not increase.
SFR, WFR = steam/water to fuel ratio.

Fig. 12.3   Water and steam injection: compressor working lines.

Unlike steam injection, described in section 12.4 below, water injection into turbines down-stream of the HP turbine is not beneficial. The large heat absorption in evaporation drastically reduces that turbine's inlet temperature, and hence also its achievable specific power. The difference for water injection into the combustor is that extra fuel is added *simultaneously*, which offsets the temperature reduction due to the water. To add extra fuel to offset the temperature reduction for injection into other turbines would require impractically high temperatures in the HP turbine, or the complexity of additional combustors between the turbines.

Parameter	Changes at		
	Fixed LP Speed	Fixed last turbine inlet temperature	Fixed power or thrust
LP speed	None	Increase	Negligible
HP speed	Negligible	Increase	Negligible
Mass flow	Decrease	Increase	Decrease
Fuel flow	Increase	Increase	Increase
Output power	Increase	Increase	None
Net thrust	Increase	Increase	None
SFC	Increase	Increase	Increase
Temperatures	Decrease	See Note 1	Decrease
Pressures	Increase	Increase	Increase

Notes:

(1)   Turbine temperatures decrease apart from last turbine inlet.
(2)   Changes shown are for physical, not referred parameters.
(3)   The effects shown assume no use of methanol for antifreeze.

Fig. 12.4   Combustor water injection: effects on leading performance parameters.

12.3.3   *Operability and control philosophy*

Compressor entry

Water injection at entry to the compressor raises few issues as it has no impact on compressor running lines, only resulting in a higher referred speed for a given SOT. The aim is to increase power whilst keeping within the same operating limits; compressor *referred* speed is increased only on occasions where it would otherwise be well below its limit. The normal control strategies of temperature and speed limits remain applicable.

Combustor

Steady state the baseline control strategy is invariably constant water fuel ratio at all power levels, and the rated power governed to the same SOT as the dry engine. However on hot and cold days the compressor mechanical or referred speed limits are reached first and the power boost is limited. In practice very cold days often already have referred speed as the limiting parameter even for the dry engine, and no power boost is possible.

Transiently however there are two major consequences of water injection which must be addressed by control strategies:

(1)   Flame out avoidance: During engine deceleration water flow is normally reduced or shut off completely. This is because the additional heat required to evaporate the water would otherwise increase the tendency for weak extinction.
(2)   Surge avoidance: During engine acceleration water flow is again reduced or shut off completely. Otherwise the loss in HP compressor surge margin due to the the overfuelling for acceleration would be too great, given the surge margin reduction due to water injection. Indeed, only by this means can the lower steady state surge margin be accepted.

12.3.4 *Specific engine design for high injection rates into the combustor*

Water injection into the combustor beyond around 2:1 WFR would usually require reselection of turbine capacities, to prevent unacceptable surge margin or overspeed as described above. Such modification is rarely if ever undertaken, since most engines must also operate 'dry'. However both 'wet' and 'dry' cases may be considered in the design phase. Another item to address is increased power turbine entry temperature (PTET) at fixed SOT, due to the reduction in temperature drop in the turbine(s) driving the compressor(s).

The need for turbine capacity changes with high water injection rates arises because for the same compressor operating points compressor work is fixed, but turbine mass flow and specific heat are higher. For example the requirements for a two shaft gas generator plus power turbine are illustrated below:

- HP turbine capacity must be increased. This accommodates the extra injected flow, preventing the rise in the HP compressor working line. It also reduces the HP turbine expansion ratio, and hence power output, to prevent the increase in HP compressor speed described in section 12.3.2.
- LP turbine capacity must be decreased despite the higher mass flow, by a relatively modest amount. This forces a higher LP turbine inlet pressure, hence reducing the HP turbine expansion ratio further.
- Power turbine capacity must also be decreased, to reduce LP turbine expansion ratio, and prevent LP compressor overspeed. The lower HP and LP turbine expansion ratios are manifested as a higher power turbine inlet pressure and temperature, which give the increased engine power output.

By restagger of turbine aerofoils within casting allowances power boosts of up to 30% may be achieved for 4 WFR. As an example, at 6 WFR to retain acceptable compressor working lines and speeds the engine changes would be approximately:

- Power increase: 70%
- SFC deterioration: 10–15%
- HP turbine: 20% capacity increase, and 20% reduction in expansion ratio
- LP turbine: 5% capacity reduction, and 20% reduction in expansion ratio
- Power turbine: 20% capacity reduction, and 50% increase in expansion ratio

12.4   Steam injection

12.4.1 *Description*

Steam injection is employed on industrial engines to reduce NO_x emissions and boost power. SFC also improves, as steam is usually raised using the engine exhaust heat, in a *heat recovery steam generator* (HRSG). The chilling effect of steam in the combustor is vastly less and more uniform than that of liquid water, hence there is little or no increase in CO emissions or combustion noise.

Steam injection into other turbines gives around half the SFC improvement and 1/5 of the power boost relative to the combustor, though it costs less in terms of compressor operability. Current systems have various proprietary names, such as *STIG* (steam injected gas turbine) or the *Cheng cycle*. *Supplementary firing* could also be employed, to increase steam production using a *boiler* with its own dedicated combustion.

Besides the need for steam plant and injection hardware, one significant issue for steam injection, like water injection, is the substantial need for purified water. Higher flow rates are considered than for water injection, as steam injection improves SFC. Owing to the cost and complexity of cooling towers etc. almost all systems are *total loss,* with the used steam simply passing up the exhaust stack. For all these reasons steam injection is employed almost solely on industrial engines rather than in mobile applications.

Parameter	Changes at		
	Fixed LP Speed	Fixed last turbine inlet temperature	Fixed power or thrust
LP speed	None	Increase	Negligible
HP speed	Negligible	Increase	Negligible
Mass flow	Decrease	Increase	Decrease
Fuel flow	Decrease	Increase	Decrease
Output power	Increase	Increase	None
Net thrust	Increase	Increase	None
SFC	Decrease	Decrease	Decrease
Temperatures	Decrease	See Note 1	Decrease
Pressures	Increase	Increase	Increase

Notes:
(1)   Turbine temperatures decrease apart from last turbine inlet.
(2)   Changes shown are for physical, not referred parameters.

Fig. 12.5   Steam injection: effects on leading performance parameters.

12.4.2   *Effects on engine performance*

Figure 12.5 summarises the effects of steam injection on an engine cycle. Typically an aero-derivative industrial engine at constant SOT with 2:1 steam fuel ratio (SFR)would achieve:

- 15–20% power boost
- 10% SFC improvement

Raising steam at the highest possible temperature, dictated by the engine exhaust temperature and HRSG design, gives the best SFC; this reflects a maximum efficiency of energy recovery from the gas turbine exhaust.

In terms of component rematching the effects of steam and water injection are similar, additional power being available from the turbines due to their increased gas mass flow and specific heat. The HP turbine must pass extra flow, as for water injection, resulting in a higher HP compressor working line as per Fig. 12.3. At fixed geometry the highest power boost occurs with the highest SFR that can be accepted within surge margin limits, which is normally not higher than 2.

12.4.3   *Operability and control philosophy*

Comments here almost exactly match those for water injection in section 12.3.3 above, though the tendency to flame out during a decel is less for steam injection as without the need for evaporation it produces less chilling in the combustor.

12.4.4   *Specific engine design for high injection rates into the combustor*

Almost all the same comments apply as for water injection, except that SFC improves rather than deteriorates. For a full turbine redesign, for 6:1 SFR the engine changes would be:

- Turbine capacity changes as per the 6 : 1 WFR water injection example in section 12.3.4
- Power increase: 70%, again as for 6 : 1 WFR
- SFC improvement: 20%

The one difference relative to water injection is that due to the *improved* SFC going to high steam to fuel ratios (SFR) is more attractive, however as with water injection actually doing this is almost unknown in practice due to the compromise to 'dry' performance. References 1 and 2 provide further examples and discussion.

In terms of engine cycle selection there are two other notable effects which impact the ability to produce steam:

(1) Increasing SFR raises engine exit temperature for constant SOT and pressure ratio, as the temperature drop within the turbines reduces due to increased flow and specific heat.
(2) Higher pressure ratio cycles allow less steam to be produced at constant SOT, as engine exit temperature is lower.

12.5   Condensation

12.5.1   *Description*

Condensation is confined to the 'cold', compressor end of the engine and is due only to ambient humidity. The resultant water simply re-evaporates further along the gas path. Condensation *may* occur once flow conditions cause the local static temperature to be below the saturation value for the local water vapour partial pressure. This is determined by local static pressure and water vapour concentration; saturation corresponds to 100% relative humidity as described in Chapter 2. The *actual* rate of condensation is determined by residence time, and by minute particles in the air which act as nucleii for water droplets. For background, section 12.7 discusses the fundamental properties of the phases of water.

In the cold end there are two regions where water vapour may condense:

(1) Intake: Flow acceleration upstream of the first compressor reduces the static temperature, increasing the tendency for condensation. Chart 12.3 shows the range of humidity and local Mach number at which condensation may occur, at ambient pressure. At 40% ambient relative humidity (RH) the threshold Mach number is 0.5, and at 80% ambient RH it is 0.24.
(2) Intercooler: The pressure increase in the first compressor raises the water vapour's partial pressure. This may exceed the saturated value once the intercooler reduces the temperature. The flow velocities are low within the intercooler heat exchanger matrix to achieve the required heat transfer; this gives adequate residence time for condensation, which occurs on the cold passage surfaces. If this produces water droplets, erosion of the downstream compressor blades may be a serious problem. The important difference from water injection or rain ingestion is that intercooler condensation may continue almost indefinitely in some geographical locations, unless the degree of intercooling is reduced.

12.5.2   *Effects on engine performance*

Intake

Intake condensation can cause a significant change in engine performance on a test bed. The apparent loss in power or thrust and increase in SFC at fixed fuel flow may be up to 2% for hot ambient conditions, and nearer 0.5% for temperate climates. The reason for this loss in performance may not be immediately apparent, being due to pressure and temperature changes occurring downstream of test bed inlet pressure and temperature measurements. Were it possible to measure accurately the pressure and temperature actually entering the compressor then there would be negligible impact on engine referred performance relative to that without condensation. In the absence of perfect measurements or correction methods, performance testing should not be undertaken during conditions of high humidity or precipitation.

The effects of condensation are an increase in air inlet temperature and a reduction in total pressure, as well as reduced specific humidity in the initial compressor stages. The increased temperature results from the latent heat release while the pressure loss reflects the momentum change in flow acceleration due to reduced density. Once within the engine the condensed water evaporates shortly downstream.

Any engine on the test bed (or even in service) will experience a real difference in performance as referred to the engine intake lip.

Intercooler

For an intercooler system some condensation control algorithm is often employed to prevent condensation causing blade erosion downstream. The coolest air temperature must be maintained above the calculated saturation value by control of the degree of cooling. For example, for a system with an intermediate transport loop between the cold sink and the engine this may involve a partial bypass of the cold sink. The intercooler exit temperature is thereby increased, with two main consequences for engine operation:

(1)   Reduced LP compressor surge margin due to decreased mass flow: The reduced inter-cooling means the HP compressor referred speed and hence referred flow are reduced by the higher inlet temperature. In addition for a given referred HP compressor flow the actual mass flow is reduced even further.

(2)   Increased turbine temperatures and actual HP shaft speed: The power required to drive the HP compressor is directly proportional to its absolute inlet temperature. If this power increases, the engine must run hotter and faster to produce the same output power.

12.6   Rain and ice ingestion

An aircraft will encounter large amounts of rain and hail when flying through storms. Furthermore in cold conditions ice may form on the intake surfaces, occasionally breaking off to pass into the engines. Significant ingestion is normally confined to turbofans and turbojets, which must pass ingestion tests for certification as described in section 11.4.3. For aero turboprop and turboshaft engines all these phenomena may be reduced by *inertial separators* in the intake ducts. Here air flow is drawn in around a tight bend; heavier particles fail to take this bend, passing straight on and then overboard. Industrial, automotive and marine engines are normally protected by intake filters.

12.6.1   *Description – rain ingestion*

The highest rain ingestion occurs when flying through monsoon conditions. For a turbofan the fan *centrifuges* a significant proportion of the liquid water into the bypass duct, i.e. the rotation of the blades forces particles radially outwards. This effect is dependent on residence time and hence is increased by wide chord fan blades and reduced by high forward speed. However some water always passes into the core compressors.

In extreme circumstances, should liquid water persist into the combustor the heat it aborbs in evaporation has a major chilling effect. This can be a serious problem, and on the CFM56 turbofan rain ingestion did result in several run downs and at least one forced landing.

12.6.2   *Description – ice and hail ingestion*

Icing occurs when an aircraft encounters cold cloud conditions corresponding to 'freezing fog', where air initially containing water vapour has dropped below 0 8C. This causes ice to build up on the wings and engine intakes or nacelles. The ice may break off naturally when either it reaches some critical thickness, warmer air is encountered, or an *anti-icing* system is switched on. For engines, such a system consists of electrical or bleed air heating of the intake surfaces.

The fan spinner is normally made of rubber to ensure flexibility which results in ice being shed before it has reached a dangerous thickness.

Hail is ingested simply by flying through hailstorms, the worst often being produced by a thundercloud. In either case the main issues are of mechanical integrity.

12.6.3   *Effects on engine performance*

Mechanical issues aside, the effects of water and ice ingestion are basically similar. Control strategies must increase fuel flow to avoid engine run down or even flame out. Though ice has a greater chilling effect than liquid water, in practice the overall concentration is less than that of liquid water and hence this is not a major concern.

Once fuel flow is increased the thermodynamics are closer to those for combustor water injection, described in section 12.3, than to compressor entry water injection. However within the core compressors there are additional phenomena which cannot be precisely predicted:

- Water progressively evaporates through the compressors, affecting gas properties, temperature levels, and gas mass flow. It also produces an intercooling effect which is specific to the compressor type. Evaporation begins when the partial pressure and temperature cross the saturated liquid line shown at the left-hand side of Fig. 12.6. It continues until the saturated vapour line is reached on the right-hand side. The rate of evaporation is also specific to a compressor type, being also strongly dependent on residence time to transfer heat to the water, and hence can only be determined empirically.
- The surge line may be lowered, due to liquid water being centrifuged outwards which affects aerodynamics at the tip section.
- Mechanical power is absorbed, by the work done in accelerating liquid water through a change in swirl velocity in each compressor stage.

Because of the various effects, rain normally results in aero-engine testing being suspended if an important performance curve is planned. The difference between rain injection and water injection into compressors is that there the intention is to drop the inlet temperature by evaporation, and not normally to pass large quantities of liquid water through the compressors.

12.7   The thermodynamics of water

As is well known, water may exist in both liquid and vapour phases over a wide temperature range. At a pressure of 1 bar and temperature of 100 8C, water evaporates to steam if energy is supplied. This energy requirement is *the latent heat of evaporation*, and the process is known as *boiling*. At these conditions the volume increases by a factor of 1600. The boiling temperature rises as pressure increases, since a different change in volume then occurs. The inability to make good tea with water boiled up a mountain, which is then too cold, is well known. Conversely, if water is heated in a closed vessel the pressure and temperature both increase once boiling has begun.

Figure 12.6 shows schematically the properties of water in its three phases on a traditional temperature–entropy (T–S) diagram. Such a diagram is discussed in Chapter 3; for water it is known as the *Mollier diagram*. The characteristic 'dome' shape defines the region through which evaporation may occur; inside this water exists as *wet steam*, a mixture of liquid and vapour. To the left of the dome only liquid water can exist, and to the right and above, only *dry steam*. Below 0 8C liquid water cannot exist, only ice and an almost negligible vapour pressure due to *sublimation*. A temperature of 0 8C is known as the *triple point*, where water may exist as ice, liquid, and steam. A pressure of 220.9 bar at 373.7 8C is the apex of the 'dome', known as the *critical point*.

Figure 12.6 includes several lines of constant pressure. At the left-hand end of each line there is the *saturated liquid* line, onto which lines collapse for a substantial pressure range; pressure changes have only a small effect, as liquid water is almost entirely incompressible.

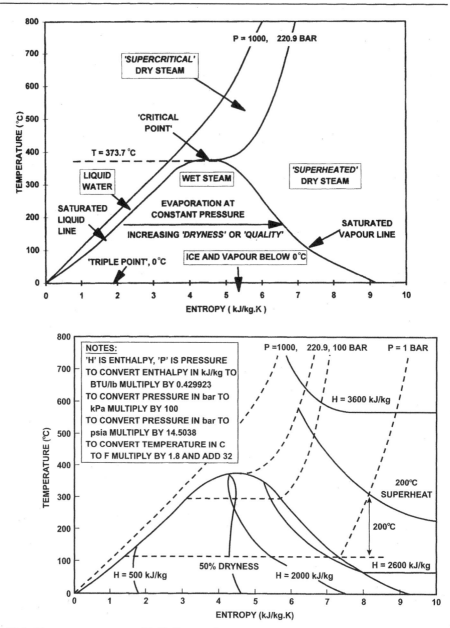

Fig. 12.6   Temperature–entropy (T–S) diagram for water.

At fixed pressure, heat input to liquid water increases the temperature until it crosses the saturated liquid line, when the boiling temperature is reached. Further heat input causes evaporation at constant temperature, and hence movement to the right within the dome; the degree of evaporation is described by the *dryness fraction* or *quality,* which is the ratio of vapour to total by mass. The right-hand edge of the dome is the *saturated vapour* line, where evaporation is complete and the steam is dry. Beyond the dome still further heat input results in a temperature increase, and the dry steam becomes *superheated. Superheating* is applied as a term in various forms to the temperature difference from the saturation line at that pressure. If the pressure is raised above the critical point (220.9 bar) it is *supercritical.* Here evaporation from liquid water to steam cannot be distinguished by any step change in density, which instead decreases uniformly as temperature increases.

Steam is not a perfect gas in that enthalpy is dependent on pressure as well as temperature, due to attraction forces between the polarised water molecules. This effect is least at low density and high temperature.

For illustration, Fig. 12.6 also includes the following:

- A line of 50% dryness
- Lines of constant enthalpy, showing the large energy input required for evaporation
- A line of constant superheating

Reference 3 gives a full numerical version of the T–S diagram for steam.

Use of steam in power cycles utilises the pressure increase when heating liquid water in an enclosed volume. There is no need for large work input, unlike for gas turbine cycles, as pressurising incompressible feed water requires very little power. It is mainly for this reason that steam cycles were developed much earlier than gas turbines. Injected steam in a gas turbine remains in the superheated region all through the engine, whereas in some purely steam cycles it condenses. Examples include the back stages of a steam turbine running 'wet', and some reciprocating cycles utilising the pressure reduction when steam is chilled.

Figure 4 in the Configurations section illustrates possible steam plant arrangements downstream of a gas turbine, including *combined cycle*, where a steam turbine produces additional shaft power. Reference 4 provides an introduction to the various possible steam cycles; these are beyond the scope of this book.

The following formulae are relevant:

- Formula F12.4 gives the enthalpy of saturated steam as a function of temperature.
- Formula F12.5 gives the evaporation temperature as a function of pressure.
- Formula F2.10 gives saturation pressure as a function of temperature.
- Formula F12.6 gives the partial pressure of water vapour as a function of its concentration.
- Formula F12.7 gives the enthalpy of liquid water as a function of temperature.
- Formula F12.10 gives the enthalpy of superheated and supercritical steam as a function of temperature and pressure.

12.8 Gas turbine performance modelling and test data analysis

The following sections present methods to model the individual effects occurring when water is encountered. These techniques are suitable both for predictions and test data analysis. Figure 12.7 presents a flowchart of how they interact; clearly many effects are common to several situations.

12.8.1 *Ice and melting*

Accurate theoretical modelling is not practical, however empiricism and approximation are sufficient to establish the impact on compressor efficiency and surge lines. Whether ice melts in the combustor or the compressor depends on residence time and particle size; extremes may be established by assuming each in turn; the difference in terms of the overall cycle may be small. Once ice has melted it becomes water droplets, which are discussed below. Formula F12.8 enables the heat absorption in melting to be calculated, from which the resulting air temperature reduction may then be simply evaluated based on the formulae in Chapter 3.

12.8.2 *Liquid water and evaporation*

For the compressor the existing characteristic may still be used for the air flow, a parallel water stream must also be considered, and the compressor drive power increased accordingly.

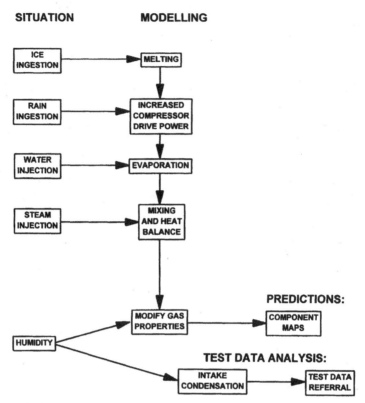

Fig. 12.7 Performance modelling: main elements and sequence.

Accelerating liquid water to some swirl velocity in each stage absorbs power, around 40–60% of the work that would be required if it reached the full mean blade speed; this power absorption may be approximated by applying Formula F12.9 for each stage before evaporation has occurred. Some evaporation may occur in the compressor, though for large flows much will pass through to the combustor; judgement or empiricism must be utilised regarding the relative quantities, as discussed in section 12.6.3.

During evaporation in either the compressor or the combustor the heat absorbed must be calculated from the difference between Formula F12.7, and F12.10 with the total temperature and partial pressure. The gas path temperature reduction due to this must then be calculated via a mixing and heat balance calculation as described in section 12.8.4. Sample calculation C12.1 illustrates the approach.

12.8.3 *Steam injection*

The mixing and heat balance calculation is as per section 12.8.4. For higher steam flows giving more than 10% WAR the turbine calculations should employ steam formulae, described in section 12.7. Modelling the HRSG must consider several key points:

- Steam formulae must be used on the steam side, including the evaporation of the liquid water supplied.
- Steam formulae may also be required on the gas side, depending on the steam concentration.
- Enthalpy changes on the steam and gas sides must be equal.
- Design minimum levels of the main temperature differences must be maintained.

- Steam delivery pressure must be some specified amount above that at the engine injection point.

To meet the last two criteria the steam delivery temperature may vary as matching guesses in a matching model, and/or the rate of steam production.

12.8.4   *Mixing and heat balance*

Prior to evaporation the ice/water and air streams are considered in parallel. Afterwards, once the latent heat absorption has been accounted as per section 12.8.2, the streams must be combined via a mixing and heat balance calculation:

- The changes in enthalpy for each stream must be equal and opposite. Sample calculation C12.1 illustrates this process.
- Total pressure is normally assumed not to change during mixing, as the steam is normally a small fraction of the total flow.
- Gas properties downstream of the mixing plane should be adjusted by one of the methods described in section 12.8.5.

Such calculations may require iteration, if so they become simply an extension of a 'matching' model of the type described in Chapter 7.

12.8.5   *Gas properties*

Three methods for adjusting gas properties to account the presence of water vapour for thermodynamic calculations are shown below. They are listed in order of increasing accuracy:

(1)   Using Formulae F12.1–F12.3, which represent the factors to be applied to gas properties versus water vapour concentration as per Fig. 12.1. This method assumes that pressure does not change enthalpy (i.e. that water vapour is a perfect gas), and that the gas property ratios are not affected by temperature; it is adequate for first-order calculations.

(2)   Using the various formulae for specific heats given in Chapter 3. For kerosene or diesel fuel, Formulae F3.23 and F3.24 should be used for dry air and combustion products respectively, and F3.23 used again for the additional water vapour. Formula F3.25 covers a sample natural gas; for other compositions Formula F3.23 should be used for each of the constituents. In all cases the overall specific heat should be obtained by averaging the gas constituents and additional water vapour on a molar basis, and the ratio of specific heats gamma is then easily derived. This method still assumes that water vapour is a perfect gas, however where vapour content is less than 10% the error incurred is negligible; this method is the most widely used.

(3)   Evaluating CP for the additional water from Formula 12.11, which does not assume a perfect gas and accounts for the effect of pressure on steam properties. CP calculated in this fashion will be as per the steam tables presented in Reference 3. The gas constant and ratio of specific heats are then derived as per method 2. This is the most rigorous method, and should be used for water content exceeding 10%.

12.8.6   *Component maps*

Figures 5.4 and 5.30 illustrate the form of compressor and turbine maps respectively. Where there is water vapour but no liquid or ice present, these component maps are normally read via the simplified quasi non-dimensional groups listed in Chart 4.1. These groups relate flow, speed, pressure and temperature and use implicit 'dry air' values of the fundamental gas properties, gamma and R. As described in section 12.1, water vapour changes these properties, hence with water vapour present the full forms of the non-dimensional groups must be recognised, as illustrated on Fig. 5.7.

When water or ice are present, the use of the maps must incorporate gamma and R corresponding to 100% RH, or more often their ratios to the values for dry air. This is in addition to the calculations described in sections 12.8.1 and 12.8.2, where the water/ice is considered as a separate parallel stream. The changes in gas properties accounts the effect on both the independent parameters against which the maps are tabulated, and on the dependent parameters returned from the map read. Evaporation within compressors produces an intercooling effect which is specific to the compressor design. Given the uncertainty of evaporation location, it is usual to simply retain the same compressor map.

12.8.7   *Test data*

To understand engine behaviour on test the effects of humidity must be accounted. These are due to the variation in gas properties, and potentially also to condensation. The first step is accurate humidity measurement at test conditions, methods for which are described in section 11.2.10. As outlined in section 11.8 there are then two main approaches to test data analysis.

Non-dimensional corrections
Engine test data is referred to standard conditions via non-dimensional groups, to allow valid comparison of prediction and other test data. For humid conditions revised gas properties CP, gamma and R are evaluated, via any of the methods described in section 12.8.5. The full forms of the non-dimensional groups from Chart 4.1 are then used for the referral to standard conditions.

Rigorous modelling
No attempt is made to correct test data to standard conditions. Instead it is compared with expectation via a rigorous thermodynamic prediction model, as described in Chapter 7, run at the tested conditions. The modelling accounts for effects of humidity on gas properties and hence on component maps, thermodynamic processes, etc. as described in sections 12.8.5 and 12.8.6. This applies equally to analysis of component performance changes via matching methods, as described in section 11.8.6.

Intake condensation
As stated the concentration and properties (hydrophobic, charged, etc.) of particles which may cause intake condensation cannot be determined. A theoretical, maximum rate at which it *could* occur may be calculated from relative humidity and intake Mach number. Reference 5 illustrates this and F12.12 provides an approximate method. There are then two possible approaches, listed in order of increasing accuracy:

(1)   Assume condensation occurs at half the theoretical rate.
(2)   Correlate measured engine performance versus theoretical maximum rate. This would be done over a series of different engines, which also have build and manufacturing variabilities. Coefficients would be determined for the effects on main measured parameters, as the mean slopes of plots versus the theoretical temperature rise for measured ambient conditions. Comparison of engines would involve correction back to the same theoretical temperature rise using these coefficients.

Directly measuring the actual rate of condensation during a test would clearly be preferable, however at the time of writing no absolutely successful technique has been demonstrated.

Intercooler condensation
The control algorithms should adjust the degree of intercooling to prevent this. The requirement is therefore accurate modelling of the control action, which depends on the specific system. There should be no need to make adjustments to test data.

Formulae

F12.1 Specific heats factor for moist air = fn(water air ratio)

CPfac = (WAR.molar * CPW + (1 − WAR.molar) * CPA)/CPA
WAR.molar = WAR * 28.96/18.015

(i)   CPfac is the ratio of CP for moist air to that for dry air.
(ii) WAR.molar is the ratio of water to dry air, by number of moles.
(iii) CPW, CPA are the specific heats of water and dry air, from Formula F3.23.
(iv) WAR is the ratio of water to dry air, by mass.
(v)   18.015, 28.96 are the molecular weights of water and dry air respectively.

F12.2 Gas constant factor for moist air = fn(water air ratio)

Rfac = Ro/(MW * RA)
MW = 1/((WAR/18.015) + ((1 − WAR)/28.96))

(i)   Rfac is the ratio of R for moist air to that for dry air.
(ii) Ro is the universal gas constant, 8.31 kJ/mole K.
(iii) MW is the molecular weight of the moist air.
(iv) RA is the gas constant of dry air, 0.28705 kJ/kg K.
(v)   WAR is the ratio of water to dry air, by mass.
(vi) 18.015, 28.96 are the molecular weights of water and dry air respectively.

F12.3 Gamma factor for moist air = fn(water air ratio)

GAMMAfac = (WAR.molar * GAMMAW
            + (1 − WAR.molar) * GAMMAA)/GAMMAA
WAR.molar = WAR * 28.96/18.015

(i)   GAMMAfac is the ratio of GAMMA for moist air to that for dry air.
(ii) WAR.molar is the ratio of water to dry air, by number of moles.
(iii) GAMMAW, GAMMAA are the ratio of specific heats for water and dry air, from Formulae F3.4, F3.23 and F3.7.
(iv) WAR is the ratio of water to dry air, by mass.
(v)   18.015, 28.96 are the molecular weights of water and dry air respectively.

F12.4 Enthalpy of saturated steam (kJ/kg) = fn(temperature (8C))

H = −7.352E-06 * T^3 − 2.333E-03 * T^2
    + 2.437 * T + 2492 + 6349/(T − 387.5)

(i)   Accuracy is within 0.6%, range is 0–370 8C.

F12.5 Temperature level of evaporation (8C) = fn(pressure (bar))

For P = 1 to 25 bar:

T = −1.811E-05 * P^6 + 0.0014006 * P^5 − 0.043 * P^4
    + 0.67482 * P^3 − 5.9135 * P^2 + 33.2486 * P + 72.1585

For P = 25 to 210.5 bar:

T = −1.11726E-11 * P^6 + 8.97543E-09 * P^5 − 2.9476E-06 * P^4
    + 0.00051476 * P^3 − 0.05329436 * P^2 + 3.933136 * P + 152.0676

(i)   Accuracy is within 0.5 8C.

F12.6   Partial pressure of water vapour in moist air (bar) = fn(specific humidity, pressure (bar))

$Pw = P/((0.622/SH) + 1)$

(i)   Pw is water vapour partial pressure.
(ii)   SH is specific humidity, kg water vapour per kg dry air.
(iii)   0.622 is the ratio of the molecular weights of water and dry air.

F12.7   Enthalpy of liquid water (kJ/kg) = fn(temperature (8C))

$H = 3.1566E\text{-}12 * T^6 - 2.9348E\text{-}09 * T^5 + 1.0407E\text{-}06 * T^4$
$- 0.16703E\text{-}03 * T^3 + 0.0120915 * T^2 + 3.87675 * T + 0.74591$

(i)   Average accuracy is within 0.7%.
(ii)   Range is 0–370 8C.

F12.8   Latent heat of melting of ice (kJ/kg)

$DHif = 333.5$

F12.9   Work on liquid water in compressor (W) = fn(water flow (kg/s), mean blade speed (m/s))

$DPW = 0.5 * Wwater * U^2$

(i)   For an axial compressor this work is done in each stage.

F12.10   Enthalpy of dry steam (kJ/kg) = fn(temperature (8C), pressure (bar))

Superheated:

$H = 2.98E\text{-}04 * T^2 + 1.83 * T + 2500 - (5.14207E08 * P/(T + 276)^3$
$- (1.03342E37 * P^3 - 6.42613E31 * P^5))/(T + 276)^{14.787}$

(i)   Average accuracy is 0.2%.
(ii)   Range is 0.01–210 bar, and saturation to 800 8C.

Supercritical:

$H = P * (T^3 * (1 + 0.001634 * P) + T^2$
$* (1094.941 - 1.663087 * P) - 8169907)/5.37E + 07 + 0.004505$
$* T * (738.0074 + 31929.78/P - 0.30077 * T) + 611.736 + 39.63429$
$* (9.551098 - 0.002642 * P) * \arctan((T + 0.005184 * P$
$* T - 0.009188 * P - 8.630696)/(1 + 0.005184 * P)$

(i)   Average accuracy is 0.5%.
(ii)   Range is 220–400 bar, and 100–600 8C.

F12.11   Specific heat of steam (kJ/kg K) = fn(temperature (8C), enthalpy (kJ/kg))

$CP = dH/dT = (H(T + DT) - H(T - DT))/(2 * DT)$

(i)   H is enthalpy, evaluated at temperatures shown via Formula F12.10; T is temperature, 8C.
(ii)   DT is a small temperature increment, e.g. 5 8C.

F12.12 Theoretical maximum rate of condensation = fn(temperature, pressure, Mach number, humidity)

The following outlines an approximate method:
- Calculate initial total enthalpy for air and water vapour from Formulae F 3.14 and 3.23, using T0 and SH.
- For case of no condensation, calculate local static temperature after flow acceleration from Formula F3.31 (rearranged), using T0 and M.
- Calculate increased local static temperature TS after condensation as above + TRISE.
- Calculate local static pressure after flow acceleration from Formula F3.32 (rearranged), using P0 and M.
- Calculate partial pressure of water vapour from Formula F12.11.
- Calculate saturated value of water vapour partial pressure from Formulae F2.10 and F12.6. This also gives specific humidity at saturation, SHsat.
- Calculate amount of condensed water to reduce the partial pressure of water vapour to the saturated value.
- Calculate enthalpy of air and remaining water vapour from Formulae F3.14 and 3.23, using T0 + TRISE and SHsat.
- Calculate enthalpy of condensed liquid water from Formula F12.7.
- Iterate on TRISE until enthalpies balance before and after condensation.

(i) The main approximation is to neglect the loss in total pressure that occurs due to the momentum change when heat release causes an increase in flow velocity.

Sample calculations

C12.1 Ingested rain evaporates within a compressor. Determine (i) the possible range of power absorption due to varying amounts of work done on the liquid water, (ii) gas conditions downstream and (iii) the variation of exit temperature resulting from changing the assumptions about how many stages the liquid water absorbs work in.

Conditions are as below:

Blade speed (m/s)	700	Compressor isentropic efficiency	0.86
Compressor PR	5	Number of compressor stages	6

	Water	Air
Temperature (K)	293	293
Pressure (kPa)	100	100
Mass Flow (kg/s)	1.0	100

(i) Work done on liquid water
From Formula F12.9 the work done in each compressor stage is DPW = 0.5*Wwater * U^2:

DPW = 0.5 * 1 * 700^2
DPW = 245 kW per stage

Since evaporation requires some temperature increase, work will be done in one stage at the very least. The very maximum number of stages would be the full 6:

DPW = 245 kW minimum, and 6 * 245 = 1470 kW maximum

(ii) Gas conditions at compressor exit

Compressor temperature rise for dry air. Ignore changes in compressor performance due to intercooling.
From Formula F5.1.4:

$$T3 - T2 = T2 * (P3Q2^{\wedge}((\gamma - 1)/\gamma) - 1)/ETA2$$
$$T3 - T2 = 293 * (5^{\wedge}(2/7) - 1)/0.86$$
$$T3 = 492\,K = 219\,8C$$

Evaporation calculation as per section 12.8.2

Calculate enthalpy of ingested water using Formula F12.7:

$$Hwater = 3.1566E\text{-}12 * 20^{\wedge}6 - 2.9348E\text{-}09 * 20^{\wedge}5$$
$$+1.0407E\text{-}06 * 20^{\wedge}4 - 0.16703E\text{-}03 * 20^{\wedge}3$$
$$+0.0120915 * 20^{\wedge}2 + 3.87675 * 20 + 0.74591$$
$$Hwater = 81.94\,kJ/kg$$

Calculate partial pressure of steam at compressor exit using Formula F12.6:

$$Pw = P/((0.622/SH) + 1)$$
$$Pw = 5/((0.622/0.01) + 1)$$
$$Pw = 0.0791\,bar$$

Calculate enthalpy of steam using Formula F12.10 for superheated steam. Add work done on liquid water in first compressor stage.

$$Hsteam = 2.98E\text{-}04 * T^{\wedge}2 + 1.83 * T + 2500 - 5.14207E08$$
$$* P/(T + 276)^{\wedge}3 - (1.03342E37$$
$$* P^{\wedge}3 - 6.42613E31 * P^{\wedge}5)/(T + 276)^{\wedge}14.787$$

$$QUwater = (Hsteam - 81.94) * 1.0 + 245$$

Now calculate change in air enthalpy due to heat absorbed by the water:

$$QUair = 100 * 1.005 * (219 - Tmix)$$

Mixing sum

Iterate to make QUair = QUwater, using either inbuilt Spreadsheet functions or 'manual' updates. Converged solution gives Tmix = 193.5 8C.

(iii) Effect of variation in mechanical power absorption

Recall work done on liquid water per stage = 245 kW.
Repeat above iteration with QUair = QUwater + 245:

$$Tmix = 195.9\,°C$$

Hence a difference of one stage changes the mixed temperature by 2.4 K; five stages would change it by 12 K.

Note: For this small water concentration the presence of the water could have been neglected and the temperature change found by considering the air alone.

Charts

Chart 12.1   Variation of gas properties with water vapour content.

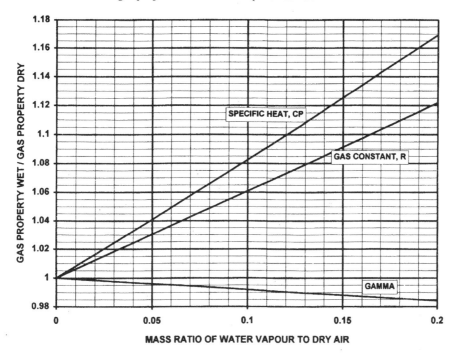

Chart 12.2   Effects of humidity on engine performance: generic exchange rates.

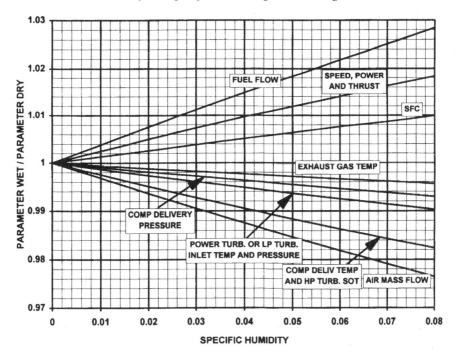

Chart 12.3 Intake condensation: threshold Mach number versus ambient relative humidity.

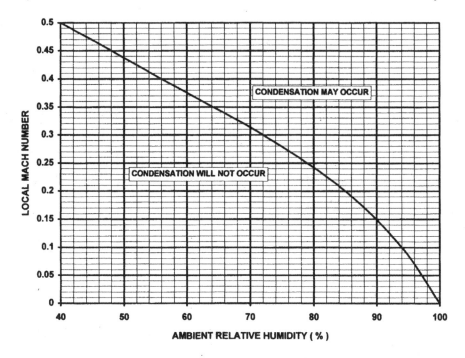

References

1. M. S. Chauhan (1995) *Effects of steam injection into an industrial gas turbine*, MSc Thesis, Cranfield University, Cranfield.
2. P. D. Noymer and E. G. Wilson (1993) *Thermodynamic Design Considerations for Steam-Injected Gas Turbines*, ASME 93-GT-432, ASME, New York.
3. Y. R. Mayhew and G. F. C. Rogers (1967) *Thermodynamic and Transport Properties of Fluids*, 2nd edn, Blackwell Science, Oxford.
4. Rogers and Mayhew (1957) *Engineering Thermodynamics Work and Heat Transfer*, Longman, London.
5. J. C. Blake (1975) *Effects of Condensation in the JT9D Engine Bellmouth Inlet*, AIAA Paper No 75-1325, AIAA, New York.

Chapter 13
Fuel and Oil Properties and their Impact

13.0   Introduction

Fuel and oil properties are required for engine performance calculations including design point, off design, windmilling, starting, transient and test data analysis. This chapter describes the parameters of relevance and provides a data base to cover all needs for such calculations.

The more general topic of properties of fuels and oils in relation to all gas turbine engine design disciplines is exhaustive and cannot be covered here. References 1–6 provide comprehensive coverage.

13.1   The combustion process and gas turbine fuel types

13.1.1   *The combustion process*

The combustion process primarily entails the exothermic reaction of a hydrocarbon fuel with oxygen, to produce carbon dioxide and water. The number of moles of each is dictated by the fuel composition. For illustration the reaction for one of the hydrocarbons present in kerosene is described below.

$C_{10}H_{20}$	$+\ 15O_2$	$=$	$10CO_2$	$+$	$10H_2O$	Chemical equation
1	15		10	10	By mole
$10*12+20*1$	$15*32$		$10*(12+32)$	$10*(2+16)$	By mass
140	480		440	180	By mass

The *stoichiometric* fuel air ratio, where all the atmospheric oxygen would be consumed, is then easily derived. The required number of moles of air is evaluated, and then converted to mass. This is shown below using the molar oxygen content of air, and then the molecular weight of air, from Chapter 3:

$15*(1/0.2095)=71.6$ Number of moles of air
$71.6*28.964=2073.8$ Mass of air
$140/2073.8=0.0675$ FAR by mass

In fact the chemical reaction is far more complex with many reactions taking place simultaneously. This includes the above for the whole range of hydrocarbons present, as well as formation and consumption of carbon monoxide and oxides of nitrogen (NO_x). The last is due to dissociation of atmospheric nitrogen.

13.1.2   *Direct firing and indirect firing*

The vast majority of gas turbine engines employ *direct firing* where the fuel is injected into the engine combustion chamber and then burnt. This is the case for the three primary fuels: kerosene, diesel and natural gas, and also for a number of less common fuels.

However, for certain obscure fuels considered for industrial engines direct firing may not be practical due to their corrosive or erosive nature. The only manner in which such fuel may be utilised is via *indirect firing* where it is burnt external to the gas turbine, with a heat exchanger to transfer heat to the compressor delivery air. Another indirect firing case of interest is a nuclear powered closed cycle, as described in Chapters 1 and 6.

For indirect firing the following changes must be made to performance calculations:

- No mass of fuel is added in the combustor calculations.
- CP downstream of the combustor is that for air at the given temperature, as opposed to that of combustion products.

The impact of both of the above is to lower power output and thermal efficiency since the mass flow and CP in the turbines are both lower. The exact power loss depends upon the engine cycle, and is between 4 and 8%.

13.1.3   *Kerosene*

Kerosene is a faction of crude oil primarily comprising a band of hydrocarbons with an average composition of $C_{12}H_{23.5}$ and molecular weight of 167.7. There are a number of commercial grades available such as JP4, JP5, Jet A1 and AVTUR which are refined to a tight specification. The proprietary high cost JP10 is a high density fuel used for certain military applications such as missiles. The term kerosene is usually also taken to encompass the lighter faction aviation gasoline, or AVGAS.

Aero-engines almost exclusively use kerosene as its high calorific value minimises fuel weight, and because it is free of corrosive elements such as sulphur which is essential for such high cost engines. The premium in fuel price is warranted by these gains.

13.1.4   *Diesel*

Diesel fuel is a heavier faction of crude oil than kerosene, again comprising a band of hydrocarbons, with an average composition of $C_{12.9}H_{23.9}$ and molecular weight of 178.6. It is less refined and hence has a lower cost, but therefore also contains small percentages of other elements such as the corrosive agent sulphur. The terms diesel and *fuel oil* are commonly used interchangeably. There are a number of *grades* spanning *Number 1 fuel oil* to *Number 6 fuel oil*, the tolerance relative to the nominal specification for each grade is significantly wider than for the kerosenes. The lower *Numbers* are closer to kerosene, have lower amounts of sulphur (typically up to 1.5% by weight being acceptable), and hence usually only Numbers 1 and 2 are used for gas turbine engines.

For cost considerations, diesel is almost exclusively used for marine engines, and in military applications it presents a lower risk of explosion. Its corrosive sulphur content is less of an issue, since marine engines must be designed to withstand a corrosive atmosphere anyway due to the sea water environment.

Though landbased power generation engines usually burn natural gas, a back up liquid fuel capability is often required to guard against an interrupted gas supply. In addition some niche applications use only liquid fuel. Again, due to cost, fuel oil Numbers 1 or 2 are employed, however life is shorter than for natural gas.

13.1.5   *Natural gas*

Natural gas comprises over 80% methane with minor amounts of ethane, propane, butane and heavier hydrocarbons. It may also include carbon dioxide, nitrogen and hydrogen. There are a plethora of blends of natural gas available world-wide. The composition on a molar basis of a common blend, used as a 'typical' natural gas for deriving gas properties of combustion products in Chapter 3, is listed below:

Methane:	95%	I-Pentane:	0.1%
Ethane:	1.9%	N-Pentane:	0.1%
Propane:	0.5%	Hexane	0.1%
I-Butane:	0.5%	Nitrogen:	1.5%
N-Butane:	0.1%	Carbon dioxide:	0.2%

As stated, industrial engines for power generation predominantly use natural gas. This is because of its abundance, competitive cost, and negligible content of corrosive elements such as sulphur. The benefits include lower emissions of carbon dioxide and good engine life, and the large volume required is not a logistics issue as it is for aircraft or marine propulsion. The lower emissions of carbon dioxide are because the lighter hydrocarbons contain more hydrogen and produce more water, and less carbon dioxide when burnt. As described later the higher water content of the combustion products leads to a higher power output and thermal efficiency for a given engine and SOT level. Engines for natural gas pumping burn solely the gas tapped off the pipeline.

Liquefied petroleum gas (LPG) may be burnt in gas turbines injected either as a liquid or as a gas. Refinery tail gases are also of interest, and are produced as a by-product of oil refining. Other gas fuels occur naturally in lower quantities, such as *condensates* and *naptha*. Their composition does not comprise the high percentage of methane present in 'natural gas' and their properties provide significant difficulties for satisfactory fuel injection. For instance condensates condense into a liquid at ambient temperature under the action of pressure.

13.1.6   *Other fuels*

For industrial engines other fuels have been considered for specific applications/projects, none have managed to take a significant share of the market.

Extensive efforts over many decades have been spent on developing coal as a fuel for gas turbine engines, however to date success has been limited. Coal gasification involves pyrolysis of coal in an atmosphere of steam and usually oxygen (as opposed to air). The resultant fuel comprises primarily carbon monoxide and hydrogen, with minor amounts of carbon dioxide, methane and sulphur compounds. Other attempts have also been made to coarse mill the coal to around 1 mm particles and then chemically clean it to remove the corrosive and erosive elements such as sodium, potassium, vanadium and ash (silicon based compounds). It is fine milled to sub 10 μm particle size such that it has characteristics of a gas for direct firing.

Hydrogen has been considered on a research basis for aircraft engines due to its high calorific value by weight. However the volume of fuel required is prohibitive, requiring complex tank arrangements. Hydrogen is present in many blends of natural gas; too high a content leads to problems with fuel injector flow number.

Biomass is a gas fuel produced from natural vegetation. While it has been considered more recently as a *renewable* fuel it is not widely used due to the large area of cultivated land required to fuel one engine continuously.

13.2   Data base of key fuel properties for performance calculations

This section presents descriptions of the key fuel properties for performance calculations, together with a data base. There are many other properties essential to gas turbine design, such as volatility and cloud point, which are outside the scope of this book.

13.2.1 *Calorific or heating value on a mass or volumetric basis*

Calorific value is the heat energy released per unit mass of fuel burned as shown by Formula F13.1. *Gross calorific value*, or *higher heating value* (*HHV*), is the total heat released making no allowance for the latent heat required to vaporise the liquid water produced by the combustion process. Hence it is a theoretical parameter and is rarely used. The *net calorific value*, or *lower heating value* (*LHV*), does account the latent heat of vaporisation and hence is the parameter most commonly used for gas turbine performance calculations. Strictly LHV is quoted as *the heat released under pressure in a constant volume when the combustion products are cooled to the initial temperature of 25 8C*. Hence for exhaustively rigorous combustion calculations the enthalpy balance should include:

- Heat release/absorption to raise fuel to 25 8C
- Heat release due to reducing compressor delivery temperature to 25 8C
- Heat release from fuel
- The resultant heat from the above is that available to raise combustion products from 25 8C to combustor outlet temperature

Occasionally for gas fuels calorific value is quoted on a volumetric basis (Formula F13.2) having units of either kJ/m^3 or kJ/scm. The scm (standard cubic metre) is the amount of gas that would occupy $1\,m^3$ at the ISO pressure and temperature of 101.325 kPa and 288.15 K respectively. Formula F13.3 shows how the number of scm present in a volume of gas fuel at any given pressure and temperature may be calculated.

LHV is used in most performance calculations such as design point, off design and test data analysis. Nominal values for the three primary gas turbine fuels are presented below, followed by methodologies for deriving values for a specific batch of fuel used during an engine test.

- Kerosene: 43 124 kJ/kg
- Diesel: 42 600 kJ/kg
- Natural gas: 38 000–50 000 kJ/kg
- Sample natural gas: 48 120 kJ/kg

Liquid fuels
For engine testing Formula F13.4 enables LHV to be calculated for kerosene if its specific gravity has been measured. Formula F13.5 enables LHV to be calculated for kerosene and light diesel fuels, knowing the carbon to hydrogen ratio and sulphur content. As described in Chapter 11, the most accurate method is to measure LHV by testing a sample in a bomb calorimeter, and is normal practice for performance testing.

Natural gas
For key performance tests using natural gas the fuel composition must first be determined using a gas chromatograph. The formula for then calculating LHV is complex, being a function of the wide range of constituents. It is presented in Reference 5 together with all data required, and is accurate to within ±0.1% once composition is known. Natural gas LHV can also be measured using a bubble calorimeter, however this is only undertaken for 'guarantee' performance tests as a further check on the calculated value.

13.2.2 *Density and specific gravity*

As described in Chapter 11, *fuel density*, the mass of fuel in a given volume (Formula F13.6), is important during engine performance testing. Invariably volumetric flow rate is measured and density must be derived such that heat release may be calculated. It is not of interest for other gas turbine performance topics.

Specific gravity is commonly used for liquid fuels and is the ratio of the fuel density to that of water at 4 8C as shown by Formula F13.7. Chart 13.1 presents SG versus fuel temperature for commonly used kerosenes and diesels. At 15 8C the SG for AVGAS is 0.704, the rest of the kerosenes are between 0.76 and 0.818, and diesels range from 0.82 to 0.88. For a given kerosene type the tolerance is around ±0.1 and for diesel it is even wider. Hence Chart 13.1 can only be used as indicative of SG level and for key engine performance tests it must be measured. As described in Chapter 11, this is accomplished using a fuel sample in a laboratory with a calibrated hydrometer. The fuel sample temperature must be measured simultaneously and Formula F13.8 used to derive SG at the fuel temperature measured during the engine performance test.

Gas fuel density may be calculated from Formula F13.9 where the 'compressibility' term z is used to modify the perfect gas law such that it may be applied to non-perfect gases. Gas fuel pressure and temperature must be measured in the vicinity of the turbine flow meter. References 7 and 8 show how R and z may be derived once the fuel gas composition is known. The calculation process for the latter is complex, however if it is not followed then significant errors will result.

13.2.3 *Viscosity, dynamic and kinematic*

Dynamic viscosity is the resistance to movement of one layer of a fluid over another and is defined by Formula F13.10. *Kinematic viscosity* is dynamic viscosity divided by density (Formula F13.11) and is the ratio of viscous forces to inertia forces. Dynamic viscosity for liquid fuels is occasionally required for performance calculations in determining fuel pump power requirements such as during start modelling. Kinematic viscosity is a second-order variable in the calibration of bulk meters and turbine flow meters for measuring liquid fuel volumetric flow rate.

Chart 13.2 presents kinematic viscosity for gas turbine liquid fuels. Dynamic viscosity can be derived by combining Charts 11.1 and 11.2 using Formula F13.11. It is common practice to arrange fuel dynamic viscosity as an input value to a performance computer program. Hence suitable values may be taken from Charts 13.1 and 13.2. If necessary the reader may 'fit' a polynomial to the data presented in these charts to embed the properties within the computer program.

Liquid fuels are not 'pumpable' with a kinematic viscosity of less than 1 cSt, and atomisation will be unsatisfactory above 10 cSt. Chart 13.1 shows that kerosenes generally have a viscosity of less than 10 cSt even at −50 8C and hence are always acceptable for atomisation. However diesels do exceed the limit and hence fuel heating is required if its temperature is allowed to fall below the threshold level.

None of the above effects are of importance for gas fuel as its viscosity is an order of magnitude lower, hence a data base is not provided here.

13.2.4 *CP of fuels*

For fully rigorous modelling fuel CP is required to calculate enthalpy input of the fuel prior to combustion. Approximate levels of CP for the primary gas turbine fuels are:

- Kerosene: 2.0 kJ/kg K
- Diesel: 1.9 kJ/kg K
- Natural gas: 2.0–2.2 kJ/kg K
- Sample natural gas: 2.1 kJ/kg K

13.2.5   *CP of combustion products*

CP of the products of combustion is required for all performance calculations involving combustion including design point, off design and engine testing. Chapter 3 presents formulae for calculating the CP of the products resulting from the combustion of kerosene or diesel fuel in air. Because these fuels are refined to a repeatable composition these formulae may be used universally to a high degree of accuracy.

However this is not so for natural gas which may come in a huge variety of blends, often with more than ten constituents. Chapter 3 presents formulae for the 'typical composition' given in section 13.1.5 above. However where second-order loss of accuracy is unacceptable then the CP of the products of natural gas combustion must be evaluated rigorously for the given natural gas composition. This complex calculation process is summarised in sample calculation C13.1.

13.2.6   *Combustion temperature rise*

Charts and formulae for evaluating temperature rise for the three primary gas turbine fuels are provided in Chapter 3.

13.3   Synthesis exchange rates for primary fuel types

Chapter 7 describes synthesis exchange rates. Industrial engines may often operate on more than one fuel type, in fact many engines are capable of switching between natural gas and diesel fuel in the field. This allows operators to pay a lower tariff for the gas fuel due to an *interruptable supply*, but not lose any availability. Also engines may often undergo a production pass off test (see Chapter 11) on kerosene at the manufacturer's facility, even though they will primarily operate on natural gas in the field.

Hence the second-order effect on engine performance of the different fuel types is often of great interest. The major contributor to performance change is the change in gas properties downstream of the combustor. Any difference in fuel mass flow resulting from different calorific values, and change in referred parameter turbine maps (full non-dimensional maps are unchanged), produces only tertiary effects.

13.3.1   *Kerosene to diesel*

The gas properties of the combustion products of kerosene and diesel, as well as their lower heating values are very similar. Hence there is negligible change in performance when burning either of these fuels.

13.3.2   *Liquid fuel to natural gas*

As described in Chapter 3, the CP of the combustion products of the natural gas is around 2% higher than for liquid fuels. This has a noticeable impact upon performance increasing both the fuel energy flow required for a given combustor temperature rise, and the shaft power output. The net effect is an improvement in SFC. The exact change in performance depends upon the engine cycle but at constant SOT the changes when burning gas relative to a liquid fuel are in the range:

- Power output: +4 to +6%
- Rotational speed: 0.5 to +1.5%
- Inlet mass flow: +2 to +3%

- HPC delivery pressure: +2 to +3%
- Fuel energy flow: +3 to +4%
- SFC: −1 to −2%

13.4   Oil types and data base of key properties

13.4.1   *Oil types*

Oils used for gas turbine engines fall into two major categories: mineral oils and synthetic oils. Mineral oils result from the refining of crude oil. Usually anti-oxidant and anti-corrosive additives are employed for gas turbine use. Synthetic oils are ester based and their cost is approximately ten times that of mineral oils.

Synthetic oils are used almost exclusively in gas generators due to their vastly higher auto-ignition temperature. Mineral oils are usually used for industrial power turbines where bearing chamber temperatures are sufficiently low. This is particularly true where journal as opposed to ball bearings are used, because the former requires approximately twenty times the oil flow to dissipate the heat resulting from significantly higher friction, hence the cost of using synthetic oil is prohibitive. Journal bearings are often required in large power turbines as they are capable of reacting far higher thrust loads.

13.4.2   *Density*

Formula F13.12 enables oil density for a typical synthetic oil with a density of $1100 \, kg/m^3$ at 15 8C to be calculated as a function of oil temperature. The same formula may be factored to give the typical density of a mineral oil at 15 8C, and then used for other oil temperatures. Typical values for oil density at 15 8C are:

- Light mineral: $850 \, kg/m^3$
- Medium mineral: $860 \, kg/m^3$
- Synthetic: $1100 \, kg/m^3$

13.4.3   *Viscosity, kinematic and dynamic*

Kinematic and dynamic viscosity are defined in section 13.2.3 above. Oil viscosity is important for calculating bearing and gearbox losses as described in section 5.17.

Chart 13.2 presents kinematic viscosity versus temperature for oils typically used in gas turbines. Light mineral oils are at the lower end of the band, synthetic oils in the middle and medium mineral oils at the top. Formula F13.13 provides kinematic viscosity versus temperature for the median line through the band on Chart 13.2. This may be factored to align with kinematic viscosity measured at a given temperature for an oil sample.

Dynamic viscosity may be calculated via Formula F13.13 and the values for density provided above. The formulae presented provide sufficient accuracy for all performance calculations.

Formulae

F13.1   Fuel lower heating value (kJ/kg) = fn(heat release (kJ), fuel mass (kg))

LHV = Q/Mass

(i)   See section 13.2.1 for difference between lower and higher heating value.
(ii)   Lower heating value is also commonly called calorific value.

F13.2   Fuel lower heating value by volume (kJ/m³) = fn(heat release (kJ), number of standard cubic metres of gas fuel (m³))

$$LHV = Q/SCM$$

(i)   See section 13.2.1 for difference between lower and higher heating value.
(ii) Lower heating value is also commonly called calorific value.

F13.3   Standard cubic metres (m³) = fn(volume (m³), gas fuel pressure and temperature (kPa, K))

$$SCM = Volume * (288.15 * P)/(101.325 * T)$$

(i)   This calculates the number of standard cubic metres that are equivalent to the amount of gas occupying the current volume at the prevailing P and T.
(ii) Hence to calculate heat release if LHV by volume is known then multiply the result of F13.3 with the output of F13.2.

F13.4   Kerosene fuel heating value = fn(FSG at 15 8C)

$$FHV = 48142.3 - 548.05 * FSG - 6850 * FSG^{\wedge}2$$

(i)   To convert from FSG at 15 8C to actual density multiply by 1000 kg/m³ and use F13.8.

F13.5   Fuel heating value (kJ/kg) = fn(hydrogen content (% mass), sulphur content (% mass), FSG at 15 8C)

$$FHV = 37290 + 566 * Hyd - 330 * S - 2300 * FSG$$

(i)   The remainder of the composition is the percentage carbon content by mass.
(ii)   The formula is valid for FSG values between 0.79 and 0.83, hydrogen content 13–14.1% and up to 0.3% sulphur. Over a wider range it provides indicative data only.
(iii) To convert from FSG at 15 8C to actual density multiply by 1000 kg/m³ and use Formula F13.8.

F13.6   Density(kg/m³) = fn(mass (kg), volume (m³))

$$RHO = Mass/Volume$$

F13.7   Fuel specific gravity = fn(actual density (kg/m³), standard density (kg/m³))

$$FSG = RHO/1000$$

(i)   The standard density of 1000 kg/m³ is that of water at 4 8C.

F13.8   Fuel specific gravity = fn(fuel sample SG, fuel sample temperature (8C), fuel temperature (8C))

$$FSG = FSGsample - (Tfuel - Tsample) * 0.00074$$

(i)   'Sample' refers to a fuel sample for which the SG and temperature have been measured in a laboratory.
(ii) To convert from FSG to actual density multiply by 1000 kg/m³.

F13.9 Gas fuel density = fn(static pressure (kPa), gas constant (J/kg K), temperature (K), compressibility)

$$RHO = P/R * T * z$$

F13.10 Dynamic viscosity (N s/m^2) = fn(shear stress (N/m^2), velocity gradient (m/s m), static temperature (K))

$$VIS = Fshear/(dV/dy)$$

(i)   Fshear is the shear stress in the fluid.
(ii)   V is the velocity in the direction of the shear stress.
(iii) dV/dy is the velocity gradient perpendicular to the shear stress.

F13.11 Kinematic viscosity (cSt) = fn(dynamic viscosity (N s/m^2), density (kg/m^3))

$$VISkinematic = 1\,000\,000 * VIS/RHO$$

F13.12 Oil density (kg/m^3) = fn(oil temperature (K))

$$RHOoil = 1405.2 - 1.0592 * T$$

(i)   This is for a typical synthetic oil.
(ii) This may be factored to give density for mineral oils using the data provided in section 13.4.2.

F13.13 Oil kinematic viscosity (cSt) = fn(oil temperature (K))

$$A = 76.14233 - 86.75707 * LOG_{10}(T)$$
$$+ LOG_{10}(T) * (34.35917 * LOG_{10}(T)$$
$$+ LOG_{10}(T) * (-4.726616 * LOG_{10}(T)))$$

$$VISkinematic = 6.82 * (10^\wedge(10^\wedge A - 0.6))$$

(i)   This is for the median line from Chart 13.2.
(ii) This may be factored to align it to the kinematic viscosity of a sample of oil at the sample temperature.

F13.14 Average molecular weight of combustion products = fn(molecular weights of constituents, number of moles of constituents)

$$MWav = \Sigma(Moles * MW)/\Sigma.Moles$$

Sample calculations

C13.1 Describe the calculation process to derive the CP of combustion products resulting from the combustion of the typical natural gas presented in section 13.1.5 in a stoichiometric mixture with dry air.

Calculate the molecular weight of the fuel
As per section 13.1.1 the number of moles of combustion products will be as below:

$$MW = 0.95 * (12 + 4 * 1) + 0.019 * (2 * 12 + 6 * 1)$$
$$+ 0.005 * (3 * 12 + 8 * 1) + (0.005 + 0.001) * (4 * 12 + 10 * 1)$$
$$+ 2 * 0.001 * (5 * 12 + 12 * 1) + 0.001 * (6 * 12 + 14 * 1)$$
$$+ 0.015 * 14 + 0.002 * 44$$
$$MW = 16.677$$

Moles fuel $= 2000/16.677$
Moles fuel $= 119.93$

Calculate the the number of moles of combustion products

$CO_2 = 119.93 * (0.95 * 1 + 0.019 * 2 + 0.005 * 3$
$\quad\quad + 0.006 * 4 + 2 * 0.001 * 5 + 0.001 * 6 + 0.002)$
$CO_2 = 125.327$

$H_2O = 119.93 * (0.95 * 4 + 0.019 * 6 + 0.005 * 8$
$\quad\quad + 0.006 * 10 + 2 * 0.001 * 12 + 0.001 * 14)/2$
$H_2O = 242.98$

The number of moles of oxygen consumed are then calculated, noting that for each mole of C_XH_Y fuel burnt then $(X + Y/4)$ moles of oxygen are consumed.
The number of moles of air are then calculated as per section 13.1.1.
The number of moles of nitrogen is that in the air plus that in the fuel.

Notes:
(i)   For ease of illustration this neglects the small amounts of atmospheric carbon dioxide and inert gases (argon and neon) present. This incurs negligible loss of accuracy, however Chart 3.5 includes data for argon and neon if required.
(ii) An alternative approach is to express the fuel as an 'average' chemical formula C_nH_m, where n and m may be non-integer.

Calculate the CP of the products in kJ/mole K
Derive the CP in kJ/kg K for each constituent for the given temperature using the polynomials for the constituents presented in Chapter 3.
Multiply each by its molecular weight to derive CP in kJ/mole K.
Evaluate the molar average:

Molar average $= \Sigma$ (No. of moles of each constituent $*$ CP)/Total number of moles

Derive the CP of the products in kJ/kg K
Derive the average molecular weight of all the combustion products using Formula F13.14 and the number of moles of each calculated above.
Derive the CP in kJ/kg K by dividing the molar average Cp by the average molecular weight of the combustion products.

Charts

Chart 13.1   Fuel specific gravity and viscosity versus temperature.

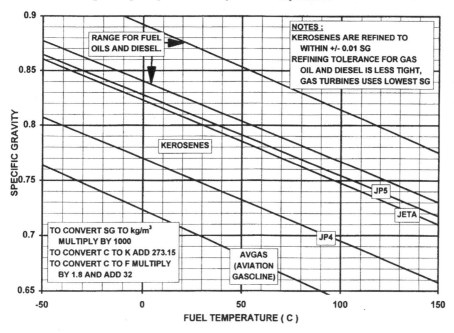

(a)   Specific gravity versus temperature

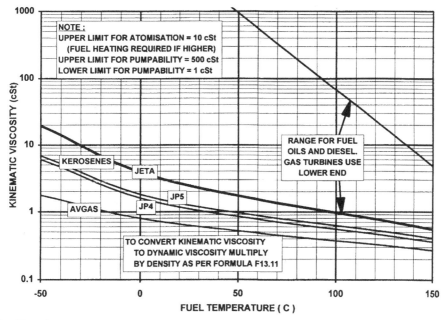

(b)   Viscosity versus temperature

Chart 13.2   Oil viscosity versus temperature.

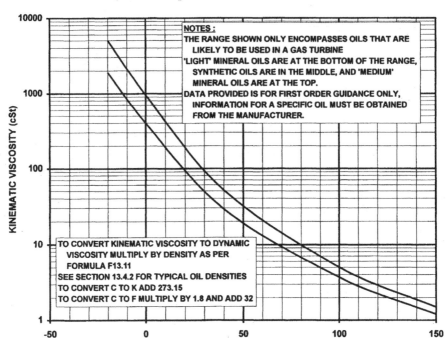

References

1. A. H. Lefebvre (1983) *Gas Turbine Combustion*, Hemisphere, New York.
2. H. M. Spiers (1961) *Technical Data on Fuel*, The British National Committee, London.
3. E. M. Goodger (1975) *Hydro-Carbon Fuels*, Macmillan, London.
4. J. A. Walowit and J. N. Anno (1975) *Modern Developments in Lubrication Mechanics*, Applied Science Publishers, Barking, Essex.
5. BSI (1993) *Liquid Fuels for Industrial Gas Turbines*, BS EN590, ASTM D975, British Standards Institution, London.
6. BSI (1994) *Liquid Fuels for Industrial Gas Turbines*, BS EN590, ASTM D2880, British Standards Institution, London.
7. ASME (1993) *Performance Test Code PTC-22*, ASME, New York.
8. ASME (1969) *Gaseous Fuels PTC 3.3*, ASME, New York.

Chapter 14
Performance of In-Service Products

14.0   Introduction

Performance of in-service products has traditionally created a wide range of technically challenging issues. However, its importance has increased significantly over recent years due to a heightened focus on life cycle costs, to which performance is central. Furthermore the internet has enabled the almost instantaneous transfer of field performance data to a personal computer in a technical centre for the cost of a local telephone call.

14.1   Instrumentation and test data analysis

14.1.1   *Instrumentation*

Field engines operate with a very sparse suite of *production instrumentation*. This is normally just sufficient for engine control and health monitoring, and is not designed for accurate diagnosis of in-service performance. Key drivers for this instrumentation are low unit cost and robustness, as opposed to the stringent accuracy demanded for development instrumentation.

Chapter 7 describes engine ratings and how they are implemented by set points in the control system; this defines the steady state instrumentation required. At production pass off the production instrumentation is often recalibrated or *trimmed* to match the more accurate data obtained by other instruments during the test. In service, drift or failure of the production instrumentation is one of the potential issues to resolve. Chapter 8 describes transient control philosophies and the performance parameters used by the control system.

Section 14.3 describes the design of a health monitoring system, and the gas turbine industry's increasing emphasis on it. Increasingly a business case can be shown for fitting some measurements dedicated to health monitoring, as opposed to the health monitoring system being forced to make the best of what is already there for engine control.

Dedicated performance tests may be scheduled for a field engine to diagnose an in-service performance issue. In this case some additional instrumentation may be fitted. However, the instrumentation level will still be significantly short of that for development testing due to field engine integrity concerns, limited data acquisition capabilities, the need to minimise the interruption to site operation, etc.

Instrument data for a field operating event such as an engine failure is usually particularly scant. This is because the historical log of production instrumentation readings is typically only at one scan per second, and for a limited duration of operation. Data storage requirements for more frequent readings are prohibitive and most events have a strong transient element. The key elements of events such as a surge or shaft failure occur within 0.5 seconds, rendering the field performance measurements of limited use.

14.1.2   *In-service performance data analysis*

The calculations for deriving performance parameters from raw in-service data, and then referring back to standard ambient conditions are similar to those discussed in section 11.8

and sample calculations C11.1 to C11.3. However, due to there often being limited measurements available, parameters may have to be derived by inference from other measurements, based on their relationship defined by either the off design model or production pass off data. This is clearly less accurate than having a full suite of measurements.

14.2   Traditional in-service performance issues

14.2.1   *Field performance guarantee acceptance tests*

As described in Chapter 11 most production gas turbines undergo a pass off test to demonstrate the compliance of each unit with the performance guaranteed to the customer. However, there are some engines for which this is not practical, such as very large (75 MW to 350 MW) heavyweight gas turbines. Here the cost of the test facility is prohibitive, also the engine cannot be transported in one piece and is delivered to the field site in modules.

In this latter case the compliance with the performance guarantees must be demonstrated when the engine is commissioned in the field. Reference 1 is an invaluable document providing detailed guidance on how this test should be carried out, including what specific instrumentation should be fitted. Also, very importantly, it states that the allocation of inaccuracies to either the manufacturer or customer/operator should be decided in advance.

14.2.2   *Performance deterioration*

Section 6.13.4 discusses performance *deterioration* of in-service engines and presents typical levels. This is distinct from performance losses due to compressor fouling, which can be recovered by compressor washing. Customers buying a new gas turbine are increasingly asking for guarantees of performance deterioration, as even a 1% loss in power, thrust or fuel efficiency has a noticeable impact on the through life cost of operation.

To combat deterioration it is common to allow measured turbine temperature to increase over time up to a limit, and only after this point will power or thrust fall. An aero engine would normally be pulled for maintenance at this point, once it had inadequate *TGT margin* relative to the value cleared on certification.

Performance deterioration can only be determined from engine measurements, however, obtaining accurate data is challenging. Experience has shown that utilising data from endurance tests during an engine development programme is invariably unrepresentative of what happens in field operation.

The best approach is to utilise new engine performance data available from the production pass off test described in Chapter 11. The engines also undergo an *as received* test when they are returned from field operation for overhaul. However, since the SFC measurement on a production bed will be only accurate to around ±1% for both tests, and the deterioration in SFC at constant SOT may only be 3% to 6%, then the measurement error in the deterioration is relatively large. Furthermore, it will usually be years between the pass off and as received tests, in which time test bed calibration and instrumentation may have been modified. Also, in many cases, it may not be clear how the engine has been operated; for example in oil and gas applications the number of operating hours will be known, but at what power level may not. Hence, to get reasonable confidence, at least 10 engines must be measured and a best-fit curve obtained of deterioration versus hours or/and engine cycles.

For engine types that do not undergo a production pass off test, such as large, heavyweight industrial gas turbines, then 'back to back' field tests must be employed. Due to the limited instrumentation and greater inaccuracy of field performance measurements, evaluating performance deterioration with confidence is fraught with difficulty.

During a *hot end overhaul* certain parts of the turbine and combustor sections are replaced. The performance expected at pass off is lower than for a new production engine, as there will still be larger tip and seal clearances in the compressors. During a *full overhaul* the remaining hardware that is subject to degradation is replaced, so that the expected performance is the same as for new production. One commonly used term is *zero lifed engine*.

14.2.3 *Diagnosis of in-service problems/failures*

Diagnosis of the root cause of in-service mechanical problems or failures is highly diverse and often presents fascinating challenges. Performance is invariably key to this process as understanding how the engine was being operated at the point of failure is critical. This may draw on a whole range of performance technology described in other chapters, as the problem or failure may have occurred while in a steady state, transient or starting regime.

Some typical problems, and potential root causes, are presented below:

- Steady state engine power or thrust output is low – control system thermocouples reading erroneously high, gas path components are damaged, control set points have been re-set erroneously after a module change, etc.
- An engine has surged in service – control system measurements are reading erroneously, the compressor has mechanical damage, an unforeseen path through the control algorithms has provided a fuel spike, etc.
- Engine won't start – a bleed valve is jammed closed, starter power output is low, thermo-couples reading erroneously are limiting start fuel flow, the start condition is at high altitude and on a cold day (which was not practical to replicate during the development programme), hence the start system needs further improvement, etc.

14.2.4 *Design modifications to in-service products*

Once problems or failures in the field have been fully understood, a modification to the engine design may be required to prevent future occurrences. As per Chapter 11, performance will be central to the processes of designing and testing these modifications.

Should a field problem be so serious that the whole fleet must be forced to operate at reduced power or thrust until a modification is developed and rolled out, then performance is pivotal to deciding how the limits will be set and implemented within the control system.

14.2.5 *Emissions tests*

For land based engines field emissions tests are important, as there is very often strict *permitting* legislation to meet regarding levels of NO_x, CO, unburnt hydrocarbons, smoke, etc. Sections 5.7 and 5.8 describe the design and operation of low emissions combustors. Should a field engine be non-compliant then performance is pivotal to understanding why, and what corrective action needs to be taken.

14.3 Unit Health Monitoring

Due to the heightened focus on the through life cost of gas turbines, engine health monitor-ing technology has become of increasing importance. As described in Chapter 15, many arrangements such as 'power by the hour' deals drive the manufacturer to minimise main-tenance requirements. Historically, maintenance has largely been *reactive*, where a problem is tackled after it has arisen, or *preventative*, where maintenance is carried out at regular scheduled intervals intended to prevent problems occurring. The latter is often over-cautious, impacting availability.

Health monitoring systems can add significant value by being able to *diagnose* what components are at the root of a change in engine performance. This information can be used to focus overhaul planning, understand performance degradation, etc.

Also, these diagnoses can enable *predictive* maintenance by analysis of data from a field engine, often in real time, to predict the onset of a mechanical problem so that a maintenance action can be taken in a timely fashion. Hence, both reliability and availability are improved. Performance is invariably at the heart of the algorithms which act on the raw data.

14.3.1 *Reliability and availability*

Reliability and availability are key product attributes for a gas turbine operator. *Reliability*, as defined by Formula F14.1, shows the percentage of time an engine is operational without unplanned events such as mechanical failures, spurious control system trips or failed starting. *Availability*, as defined by formula F14.2, is the percentage of time an engine is available to operate. This accounts for both unplanned and planned down time, the latter being due to the outages required for planned maintenance.

To put into perspective the challenge for a gas turbine manufacturer, power generation operators will often contractually demand 98% availability. This allows only 7 days per year for planned and unplanned events; this time must cover not only that required to work on the engine, but also the logistics of getting staff, equipment and parts to site. For aircraft engines, or aeroderivative industrial engines, replacement gas generators or engines are often provided to enhance plant availability, changeout being possible in a day. For large heavyweight industrial engines this is not possible and the duration of planned maintenance is significant, up to around 2 months. Historically, this has been offset by failures having been less frequent and maintenance intervals longer.

14.3.2 *The unit health monitoring system*

Figure 14.1 presents a typical health monitoring system. Raw data is logged from a field engine and then transmitted via an internet link to a data store. After processing by health monitoring algorithms, the raw data and calculated parameters are transmitted via the internet to the manufacturer's regional support and global technical centres, as well as back to the field site.

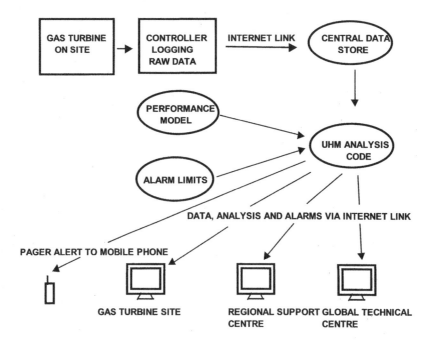

Fig. 14.1   The unit health monitoring system.

If the health monitoring predicts a forthcoming problem then alarms are automatically sent to these three locations, as well as mobile phone alerts to the personnel responsible.

If the alarm is minor or a repeat of a previous incident, then the site personnel will take the appropriate action. If it is more complex then it will be escalated to the manufacturer's regional support centre. The most challenging cases, with potentially high costs, are then referred on to the global technical centre where the full engineering team are located, to help solve the issue.

14.3.3 *Trending*

Trending with limits is where key parameters are plotted versus time and compared with upper and lower warning limits known as *tram lines*. To achieve this, data is typically expressed as a percentage of the value predicted by the cycle deck at each operating condition. The warning range is less than that between the *trip levels* locked into the engine control system for engine shut down. This enables the operator to investigate why a parameter is heading towards its trip value before shut down. Generally, one of the more complex systems below are operational, but trending is virtually always retained as the first line of defence.

Another form of trending is the CUSUM technique, as described in section 11.7.3 for production pass off. For health monitoring it is plotted against time rather than engine number. Again, data is expressed as a percentage of the value predicted by the cycle deck at each operating condition.

TGT spread is normally monitored. This is the term for circumferential temperature distortion, where multiple temperature readings start to show scatter; one of most common causes is blockage of one of the individual combustor burners. Charts are employed based on development test data which define the circumferential offset between low temperature at particular station to the reduced fuel flow in an individual burner. Warnings and trip limits are also applied to the degree of spread.

14.3.4 *Performance diagnosis techniques*

Engine and sensor diagnosis is comprised of 3 steps:

(1) First, *normality* must be defined.
    This may be via *pattern recognition* of previous performance data, or via an off design performance model. Both have to be *tuned* to the engine in question.
(2) Secondly, the threshold of difference that indicates abnormality must be defined.
(3) The final step is *identification* of the root cause.

Pattern recognition
This can be implemented in both simple and highly complex forms. For example, a number of incidences of nozzle guide vane burn through may have occurred within an engine fleet, leading to unplanned shut downs. Health monitoring data from these engines could then be manually examined and a shift in performance parameters identified prior to the failures, such as, say, exhaust gas temperature rising at 30°C per thousand hours as opposed to the normal degradation rate of 10°C per thousand hours. This threshold can then be coded into the fleet health monitoring system such that if 30°C per thousand hours is seen on an engine, then a warning message is automatically sent to pull the engine for overhaul within the next, say, 200 hours. Another simple example is for the health monitoring system to diagnose the onset of fuel injector blockage via changes in exhaust gas circumferential and radial temperature distributions.

Neural networks
These are employed for the most complex forms of pattern recognition. First, they are *trained* to learn the *footprint* of normal operation in terms of how parameters relate to each other.

Once trained they can examine performance data continually in real time and if the pattern goes outside the normal footprint then a warning is sent. It is then possible for an individual abnormal pattern to be tagged to a given component change.

The other main group of engine and sensor fault diagnosis techniques use an off design performance model to conduct gas path analysis. Advanced numerical methods may be used to both tune the average engine model to the deteriorated state of the actual engine in question, and for fault diagnosis.

Kalman filter

This is used in a number of commercialised systems and relies upon the assumption that the most likely answer to a given set of measurements is the minimum change from the expected results.

To operate the Kalman filter, exchange rates from the off design model are generated first. These show the percentage change in a health parameter (a measured parameter) for a 1% change in each of the key component performance parameters, such as efficiencies, capacities, etc. They are analogous to the synthesis exchange rates discussed in section 7.7. The Kalman filter then solves Formulae F14.3, which finds the set of component changes and measurement errors that give the minimum least squares error from both the expected component performance parameters and the actual measured parameters.

Genetic algorithms

These are computing-intensive techniques which try a range of 'solutions' as input to the off design model to minimise either Formulae F14.3, or the least absolute error function shown in Formulae F14.4. New potential solutions are 'bred' from only the best previous ones, hence the term 'genetic'. This type of solver is most suited to situations where there are discontinuities in the system, which is not the case with an off design performance model and hence a more straightforward deterministic solution of the error function is usually adequate.

Ansyn

Short for 'analysis-synthesis', this uses an extended performance model to ascribe changes in measured data to changes in engine component performance. Due to the reduced suite of measurements, Ansyn is mainly useful in service to determine whether something has changed or not, because it runs the model at the right conditions. If there are enough measurements then, in many cases, the 'signature' of specific problems can be recognised from past experience.

14.3.5   *Use of other engine measurements*

A number of non-performance parameters can also be used for health monitoring. These include engine vibration, automatic detection of debris on magnetic chip detectors in the oil system, and engine noise signatures. However, invariably these must be related to acceptable levels or previous values via engine performance parameters.

14.3.6   *Used life calculations*

Traditional practice had been to overhaul engines at the fixed number of hours or cycles that the engine had been certified to. However, very often this is a pessimistic approach as it assumes a standard usage profile. For example, naval marine engines often have a usage profile with significant time at lower power settings than those against which engine life was originally certified.

Health monitoring enables used life calculations to be implemented such that if an engine is operated at low power, cool ambient temperatures, or the cycle accelerations or decelerations are significantly lower than used in calculating the standard life, then an extended life may be evaluated for that individual engine. Care must be taken such that if modules are changed in the field, then the individual module lives are tracked.

A major issue of concern with respect to hours of operation is turbine creep and oxidation. One approach is to define bands of speed and firing temperature which are each given an *effective life*. For operation in the bands around base load or cruise, then one actual hour is also one *effective hour*. However, an hour spent in the next band up uses, say, 2 hours of *effective life*. At a lower band, an hour may be equivalent to only 0.7 hours of effective life usage. The effective life map is embedded in the UHM software. Speed and temperature are read from the engine control system and an effective life calculated. These are then added together for every actual hour of operation and an integrated used life is calculated.

Used cyclic life can also be implemented in the control system by counting engine starts and shut downs. Performance-based algorithms can be used in the health monitoring software to allow benefit for slower cycles than that on which the standard cyclic life is based, or those that only start to, or shut down from, an intermediate power. In the first instance the stress engineer will declare an effective cyclic life where if start times are, say, doubled then that only counts as, say, 0.7 of a cycle.

14.4   Other services

There are many other services that can be offered around in-service engines to which performance is pivotal.

14.4.1   *Training simulators*

Figure 14.2 illustrates a typical flight simulator used to train pilots. Clearly, this enables the trainee to undergo a significant amount of highly valuable training without the risk and cost of using a real aircraft. The cost savings for airlines or airforces can be dramatic.

The trainee sits in a hardware replica of the cockpit with all of the instrumentation, joy sticks, etc., identical to the real aircraft. Software is available to provide both cockpit instrumentation readings, and windscreen views for a whole range of airports or in-flight scenarios. A real-time model of the engine and its control system, as described in section 8.11, is utilised.

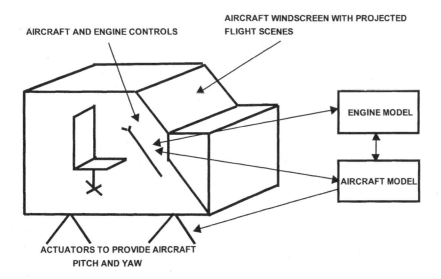

Fig. 14.2   Schematic diagram of flight simulator.

It responds to inputs from the cockpit and provides outputs to cockpit instrumentation, as well as providing thrust versus time to the real-time aircraft model. The latter also takes inputs from the cockpit and interacts with the airport/in-flight model to visualise the resulting flight path. Actuators are also used to pitch and yaw the simulator to mimic the aircraft response.

14.4.2   *Fleet management*

Chapter 15 describes how the off design performance is pivotal to evaluating through life cycle costs before a customer/operator purchases a gas turbine. This model can also be used in service to aid operational decisions on a techno-economic basis. It considers the performance and also the revenue issues, maintenance and availability impacts etc. Here the engine manufacturer is getting into the heart of the customer's business, to try and capture more value. This is an increasing trend in the industry.

Formulae

F14.1   Reliability

Reliability $= 1 -$ Unplanned downtime per year (hrs)/Planned running time per year (hrs)

F14.2   Availability

Availability $= 1 - ($Unplanned downtime per year (hrs) $-$ Planned downtime per year (hrs))/8760

(i)   Where 8760 is the number of hours in a year.

F14.3   Least Squares function $=$ fn(sum of component factors and deltas (y) divided by their standard deviation, sum of measurement errors (Z) divided by their standard deviations)

LS function $= \sum (\Delta y/\sigma)^\wedge 2 + \sum (\Delta Z/\sigma)^\wedge 2$

F14.4   Least Absolutes function $=$ fn(sum of component factors and deltas (y) divided by their standard deviation, sum of measurement errors (Z) divided by their standard deviations)

LA function $= \sum (\Delta y/\sigma) + \sum (\Delta Z/\sigma)$

References

1.   ANSI/ASME (1995) *Performance Test Code on Gas Turbine Power Plants PTC-22*, ASME, New York.

Chapter 15
Performance and the Economics of Gas Turbine Engines

15.0 Introduction

Performance is pivotal to a gas turbine engine's economic viability, both from the perspective of a manufacturer and an operator. Performance comprises the thrust, or shaft power, delivered for a given fuel flow, life, weight, emissions, engine diameter and unit cost. This is fundamentally what a gas turbine manufacturer sells and an operator buys. If a manufacturer designs an engine with poor performance then it will struggle to sell and is likely to make a loss on the project. Likewise, an operator will lose money should it purchase an engine with poor performance.

To put in perspective the levels of investment involved, a new large civil aero engine is likely to cost over a billion US dollars, and take at least 3 years to develop. Even the development of a new small 50 kW micro-turbine for land-based power generation will require in excess of ten million US dollars. Operator levels of investment are also high – a 50 MW power generation plant may have a first cost of over thirty million US dollars, with operating costs of around five million US dollars per year; engines are often in service for 25 years.

Engineers can no longer focus purely on the technical aspects of a gas turbine. In the modern economic climate it is critical that they understand the economics associated with the launch into product development, or purchase, of a new gas turbine as they are a major contributor to the preparation of the economic model or *business case*. *Techno-economic* analyses have become the 'norm' as opposed to the exception, even for government funded military products. Performance is inseparable from the business case as fuel burn, life, maintenance requirements, etc., are all driven by the performance parameters; fuel burn is often dominant. For example, in base load power generation fuel costs are typically 85% of the operating costs, and items related to engine performance represent 35% of the operating costs of a Boeing 737.

15.1 The business case for a gas turbine project

There are two main business cases that must be considered from the outset:

- Firstly, the gas turbine manufacturer must see a good return on the large investment required to develop a new product. This includes not only the cost of the engineering design and development programme, but also capital investments for manufacture or build facilities, warranty costs associated with early field units, etc. Returns come not only from the sale of new units, but also from the aftermarket support of products via spare parts, repair and overhaul services.
- Secondly, the customer who will buy and operate this product must see that the financial returns, such as passenger revenue, comfortably exceed the first cost and through life operational costs.

Both business cases must be considered from the outset of the concept design phase discussed in Chapter 6. For a product to be successful, both must be robust, providing a 'win–win' for the manufacturer and the customer/operator. References 1 and 2 provide comprehensive coverage of the issues faced by a business when deciding whether to develop a new product or not.

15.1.1 *Accounting terminology*

Reference 3 provides an introduction to company accounts. Many readers will be aware of the main instruments that a company uses to manage its financial performance.

- The *profit and loss account (P&L)* summarises and compares transactions in the accounting period, usually a year, which relate to either income from the sale of products/services, or expenses related to their production.
- The *balance sheet* is a statement of the financial position of a company at a point in time. This comprises: on one side what it owns including tangible assets such as land, buildings, machinery, as well as stock, debtors and cash; on the other side what it owes, including creditors, loans and share capital issued.
- The *cash flow statement* shows the change in cash owned by the company over a period, such as a year.
- *Capital cost* relates to purchase of an item which will contribute to profit over a number of years, and often over a number of projects, for example, a manufacturer setting up a new test facility.

All these describe the financial performance of a complete company over a year, or at a point in time. However, when a company is deciding whether it should invest in developing, or purchasing, a new gas turbine then it needs to examine how this individual project will contribute financially over, typically, at least ten years.

15.1.2 *The 'time value' of money*

A business case for an individual product is generated by modelling all forecast incomes and expenses throughout its life, then deducting one from the other to calculate the net cash flow for each year. In the early years the cash flow will be negative, then positive in the later years.

However, just adding up all of the annual cash flow returns over the project life would be very misleading. It is essential that the *time value of money* is considered, utilising two key parameters:

(1) *Discounted cash flow rate (DCFR)* or *the cost of money* can equally be considered as the interest rate at which a company could either borrow money at, or invest it in a bank, in bonds or on the stock market; it is typically of the order of 10%. Clearly for all the risk and challenges involved in a gas turbine project the manufacturer or operator is looking for its investment to comfortably exceed the DCFR that it could have earned from, say, a stock market tracker fund.

(2) *Discounted cash flow* as shown in Formula F15.1 is the cash flow in a given year discounted by the DCFR back to its *present value*.

To illustrate this, consider the simple example of one million dollars being invested in a product now, which returns a single cash flow of two million dollars in 10 years time. Without considering the time value of money, then this project makes a profit of one million dollars. However, if the two million dollars is brought back to its present value using Formula F15.1, with a DCFR of 10%, then it would be worth only 0.77 million dollars. Hence, the product is considered to have made a loss.

15.1.3 *Business case outputs*

There are a number of key business case output parameters which fully account the time value of money:

- *NPV – net present value* is the profit in dollars, in today's value, over and above what would have been made from the same investment at the DCFR. Formula F15.2 defines NPV, and sample calculation C15.1 illustrates its use. It is also a standard function available in most spreadsheets such as Microsoft Excel.
- *IRR* – the *internal rate of return* is the interest rate, in percent, that gives a net present value of zero; this is another way of evaluating how the investment will exceed the DCFR. If it is equal to the DCFR then the company may as well not bother. However, if it is, say, 25% when the DCFR is 10% then the investment is looking more interesting. Formula F15.3 and sample calculation C15.1 show how it is used; again, it is a standard function in most spreadsheets.
- *Worst annual negative cash flow* is the cash flow in the year, during the early part of the project, where the project makes the highest annual loss. This can be compared to the typical level of annual profit that the overall company makes and a judgement made as to whether this is financially viable.
- *Break even time* is the number of years until the cumulative cash flow gets to zero, and hence shows the point in time when the project will begin to contribute positively to overall company profit.

15.1.4 *Operator business case inputs*

There are a number of attributes such as life, emissions, noise, transient capability and safety, as per Chapter 1, that are essential for a product to be even considered by an operator. After this, business cases are evaluated for competing product offerings. Sample calculation C15.2 presents a business case evaluation for a customer considering purchasing a 100 MW combined cycle gas turbine plant for power generation; it is clear that all of the key drivers relate strongly to Performance. These are largely set by the concept definition and off design processes described in Chapters 6 and 7.

The sources of *revenue* depend upon the application. For example, in power generation they are MW hours generated and the market price of electricity. For civil aero applications, an aircraft business case must be generated in which the engine's contribution to the number of passenger miles is evaluated. Gas turbine performance is clearly critical to operator revenues. For example, more power or thrust within a market block, enabling extra revenue, can be a significant differentiator.

The *operational costs* are often combined into terms such as cost of electricity (CoE) in US cents/kWh, or US cents per passenger seat mile. Traditionally the main contributors have been:

- Fuel cost – this is usually the most critical parameter to the business case. This can be seen from sample calculation C15.2 where fuel costs represent 85% of the total annual operating cost. Its main drivers are the engine performance and the cost per kWh of the fuel.
- First cost – this is equivalent to the manufacturer's price and is strongly driven by engine performance. Specific cost in \$/kW or \$/kN is important; these are strongly related to specific power and specific thrust covered in Chapter 6. Generally the levels of specific cost reduce with increasing rated power or thrust.
- Overhaul and repair costs – these are typically circa 15% of the fuel costs.
- Engine weight – this is mainly an issue for aero applications as it displaces passengers or military payload.

More recently there has been a move towards operators buying sustained performance, offloading the risk of unreliable equipment to the manufacturer. These arrangements change the impact of operating costs:

- Under *long term service agreements* (LTSAs) the manufacturer guarantees availability after engine purchase in exchange for up front fees.
- Under 'power by the hour' deals engines are effectively leased.
- Some gas turbine manufacturers are moving into service provision at all levels in the operator's organisation, using the gas turbine to leverage increased business volumes. GE capital famously financed development of the Boeing 777–300 ER, securing exclusive use of the GE–90 – 115B.

15.1.5   *Manufacturer business case inputs*

In this instance revenues are generated by first predicting sales volumes per annum and then multiplying this by the forecast unit price:

- New unit *sales forecast* – the *market size* for the product will be forecast year on year. By its nature this is an imprecise process relying on extrapolating the historical market size, and then superimposing changes in market trends established via customer surveys, etc. A detailed comparison of the product to its competitors must then be made to come up with what portion of the market will be captured – the sales forecast. In doing this *competitor analysis* it is the performance issues of power/thrust, fuel efficiency, unit cost, weight, etc., that dominate.
- New unit price – this is the amount a customer is prepared to pay for the product and will be driven by the customer having a satisfactory business case, and how well the product compares with what the competition are offering. *Gross margin* is the difference between price and cost described below.
- Aftermarket sales forecasts and prices – these revenues are often the difference between a product business case showing a profit or loss. Revenues may be generated by the sale of spare parts and overhaul and repair services, or by LTSA payments as described above.

Historically, engines were often sold at a loss and spares at a large profit, meaning it took a long time to break even. In some instances, improved engine life and reliability over previous generations threatened the ability to return a profit at all. More recently the up front LTSA payments have helped this situation. They have also made high life and reliability even more essential for the manufacturer.

The major expenditures for a manufacturer are:

- Research and development (R & D) programme costs – section 15.0 indicates typical magnitudes. It includes salary and overheads for the engineers, purchase of development hardware, running component rigs as described in Chapter 5 and development engine testing as described in Chapter 11. One way to offset some of the testing costs is via a commercial demonstrator where early engine(s) complete endurance running at a customer site and earn revenue, rather than just burning fuel at the manufacturer's plant. Again, engine performance issues drive the magnitude of the overall R & D cost, as it strongly relates to the degree that emissions, SOT, pressure ratio, component efficiencies, etc., are pushed beyond levels the company has previously demonstrated.
- Cost of a production unit – again, as described in Chapter 6, engine performance is a major contributor. If a high pressure ratio has been chosen then this will drive up the number of compressor and turbine stages; if a high SOT has been set then more expensive materials with complex cooling geometries will be required. Conversely driving up specific power or thrust will reduce the engine size and, hence, cost. The cost is also strongly influenced by the production volume.

- Cost of providing aftermarket spares, repair and overhaul or maintenance – if the manufacturer is new to this market then there will be additional investments for the provisioning of spares and overhaul services to set up the customer support infrastructure.
- Test, manufacturing and production facilities – if they are to be shared with other projects then a percentage of their cost may be apportioned. As described in Chapter 11 performance drives the magnitude of these costs, by defining engine test facility requirements. The degree of advancement in components will drive what level of rig testing will be required, as described in Chapter 5.
- In field problems – costs are likely to be incurred due to fixing in-service problems, both in terms of re-engineering and compensation to customers. The greater the step change in performance, the higher these costs are likely to be.

15.2   Coupling the business case to the performance model

15.2.1   *Operator*

As summarised in Fig. 15.1 the operator's business case calculations can easily be coupled to the off design performance model. This enables the operator to input an estimated annual usage profile, in terms of time, at given ambient temperatures, pressure altitudes, humidity and power settings. Hence the IRR, NPV, etc., can be calculated quickly for this profile, as opposed to assuming single point operation.

This type of model may also be used by the manufacturer to evaluate customer NPV etc., for competing concept designs of a new product. As described in Chapter 6, off design models

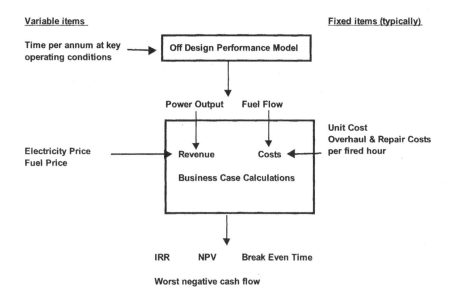

Notes:
Revenue and cost calculation repeated for all key operating conditions to capture performance and revenue variation
Business calculation runs over typically 10–25 years to capture the time value of money
Formula F15.3 defines IRR
Formula F15.2 defines NPV

Fig. 15.1   Coupling the off design performance model to the operator's business case.

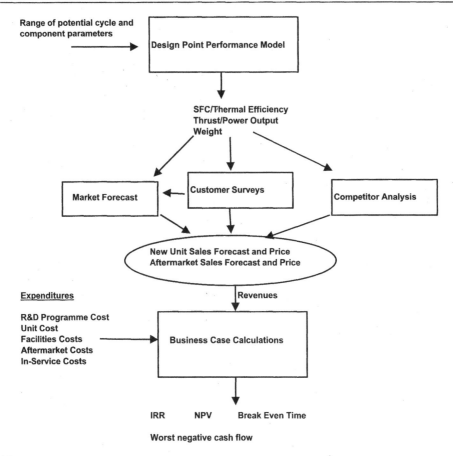

Fig. 15.2   Coupling the design performance model to the manufacturer's business case.

Notes:
Revenue and cost calculation repeated for all key operating conditions to capture performance and revenue variation
Business calculation runs over typically 10–25 years to capture the time value of money
Formula F15.3 defines IRR, Formula F15.2 defines NPV

are created for a number of engine designs before any down select can be completed. Just as SFC and thrust can be plotted versus the cycle parameters, so can financial parameters such as operator NPV, enabling techno-economic evaluation of competing cycles. Reference 4 illustrates the use of such a model to trade thrust and weight against customer NPV.

15.2.2   *Manufacturer*

Figure 15.2 shows how the manufacturer can extend the coupling described in section 15.2.1 to also include the manufacturer's own business case.

The NPV, IRR, etc., from the manufacturer's business case, as well as that of the operator, can again be plotted versus the cycle parameters in the same way as engine performance parameters such as thermal efficiency, SFC, specific power or thrust, as described in Chapter 6. Hence, the full techno-economic loop is closed right from the outset of a project.

15.3   Operational planning using in-service models

15.3.1   *Utilising the operator's business case model in-service*

Section 15.2.1 describes how the off design performance is pivotal to evaluating the through life business case before a customer/operator purchases a gas turbine. This through life business case model can also be employed to great advantage in-service. It can be used to show the operator whether, under the prevailing market conditions such as price of natural gas and price of electricity, it is worth running a plant, when to do scheduled maintenance, when to order spare parts, etc. The model can be extended to cover a string of engines, such as those along a stretch of natural gas pipeline or a fleet of aircraft.

15.3.2   *Power plant models*

The next generation of in-service models will combine online the performance and cost of operation via the business case model, with the reliability and availability models. For example, if an oil and gas operator has three engines at a pipe line station then one model will be run to evaluate both the economic impact of shutting one engine down for maintenance, and the reduced reliability of being able to compress the minimum necessary amount of gas along the pipeline. This model will also provide the statistical likelihood, and financial impact, of another engine shut down due to an unplanned failure.

15.4   Business case exchange rates

It is critical to test the sensitivity of the business case to variation in any of the input parameters. This is done in the same way as for engine design point, and off design, exchange rates described in Chapters 6 and 7. The business case model is run with a percentage change to each of the inputs, and the percentage change to the key outputs is tabulated. Invariably the business case is most sensitive to the engine performance-related parameters such as fuel efficiency or price, power, and the imposition of emissions taxes.

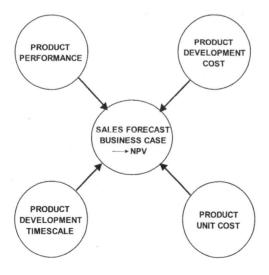

Notes:
Effect of each on NPV is generated in turn by assessing a 10% change
Decisions involving changes to several parameters are assessed by adding up the changes to NPV

Fig. 15.3   Product development exchange rates.

15.5   Product development exchange rates

Throughout a gas turbine product development programme, decisions must be made about whether to introduce additional product development cost and time to further enhance product attributes, such as engine performance and unit cost. To facilitate expedient business case based decisions *product development exchange rates* are generated for each of the four parameters. For example, if there was a 10% improvement in SFC of the product, then the number of extra sales could be estimated, due to this and the change in NPV generated.

Figure 15.3 summarises how trade-offs between these four parameters are achieved. A typical decision that may be faced would involve two or more of them changing simultaneously. For example, for an extra 5% product development cost, SFC and thrust may be improved by 3%, but unit cost increases by 1%. The development programme timescale is unaffected. The changes in NPV for performance, product development spend and unit cost can be quickly added up to see whether the result is positive or negative.

Formulae

F15.1   Discounted Cash Flow in year n ($) = fn (cash flow in year n ($), discounted cash flow rate (%), n)

DCF = (Actual cash flow in year n)/(1 + DCFR/100)^n

F15.2   Net Present Value ($) = fn (cash flow in years zero to n ($), discounted cash flow rate (%), n)

NPV = SUM (DCF) for years zero to n

F15.3   Internal Rate of Return (%) = fn (Upper DCF (%), Lower DCF (%), NPV at upper DCF ($), NPV at lower DCF ($))

IRR = a + NPVa/(NPVa − NPVb)/(b − a)

(i)   Where a refers to Upper DCF and b refers to lower DCF.
Note: this is an approximate formula using linear interpolation and the upper and lower likely bounds of DCF.

Sample calculations

C15.1   Calculate the NPV, IRR, maximum negative cumulative cash flow and the break even time for a potential investment with the following cash flows. The residual value at year 10 is considered to be zero and the DCFR is 10%. Comment as to whether this is a good investment.

Start year 1 = −$1.0M, End year 1 = −$2.0M, End year 2 = −$1.0M,
End year 3 = −$0.2M,   End year 4 = $0.5M,   End year 5 = $0.75M
End year 6 = $1.0M,   End year 7 = $1.25M, End year 8 = $2.0M
End year 9 = $2.0M,   End year 10 = $2.0M

(i) Calculate discounted cash flow for each year using Formula F15.1 with DCFR = 10%:

Year	Cash flow ($M)	Discounted Cash Flow ($M)
0	−1.0	−1.0
1	−2.0	−1.82
2	−1.0	−0.83
3	−0.2	−0.15
4	0.5	0.34
5	0.75	0.47
6	1.0	0.56
7	1.25	0.64
8	2.0	0.93
9	2.0	0.85
10	2.0	0.77

(ii) Calculate NPV using formula F15.2:

NPV = Sum of Discounted Cash Flows = 0.77$M

(iii) Calculate NPV for 5% and 20% DCFR – likely range:

Year	Cash flow ($M)	DCF at 5% ($M)	DCF at 20% ($M)
0	−1.0	−1.0	−1.0
1	−2.0	−1.9	−1.7
2	−1.0	−0.91	−0.69
3	−0.2	−0.17	−0.12
4	0.5	0.41	0.24
5	0.75	0.59	0.30
6	1.0	0.75	0.33
7	1.25	0.89	0.35
8	2.0	1.35	0.47
9	2.0	1.29	0.39
10	2.0	1.23	0.32
NPV		$2.52M	$−1.07M (−1.08 if rounded numbers are added)

(iv) Calculate IRR using Formula F15.3:

IRR = 5 + 2.52/(2.52 + 1.08) * (20 − 5)
IRR = 15.5%

(v) The project makes an NPV of $0.77 M for a total investment of $4.2M. The worst annual cash flow is a loss of $2M and the project has a break even time of 9 years. As would be expected with a positive NPV the IRR of 15.5% is greater than the DCFR of 10%. Most companies would only proceed for these levels of return if the project was very low risk or strategic in nature.

C15.2 **An electrical utility is in the process of buying a combined cycle gas turbine plant; it can export up to 100 MW. It is offered engines from 'manufacturer A':**

Fully installed plant price (US$M)	**72**
Thermal efficiency (%)	**54.0**
Power output (MW)	**98**
Overhaul/maintenance cost (US$/MW.hour)	**4**

Calculate the NPV for this product offering with the following assumptions:

Running hours per year	**8000**
Average electricity price	**55 US\$/MW.hour**
Fuel price	**0.33 UScents/MJ**
Evaluation period	**10 years**
DCFR	**10%**

(i)   Calculate costs per annum for manufacturer A's offering:

Fuel cost = power output/thermal efficiency * 3600 * hours per year * fuel cost per MJ
Fuel cost = 98/0.54 * 3600 * 8000 * 0.0033
Fuel cost = US\$17.3M
O & M cost = power output * hours per year * O & M cost per MW.hour
O & M cost = 98 * 8000 * 4
O & M cost = US\$ 3.1M

(ii)   Calculate revenue per annum from manufacturer A's offering:

Revenue = power output * hours per year * price of electricity
Revenue = 98 * 8000 * 55
Revenue = US\$43.1M

(iii)   Calculate discounted cash flow for per annum from manufacturer A's offering by adding the revenue and costs and then using Formula F15.1 with DCFR = 10%:

Year	Cash flow (\$M)	Discounted Cash Flow (\$M)
0	−72	−72
1	22.7	20.6
2	22.7	18.8
3	22.7	17.1
4	22.7	15.5
5	22.7	14.1
6	22.7	12.8
7	22.7	11.6
8	22.7	10.6
9	22.7	9.6
10	22.	8.8

(iv)   Calculate NPV for manufacturer A:

Apply F15.2 to add up the DCF column,

NPV = US\$67.5M

References

1.   M. Robert (1995) *Product Innovation Strategy*, McGraw-Hill, New York.
2.   P. G. Smith and D. G. Reinersten (1995) *Developing Products in Half the Time*, Van Nostrand Reinhold, New York.
3.   W. Reed and D. R. Myddleton (1997) *The Meaning of Company Accounts*, 6th edn, Gower, Bournemouth.
4.   J. Hartsel (1998) ASME 98-GT-182, ASME, New York.

Appendix A
Engine Station Numbering and Nomenclature

A.0 Introduction

This appendix provides in full the international standard for engine station numbering and nomenclature, which is used throughout this textbook. At first this topic may seem mundane and of secondary importance. However in industry the ability to unambiguously transfer performance data world-wide provides substantial cost savings due to efficiency gains and avoidance of misinterpretation. Virtually every gas turbine company has some form of alliance or joint venture with other companies, due to the huge cost of developing new engines. Furthermore customers, such as airframe manufacturers, often have large departments of engineers dealing with gas turbine performance.

A.1 International station numbering and nomenclature standards

Reference 1, ARP 755A, is the main internationally recognised standard for gas turbine engine station numbering and nomenclature. ('ARP' means 'Aerospace Recommended Practice'). This was developed within the aerospace industry to improve efficiency and quality control when data passed between companies. Indeed, prior to the introduction of ARP 755A different station numbering and nomenclature often existed even within a company's own product range, creating confusion for engineers moving between engine programmes. ARP 755A is now used almost universally by aero gas turbine engine companies, airframe manufacturers, etc.

Reference 2, AS681, defines more rigidly the requirements for engine steady state performance representation on digital computer, via *customer decks*. (This term dates from when computer programs and data were supplied as a 'deck' of cards.) The rigid definition ensures interchangability of decks supplied by different companies covering engines for the same application. Section A.4 describes the main features. Many gas turbine companies use the AS681 system in house, to avoid the effort of translation when supplying decks externally.

Engine performance standards evolve as the design and development progresses. AS 681 mentions *preliminary design* decks, where the engine and deck are basically unfinished, and at the other extreme *specification* decks which model the production engine developed to meet the specification. In addition there are *status* decks, which cover the evolving status of the engine performance during development; Reference 3 defines these more fully. One final category is decks for engine test data reduction, defined by Reference 4.

Reference 5, ARP 1257, covers transient performance decks. The requirements are an extension of those in AS681, with additional necessary parameters relating to the simulation timebase.

Reference 6, PTC 22 includes an industrial gas turbine nomenclature. ARP 755A is more commonly used however, as many industrial engines are aero-derived, and also most heavy-weight engine manufacturers are now partnered with an aero gas turbine company for technology transfer.

Owing to its widespread use throughout industry, ARP 755A station numbering and nomenclature have been employed throughout this textbook.

A.2   ARP 755A station numbering

This section describes the basis for station numbering. The fundamental station numbers are marked on Figs 1–4 describing the major engine configurations, which are located in the Gas Turbine Engine Configurations section, just before Chapter 1. Station numbers are appended to symbols, such as for total pressure, to identify exactly at what point in the engine that value of pressure occurs.

A.2.1   *Fundamental station numbers*

The fundamental station numbers for the core stream of an engine are as below:

AMB	Ambient conditions
0	Ram conditions in free stream
1	Engine intake front flange, or leading edge
2	First compressor/fan front face
3	Last compressor exit face
4	Combustor exit plane
5	Last turbine exit face
6	Front face of mixer, afterburner etc.
7	Propelling nozzle inlet
8	Propelling nozzle throat
9	Propelling nozzle or exhaust diffuser exit plane

A.2.2   *Intermediate station numbers*

Stations between the fundamental ones are numbered using a second digit suffixed to the upstream fundamental station number. In general this is not formally defined, hence companies have their own practices. For example T4 is the combustor exit/turbine nozzle guide vane leading edge temperature, and T41 is usually employed for the first stator outlet temperature (see section 6.2.2 for its definition).

Where more than ten intermediate stations are required a third digit is used. Continuing the above example the first nozzle guide vane throat, which occurs between station 4 and 41, is usually numbered 405.

A.2.3   *Turbojets*

Intermediate station numbers most commonly used for a two spool turbojet are listed below. Additional station numbers would be created to deal with the mixing of cooling air flow back into the main stream.

24	First compressor exit
26	Second compressor front face
31	Compressor outlet diffuser exit/combustor inlet
405	First turbine nozzle guide vane throat
41	Stator outlet temperature
44	First turbine exit
45	Second turbine nozzle guide vane leading edge

A.2.4   *Turbofans*

Here the fundamental station numbers are prefixed with a 1 for the bypass stream, and the core numbering is as per sections A.2.1–A.2.3. For a turbofan with separate jets, common bypass duct station numbers include:

12	Fan tip front face, if conditions are different from the fan root front (station 2)
13	Fan exit
17	Cold propelling nozzle inlet
18	Cold propelling nozzle throat

In the more complicated instance of mixed streams and an afterburner the following numbers are usually used through these components:

16	Cold mixer inlet
6	Hot mixer inlet
65	Mixer outlet/afterburner inlet
7	Afterburner outlet/propelling nozzle inlet

For a three spool turbofan common additional stations are 24 for the second compressor entry, and 26 for the third.

A.2.5 *Shaft power engines*

For a simple cycle shaft power engine the key station numbers are as per sections A.2.1–A.2.3, however stations 6, 7 and 8 are normally redundant as there will only be an exhaust diffuser between stations 5 and 9. In the case of an industrial engine station 1 would be the engine inlet flange, and station 9 the engine exhaust flange. Station 0 would be used for the plant intake flange, and station 10 for the plant exhaust flange.

For the more complicated intercooled, recuperated shaft power cycle the following intermediate station numbers would typically be employed for these components:

21	First compressor exit face
23	Intercooler inlet face
25	Intercooler exit face
26	Second compressor inlet face
307	Recuperator air side inlet face
308	Recuperator air side exit face
31	Combustor inlet
6	Recuperator gas side inlet
601	Recuperator gas side exit

A.2.6 *Spool rotational speeds, inertias, etc.*

These are numbered as per that at the inlet to the first compressor on the given spool. For example, for a two spool turbojet the polar moments of inertia are XJ2 for the first spool and XJ26 for the second.

A.3 Nomenclature

A.3.1 *Fundamental parameters and commonly used ratios, functions, etc.*

The nomenclature listed below for fundamental parameters, commonly used ratios etc. is taken from ARP 755A and is used throughout this textbook. Some symbols listed do not occur in the text but are provided here for reference purposes. In some instances an *alternative* as well as a *recommended* symbol is provided; either of these may be employed. An asterisk denotes where the alternative symbol has been used herein. The 'recommended' symbols are regimented to allow their use in FORTRAN computer programs. For instance 'XM' is recommended for Mach number, as FORTRAN would interpret the alternative 'M' as an integer variable ('M' is more commonly used outside of computer code, however).

Further parameter names can be derived from the fundamental parameter symbols by applying the rules provided in sections A.3.2–A.3.5, a range of examples is provided in each section. Some of the symbols for commonly used functions listed below, such as SFC, do not adhere to these rules, however they are retained within ARP 755A due to their widespread use throughout industry. For clarity this text also uses the common terms 'SOT' (HP turbine stator outlet temperature) and 'PR' (pressure ratio); these are in widespread use, though not listed below.

	Recommended	Alternative	
Area, geometric	A		
Area, effective	AE		
Altitude, geo-potential	ALT		
Angle	ANG	α, β, γ, etc.	*
Blow out margin	BOM		
Bypass ratio	BPR		
Coefficient of discharge	CD		
Coefficient of velocity or thrust	CV		
Delta (pressure/101.325 kPa)	DEL	δ	*
Diameter	DI		
Density	RHO	ρ	
Drag	FD		
Efficiency, adiabatic[†]	E	η	
Efficiency, polytropic	EP		
Enthalpy, total per unit mass	H		
Entropy, total per unit mass	S		
Entropy function	PHI	ϕ	
Force	F		
Fuel air ratio	FAR		
Fuel flow	WF		
Fuel lower heating value	FHV		
Fuel specific gravity	FSG		
Frequency	FY	f	
Gas constant (per unit mass)	R		
Heat transfer rate	QU	Q	
Inertia, polar moment	XJ	J	
Length	XL	L	
Light off margin	XLOM	LOM	
Mach number	XM	M	*
Mass	GM	m	*
Mass flow rate	W		
Molecular weight	XMW	MW	*
Power lever angle	PLA		
Power	PW		
Pressure, static	PS		
Pressure, total	P		
Radius	RAD		
Ratio of specific heats, gamma	GAM	γ	*
Relative humidity	RH		
Reynolds number	RE		
Reynolds number index	RNI		
Rotational speed	XN	N	*
Specific fuel consumption	SFC		
Specific gravity	SG		

	Recommended	Alternative	
Specific heat, constant pressure	CP		
Specific heat, constant volume	CVOL		
Surge margin	SM		
Tangential wheel speed	U		
Temperature, static	TS		
Temperature, total	T		
Theta	TH	θ	*
Thrust, gross	FG		
Thrust, net	FN		
Time	TIME	t	
Torque	TRQ		
Velocity	V		
Velocity, true air speed	VTAS		
Velocity, equivalent air speed	VEAS		
Velocity, calibrated air speed	VCAS		
Velocity, of sound	VS	a	
Velocity, dynamic head	VH	q	
Viscosity	VIS		
Volume	VOL	v	
Weight	WT	w	

†The strict scientific term would be *isentropic*, which means adiabatic plus reversible. In addition, 'ETA' is also commonly used for efficiency.

A.3.2 *Combining station numbers and symbols for fundamental parameters*

The station number from section A.2 is applied as a suffix to the symbol from section A.3.1. For example:

P2	Total pressure at the first compressor front face
PS5	Static pressure at the last turbine exit face
W31	Combustor inlet mass flow rate
T1	Total temperature at the engine inlet
FAR8	Fuel air ratio at the propelling nozzle throat
FG19	Cold nozzle gross thrust

A.3.3 *Operating symbols*

Apart from 'AV', which would normally be a suffix, the following operating symbols would normally be embedded in compound groups.

	Recommended	Alternative
Average	AV	
Derivative with respect to time	U	d/dt
Derivative with respect to X	UX	d/dX
Difference	D	$-$ or Δ
Quotient or Ratio	Q	/
Square Root	R	

Some examples of the use of these operands are:

P3Q2	Overall pressure ratio
WRTQP26	W * SQRT(T)/P at station 26
T4D5	The difference between total temperature at stations 4 and 5
UN*	Rate of change of rotational speed with time
DTRQ	Unbalanced torque

Note: Operating symbols may be omitted if length restrictions would be exceeded, e.g. 'WRTQP26' could become 'WRTP26' to achieve the six characters permitted in Fortran 66.
* The term NDOT is commonly used.

A.3.4   *Fluid description*

It is often useful to employ an additional symbol immediately after that for a fundamental parameter to denote the type, or use, of fluid. This is rarely needed within the main gas path. The following symbols are recommended in ARP 755A:

Air	A
Bleed	B
Boundary Layer	BL
Coolant	CL
Fuel	F
Leakage	LK
Water	W

Some examples of these fluid description symbols are:

WB3405	Bleed flow rate from station 3 to station 405
WF	Fuel flow rate
WW3	Water injection flow rate at compressor delivery
PSB3	Static pressure of bleed air flow at compressor delivery

A.3.5   *General descriptive symbols*

The following symbols may be used as suffixes to parameter names to aid understanding:

Ambient conditions	AMB
Controlled parameter	C
Distortion	DIST
Effective	E
Extraction	X
Gross	G
High (maximum)	H
Ideal	I
Installed	IN
Low (minimum)	L
Map value	M
Net	N
Parasitic	PAR
Polytropic	P
Referred (corrected)	R
Relative	REL
Sea level	SL
Shaft delivery	SD

Standard	STD
Static	S
Swirl	SW
Tip	TIP

Some examples of the use of these suffixes are provided below:

UTIP	Tangential blade speed at the tip section
ANGSW5	Swirl angle at exit from the last turbine stage
P3Q2M	Compressor pressure ratio from compressor map before any factors or deltas are applied
TAMB	Ambient temperature
W3R2	Flow rate at compressor delivery referred to conditions at compressor front face
PWPAR	Parasitic power
PWSD	Engine output power
NSD	Engine output speed

A.4 Customer deck requirements

As stated, References 2–5 make various stipulations which aim to ensure compatibility between customer decks supplied by different engine companies. The main points are as follows.

- A very comprehensive range of engine features must be covered, including such items as customer bleed, power extraction, inlet distortion, ram recovery and windmilling.
- Power or thrust level must be selectable either by running to a specified level of it, of some other engine parameter, or via a *rating code* value.
- Thermodynamic relationships must be as defined, such as standard air and fuel properties.
- The content and structure of the deck user guide must be as specified.
- Nomenclature must follow that of ARP 755A, with specified extensions to assist in 'case' (point) definition and identification. For example prefixing a parameter name with 'Z' denotes an input value, and items such as 'TITLE' are added. The form in which to print out various units is also specified; doing this properly was difficult in the 1960s without superscript, lower case, etc.
- Computer capabilities and language are specified. FORTRAN 66 must be used, though with later enhancements if agreed with the customer.
- The *engine routine* must work as a subroutine, *ER*. Several FORTRAN *common* blocks form the *complete* interface for data transfer. FIXIN and FIXOUT have fixed items, VARIN and VAROUT have items dependent on the engine configuration, and EXPIN and EXPOUT have items to be agreed between user and supplier. (The names come from 'fixed', 'variable', 'expanded', 'input', 'output'.)
- Deck execution errors must be indicated by output messages known as *numerical status indicators* (*NSIs*). These have four digits, each of which has a specified meaning.

On the computing side much appears dated. Partly for this reason, the standards do ensure that compatibility is achieved between decks and users' host systems.

References

1. SAE (1974) *Gas Turbine Engine Performance Station Identification and Nomenclature*, Aerospace Recommended Practice, ARP 755A, Society of Automotive Engineers, Warrandale, Pennsylvania.

2. SAE (1989) *Gas Turbine Engine Steady-State Performance Presentation for Digital Computer Programs*, Aerospace Standard, AS681 Rev. E, Society of Automotive Engineers, Warrandale, Pennsylvania.

3. SAE (1974) *Gas Turbine Engine Status Performance Presentation for Digital Computer Programs*, Aerospace Recommended Practice, ARP 1211A, Society of Automotive Engineers, Warrandale, Pennsylvania.

4. SAE (1996) *Gas Turbine Engine Interface Test Reduction Computer Programs*, Aerospace Recommended Practice, ARP 1210A, Society of Automotive Engineers, Warrandale, Pennsylvania.

5. SAE (1989) *Gas Turbine Engine Transient Performance Presentation for Digital Computer Programs*, Aerospace Recommended Practice, ARP 1257, Society of Automotive Engineers, Warrandale, Pennsylvania.

6. ASME (1993) *Performance Test Codes, Gas Turbines*, PTC 22, The American Society of Mechanical Engineers, New York.

Appendix B
Unit Conversions

B.0   Introduction

This appendix presents, in alphabetical order, all unit conversions likely to be required for gas turbine performance calculations. For all tables, except that covering pressure or stress, a quantity in given units is multiplied by the value in the next column to the right to convert it to the units in the column to the right of the conversion factor. The conversions presented may be combined, for example for acceleration mile/h s may be converted to ft^2/s by multiplying by 0.447 and then by 3.28084.

Almost all conversion factors are taken from the references quoted at the end of this appendix.

B.1   Acceleration

From	Multiply by	To	From	Multiply by	To
ft/s^2	0.3048	m/s^2		3.28084	ft^2/s
km/h s	0.27778	m/s^2		3.6	km/h s
mile/h s	1.609344	km/h s		0.621371	mile/h s
mile/h s	0.447	m/s^2		2.23714	mile/h s

B.2   Area

in^2	645.16	mm^2		0.00155	in^2
in^2	6.4516E-04	m^2		1550.0	in^2
ft^2	0.092903	m^2		10.7639	ft^2

B.3   Density

lb/ft^3	16.0185	kg/m^3		0.062428	lb/ft^3
lb/in^3	27 679.9	kg/m^3		0.0000361273	lb/in^3
lb/UKgal	0.0997763	kg/m^3		10.0224	lb/UKgal
lb/USgal	0.0830807	kg/m^3		12.0365	lb/USgal

B4   Emissions (approx.)

CO mg/Nm^3	0.8	vppm	1.25		mg/Nm^3
CO mg/kWh fuel	0.278	vppm	3.6		mg/kWh fuel
CO mg/kg fuel	0.0204	vppm	49		mg/kg fuel
UHC mg/Nm^3	1.41	vppm	0.71		mg/Nm^3
UHC mg/kWh fuel	0.455	vppm	2.2		mg/kWh fuel
UHC mg/kg fuel	0.036	vppm	28		mg/kg fuel

From	Multiply by	To	From	Multiply by	To
NO_x mg/Nm^3	0.487	vppm	2.054		mg/Nm^3
NO_x mg/kWh fuel	0.164	vppm	6.1		mg/kWh fuel
NO_x mg/kg fuel	0.0125	vppm	80		mg/kg fuel

Notes: Natural gas fuel $= 47$ MJ/kg, 0.9 kg/Nm^3.

vppm are quoted with water content neglected and equivalent for 15% oxygen (this specifies fuel air ratio and forbids dilution).

UHC (unburnt hydrocarbon) is taken here to be CH_4 – a reasonable assumption for most natural gas.

All oxides of nirogen are taken to be NO_2 – usual practice for emissions legislation.

Nm^3 is a normal cubic meter i.e. 1 atmosphere, $20\,^\circ$C.

B.5 Energy

From	Multiply by	To	Multiply by	To
Btu	0.555558	Chu	1.8	Btu
Btu	778.169	ft lbf	1.28507E-03	Btu
Btu	1.05506	kJ	0.947817	Btu
calorie	4.1868	J	0.238846	calorie
calorie	0.0022046	Chu	453.597	calorie
Chu	1899.105	J	0.0005265	Chu
Chu	1400.7	ft lbf	7.1393E-04	Chu
hp h	1.98E+06	ft lbf	5.0505E-05	hp h
hp h	2.68452	MJ	0.372506	hp h
kW h	2.65522E+06	ft lbf	3.76617E-05	kW h
kW h	3.600	MJ	0.277778	kW h

B.6 Force

From	Multiply by	To	Multiply by	To
kgf	9.80665	N	0.101972	kgf
lbf	0.4535924	kgf	2.20462	lbf
lbf	32.174	lb/ft s^2 (pdl)	0.031081	lbf
lbf	4.44822	N	0.224809	lbf
tonf (Imperial)	9964.02	N	1.00361E-04	tonf

B.7 Fuel consumption

From	Multiply by	To	Multiply by	To
mile/UKgal	0.354006	km/litre	2.82481	mile/UKgal
mile/UKgal	0.83267	mile/USgal	1.20096	mile/UKgal
mile/USgal	0.29477	km/litre	3.39248	mile/USgal

B.8 Length

From	Multiply by	To	Multiply by	To
ft	0.3048	m	3.28084	ft
in	0.0254	m	39.3701	in
in	25.4	mm	0.0393701	in
mile	1.609344	km	0.621371	mile
nautical mile	1.852	km	0.539957	nautical mile
yard	0.9144	m	1.09361	yard

Note: The above is for the international nautical mile; the UK nautical mile is obsolete.

B.9   Mass

From	Multiply by	To	From	Multiply by	To
lb	0.45359237	kg		2.20462	lb
lb	0.031056	slug		32.2174	lb
ounce	28.3495	g		0.035274	ounce
tonne	1000	kg		0.001	tonne
UK ton	1016.05	kg		9.84207E-04	UK ton
UK ton	1.01605	tonne		0.984207	UK ton

B.10   Moment of inertia

$lb\,ft^2$	0.0421401	$kg\,m^2$		23.7304	$lb\,ft^2$
$lb\,in^2$	2.9264E-04	$kg\,m^2$		3417.17	$kg\,m^2$

B.11   Momentum – angular

$lb\,ft^2/s$	0.0421401	$kg\,m^2/s$		23.7304	$lb\,ft^2/s$

B.12   Momentum – linear

$lb\,ft/s$	0.138255	$kg\,m/s$		7.23301	$lb\,ft/s$

B.13   Power

Btu/s	0.555558	Chu/s		1.799992	Btu/s
Btu/s	778.169	ft lbf/s		1.28507E-03	Btu/s
Btu/s	1.05506	kW		0.947817	Btu/s
Chu/s	2.54674	hp		0.39266	Chu/s
Chu/s	1.899105	kW		0.5265	Chu/s
ft lbf/s	1.35582	W		0.737562	ft lbf/s
hp	550	ft lbf/s		1.81818E-03	hp
hp	0.7457	kW		1.34102	hp
PS	0.98632	hp		1.01387	PS
PS	75	kgf m/s		0.0133333	PS
PS	735.499	W		1359.62E-06	PS

Note: The PS is also called a 'metric horsepower'.

B.14   Pressure

See Table B.1 at the end of this appendix.

B.15   Specific energy

Btu/lb	2.326	kJ/kg		0.429923	Btu/lb
Chu/lb	45066.1	ft^2/s^2		2.219E-05	Chu/lb
Chu/lb	4.1868	kJ/kg		0.238846	Chu/lb
ft lbf/lb	2.98907	J/kg		0.334553	ft lbf/lb

B.16   Specific fuel consumption (SFC)

From	Multiply by	To	From	Multiply by	To
kg/kW h	0.735499	kg/PS h		1.35962	kg/kW h
lb/lbf h	0.10197	kg/N h		9.80665	lb/lbf h
lb/lbf h	1.0197	kg/daN h		0.980681	lb/lbf h
lb/hp h	0.60828	kg/kW h		1.64399	lb/hp h
lb/hp h	0.447387	kg/PS h		2.2352	lb/hp h

See B.21 for conversions to thermal efficiency.

B.17   Specific heat

Chu/lb K	1	Btu/lb R		1	Chu/lb K
Chu/lb K	4186.8	J/kg K		2.38846E-04	Chu/lb K
ft lbf/lb R	5.38032	J/kg K		0.185863	ft lbf/lb R
HPs/lb K	1643.99	J/kg K		6.08277E-04	HPs/lb K

B.18   Specific thrust

lbf s/lb	9.80665	N s/kg		0.1019716	lbf s/lb

B.19   Stress

See Table B.1 at the end of this appendix.

B.20   Temperature

Conversion shown is to convert a quantity *from* units after the = sign *to* units before it.

C = K − 273.15	F = 1.8 * K − 459.67	R = 1.8 * K	K = C + 273.15
C = (R − 491.67)/1.8	F = R − 459.67	R = 1.8 * (C + 273.15)	K = (F + 459.67)/1.8
C = (F − 32)/1.8	F = 1.8 * C + 32	R = F + 459.67	K = R/1.8

B.21   Thermal efficiency of a turboshaft engine – conversion to or from SFC.

See Table B.2 at the end of this appendix.

B.22   Torque

lbf ft	0.138255	kgf m		7.23301	lbf ft
lbf ft	1.35582	N m		0.737562	lb ft
lbf in	0.112985	N m		8.85075	lbf in

B.23   Velocity – angular

From	Multiply by	To	From	Multiply by	To
deg/s	0.0174533	rad/s	57.2958		deg/s
rev/min (rpm)	0.104720	rad/s	9.54930		rev/min (rpm)
rev/s	6.28319	rad/s	0.159155		rev/s

B.24   Velocity – linear

ft/s	0.59248	kt	1.68782		ft/s
kt	1.852	km/h	0.539957		kt
kt	0.514444	m/s	1.94384		kt
mile/h	1.46667	ft/s	0.681818		mile/h
mile/h	1.609344	km/h	0.621371		mile/h
mile/h	0.86896	kt	1.1508		mile/h
mile/h	0.44704	m/s	2.23694		mile/h

Note: The above is for the international knot; the UK nautical mile is obsolete.

B.25   Viscosity – dynamic

lb/ft s	1.48816	kg/m s	0.671969		lb/ft s
lb/in s	17.858	kg/m s	0.055997		lb/in s
$lbf\,h/ft^2$	0.172369	$MN\,s/m^2$	5.80151		$lbf\,h/ft^2$
$lbf\,s/ft^2$	47.8803	kg/m s	0.0208854		$lbf\,s/ft^2$
Pa s (kg/m s)	1000	cP	0.001		kg/m s
Pa s	1.0	$N\,s/m^2$	1.0		Pa s

B.26   Viscosity – kinematic

cSt	10^{-6}	m^2/s	10^6		cSt
ft^2/s	0.092903	m^2/s	10.7639		ft^2/s
in^2/s	6.4516	cm^2/s	0.155		in^2/s

B.27   Volume

in^3	16.3871	cm^3	0.0610237		in^3
ft^3	28.3168	litre	0.0353147		ft^3
UK gallon	4.54609	litre	0.219969		UK gallon
UK gallon	1.20095	US gallon	0.832674		UK gallon
US gallon	3.785	litre	0.2642		US gallon
$yard^3$	0.764555	m^3	1.30795		$yard^3$

References

1. BSI (1967, 1974) *BS 350 Parts 1 and 2:   Basis of Tables, Conversion Factors and Detailed Conversion Tables*, British Standards Institution, London.
2. BSI (1993) *BS 5555 Specification for SI Units and Recommendations for the Use of their Multiples and of Certain Other Units*, BSI, London.
3. J. L. Cook (1995) *Conversion Factors*, Oxford Scientific Publications, Oxford.

Table B.1: Unit conversions for pressure and stress

	atm	bar	in Hg	in H₂O	kgf/cm²	mm Hg	mm H₂O	ibf/in² (psi)	kPa
atm		1.01325	29.9213	406.782	1.03323	760.0	10332.3	14.6959	101.325
bar	0.986923		29.53	401.463	1.01972	750.062	10197.2	14.5038	100
in Hg	0.0334211	0.0338639		13.5951	0.0345316	25.4	345.316	0.491154	3.38639
in H₂O	0.0024583	0.002491	0.073556		0.00254	1.86832	25.4	0.036127	0.249089
kgf/cm²	0.967841	0.980665	28.959	393.701		735.559	10000	14.2233	98.0665
mm Hg	0.0013158	0.0013332	0.03937	0.53524	0.0013595		13.5951	0.0193368	0.133322
mm H₂O	0.0000978	0.0000981	0.002896	0.0393701	0.0001	0.073556		0.0014223	0.009807
lbf/in² (psi)	0.068046	0.0689476	2.03602	27.68	0.070307	51.7149	703.07		6.89476
kPa	0.0098692	0.01	0.2953	4.01463	0.0101972	7.50062	101.972	0.145038	

Notes:
To convert a value in the units in the left-hand column to those in the top row multiply by the number at the junction of the row and column.
The conversion factors for columns of H₂O are for water at a uniform density of $1000 \, \text{kg/m}^3$ under the standard gravity of $9.80665 \, \text{m/s}^2$.
The conversion factors for columns of Hg are for mercury at a uniform density of $13.590 \, \text{kg/m}^3$ under the standard gravity of $9.80665 \, \text{m/s}^2$.
1 kPa is equivalent to $1 \, \text{kN/m}^2$.

Table B.2: Conversion between turboshaft engine thermal efficiency, SFC and heat rate

$$ETATH = 1413.6/(SFC*LCV)$$
where SFC is in lb/hp.h and LCV is in CHU/lb

$$ETATH = 3413/Heat \ Rate$$
where Heat Rate is in BTU/kW.h

$$ETATH = 2545/Heat \ Rate$$
where Heat Rate is in BTU/hp.h

Index

Printed and bound by CPI Group (UK) Ltd, Croydon, CR0 4YY

16/04/2025

14658830-0005